The biology of birds is diverse and frequently differs significantly from other vertebrates. Many birds migrate or fly at high altitudes, while egg-laying and feather production places high demands on nutrient uptake and storage. This book is the only comprehensive and up-to-date survey of avian biochemistry and molecular biology available. It emphasises the similarities and differences between birds and other vertebrates, concentrating on new developments.

The first section deals with protein, lipid and carbohydrate metabolism, its hormonal control and the adaptations that occur in birds. The second covers the avian genome, gene expression and avian immunology. Growth and embryological development are also discussed.

Avian Biochemistry and Molecular Biology will be of interest to all those working on birds, especially postgraduate students and researchers.

AVIAN BIOCHEMISTRY
AND
MOLECULAR BIOLOGY

AVIAN BIOCHEMISTRY AND MOLECULAR BIOLOGY

LEWIS STEVENS

Senior Lecturer
Stirling University, Scotland

CAMBRIDGE
UNIVERSITY PRESS

PUBLISHED BY THE PRESS SYNDICATE OF THE UNIVERSITY OF CAMBRIDGE
The Pitt Building, Trumpington Street, Cambridge, United Kingdom

CAMBRIDGE UNIVERSITY PRESS
The Edinburgh Building, Cambridge CB2 2RU, UK
40 West 20th Street, New York NY 10011–4211, USA
477 Williamstown Road, Port Melbourne, VIC 3207, Australia
Ruiz de Alarcón 13, 28014 Madrid, Spain
Dock House, The Waterfront, Cape Town 8001, South Africa

http://www.cambridge.org

First published 1996
First paperback edition 2004

A catalogue record for this book is available from the British Library

Library of Congress cataloguing in publication data

Stevens, Lewis.
Avian biochemistry and molecular biology / Lewis Stevens.
p.   cm.
Includes bibliographical references and index.
ISBN 0 521 45510 3 (hardback)
1. Biochemistry.   2. Molecular biology.   3. Birds–Molecular
aspects.   I. Title.
QP514.2.S64   1996
598.219′2–dc20   95-15166 CIP

ISBN 0 521 45510 3 hardback
ISBN 0 521 61211 X paperback

To Evelyn for her constant support

Contents

Contents

Preface

This book aims to give a perspective view of avian biochemistry and molecular biology. The main topics discussed are either those which are particularly important in birds, e.g. the metabolism associated with energy provision for flight and the expression of keratin genes necessary for feather growth, or those in which a process occurs in a different manner in birds as compared with other vertebrates, e.g. the generation of antibody diversity and the excretion of nitrogen as uric acid. Areas where there are only slight differences between birds and mammals, and which are well covered in standard textbooks of biochemistry and molecular biology, are either not considered or discussed only briefly.

The book is divided into two parts; the first is concerned broadly with metabolism and its control, and the second is concerned with the organisation of the avian genome and its expression.

Much of the work cited has been carried out in the 1980s, and a significant proportion has not been previously reviewed. For this reason an extensive list of references is given to the primary literature so that readers can locate the sources of information. A basic knowledge of biochemistry and molecular biology equivalent to that which might be covered in a one year undergraduate course is assumed.

Lewis Stevens
Stirling, Scotland

Nomenclature

Taxonomy

Common English names as given in *A Complete Checklist of Birds of the World*, 2nd edn, (1991) ed. R. Howard & A. Moore. London: Academic Press are used throughout the text, the Latin binomial equivalents are given in the Appendix. Avian classification and the standardisation of English names is still hotly debated (see Crosby, 1994), but this is generally of lesser importance for the biochemical and molecular biology studies described in this book. Domesticated species recur frequently throughout the text. The reader may assume the following applies: domestic fowl, *Gallus gallus domesticus*; duck, or domesticated duck, *Anas platyrhynchos*; goose, *Anser anser*; turkey, *Meleagris gallopavo*; quail, *Coturnix coturnix*; pigeon, *Columba livia*; and pheasant, *Phasianus colchicus*.

Developmental stages of embryos and growing birds

Unless otherwise indicated, where an age is given in days or weeks it refers to the age post-hatching. For the most part, the stage of embryonic development is given in days prefixed by 'E'. Therefore, E5 means an embryo incubated for 5 days. For much of the biochemical data cited in the book, this is sufficiently precise. The rate of development of an embryo varies with the conditions of incubation, the time interval between oviposition and incubation and the strain or species of bird. For example, White Leghorn embryos develop more rapidly than Barred Plymouth Rocks and hatch approximately a day earlier. For experiments that focus primarily on stages in embryological development, greater accuracy is required. Hamburger & Hamilton (1951) made a detailed analysis of the morphological stages of embryonic development dividing the incubation period (in the case of the domestic fowl usually 21 days) before hatching into 46 stages. Each of these stages can be unambiguously identified by comparison with an illustration. An approximate conversion for the domestic fowl is given in Table A (overleaf).

Enzyme nomenclature

The nomenclature adopted is that given in Enzyme Nomenclature: Recommendations (1992) of the Nomenclature Committee of the International Union of Biochemistry and Molecular Biology.

Genes and gene products

Genes are designated by three letters written in italics. The protein products of gene expression are abbreviated to the same three letters not italicised, with the first letter in upper case.

Table A. *Relationship between stage of development and incubation period for the domestic fowl*

Stage	Incubation time (h)	Stage	Incubation time (days)	Stage	Incubation time (days)
1	<6	21	≈3.5	41	≈15
2	6–7	22	≈3.5	42	≈16
3	12–13	23	3.5–4.0	43	≈17
4	18–19	24	4	44	≈18
5	19–22	25	4.5	45	≈19–20
6	23–25	26	4.5–5.0	46 (newly	≈20–21
7	23–26	27	≈5	hatched chick)	
8	26–29	28	≈5.5		
9	29–33	29	≈6		
10	33–38	30	≈6.5		
11	40–45	31	≈7		
12	45–49	32	≈7.5		
13	48–52	33	≈7.5–8.0		
14	50–53	34	≈8		
15	≈50–55	35	≈8–9		
16	≈51–56	36	≈10		
17	≈52–64	37	≈11		
18	≈65–69	38	≈12		
19	≈68–72	39	≈13		
20	≈70–72	40	≈14		

1

Introduction

A scientific interest in birds dates back at least to the time of Aristotle. In his book *Historia Animalium* he describes over 140 species, their habits and physiology. At the present day, the number of people interested in ornithology, with and without scientific backgrounds, exceeds those interested in other vertebrates except perhaps humans, dogs, cats and horses. An indication of an affectionate attitude towards birds by scientists can be detected from the names of research publications. *Auk, Condor, Ibis, Emu* are important avian biology journals, not solely devoted to a single avian genus, but there are at present no journals with titles like *Rat, Mouse, Guinea-pig, Hamster* or *Monkey*, in spite of very substantial amount of research on these species.

Birds have been classified into nearly 10 000 different species. Nevertheless, when compared with the other classes of vertebrate, birds show a greater degree of uniformity of shape and size, while there is less diversity within an order of birds than within a family of the other classes of vertebrate. The latter is usually attributed to the selective pressure that the evolution of flight has imposed on them. The vast majority of present-day birds are capable of flight. The class of vertebrate to which birds are most closely related are reptiles. They have a number of features in common with reptiles, including egg-laying, producing structures derived from the integument, such as feathers or scales, possessing nucleated erythrocytes and possessing a urogenital system that excretes urates as the principal nitrogenous excretion product.

The particular anatomical characteristics common to almost all birds are: (i) presence of feathers, (ii) lack of teeth, (iii) bipedism, (iv) a bone structure which is specially adapted for flight, and (v) the specialisation of the fore limbs for flight, including strong pectoral muscles. Adaptation of the bone structure involves both the fusion of certain bones, e.g. the pelvis and

pectoral girdles, giving greater rigidity per given bone mass for muscle attachment, and the presence of air sacs within the bone structures.

1.1 Classification and evolutionary origins of birds

The most recent classification of birds lists 9672 species (Sibley & Monroe, 1990). This is a larger number of species than for any other class of vertebrate, except the teleost fishes. The earliest comprehensive classification using the Linnaean system was by Gadow (1892), and this formed the basis of the widely used contemporary system of Wetmore (1960), which divides Aves into 27 orders and 170 families (Fig. 1.1). For many years there has been much debate on the principles underlying taxonomic classifications in general (for a detailed discussion, see Ridley, 1993; Proctor & Lynch, 1993). The principles underlying the classification methods are: (i) the *phenetic* or phenotypic method, (ii) the *phylogenetic* or *cladistic* method, and (iii) the *evolutionary taxonomic* method. The last of these three is a synthesis, or mixture, of the first and second, attempting to overcome some of their shortcomings, but in so doing inevitably it has an element of compromise.

In the phenetic method, species are grouped on the basis of their phenotypic attributes. Numerous parameters taken from different characters (e.g. morphological, physiological, biochemical, behavioural, etc.) are compared between different species, and a quantitative difference between each is measured. The differences between each are averaged and the classification is based on the magnitude of the differences. A number of different algorithms have been used to generate phylogenetic trees from these differences (Avise, 1994). In this method, a large number of characters are used, and each is given equal weighting. The phylogenetic or cladistic

1

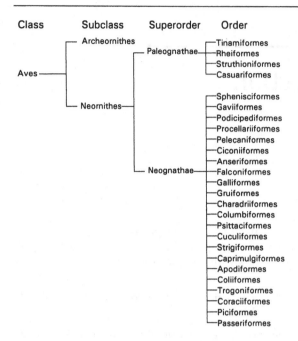

Class	Subclass	Superorder	Order

Fig. 1.1. Morphological classification of birds (Wetmore, 1960).

(meaning 'branching') method, by contrast, is based on the supposed evolutionary relationships and depends on devising an evolutionary branching pattern. Every significant evolutionary step makes a dichotomous branch that produces two sister taxa equal in rank. The phenetic method is, therefore, dependent on the differences of the phenotype for the parameters chosen, and the cladistic method is dependent on evolutionary relationships.

Apart from the underlying principles, there are the questions of how the data are obtained and their limitations. With the phenetic method, characters used are those for which it is possible to make the necessary quantitative measurements. With the cladistic method, it is rarely possible to determine each branch point, especially where convergent evolution has occurred. The phenetic method produces a practical classification but lacks the deeper philosophical justification of the cladistic method. However, both methods often give similar results, but differences arise, as in the example in Fig. 1.2. Although the crocodile and lizard show greater morphological similarity to one another than to the bird, birds and crocodiles have more recent common ancestors than crocodiles and lizards.

Much of the earlier evidence used in classification was morphological, but since the 1970s molecular evidence has been used increasingly. The morphological classification given in Fig. 1.1 is, therefore, mainly phylogenetic. The two molecular methods most used in avian classification are DNA:DNA hybridisation, and restriction endonuclease analysis of mitochondrial DNA (for details of both, see Avise, 1994; Li & Graur, 1991). The former has been used as the basis for an overall classification of birds by Sibley and co-workers, whereas the latter is more useful in examining differences between closely related species and subspecies. Since the mid 1970s, Sibley and co-workers have used the DNA hybridisation technique and have made an enormous number of pairwise comparisons between species. The results form the basis of their biochemical classification of birds (Fig. 1.3) (Sibley & Alquist, 1990). The DNA:DNA hybridisation method is based on the principle that the more closely related two complementary DNA strands from different species are the greater will be the stability of the hybrid double helix. This stability is measured as the temperature at which the double helix disrupts or 'melts'. This method compares the DNA of the whole genome and, therefore, gives an average of differences in different parts of the genome. Whilst it is generally recognised that their labour is a *tour de force* and has involved an enormous amount of work, it is still controversial (see Sheldon & Bledsoe, 1993); however, it should be regarded as the first stage towards a classification based on the genome. It is already undergoing refinement, and this will undoubtedly continue in the years to come. Some of the criticism of the work is directed at the large experimental errors in the measurement of the melting temperature, and at the way that genetic distances are calculated from the data. More recent measurements, using independent calibration and refinements in the data handling, which have been made by other workers, have lent support to Sibley & Alquist's classification (for discussion, see Mooers & Cotgreave, 1994).

Since 1975, the ability to sequence DNA has become much more routine; nevertheless it would be an enormous task to sequence large stretches of DNA from a large number of different species. So although the quality of information from DNA sequencing is much better than from DNA hybridisation, the work involved is orders of magnitude greater. It is unlikely that DNA sequencing will be used for the majority of avian species, but it may be used to resolve important controversial areas. For this, only small selected areas of the genome are likely to be compared.

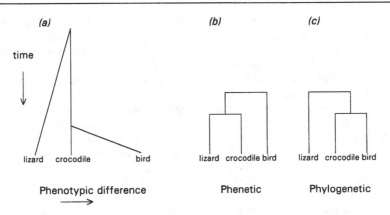

Fig. 1.2. An example of phenetic and phylogenetic principles of classification. (*a*) The probable evolutionary tree for lizard, crocodile and bird. The distances separating the three are a measure of their phenotypic differences. (*b*) Phenetic relationships. (*c*) Phylogenetic relationships.

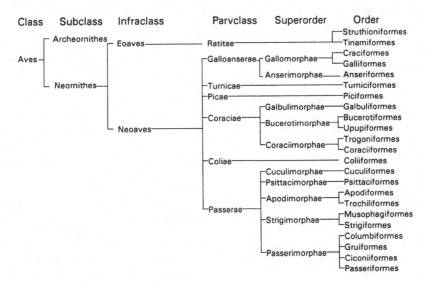

Fig. 1.3. Biochemical classification of birds (Sibley & Alquist, 1990).

The main differences between the morphological classification and the biochemical classification (Figs. 1.1 and 1.3) are: (i) the larger number of taxonomic levels of the latter, and (ii) the differences in the order Ciconiiformes. In the morphological classification, this comprises herons, ibises, storks and flamingos, but in the biochemical classification it also includes penguins, loons, grebes, albatrosses, shearwaters, petrels, skuas, gulls, terns, auks, vultures, hawks, falcons, pelicans, gannets and cormorants. The morphological classification divides these into the separate orders: Sphenisciformes, Gaviiformes, Podicipediformes, Procellariiformes, Charadriiformes,

Falconiformes, and Pelecaniformes (for details, see Proctor & Lynch, 1993; Sibley & Alquist, 1990). In morphological classification systems, the divisions into hierarchies comprising phyla, classes, orders and families are somewhat arbitrary. Using the DNA:DNA hybridisation data, Sibley & Alquist (1990) define hierarchical divisions more precisely in terms of the differences in melting temperatures, e.g. classes differ by 31–33 °, orders by 20–22 ° and families by 9–11 °, thus making the hierarchies less arbitrary.

Restriction endonuclease analysis of mitochondrial DNA (mtDNA) has been particularly useful in establishing evolutionary relationships over micro-evolutionary times. For example, it has been used to distinguish closely related species, subspecies and phylogeographic patterns within species (Avise *et al.*, 1987). The advantage of mtDNA over nuclear DNA is the faster rate of evolution of mtDNA. Based on the differences in restriction endonuclease patterns of mtDNA from five species of geese, whose divergence time has been independently assessed from fossil evidence, the rate is ≈ 2% divergence in mtDNA per million years (Shields & Wilson, 1987). This is much higher than for avian nuclear DNA, for which two estimates are 0.22% and 0.34–0.4% per million years (Sheldon & Bledsoe, 1993).

It is generally accepted that birds evolved from some form of reptile more than 150 million years ago in the Mesozoic era. The fossil toothed reptile having avian features, *Archaeopterix lithographica*, is generally regarded as an evolutionary link between reptiles and birds. It is not clear which group of prehistoric reptiles was its ancestor, but the most likely alternatives appear to be either small theropod dinosaurs, or thecodont ancestors of the crocodile. Modern bird orders may have diverged from each other as recently as about 65 million years ago (Wyles, Kunkel & Wilson, 1983). Geological evidence of a 'catastrophe' about 65 million years ago allows for the possibility that the divergence of modern bird orders arose since then, but from which ancestral form is still a matter for debate (Proctor & Lynch, 1993).

The avian genomes are generally smaller than those of most mammals and of many other vertebrates. This is coupled with a large chromosome number, including several microchromosomes. The latter are difficult to identify. The chromosome number of some species has only recently been established, and many are still to be established.

1.2 Special features of avian biochemistry and molecular biology

During the 1930s to 1950s, the details of the major metabolic pathways were resolved, and the structures of many macromolecules, particularly proteins, were determined. From this work, the concept of unity and diversity of living organisms emerged. It became clear that there are basic structures or building blocks and metabolic pathways common to all living organisms, but there is also diversity. For example, haemoglobin from horse and from the Andean goose have many features in common, including tertiary and quaternary structure, but at the same time they have subtle differences such that haemoglobin from

the Andean goose has a higher oxygen affinity, enabling it to bind oxygen at low partial pressures of oxygen. Histones are highly conserved proteins; the five major groups (H1, H2a, H2b, H3 and H4) are found in all vertebrates, each having very similar structures. A sixth type, H5, is found only in avian species and is present in their erythrocytes.

The approach adopted by many research workers in biochemistry and molecular biology is problem orientated rather than species orientated. The organism that they use to study a problem is often determined by practical considerations, such as which organism will provide the most active enzyme in the largest quantity. For many biochemists, organisms of choice have often been the laboratory rat or the bacterium *Escherichia coli*. Then, once experience has been built up using one organism, it is often an advantage to use the same organism for solving another problem, since a data base and experimental procedures have already been established. At the advent of gene cloning technology, it was easiest to use *E. coli* since many of its genes had been mapped and a wide range of plasmid and viral vectors for which *E. coli* was host were known.

Sometimes birds, and particularly the domestic fowl, have been the species of choice in which to solve a particular problem. For example, when the molecular basis of steroid hormone action was first studied, hen oviduct was regarded as the tissue *par excellence* since it could be induced to synthesise large amounts of egg-white proteins in response to oestrogen administration. Pigeon breast muscle was used for much of the work that led to the discovery of the tricarboxylic acid cycle, carried out by Szent-Györgi and co-workers, and by Krebs and Johnson in the mid-1930s (see Fruton, 1972). It was selected because of the rapid rate of respiration of this tissue. However, there are also times when a problem has been solved in one species, and it is important to know whether the same mechanism occurs in a wide range of species. The discovery of 'essential amino acids' was worked out in a classic series of experiments using laboratory rats. It was then important to know whether humans depended on the same set of amino acids. It was only after this that the protein nutritional requirements of a wider range of vertebrates, including the domestic fowl, were studied. The latter has assumed economic importance as the diet of the domestic fowl has been fine tuned to maximise meat production (see Chapter 2). By comparison, the importance of vitamins B_1 (thiamin) and vitamin K were first noted in avian species and subsequently studied in other vertebrates.

In this book, most emphasis is given to areas where some aspect of the biochemistry or molecular biology is either unique or has particular importance in birds. Areas in which there is little to add to what is well known for mammals have been given either less emphasis or have not been included, since these are included in most standard textbooks (see Mathews & van Holde, 1990; Zubay, 1993). Some of the special features of birds that have been the focus of research are outlined below and then discussed in more detail in subsequent chapters.

Many of the distinctive features of the physiology and anatomy of birds arise from adaptations for flying. Flight imposes constraints on the body mass and the overall shape, and it requires a greater rate of energy output than any other form of motion. The growth and development of feathers is a further adaptation. There are also features that arise from oviparity. These include development of an ovum so that at the point of laying it contains the food reserves for endogenous metabolism needed to sustain it until hatching. There are also features of avian anatomy that may have some special advantage, which is not at present clear, for example the presence of the bursa of Fabricius. The bursa is responsible for the maturation of B lymphocytes, which are an integral part of the immunological system. A very diverse repertoire of antibodies is generated by the B lymphocytes using a mechanism peculiar to birds.

Birds in general have very high energy requirements compared with other vertebrates. Flight is more energy demanding than running, walking or swimming. Flight at high altitudes requires the use of oxygen at lower partial pressures than at sea level. Migratory birds need to be able to store large food reserves, and to store them in the most energy efficient form, i.e. as fats. Some birds, for example penguins and geese, are able to withstand very low temperatures without feeding for periods of months and at the same time maintain a body temperature much higher than the ambient temperature. This requires both an effective insulating system and a large source of potential energy. Birds have nucleated erythrocytes, in common with reptiles and amphibians. It would be convenient to suggest that the ejection of the nucleus of mammalian erythrocytes during terminal differentiation left the cell containing more haemoglobin, thereby maximising the potential for oxygen transport. However, there is no evidence that retaining the nucleus, as in the case of birds, leads to lower efficiency.

A number of biochemical and molecular biological studies have taken advantage of the readily available protein source in eggs. In the early work on protein purification and characterisation, egg-white proteins presented fewer problems than intracellular proteins. They are more abundant and require no cell disruption, or protection from intracellular hydrolases. They are amongst the best characterised of proteins (see Stevens, 1991b). Egg-white lysozyme was the first structure for which an atomic resolution X-ray structure was published (Phillips, 1967), and from this its mechanism of catalysis was revealed. It is perhaps ironical that its function in egg-white is still a matter for conjecture (see Chapter 10). The genes encoding all the major egg-white proteins have been mapped and the control of their expression by steroid hormones is a paradigm for the study of control of gene expression. The oviparous nature of the avian embryo has meant that it is easy to obtain embryos at all stages of development, simply by incubation for the appropriate length of time and at the appropriate temperature. The morphological aspects of avian embryology have been described in much detail. Chick embryo is particularly suited to studying limb bud development, and it was in this tissue that it was first demonstrated that retinoic acid acted as a morphogen (Tickle et al., 1982). In contrast, the development of transgenic birds turns out to be more difficult than transgenic mice. This is because at the time of oviposition the embryo has already reached about 60 000 cells, and also because technical manipulation of the developing blastoderm within the large egg mass is more difficult, although possible (Sang et al., 1993). In common with reptiles, birds produce cleidoic eggs, i.e. eggs with a protective shell that largely isolates the contents from their surroundings but permits gaseous exchange and some water loss or gain. The eggs contain the full complement of nutrients (excluding oxygen) to enable the embryo to develop until hatching. Since metabolism occurs throughout embryonic development, this will generate carbon dioxide, water and nitrogenous waste. Carbon dioxide diffuses out of the shell, and the water can be utilised. However, the primary nitrogenous end-product of metabolism is ammonia. Because of its high toxicity, it has to be removed and is converted to the poorly soluble uric acid. Urea, the end-product of nitrogen metabolism in mammals and elasmobranch fishes, is relatively non-toxic, but if it were present in large amounts it would create osmotic difficulties. The slightly soluble uric acid is deposited as crystals in the allantois, and this releases water to be reabsorbed by the embryo. It is widely believed that the excretion of nitrogen in the form of urate enables birds to conserve water by excreting a semi-solid urine. However, it has been calculated that the excretion of uric acid by the domestic fowl would entail the use of 200 ml water per gram of nitrogen, whereas the excretion of urea by mammals could use 150 ml per gram of nitrogen (King & McLelland, 1984), so the importance of uric acid excretion to the developing embryo appears more significant.

The uropygial gland or preen gland is the principal cutaneous gland of birds. It is largest in aquatic species but is absent in some species, including ostrich, emu, cassowaries, bustards, frogmouths, many pigeons, woodpeckers and parrots. It produces a lipid secretion that is spread over the feathers during preening. Being an oil producer, it is particularly active in the synthesis of fatty acids. The goose uropygial gland is a particularly rich source of fatty acid synthetase, where it constitutes over 10% of its total protein, and the gland has been used extensively to study the enzyme.

An anatomical feature unique to birds is the bursa of Fabricius, a diverticulum or pouch, usually pear-shaped, located in the proctodeum of the cloaca. It was first described by Hieronymous Fabricius in the seventeenth century. The first bursectomy on a bird was performed on a pigeon by Riddle and Tange (1928), but its immunological function was not discovered until 1956 by Glick (see Glick, 1988). The two principal antibody-producing cells are known as B cells and T cells, the former proliferating in the bursa and the latter in the thymus. (Although there is no bursa in mammals, the cells serving an analogous function are still known as B cells.)

1.3 The plan

The distinction between biochemistry and molecular biology is difficult to define. There is a considerable area of overlap between the two, although there are areas that are generally accepted as belonging to one rather than the other. Part 1 of this book covers the area of intracellular metabolism and its control, generally regarded as the province of biochemistry. Part 2 begins with the genome and progresses to the organisation of genes and their expression, and it may be regarded as having more of a molecular biology flavour.

The first chapter in Part 1 is on avian nutrition. This includes the organic compounds that are required for growth and maintenance and that provide the substrates for many of the metabolic pathways discussed in Chapters 3 to 5. Chapter 6 is concerned with how metabolic processes are adapted to the special needs of avian lifestyle, e.g. flight at high altitude, migration, polar environments and egg-laying. Part 1 concludes with a chapter on the hormonal control of metabolism. Part 2 begins with an account of the avian genome (Chapter 8) and how many genes are organised into multigene families (Chapter 9). Chapter 10 considers how gene expression is regulated by steroid hormones, the best studied example of which is how oestrogens regulate protein synthesis during oogenesis. Oncogenes have been better characterised in the domestic fowl than any other species and are considered in Chapter 11. The key to understanding developmental biology lies is understanding the temporal and spatial gene expression during embryonic development (Chapter 12). Finally the unusual mechanism by which birds generate their repertoire of antibodies forms much of the final chapter (Chapter 13). Throughout the book, wherever possible, as wide a range of avian species as possible is considered. However for many aspects, particularly in molecular biology, the domestic fowl is the only species that has been extensively studied. Many aspects that relate to gene expression are likely to be similar in most avian species. In contrast, areas that relate to adaptive mechanisms are likely to show greatest diversity and have been studied in the widest range of species.

Part 1

Metabolism

2

Avian nutrition

2.1 Introduction

The range of avian diets is quite varied, and amongst the avian species there are omnivores, carnivores and herbivores. In common with other vertebrates, and indeed all nutritional heterotrophs, the diet provides both a source of energy and the necessary building blocks with which to generate new cell materials and to replace existing materials. The nutrients required by an organism fulfil these roles. Much of the study of nutrition is aimed at defining the nutrients in chemical terms; some of these are required in large quantities whereas others are micronutrients. The study of vertebrate nutrition, which was the focus of much biochemistry for the early part of this century, was carried out mainly using laboratory mammals and humans; interest in other vertebrates, especially birds, generally arose later. However, this was not always so, for example the discovery by Christiaan Eijkman in 1887 of polyneuritis in chickens fed a diet of polished rice that lacked thiamin (see Carpenter & Sutherland, 1994) and the discovery of vitamin K by Dam in 1935 as the factor that could overcome the slow blood clotting in chickens fed certain diets are cases where the avian work provided the lead.

Since the 1940s, nutritional science has been neglected by the majority of biochemists; the basic nutrients of the diet have been defined and in most cases their biochemical role is known, but, as Kornberg (1989) points out, we still lack detailed knowledge on (i) the optimum amounts of the constituents of food necessary to maintain health, (ii) the effects of dietary imbalances, and (iii) the effects of diet on disease. Although with humans the emphasis is on improving health and longevity, in the domestic fowl and the other domesticated birds, the emphasis of nutritional studies has been largely directed towards optimising growth, and longevity has rarely been a consideration. The focus has been directed towards 'feed efficiency'

in the domestic fowl, in order to 'fine tune' the diet to obtain the maximum rate of synthesis of animal protein, in the form of meat and eggs, at minimum cost. There are more detailed quantitative data on the effects of diet on growth in the domestic fowl than in any other avian species. In wild or non-domesticated birds, there are many interesting nutritional problems that have only begun to be studied at a biochemical level. These include nutrition in relation to breeding and reproductive capacity, feather production, migration, and adaptation to extremes of environment such as polar conditions.

The major nutrients of the diet are protein, carbohydrate and fat, and the minor nutrients include the vitamins and minerals. Major and minor are used in relation to the quantities required, rather than their importance to the well-being of the organism. The major nutrients provide for body protein and energy and will be considered under those headings in subsequent sections. The complexity of the nutritional requirements of an organism is inversely related to its biosynthetic capabilities. It is generally believed that some nutritional requirements have evolved as a result of the loss of some essential enzyme or enzymes during the course of evolution. Although man has a dietary requirement for ascorbic acid (vitamin C), being unable to convert the hexose precursor to ascorbic acid, the domestic fowl and many other birds require no exogenous supply of ascorbic acid because they have retained the ability to synthesise it in adequate amounts. Sometimes the loss of an enzyme has a different effect. A number of birds in the Thrush (Muscicapidae) and Starling (Sturnidae) families appear to have lost the ability to hydrolyse sucrose in the small intestine, and this leads to a sucrose intolerance causing osmotic diarrhoea and consequent feeding aversion (Malcarney, Martinez del Rio & Apanius, 1994).

This chapter aims at defining the basic nutrients of

an avian diet in relation to their metabolic role in avian cells. Modification of the basic requirements for egg-laying, migration and cold adaptation are discussed in Chapter 6.

2.2 Protein and amino acid requirements

Of the three major dietary requirements, protein may be considered the most important. This is because dietary protein is able to act as a source of body protein, carbohydrate or fat, whereas neither fat nor carbohydrate alone can give rise to a net increase in body protein. Lack of dietary protein has the most serious consequences, since it is the source of synthesis of body protein. Dietary protein is broken down in the gut, and evidence for this breakdown in birds dates back to the eighteenth century, when Réaumur and Spallanzani independently provided evidence of proteolytic activity in the digestive juices of vultures, hawks, owls, ducks, geese and pigeons (Brown, 1970). The products of proteolysis are absorbed in the small intestine in the form of amino acids or small peptides, the majority being in the latter form (Matthews, 1975). The passage of food through the digestive system is generally more rapid in birds than in other vertebrates. Storage of heavy nutrients has an adverse effect on flight and this is an important factor, except for ground birds. Proteins differ in their amino acid composition, some containing all 20 different amino acids, but others lack certain amino acids. Even those containing all 20 amino acids may not be balanced in the same proportions as body protein and, therefore will have certain amino acids that are limiting. Natural foodstuffs contain a mixture of a large number of different proteins, but it is often the case that diets made up of plant proteins have limiting amounts of methionine, lysine and tryptophan. Lysine is often a major limiting amino acid in seeds of Gramineae, which are the predominant winter food of many granivorous passerines (Parrish & Martin, 1977).

During the period 1910 to 1950, a general understanding of the protein and amino acid requirements of humans and the rat was established. Amino acids can be divided into essential and non-essential. *Essential amino acids* are those required to maintain a growing animal in positive nitrogen balance, or an adult in nitrogen equilibrium. An animal requires all 20 amino acids to make its complement of body proteins, but some of these can be synthesised from the 'essential' amino acids. For this reason, some authors prefer the terms 'indispensible' and 'dispensible' amino acids. After establishing the essential amino acids for humans

and the laboratory rat, work began to see whether birds, and in particular the domestic fowl, fitted the same pattern. Experiments to determine the amino acid requirements for the chick were carried out by Almquist and co-workers and Luckey and co-workers (see Fisher, 1954). Arginine, isoleucine, leucine, lysine, methionine, phenylalanine, threonine, tryptophan, valine and histidine (see below) are essential for maintainence of nitrogen balance in adult domestic fowl, and, in addition, glycine is necessary for growing chicks to maintain maximum growth rate (Table 2.1). For chicks, arginine is essential for any growth to occur, whereas in growing rats it is lack of arginine that results in submaximum growth rate. Glycine is not essential in either rats or humans. Broiler chicks that have been selected over a number of generations for rapid growth are able to double their weight from 100 to 200 g in less than 5 days, compared with the laboratory rat doubling from 50 to 100 g in about 10 days (Baker, 1991). Because of the more rapid growth rate shown by chicks than rats, any deficiencies in amino acid requirements are likely to be more pronounced. There are a number of amino acids that the chick is capable of synthesising, but unless exogenous supplies are provided, they become rate limiting during the rapid growth phase.

2.2.1 Functions of specific amino acids

Glycine

This amino acid is required for growth but it is not essential for maintenance of nitrogen balance in the adult fowl (Fisher, 1972). Glycine residues account for ≈ 30% of amino acid residues in collagen and ≈ 15% in feather proteins (Gregg & Rogers, 1986). Collagen constitutes about 25% of total body protein, and the total collagen content of the developing chick increases 7-fold between 12 and 19 days (Ohyama, Tokimitsu & Tajima, 1991). The main nitrogenous excretion product in birds is uric acid; the adult domestic fowl excretes 4–5 g of uric acid per day. Glycine contributes two carbon and one nitrogen atoms to the biosynthesis of uric acid (Fig. 2.1) further increasing the dietary requirement for glycine. Chicks are only able to synthesise about 60–70% of the glycine required for maximum growth (Graber & Baker, 1973).

Proline

Young chicks also have a dietary requirement for proline (Baker, 1991), and this may also be required for collagen biosynthesis, since collagen is composed

Table 2.1. *Relative amino acid requirements for domestic fowl, quail and white-crowned sparrow for growth, maintenance and egg production*

| Amino Acid | Domestic fowl | | | Quail | | White-crowned sparrow |
	Growth	Maintenance	Egg production	Growth	Egg production	Maintenance
Arginine	1.5	1.6	1.8	1.2	1.7	2.05
Glycine	0.5	0	0	0.5	0	0
Histidine	0.5	0	0.5	0.4	0.6	0.58
Isoleucine	0.9	1.0	1.4	1.0	1.2	1.42
Leucine	1.8	1.7	1.9	1.7	1.9	2.00
Lysine	1.4	0.4	1.4	1.1	1.3	1.53
Methionine	0.5	1.0	0.7	0.4	0.6	2.0
Phenylalanine	0.8	0.4	1.2	0.9	1.1	2.05
Threonine[a]	1.0	1.0	1.0	1.0	1.0	1.00
Tryptophan	0.2	0.2	0.3	0.2	0.3	0.37
Valine	1.3	0.8	1.5	0.9	1.3	1.26

[a] Standardised relative to threonine.
Source: From Fisher (1972); Shim & Vohra (1984); Murphy (1993a).

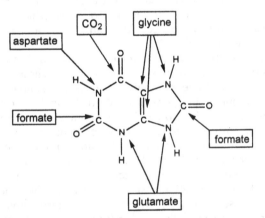

Fig. 2.1. The metabolic precursors providing the carbon and nitrogen atoms for purine and uric acid formation.

of about 30% proline and hydroxyproline residues. Hydroxyproline is synthesised from proline as a post-translational hydroxylation (see Section 2.4.5). Although proline is not an essential amino acid, chicks fed a diet containing an amino acid mixture lacking proline show a reduced growth rate over the first seven weeks, when rapid proline synthesis is required (Graber & Baker, 1973).

Cysteine

Cysteine is also essential for normal growth, and this is probably because of the high proportions (ap-proximately 10%) of sulphur-containing proteins (Table 2.2) in feathers (Brown, 1970; Gregg & Rogers, 1986). Feathers are composed of over 90% protein, of which keratin is the major one (Murphy & King, 1982). A substantial amount of protein synthesis is required both for the initial growth of feathers and for later feathering after moulting, e.g. the Emperor penguin synthesises about 900 g of feather keratin utilising about 1300 g of body protein during the 35 days of moult fasting (Groscolas, 1978). The principal amino acids in feather keratins are alanine, cysteine, glycine, leucine, proline, serine and valine (Crewther et al., 1965; Gregg & Rogers, 1986). Therefore, although birds are able to synthesise cysteine, at times of high demand exogenous cysteine is required for maximum growth.

Histidine

The essential requirement of histidine for growing chicks has been clear for some time, but whether or not it was required for the adult was less clear. It now seems justified to include it as one of the adult essential amino acids. Skeletal muscle, which comprises 30 to 40% of body protein, has considerable concentrations of the histidine-containing dipeptide carnosine (β-alanylhistidine). The metabolic function of carnosine is not clear, although it has been suggested that it may be required as an antioxidant (Kohen et al., 1988); alternatively, it may act as an intracellular buffer, since the imidazole ring of histidine

Table 2.2. *Proportions of essential amino acids in the tissue, egg and feather proteins of the domestic fowl*

Amino acid	Tissue[a] (% of protein)	Egg (% of protein)	Whole feathers (% of protein)
Arginine	7.8	6.4	7.3
Cystine	2.6	2.2	7.4
Histidine	3.9	2.3	0.6
Isoleucine	3.9	5.0	6.4
Leucine	6.4	8.3	8.5
Lysine	9.4	7.1	1.6
Methionine	1.8	3.2	0.5
Phenylalanine	3.5	4.7	5.5
Threonine	3.4	5.0	4.7
Tryptophan	1.0	1.4	0.7
Valine	4.4	6.5	8.9

[a] Tissue refers to total carcass proteins.
Source: Scott et al. (1982).

buffers most effectively at pH 6–7. Long-term nitrogen balance studies are necessary to demonstrate the indispensable nature of histidine, since the carnosine muscle pool has first to be depleted (Robbins, Baker & Norton, 1977). During egg production there is a higher histidine requirement in the adult (Table 2.1).

Arginine

Birds appear unable to synthesise any arginine via the urea cycle (see Section 5.4), which may be because of lack of carbamoyl phosphate synthetase I in mitochondria (Baker, 1991), and, as a result, the dietary requirement for arginine is higher than in growing mammals. However, they do appear to have a carbamoyl phosphate synthetase II in the cytosol (Maresh, Kwan & Kalman, 1969). This may be part of the multienzyme protein CAD (carbamoyl phosphate synthase–aspartate carbamoyl transferase–dihydroorotase), responsible for the biosynthesis of dihydroorotate, a pyrimidine precursor, but this is bound on the multienzyme protein, and it seems unlikely that it would be available for arginine biosynthesis (see Price & Stevens, 1989).

Amino acid requirements

The most abundant body proteins in birds are muscle proteins, collagen, keratin and, during the laying period, egg proteins. It is when these proteins are being synthesised most rapidly that amino acid deficiencies are likely to become apparent. The proportions of each essential amino acid required in

the diet are given in Table 2.1. It can be seen that these vary between growing and adult birds, and also during egg-laying. This in turn can be related to the differing amino acid composition of body tissue, egg protein and feathers (Table 2.2), which are synthesised at differing rates in differing physiological states. Amino acids are degraded at different rates, e.g. lysine is degraded more slowly than methionine or isoleucine, and so there is not an exact correspondence between the proportions of amino acids required in the diet and the amino acid composition of the tissues.

L-Amino acids are the naturally occurring form both in proteins and as free amino acids, and the ability of different experimental animals to use different optical isomers of the amino acids has been studied. The chick is only able to use the L-form of threonine, isoleucine and valine but can use D- and L-leucine equally. It can use both D- and L-tryptophan, although the D-form is less effective (see Fisher, 1954). This probably relates to the presence of specific racemases that enable certain D- and L-interconversions to take place. Factors that affect the requirements for specific amino acids which are particularly pronounced in avian species are summarised in Table 2.3.

A number of dietary amino acids are required for processes other than protein biosynthesis. The amino acid methionine not only provides the source of methionine, and sometimes cysteine, in the body proteins but is also an important source of methyl groups used in the biosynthesis of compounds such as creatine and thymine. Methionine is not the only dietary source of methyl groups, another is *choline*. In rats, provided there is adequate methionine in the diet, choline is not an essential ingredient. This is not the case in chicks, where some dietary choline appears to be indispensable (Ruiz, Miles & Harris, 1983). Choline has been shown to prevent perosis in turkeys. The vitamin B_{12} requirement is also related to methionine and choline since this vitamin is required for effective methylation (Almquist, 1952). Some methionine is essential for the growth of chicks, but homocysteine plus choline can replace part of the requirement. The choline requirement in the chick can be replaced by methylethanolamine or dimethylethanolamine, both of which can be methylated to choline. The results suggest a degree of exchangeability of methionine and choline, but that the choline cannot be synthesised *de novo* at a fast enough rate to maintain growth.

A significant proportion of dietary histidine is used in the synthesis of the dipeptide *carnosine*,

Table 2.3. *Factors of particular importance in birds that affect amino acid requirements*

Amino acid	Factor
Arginine	Lack of urea cycle enzymes in liver, in particular carbamoyl phosphate synthetase I
Glycine	Required for collagen, feather protein and uric acid biosynthesis
Cysteine	High content in feathers and, therefore, in high demand during initial feathering and after moulting
Histidine	High requirement during egg-laying, since high histidine content in egg protein
Lysine	High requirement during egg-laying for egg protein

Table 2.4. *Essential amino acid composition of avian carcass proteins*

Amino acid	Domestic fowl (Leghorn)	Ring-necked pheasant	Bob-white quail	Turkey	Goose	Budgerigar	White-crowned sparrow
Arginine	7.8	7.8	7.4	7.6	7.2	6.7	7.1
Cystine	2.6	2.8	2.8	2.4	1.1	(2.0)	2.5
Histidine	3.9	4.5	3.9	3.9	2.8	2.5	2.5
Isoleucine	3.9	4.0	3.9	4.2	3.7	4.4	4.3
Leucine	6.4	6.3	6.7	6.7	7.6	7.1	9.7
Lysine	9.4	10.1	10.0	10.2	7.1	8.1	8.7
Methionine	1.8	1.8	2.0	1.9	1.9	2.5	2.2
Phenylalanine	3.5	3.5	3.5	3.6	4.2	4.4	4.5
Threonine	3.4	3.4	3.5	3.5	3.1	4.6	4.8
Tryptophan	(1.0)	(1.0)	(1.0)	(1.0)	(1.0)	0.8	(1.0)
Tyrosine	2.9	3.0	3.0	2.9	3.1	3.3	3.8
Valine	4.4	4.3	4.3	4.3	4.4	5.4	5.2

The values given are the percentage of the total carcass proteins; those in parentheses are estimates.
Source: Scott *et al.* (1982); Murphy (1994).

present in large quantities in muscle. When histidine is the limiting dietary amino acid, the concentration of carnosine in muscle is depressed. If the dietary histidine is gradually increased, the level of carnosine in muscles remains depressed until the increase in protein synthesis has occurred. This suggests that histidine is used preferentially for protein synthesis, and only after this has reached its maximum rate does the intramuscular carnosine concentration increase. Cysteine is required for the biosynthesis of the tripeptide *glutathione* (γ-glutamyl-cysteinyl-glycine), which functions in protecting the body against free radicals, peroxides and superoxides (Cadenas, 1989). When the dietary levels of cysteine are limiting, the glutathione concentrations are preferentially reduced compared with protein synthesis in chicks. Tyrosine is required for the synthesis of hormones such as catecholamines and thyroxine, and for the pigment melanin. Arginine is required for the biosynthesis of polyamines and creatine.

2.2.2 Amino acid requirements in species other than domestic fowl

Although other avian species have not been as thoroughly investigated, evidence suggests that the duck, quail and ring-necked pheasant have similar essential amino acids, namely phenylalanine, tryptophan, histidine, arginine, lysine, valine, isoleucine, leucine, threonine, methionine and glycine (Scott, Holm & Reynolds, 1963; Elkin, 1987; Shim & Vohra, 1984). Certainly there is very little difference in the amino acid composition of the carcasses (Table 2.4) or of the feathers (Murphy & King, 1982) of domesticated or non-domesticated species, and the relative requirements for essential amino acids for growth, maintenance and egg-laying are very similar in domestic fowl, quail and white-crowned sparrow (Table 2.1). The amino acid requirements that increase most at the onset of egg-laying are phenylalanine and tyrosine, followed by lysine and valine (Murphy,

1994). In the quail, young growing birds (0–3 weeks) have higher amino acid requirements, particularly for lysine, methionine and glycine, than at 4–5 weeks (Svacha, Weber & Reid, 1970). The lack of adequate dietary lysine causes the deficiency disorder achromatosis in turkeys and certain breeds of domestic fowl, in which the feathers lose pigmentation (Fisher, 1972). The amino acid composition of the achromatose feathers is not significantly different from that of normal feathers.

2.2.3 Quantitative requirements for proteins and amino acids

The early studies on amino acid requirements focused primarily on the qualitative aspects of which amino acids were essential; the more recent studies on domestic fowl have focused on the quantitative aspects of efficiency of utilisation of the protein source and the balance of amino acids. The amounts of proteins and amino acids required in an avian diet depend on the following factors: (i) the physiological state, (ii) the digestibility of dietary protein, and (iii) the biological value of dietary protein. Several factors affect the physiological state: whether the bird is actively growing or has reached adulthood; in the case of a hen, whether it is laying or about to come into lay; and feather growth and moulting.

Early experiments on assessing the dietary protein requirements relied on the simple measurement of the protein content of a diet, usually by measuring its nitrogen content, and then assuming an average figure for the nitrogen content of an average protein. This does not take into consideration the differing digestibilities of proteins in different diets, nor their amino acid contents. Nowadays the amino acid content is usually measured and its digestibility assessed in nitrogen balance experiments. If only the essential amino acids, in the form of an amino acid mixture, are provided in the diet, a high amount of each is required to attain maximum growth rate. This is because the essential amino acids have to be converted to provide the 20 amino acids present in body proteins, and this results in lowered efficiency. However, in some cases, the conversion is highly efficient; phenylalanine is as efficient as tyrosine as a source of nutrient tyrosine for young chicks (Sasse & Baker, 1972). When poultry are considered, the efficiency of feed conversion is of paramount importance in profitability. Using balanced diets for domestic fowl, efficiencies in utilisation of dietary protein in the region of 60 to 70% are generally achieved (Scott, Nesheim & Young, 1982). A more

efficient utilisation of nitrogen can be achieved in a diet that otherwise contains only essential amino acids by adding other sources of non-essential nitrogen. The most efficient other sources are glutamate, diammonium citrate, proline and aspartic acid. Excesses of essential amino acids are less efficiently utilised. Dietary urea, adenine and uracil can also be used as a source of nitrogen for amino acid biosynthesis, but they are less efficiently used; for example, only about 25% of dietary urea is used (Baker, 1991). Dietary urea is degraded to ammonia in the caecum. Evidence suggests that domestic fowl fed a low-nitrogen diet utilise urea more efficiently than those fed a high-nitrogen diet (Karasawa & Maeda, 1994).

A significant proportion of the uric acid produced in the urine of birds is transported from the cloaca up into the caeca. A high content of uric acid-utilising bacteria have been found in the caeca of domestic fowl, ducks, pheasants, guinea-fowl and Willow grouse. The metabolism of radioactively labelled uric acid in the Willow grouse caeca has been studied by Mortensen & Tindall (1981). It has a half-life of $\approx$ 26 min, being rapidly degraded to carbon dioxide. However, it is not clear whether the nitrogen is recycled, or whether any organic carbon compounds are reabsorbed.

For the maximum growth rate of young chicks, in relation to feed intake, it is also important to get the correct balance between protein and 'energy'. This balance varies with the age of the bird, and in the case of domestic fowl and related species the optimum is about 28% protein (as a percentage of the whole diet) during early growth; this falls to $\approx$ 10% for adult maintenance (Scott et al., 1982). The optimum percentage of protein depends on age and strain, in the case of the domestic fowl, and also on the quality of the protein in the diet. Between 7 and 9% dietary protein is required to maintain passerines such as the white-crowned sparrow, tree sparrows, house sparrow and dark-eyed junco in positive nitrogen balance (Parrish & Martin, 1977; Murphy, 1993b). The daily requirement for maintenance of adult non-breeding hummingbirds is as little as 1.5% protein, and they can subsist for periods up to 10 days without adverse effects on a diet of sucrose that is protein-free (Brice, 1992). The emu is also able to subsist on a diet with very low nitrogen (Dawson & Herd, 1983).

The biological value of corn can be improved by changing the relative levels of the different proteins present in the seed. Normal corn has a low biological value because of the high content of the protein zein, which is low in both lysine and tryptophan contents.

Through genetic selection a variety of corn (Opaque-2 corn) has been obtained having a lower content of zein and a higher content of glutelin. Glutelin has a higher percentage of both lysine and tryptophan than zein. Soya bean, by comparison, has a high amount of the protein glycinin, which provides a rich source of essential amino acids. Its main drawback is that it also contains proteins, such as tryptic soya bean inhibitor, which inhibit protein digestion and cause enlargement of the pancreas (Scott *et al.*, 1982).

Many studies have been made to estimate how the requirement of particular essential amino acids relates to the percentage protein in the diet required for optimum growth and feed efficiency (see Almquist, 1952; Scott *et al.*, 1982; Jensen, 1989). The results generally show that if the protein content of the diet is low the amino acid requirement increases in direct relation to the protein concentration of the diet. An imbalance in the proportions of amino acids also leads to reduced efficiency; for example, an excess of lysine reduces the efficiency of utilisation of arginine. Such an imbalance of lysine:arginine may exist in diets based on rapeseed meal. A possible explanation is that increased kidney arginase activity is induced by either dietary lysine or arginine. Thus a high lysine:arginine ratio in the diet will cause the degradation of arginine to ornithine and urea, causing a deficiency of arginine (Latshaw, 1993). There is a similar interdependence between the aliphatic amino acids leucine, isoleucine and valine (D'Mello, 1988), where an excess of one impairs the utilisation of another. However, it is clear that moderate excesses (approximately 1%) of single amino acids such as lysine, methionine, threonine or tryptophan, which are common supplements in poultry diets, do not adversely affect performance (Koelkebeck *et al.*, 1991).

2.3 Carbohydrate and fat requirements

The dietary requirements for carbohydrates and fats are less exacting than those for proteins. Fats and carbohydrates form the main energy sources in a typical avian diet, and although protein can also be an energy source, this only occurs in situations in which it is present in excess or where there is a shortage of fat and/or carbohydrate. The energy yields of the major nutrients, as determined by bomb calorimetry, are carbohydrate (17.15 kJ/g), protein (22.59 kJ/g) and fat (38.91 kJ/g). The actual amount of energy obtainable from these nutrients by birds is somewhat less, as this would require each of the nutrients to be absorbed by the gut with 100% efficiency. Both fats and carbohydrate can be fully

oxidised in the body to carbon dioxide and water, but the nitrogen of proteins is not released in its fully oxidised state, since most is excreted as uric acid. This also reduces the energy available from protein to 17.99 kJ/g. The proportions of fat and carbohydrate in the diet vary with the species. Those species having the highest proportion of carbohydrate in the diet are granivorous birds. Cereals have starch as their principal storage material. Carnivorous birds, by comparison, will have a relatively low carbohydrate content in their diet. Most of the carbohydrate would be in the form of glycogen, present in muscle and liver. The principal source of fats is in the form of triglycerides present in the adipose tissue and to a lesser extent several other tissues of the body. Oil-storing seeds, e.g. sunflower seeds, are also a rich form of triglyceride.

The most abundant forms of carbohydrate in plant tissues are cellulose and starch. *Cellulose* is either not utilised at all or only poorly used by most birds. In order to be utilised, it has to be broken down by bacterial enzymes present in the caecum. Most birds have relatively small caecae compared with those of herbivorous mammals. The largest caecae (there are generally a pair in birds) are present in domestic species and galliform species (King & McLelland, 1984). In the Willow grouse, which feeds on particularly tough plant materials, the combined length of the pair of caecae is approximately the same as that of the small intestine. In contrast to most birds, grouse and ptarmigan are able to make effective use of dietary cellulose. The cellulose is fermented by the caecal bacteria to acetate, propionate and butyrate (collectively known as volatile fatty acids), which are then absorbed into the bloodstream and can be further metabolised by tissues. Using [^{14}C]cellulose, Gasaway (1976a) demonstrated that its digestion in the Rock ptarmigan's caecum is relatively complete. Gasaway (1976b,c) estimates that 4% and 7%, respectively, of the energy requirements of the Willow grouse and Rock ptarmigan are met from metabolism of the fatty acids obtained by the microbial digestion of cellulose in the caeca.

There are very few obligate folivorous (leaf-eating) birds. Leaves provide a much less concentrated form of energy than seeds and are, therefore, too bulky for most active fliers. An exception is the hoatzin, which inhabits the rain forests of South America. Over 80% of its diet comprises green leaves. It has a well-developed foregut, similar to that of a ruminant, in which microbial fermentation occurs, enabling it to absorb the volatile fatty acid products of cellulose fermentation (Grajal *et al.*, 1989).

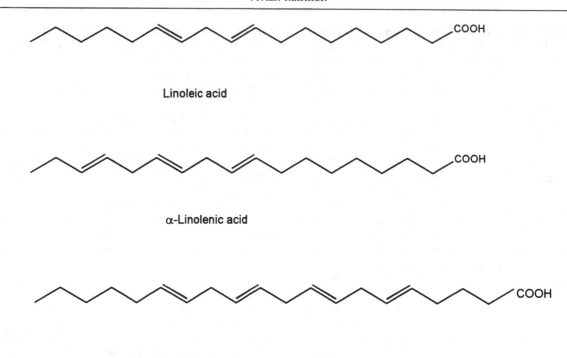

Linoleic acid

α-Linolenic acid

Arachidonic acid

Fig. 2.2. Essential fatty acids.

In general, vertebrates are more readily able to hydrolyse the α-linked than β-linked polysaccharides. Starch is broken down to oligosaccharides or glucose in the small intestine. Glucose and sucrose are efficiently utilised by domestic fowl, whereas galactose and lactose are not (Fisher, 1972). A number of birds have been found to have low sucrase activities in their small intestine, leading to sucrose intolerance. Martinez del Rio (1990) compared the activities of intestinal sucrases from three species of hummingbirds with 11 species of passerine birds from three closely related families (Muscicapidae, Sturnidae and Mimidae) and found the activities of the former were sometimes two orders of magnitude higher than the latter. Hummingbirds are unusual among birds in that they feed on concentrated sucrose-rich nectars. Not only do they have high sucrase activity in their small intestine, but they also have the highest rates of glucose carrier-mediated uptake among vertebrates (Martinez del Rio, 1990).

The domestic fowl, and probably many other birds too, are able to grow normally on a diet free of carbohydrates provided sufficient triglycerides are provided as a source of energy (Scott *et al.*, 1982). The dietary requirement for fat differs from that of carbohydrate in that it can be reduced to a low level but not eliminated completely. For carnivorous birds, fats, in the form of triglycerides will normally provide the most important energy source. Most of the dietary lipids are used as a form of energy, although they may initially be stored in the fat depots, such as adipose tissue. Triglycerides are broken down in the gut mainly to monoglycerides and free fatty acids. In both the chick and the laying hen, the fats are mainly absorbed in the jejunum (Annison, 1983). After absorption, fats enter the hepatic portal system in the form of very large low-density lipoproteins, referred to by Bensadoun & Rothfield (1972) as *portomicrons*. (Lipoproteins are discussed in detail in Chapter 4.) This is to emphasise the distinction between the route followed in the domestic fowl and probably most avian species from that of mammals; in the latter, the lipoproteins, referred to as chylomicrons, enter the lymphatic system, eventually emptying into the bloodstream via the thoracic duct. In the domestic fowl, which lacks a functional intestinal lymphatic system, the lipoproteins enter the venous system directly.

Although dietary fat can be reduced to a low level without any detrimental effects provided other energy sources are available, it cannot be eliminated entirely for prolonged periods. This was first clearly

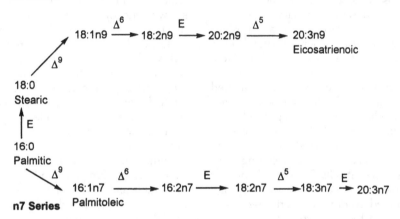

Fig. 2.3. Probable pathways of formation of polyenoic acids in avian species (Watkins, 1991). The structures of the fatty acids are indicated by their chain length, followed by the number of double bonds, followed by the series. The series relates to the position of the double bond nearest the methyl end of the chain. For example, 18:3n3 indicates a C_{18} fatty acid with three double bonds, beginning the third carbon atom from the methyl end of the chain. $\Delta^9, \Delta^6, \Delta^5$ and Δ^4, desaturation enzymes; E, elongation enzyme; PG_1, PG_2, PG_3, prostaglandin precursor fatty acids of the 1-, 2-, and 3-series, respectively.

proved to be the case in rats by Burr and Burr in 1929. They showed that if rats were fed on a fat-free diet, this led to retarded growth, dermatitis and eventual death. These deficiency symptoms could be overcome if certain polyunsaturated fatty acids were added. The acids were referred to as 'essential fatty acids' and comprised linoleic, linolenic and arachidonic acids (Fig. 2.2). The requirement for dietary polyunsaturated fatty acids in the domestic fowl was first convincingly demonstrated in the 1960s by Machlin and Gordon, and by Hopkins and co-workers (Scott et al., 1982). Linoleic acid and/or arachidonic acid are required, but there is not convincing evidence for the requirement of α-linolenic acid.

Mammals and birds possess the enzymes required to introduce a single double bond into a saturated fatty acid, such as that required to convert stearic acid into oleic acid, but they are unable to synthesise de novo polyunsaturated fatty acids. These acids are required in relatively small amounts, approximately 1% of the diet. The C18 fatty acids can be elongated by two carbon atoms at a time using elongase enzymes; further double bonds can be introduced by desaturase enzymes present in avian liver and other tissues. The probable pathways of desaturation and elongation in avian species together with the system of nomenclature is given in Fig. 2.3 (Watkins, 1991). The Δ-6 desaturase enzyme catalyses the rate-limiting step in the conversion of linoleic acid to other polyunsaturated fatty acids in mammals, but it is not clear whether this is the case in avian species. It can be seen that linoleic acid is the precursor of arachidonic acid and other polyunsaturated fatty acids via the intermediate γ-linolenic acid. Another series of polyunsaturated fatty acids arise from α-linolenic acid. Although a clear demonstration of a dietary requirement for α-linolenic acid in the domestic fowl has not been made, analyses of developing brain and retina have shown the presence of polyunsaturated fatty acids containing three double bonds ($n3$), which would be expected to be synthesised from α-linolenic acid (Watkins, 1991). Polyunsaturated fatty acids are normal components of the membrane phospholipids, and they are also biosynthetic precursors of a class of compound, the *eicosanoids*, so named because of their origin from C_{20} polyunsaturated acids (eicosanoic acids). The most important eicosanoids are prostaglandins, thromboxanes and leucotrienes. Prostaglandins were first discovered in the prostate gland but were subsequently found to be widely distributed in animal tissues. Thromboxanes were first isolated from thrombocytes (blood platelets) and leucotrienes from leucocytes. All have pronounced physiological effects at nanogram concentrations on different tissues. Prostaglandins are released from cells and have physiological effects on the secreting cells themselves, or on adjacent cells. These are known as autocrine and paracrine secretions, respectively, and they may be compared to normal endocrine secretions that are circulated via the bloodstream.

The requirement for linoleic acid is most pronounced during rapid growth and during egg formation. Deficiency symptoms in the domestic fowl include retarded growth, increased water consumption, retarded sexual development, decreased egg size and changes in the egg-yolk fatty acid composition. When the growth rate of quail on a normal and linoleic acid-deficient diet are compared, the growth was much higher on the former. In laying hens, linoleic acid deficiency reduces egg size (Griminger, 1986). The prostaglandins $PGF_{2\beta}$ and PGE_2 cause uterine contractions, resulting in egg release, and PGE_2 also facilitates oviposition by relaxing vaginal musculature (Griminger, 1986). The essential fatty acids are required both for membrane phospholipids and as precursors of eicosanoids.

2.4   Vitamins

Vitamins are organic substances that are essential for biological processes in higher organisms. They have to be supplied in small quantities in the diet for growth and normal maintenance. The requirements for vitamins have probably evolved in higher organisms through the loss of enzyme(s) required for their biosynthesis. Many prokaryotes can grow when supplied with only a single carbon source, since they are able to synthesise all the carbon compounds required for growth and multiplication; in contrast, higher organisms require a variety of carbon compounds as nutrient. The vitamin requirements for avian species are very similar to those in mammals, although there are some differences. Vitamins are subdivided into two groups, the fat-soluble and the water-soluble vitamins. Apart from the differences in solubility, there are differences in function. Most of the water-soluble vitamins, which include the B vitamins and vitamin C, have known functions as precursors of coenzymes. The functions of the fat-soluble vitamins are less-clearly understood but include both regulatory roles and redox roles. The fat-soluble vitamins can usually be stored in appreciable amounts in the body, and long-term experiments may be necessary to demonstrate deficiency symptoms. An excess dietary intake of fat-soluble vitamins can lead to toxic effects. The water-soluble vitamins, by contrast, cannot be stored in appreciable amounts and excesses are not generally harmful, since they are excreted. Tables 2.5 and 2.6 summarise the functions and deficiency diseases for fat-soluble and water-soluble vitamins in birds. Each of the vitamins is considered in turn, beginning with the fat-soluble vitamins.

2.4.1   Vitamin A

Vitamin A is the isoprenoid alcohol retinol, which is required for a number of body processes including vision, growth, maintenance of mucous membranes, reproduction and proper growth of the cartilage matrix upon which bone is deposited. Of these, the molecular basis of its role in the visual cycle is best understood. In addition to the alcohol retinol, both retinal and retinoic acid (Fig. 2.4) also act to restore some of the deficiency symptoms. In rod vision, retinol is oxidised to retinal. 11-*cis*-Retinal combines with the protein opsin to form rhodopsin. After the absorption of a photon, rhodopsin undergoes a series of changes, eventually dissociating to opsin and all-*trans*-retinal. This last compound is converted back to 11-*cis*-retinal by retinal isomerase (EC.5.2.1.3) Retinoic acid is able to replace retinol in all functions except the visual cycle and reproduction.

Vitamin A in the diet comes largely from animal

Table 2.5. *Fat-soluble vitamins in avian species*

Vitamin	Deficiency symptoms	Function
Vitamin A (retinol)	Prevents xerophthalmia, night blindness, growth retardation, bone abnormalities	Visual pigments, morphogen
Vitamin D_3 (cholecalciferol)	Malformation of bone and egg-shell, growth retardation, abnormal feather pigmentation in some breeds of fowl	Converted to 1,25-dihydroxyvitamin D_3, which acts as a hormone regulating Ca^{2+} and phosphate metabolism
Vitamin K (phylloquinone, K_1; menaquinone, K_2)	Slow blood clotting	Cofactor for glutamate carboxylase
Vitamin E (α-, β-, γ-tocopherol)	Sterility, encephalomalacia, exudative diathesis	Antioxidant, protection against peroxide radicals

Table 2.6. *Water-soluble vitamins in avian species*

Vitamin	Deficiency symptoms	Function
Thiamin (B_1)	Polyneuritis	Thiamin diphosphate is a cofactor for oxidative decarboxylation and other reactions
Riboflavin (B_2)	Growth retardation, curled toe paralysis in chicks	FAD and FMN are cofactors for several dehydrogenases
Pyridoxal (B_6)	Growth retardation	Pyridoxal phosphate is a cofactor for transaminations and deaminations
Nicotinamide (B_7)	Dermatitis, poor feathering, bowing of legs	Precursor of NAD and NADP
Pantothenic acid	Reduced growth, dermatitis, reduced egg production and hatchability	Precursor of CoA, which is a cofactor for acylation reactions
Biotin	Similar to pantothenic acid	Cofactor for carboxylases
Folic acid	Macrocytic anaemia, poor growth and feathering	Cofactor for C_1 metabolism
Cyanocobalamin (B_{12})	Slow growth, reduced hatchability	Cofactor for methylmalonyl-CoA mutase and other alkylation reactions
Choline	Poor growth, perosis	Precursor of phospholipids and acetyl choline
Ascorbic acid (vitamin C)	Only required by some avian species	Hydroxylation of proline residues in collagen

and fish tissues, the liver being a particularly good source. Plants do not contain vitamin A, but many are good sources of carotenes, which are provitamins. The enzymes of the intestinal mucosa are capable of splitting carotenes into vitamin A, with varying efficiencies. In the rat, β-carotene has 50% of the potency of retinol, but in the domestic fowl it has only 10% of the potency.

Seed-eating birds will acquire their source of vitamin A entirely in the form of the provitamin carotenes.

The requirement for vitamin A in birds has been studied most extensively in the domestic fowl. When adult domestic fowl are fed a vitamin A-deficient diet, the deficiency symptoms may take from two to five months to develop. The body reserves of vitamin A are stored in many tissues, but liver is the principal storage organ. Much of the stored vitamin A is esterified as retinyl palmitate. The developing embryo, however, does not have a large reserve (Takase & Goda, 1990). During embryonic development, the vitamin A stored in the yolk is progressively taken up into the embryo. In a tissue such as the embryonic lung, the vitamin A is present entirely in the unesterified form, and the concentration declines from about E15 until hatching and then begins to rise again once the chicks start feeding. (Development stages of embryos given in days prefixed by E, see

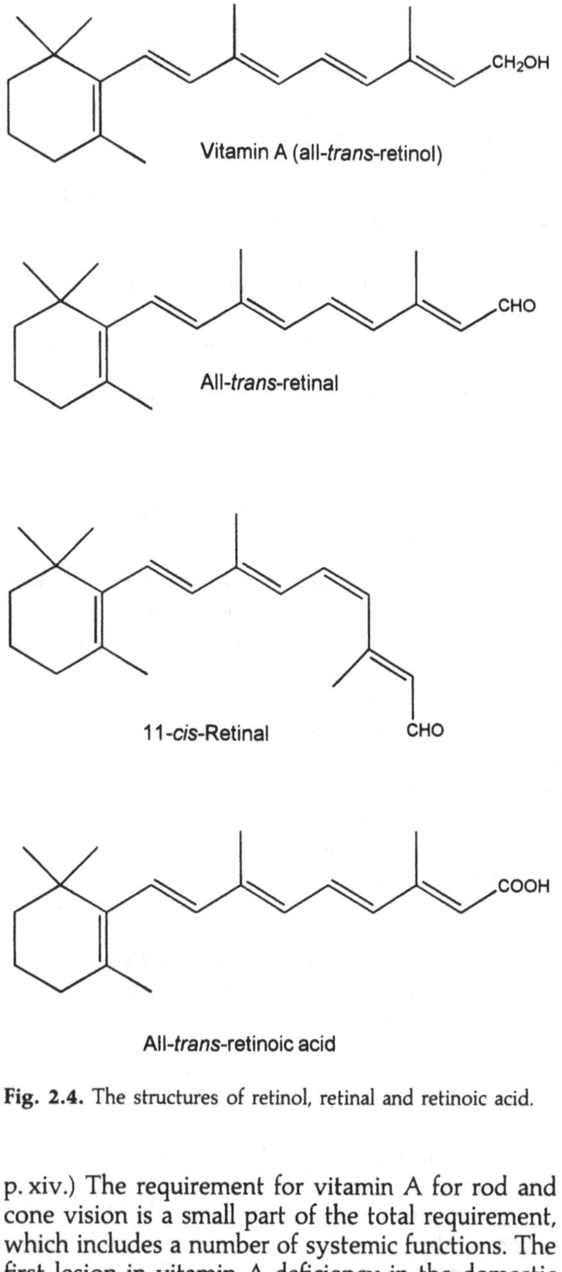

Vitamin A (all-*trans*-retinol)

All-*trans*-retinal

11-*cis*-Retinal

All-*trans*-retinoic acid

Fig. 2.4. The structures of retinol, retinal and retinoic acid.

p. xiv.) The requirement for vitamin A for rod and cone vision is a small part of the total requirement, which includes a number of systemic functions. The first lesion in vitamin A deficiency in the domestic fowl is that mucous epithelium is replaced by squamous, keratinising epithelium, and this is particularly noticeable in the oesophagus, crop and respiratory tract.

Retinoic acid is perhaps the best characterised morphogen in vertebrates and has been extensively studied in developing chick limb bud. There is a concentration gradient of retinoic acid across the limb anlage, which acts to regulate its development

(Thaller & Eichele, 1987). The expression of a chicken gene carrying a homeobox and designated *Msx-1* is closely related to limb bud formation and is discussed further in Section 12.5.

2.4.2 Vitamin D

The presence of an antirachitic factor in cod liver was discovered in cod liver oil in 1923. Cod liver oil was originally thought to contain a single vitamin, namely vitamin A, but when oxygen was bubbled through the oil at 120 °C for several hours, it was found to lose its effectiveness at curing vitamin A deficiency but not its antirachitic effect. This retention was the result of the greater chemical stability of vitamin D. At a similar time it was shown that irradiation of either deficient chicks or their food supply had a curative effect on rickets. The relationship between the vitamin and the effects of irradiation became understood soon after the chemical nature of the vitamin was known. There are two main forms of vitamin D, known as cholecalciferol (D_3) and ergocalciferol (D_2), the former occurring in animal tissues and the latter in plant tissues. They differ only by a methyl group. The original vitamin D_1 turned out to be a mixture. Cholesterol is synthesised *de novo* in all animals and is oxidised to 7-dehydrocholesterol. The latter, when present in the skin, acts as a provitamin, being converted to vitamin D_3 by ultraviolet irradiation (Fig. 2.5). Hence the requirement for vitamin D can be met either from the diet or by a photochemical reaction occurring in the skin.

Vitamin D has a range of effects, all related to proper metabolism of calcium and phosphorus. It is converted to hydroxylated secosteroids, and it is these compounds that act on target tissues by a general mechanism common to all steroid hormones. Much of the work that has led to our present understanding of the mode of action of vitamin D has come from studies using the White Leghorn cockerel and the Japanese quail (Norman, 1987). These species turn out to be very suitable for such studies since they are particularly sensitive to deficiencies of vitamin D. They have been used to study the role of vitamin D in intestinal calcium absorption, bone calcium mobilisation and the induction of vitamin D-binding protein. Although there are many similarities between vitamin D functioning in birds and mammals, there are also differences. Vitamins D_2 and D_3 have similar potencies in mammals, but, in birds, vitamin D_3 is about ten times more potent. The vitamin D-dependent calcium-

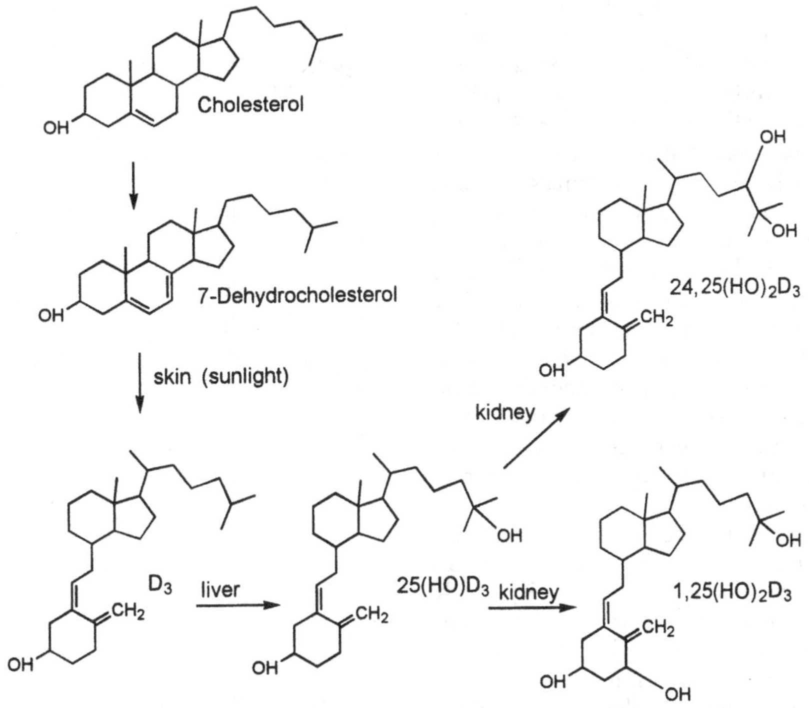

Fig. 2.5. The formation of vitamin D_3 and its hydroxylated derivatives from cholesterol.

binding protein *calbindin* is a 10 kDa protein in mammals and a 28 kDa protein in the domestic fowl.

The dihydroxylated form of vitamin D_3, $1,25(HO)_2D_3$, was first discovered by Haussler *et al.* (1968) in the intestine of the domestic fowl. Its mode of action is similar to that of other steroid hormones, e.g. oestrogens, androgens and corticosteroids; the hormone is taken up into the target tissues by intracellular binding proteins and eventually interacts with the chromatin of the cell nucleus where it modifies gene expression. Over 100 genes have been shown to come under the influence of $1,25(HO)_2D_3$ (Norman & Hurwitz, 1993). (This is discussed further in Section 10.6.) In addition to affecting gene expression, which has a long time response (hours) since it involves the synthesis of new proteins, $1,25(HO)_2D_3$ also has a fast-acting effect. There is evidence that it is able to open Ca^{2+} channels in the plasmalemma causing a rise in intracellular Ca^{2+}, which may function as a second messenger (Norman & Hurwitz, 1993). The first hydroxylation occurs in the liver and the second occurs principally in the kidney, yielding the two dihydroxylated derivatives (Fig. 2.5). Receptors for $1,25(OH)_2D_3$ in the domestic fowl have been found in intestine, kidney, bone,

parathyroid gland, pancreas, chorioallantoic membrane, egg-shell gland and pituitary (Norman, 1987). The calcium-binding protein calbindin has also been detected in all the above tissues except the pituitary.

Vitamin D is required for efficient absorption of calcium and phosphorus in the intestinal tract, and it is also necessary for calcium and phosphorus homeostasis within the body. Deficiency of vitamin D will, therefore, lead to calcium deficiency and will affect processes such as bone formation and egg-shell production. Bone mineralisation is dependent on the maintenance of Ca^{2+} and phosphate levels. The effect of vitamin D on these processes is mediated by calbindin and is discussed in Section 10.6. Aluminium toxicity is, at least in part, accounted for by disturbance of calcium homeostasis. Aluminium compounds reduce vitamin D-dependent Ca^{2+} absorption in chicks. They act by reducing the intestinal calbindin concentration (Dunn *et al.*, 1993).

2.4.3 Vitamin K

In 1929, Dam discovered that when chicks were fed a purified diet very low in lipids they developed intramuscular and subcutaneous haemorrhages, anaemia and showed prolonged blood clotting time. A few years later, Dam (1935) and other groups of workers found a fat-soluble factor present in a number of plant materials that could overcome these symptoms. He named the factor vitamin K (Koagulations vitamin). Since his initial discovery, the vitamin has been shown to be required by mammals, including humans, although the requirements can be partially or completely met in some cases through synthesis by bacteria present in the gut. Vitamin K deficiency is usually defined by impairment of blood coagulation, resulting from decreased prothrombin, factor VII, factor IX and factor X, and it is usually measured as a prolongation of a one-stage prothrombin time. Since the 1970s, the mechanism of action of the vitamin has become increasingly clear (see Vermeer, 1990), and it appears unique among fat-soluble vitamins in having a role as a cofactor.

Vitamin K includes a number of related compounds all of which have the common naphthoquinone structure. They are subdivided into K_1, K_2 and K_3 (Fig. 2.6). Vitamin K is the cofactor for the enzyme vitamin K-dependent γ-glutamylcarboxylase (Fig. 2.7). This enzyme is an integral membrane protein present in the endoplasmic reticulum of many tissues including liver, lung, testis and skin. The only tissues in which it has not been detected so far are brain and tendon. It catalyses the carboxylation of glutamate residues present in a number of secreted proteins from a variety of tissues. The effect of carboxylation of the glutamate residues is to increase considerably the Ca^{2+}-chelating ability. The first substrate to be discovered for this enzyme was prothrombin, which has 10 glutamate residues that become carboxylated (Gla). A number of substrates have been discovered from a range of tissues. At least four of the proteins involved in the blood-clotting cascade (prothrombin, factors VII, IX and X) become carboxylated. The two most important ones associated with calcified tissues are the matrix Gla protein (MGP) and osteocalcin. MGP is expressed in several tissues; osteocalcin is synthesised in the osteoblasts and secreted, where it becomes bound to hydroxyapatite. The functions of these proteins are not yet understood. Besides these Gla proteins, others have been detected in lower concentrations in other tissues.

The role of vitamin K as a cofactor is that it provides the energy necessary for the carboxylation step. Vitamin K exists in three forms, the quinone (K), the dihydroquinone (KH_2) and the epoxide (KO), as illustrated in Fig. 2.8. The carboxylation reaction is linked to the conversion of KH_2 to KO. Two reduction steps are then necessary to recycle the

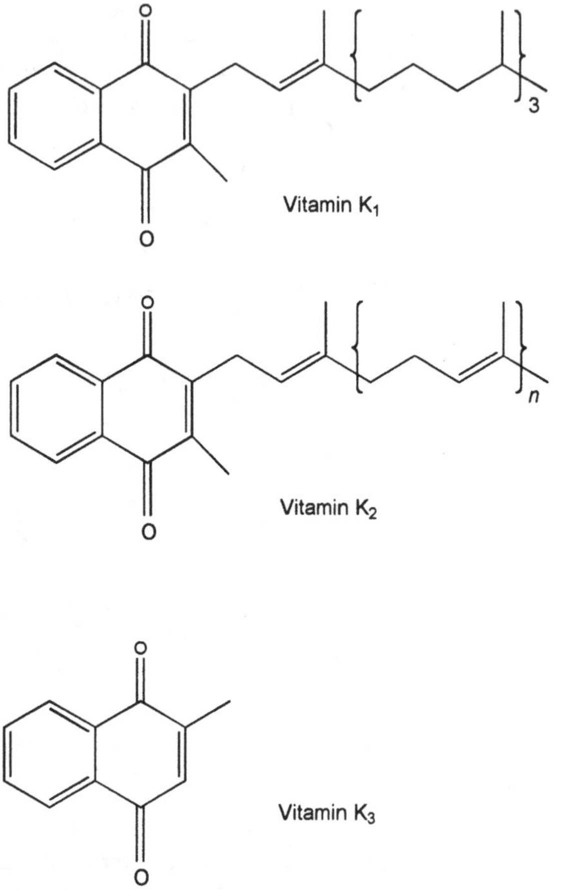

Fig. 2.6. Structures of vitamins K_1, K_2 and K_3.

cofactor. The exact nature of this linkage is not clear. Experiments carried out *in vitro* show that KO formation may exceed carboxylation by 5–10-fold (Vermeer, 1990). The vitamin K requirement for chicks is approximately ten times that required for rats, and it is easier produce the deficiency symptoms by dietary restriction in chicks than in rats. This is perhaps why vitamin K deficiency was first detected in chicks. The basis of this higher requirement is now fairly clear. During the cycle in which glutamate residues become converted to γ-carboxyglutamyl residues (Fig. 2.8), vitamin K becomes converted to its epoxide. After feeding vitamin K_1, high concentrations of vitamin K 2,3-epoxide can be detected in chick liver and serum, and also in rat serum but not liver (Will, Usui & Suttie, 1992). The epoxide is normally reduced by hepatic vitamin K epoxide reductase, in order for it to be recycled. The activity of this enzyme in chicks is only $\approx 10\%$ that in rat

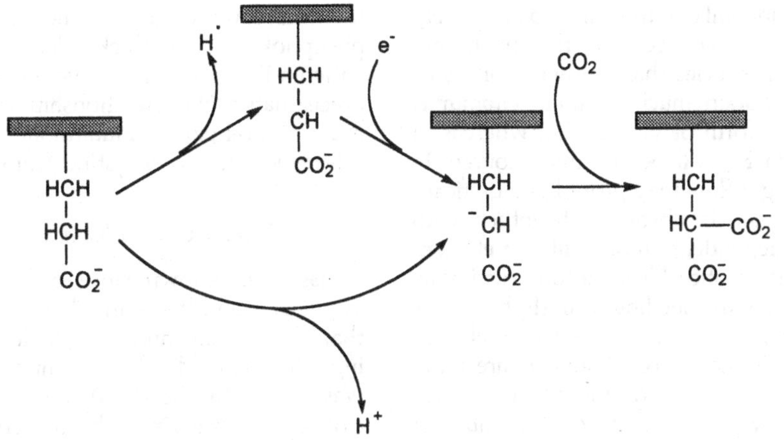

Fig. 2.7. Proposed mechanism of carboxylation of glutamate residues by vitamin K-dependent carboxylase (Vermeer, 1990).

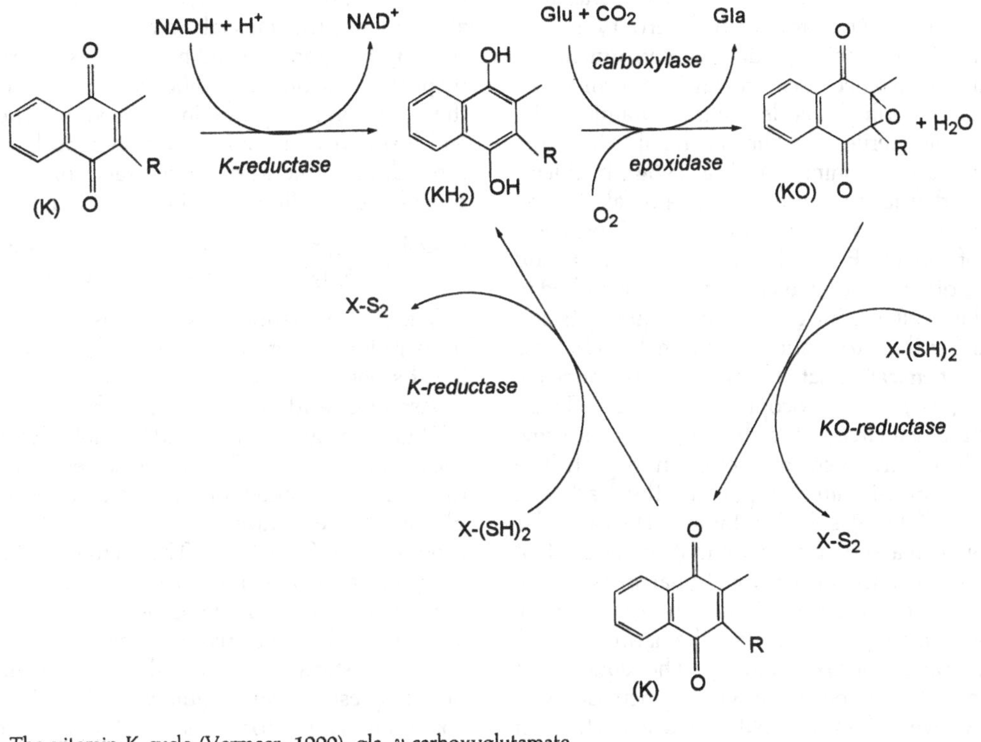

Fig. 2.8. The vitamin K cycle (Vermeer, 1990). gla, γ-carboxyglutamate.

liver, and it is the inability of the chick to effectively recycle the epoxide that accounts for the higher requirement. Avian species that are herbivorous or omnivorous will obtain much of their vitamin K requirement in the form of vitamin K_1, whereas in carnivores vitamin K_2 will be the main source. In many species, the gut flora may provide a significant source. The vitamin is absorbed together with dietary fat and is dependent on bile salts for efficient absorption. The most noticeable symptoms of vitamin K deficiency in chicks are bleeding from slight bruises and haemorrhages; bone development may also be affected. The requirements for vitamin K are more pronounced under stress, e.g. during infections such as coccidiosis or Newcastle disease. A number of vitamin K antagonists are known, such as sulphaquinoxaline and warfarin; these are structural analogues of vitamin K and compete with them.

2.4.4 Vitamin E

Vitamin E was discovered as a fat-soluble vitamin required for normal reproduction in rats. It was later shown to be required for normal fertility in the rooster and for normal reproductive performance in hens. This has also been demonstrated in quail: when fed on a semi-purified diet deficient in vitamin E, the males became sterile, and the hatchability of eggs was depressed (see Shim & Vohra, 1984). In chicks, vitamin E deficiency causes encephalomalacia, exudative diathesis and muscular dystrophy. The richest sources of vitamin E are vegetable oils, cereals and eggs. Like other fat-soluble vitamins, it is absorbed in the small intestine as micelles, which require fats and bile salts for their formation. Vitamin E includes a number of chemically related compounds, the principal ones being α-, β- and γ-tocopherols (Fig. 2.9). These are structurally related to the ubiquinones (coenzyme Q), which are involved in electron transport. The cellular role of vitamin E is perhaps the least well understood of the fat-soluble vitamins. The two roles that have been assigned to it are that of antioxidant and of maintenance of the functional integrity of membranes, the two being related. There is evidence that vitamin E can protect against the harmful effects of peroxide and superoxide radicals. The requirement for vitamin E has been shown to increase with increasing levels of unsaturated fatty acids in the diet (see Coates, 1984). The integrity of cellular membranes in ducklings involves vitamin E. Haemolysis is a feature of vitamin E deficiency. The muscular dystrophy associated with vitamin E deficiency may be related to specific muscle proteins; for example,

there is an increased turnover rate of creatine phosphokinase in chicks deficient in vitamin E. Vitamin E and selenites have related effects, but the precise nature of the relationship is not clear. Selenium is an integral part of glutathione peroxidase, and in this respect, its role is distinct from that of vitamin E.

2.4.5 Vitamin C (ascorbic acid)

It has been known since before Captain Cook's voyages round the world, between 1772 and 1775, that fresh fruit and vegetables were necessary ingredients of a diet to prevent scurvy in humans. It was not until the 1930s that ascorbic acid was isolated and shown to be the antiscorbutic agent. Since then much research has been carried out to try to understand the molecular basis of the action of ascorbic acid. The clearest roles that have been ascribed to it are as an intracellular reducing agent and as a cofactor in the hydroxylation of proline present in collagen. Collagen is the most abundant protein in the body, and it contains about 30% combined proline and hydroxyproline residues. It also contains the unusual amino acid hydroxylysine. Both hydroxyproline and hydroxylysine are formed by post-translational modification of procollagen, i.e. the hydroxyl groups are introduced into the proline and lysine residues after the polypeptide chain has formed. Ascorbic acid is necessary for the enzyme procollagen-proline dioxygenase (EC.1.14.11.2):

procollagen L-proline + 2-oxoglutarate + $O_2 \rightarrow$ procollagen *trans*-3-hydroxy-L-proline + succinate + CO_2

The hydroxylation of collagen is necessary for its proper functioning, which is why, in cases of scurvy, it is tissues rich in collagen that are most affected.

Ascorbic acid is synthesised from glucose via UDPG and glucuronic acid as intermediates, as shown in Fig. 2.10. The key enzymes that catalyse, respectively, a reduction and an oxidation step are glucuronate reductase (EC 1.1.1.19) and L-gulonolactone oxidase (EC 1.1.3.8). These enzymes have been detected in all avian species that have no dietary requirement for vitamin C, either in the liver or kidney, or in both tissues. Early reports in the 1920–30s showed that the domestic fowl, pigeon, ducks, geese, turkeys, guinea-fowl and pheasants could be reared for periods of months on diets apparently lacking vitamin C (see Pardue & Thaxton, 1986). Chaudhuri & Chatterjee (1969) have examined the biosynthetic capacity of liver and kidney of different species to synthesise ascorbic acid. The two key enzymes glucuronate reductase and gulonolactone

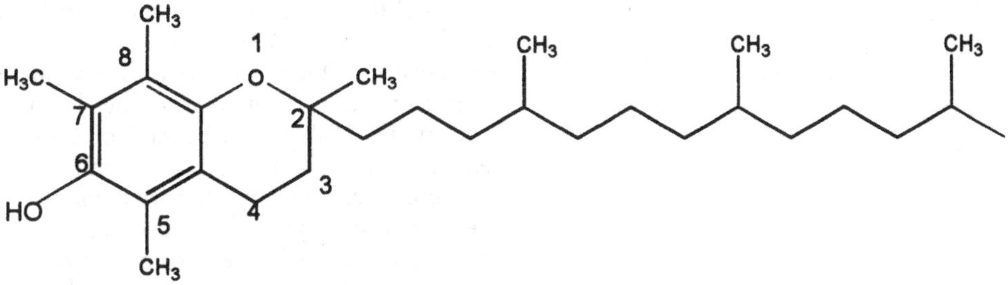

β-Tocopherol, H in position 7
γ-Tocopherol, H in position 5

Fig. 2.9. Structures of α-, β- and γ-tocopherols (vitamin E).

oxidase which were measured are present in the microsomal fractions. Their results show interesting differences within the class Aves that allowed grouping into four categories according to their biosynthetic capabilities: those in which synthesis occurs (i) in kidney only, (ii) in the liver only, (iii) in both kidney and liver, and (iv) in neither tissue. Only the last group has a dietary requirement for ascorbic acid. The tissue distribution of ascorbic acid-synthesising activity in different avian orders is shown in Fig. 2.11. Chaudhuri & Chatterjee (1969) suggest that the most primitive orders synthesise ascorbic acid in the kidney, whereas in more advanced species the process is switched to the liver or is apparently lost altogether. Passeriformes are regarded as the more advanced. However, within the Passeriformes order three different patterns emerge (Fig. 2.11), and even within families (not shown in Fig. 2.11), there is more than one pattern. Amphibians and reptiles synthesise ascorbic acid in the kidney, whereas most mammals synthesise it in the liver, except for humans, monkey, guinea-pig and the fruit bat, which cannot synthesise it. Analysis of the genes and gene products (glucuronate reductase and gulonolactone oxidase) is needed to further this hypothesis.

In normal circumstances, the domestic fowl is capable of synthesising adequate amounts of ascorbic acid, but there is evidence that under stress, e.g. temperature stress, high humidity, parasitic infestation and during egg-laying, the bird's requirement is increased (see Coates, 1984). The rate of synthesis decreases and the adrenal glands become depleted of ascorbic acid under stressful conditions (Kutlu & Forbes, 1993). Diets that are

satisfactory under normal conditions may become ascorbic acid limiting under stress. Under these conditions, addition of ascorbic acid to the diet ameliorates the stress.

2.4.6 B vitamins

One of the initial observations that led to the discovery of the B vitamins was Eijkman's discovery in 1897 that polyneuritis developed in domestic fowl after feeding a diet of polished rice (Carpenter & Sutherland, 1994). This was similar to the condition in humans known as beri-beri. Grijns, Eijkman's successor, showed that it was the polishings that contained the factor that could alleviate the symptoms, and this eventually led to the characterisation of vitamin B_1, thiamin, in the 1920s. It later became apparent that the water-soluble extracts containing the antiberi-beri factor also contained other vitamins. Thiamin, or vitamin B_1, is heat labile. After heating the water-soluble extracts, they were shown to contain a growth factor that was heat stable, and this turned out to be riboflavin or vitamin B_2. In 1930, Norris and Ringrose showed that pellagra-like symptoms occurred when chickens were fed a diet deficient in pantothenic acid. There are now known to be at least eight different B vitamins (Table 2.6). Some of these were discovered as essential nutrients in mammals and then subsequently shown to be requirements in the domestic fowl and other poultry. All those that are required for mammals have also been found to be required for the domestic fowl and poultry in general. In addition, choline, which is not

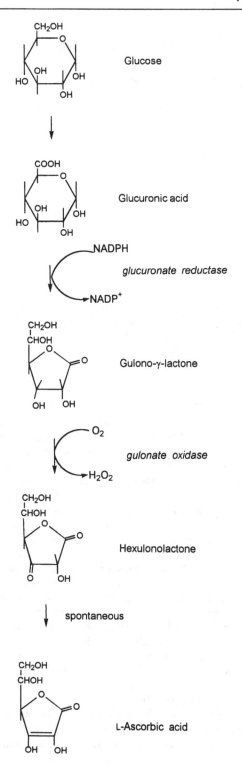

Glucose

Glucuronic acid

NADPH

glucuronate reductase

NADP⁺

Gulono-γ-lactone

O₂

gulonate oxidase

H₂O₂

Hexulonolactone

spontaneous

L-Ascorbic acid

Fig. 2.10. Biosynthesis of ascorbic acid from glucose.

essential for mammals, is required for the domestic fowl (see Section 2.2).

A feature which all B vitamins have in common is that they are required for the biosynthesis of coenzymes (Table 2.6). Several of these cofactors, e.g. thiamin, riboflavin, pantothenic acid and nicotinamide, are essential for what might be broadly described as energy metabolism, and there are common features in the deficiency symptoms. Deficiency of any of these four causes degeneration of the cells lining the crypts of Lieberkuhn and vacuolations in the pancreatic acinar cells. The minimum dietary requirements for these vary for individual vitamins from micrograms to milligrams daily, and this partly reflects the different amounts of the corresponding cofactor present in each cell. These requirements may also be affected by other factors, e.g. synthesis of the vitamin by gut bacteria reduces the dietary requirement. This is less important in the domestic fowl than in some mammals, since the domestic fowl has a comparatively short gut and the microorganisms most active are present beyond the position in the gut where maximum absorption occurs.

The dietary requirement for nicotinamide is also related to the requirement for tryptophan. Dietary tryptophan can be converted with varying efficiencies into nicotinamide, thus dietary tryptophan spares the requirement for nicotinamide. In the domestic fowl and grain-eating birds, there is rarely an excess of tryptophan. Pyridoxal phosphate (vitamin B₆) is required for the interconversion of tryptophan to nicotinamide, and so the requirements for nicotinamide are moderated by the amounts of pyridoxal and of tryptophan residues present in the diet. The ability to convert tryptophan to nicotinamide also appears to depend on the level of picolinic acid carboxylase in the liver. This enzyme converts one of the intermediates, 2-amino-3-acroleylfumaric acid into a branch path and so competes with the main pathway. The levels of this enzyme are comparatively low in the domestic fowl but are much higher in the duck, which fits with the duck having a nicotinamide requirement about double that of the domestic fowl (Scott *et al.*, 1982).

Pyridoxal phosphate is a cofactor required for several reactions involving amino acid interconversion. Its requirement is increased in relation to the amount of protein in the diet. Some of the deficiency symptoms can be readily correlated with their coenzyme function. Thiamin pyrophosphate is a cofactor for pyruvate dehydrogenase, the activity of which is decreased in the brain as a result of deficiency. Pantothenic acid is required not only for

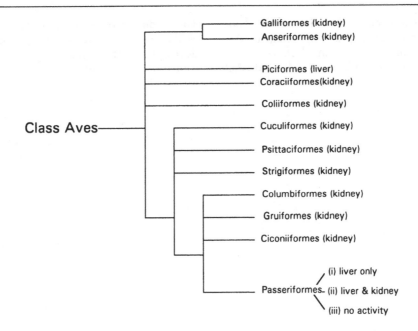

Fig. 2.11. The capacity for ascorbic acid biosynthesis in avian liver and kidney microsomes in different orders of bird based on data from Chaudhuri & Chatterjee (1969), using the biochemical classification of avian orders as shown in Fig. 1.3.

coenzyme A, but also for the acyl carrier protein (ACP) and for citrate lyase and citromalate lyase. Biotin causes a condition known as fatty liver and kidney syndrome in poultry (Coates, 1984). It occurs in broiler chicks having a high energy/protein diet. Affected birds have enlarged livers and kidneys, which also have a high fat content. They are also hypoglycaemic and free fatty acids and triglycerides are increased. Biotin is a cofactor for a number of carboxylation reactions. The liver pyruvate carboxylase is one of the enzymes most sensitive to this deficiency, and there is a consequent reduction in gluconeogenesis.

2.4.7 Vitamin requirements for the embryo

The vitamins requirements by most vertebrates are similar, but one important difference between oviparous and viviparous vertebrates is that in the former all the nutrients required until hatching have to be deposited in the cleidoic egg before oviposition. A cleidoic egg is one sealed from its surrounding by a shell that only permits exchange of gases and limited exchange of water, in contrast to many aquatic eggs, in which water, salts and some nutrient exchange occurs freely. B vitamins are stored in both the yolk and albumen, with the higher concen-

trations in the former. All the lipid stored by the embryo is present in the yolk, so also are all of the fat-soluble vitamins. The vitamin content of the egg is related to the nutritional status of the laying bird. For example, if domestic fowl are fed diets with riboflavin restricted, there is a corresponding reduction in the riboflavin content of both egg-yolk and egg-white (Squires & Naber, 1993). Several of the vitamins are transported to the developing oocyte bound to specific binding proteins, e.g. riboflavin, thiamin, biotin, cobalamin and cholecalciferol (White, 1987; Sherwood et al., 1993). Pantothenic acid is present in the highest concentrations ($\approx$ 1 mg in the yolk of a domestic fowl's egg), but so far no specific binding protein has been identified for it (White, 1987). Nicotinamide is present in very low concentrations, in spite of the high concentrations of NAD and NADP in developing embryos. Part of the nicotinamide requirement is met by synthesis from tryptophan. Many of the binding proteins have been well characterised (see White, 1987; Stevens, 1991b). The importance of the riboflavin-binding protein in the developing embryo was highlighted by the discovery of a mutant incapable of concentrating riboflavin in its eggs. The riboflavin concentration in the eggs was $\approx$ 1% normal. The developing embryos from this mutant die between E10 and E14, unless they are rescued by injection of riboflavin into the egg (see White, 1987). At this stage (E10 to E14) there is a rapid increase in riboflavin kinase activity needed for the synthesis of FMN and FAD.

The binding proteins eventually become degraded as the yolk and albumen become injested by the developing embryo and so provide an additional source of amino acids.

3

Carbohydrate and intermediary metabolism

3.1 Introduction

The principal form in which carbohydrates are stored in birds is glycogen, and the principal form in which they are transported between tissues is glucose. Much of this chapter is devoted to the metabolism of these two compounds. Glucose is an important form of energy currency that can be transported between a number of different tissues of the body through the bloodstream. A measure of its importance in birds may be gauged from the high blood glucose concentrations in a wide range of birds, which are generally higher than in mammals (Table 3.1). Although this in itself does not prove its importance, when examined in conjunction with the active gluconeogenesis occurring in avian liver and kidney, it seems very likely. Before considering glucose metabolism, therefore, the factors that are important in controlling its cellular uptake are discussed. The different metabolic pathways are then considered, followed by their intracellular control and, finally, the changes in carbohydrate metabolism that occur during embryonic development and upon hatching. An important aspect of intracellular control in eukaryotes is intracellular compartmentation. Biosynthesis of glucose and polysaccharides is generally a reductive process requiring a source of energy, whereas glucose catabolism is generally oxidative and generates ATP. Compartmentation of the enzymes involved and different redox conditions are important to enable these to occur within the same cell.

3.2 Intercellular transport of glucose

Glucose circulating in the blood is either of dietary origin, largely arising from polysaccharides that have been hydrolysed in the gut and transported across the intestinal wall, or endogenously released from tissues, mainly the liver and kidney. Blood glucose concentrations have been measured in quite a wide range of birds by a number of different workers (Table 3.1). Since the conditions under which the measurements have been made are not always equivalent, probably only the larger differences between species given in Table 3.1 are significant. The dietary intake of carbohydrate, and hence glucose, will vary considerably between seed eaters at one extreme and carnivores at the other. It is interesting to note that even those birds with relatively low carbohydrate intakes nevertheless maintain high blood sugar levels. Although blood glucose has been measured in several species, it is only in a few species that it has been studied under differing physiological conditions. Blood glucose concentrations are maintained during 3 days fasting in domestic fowl and quail (Davison & Langslow, 1975; Didier, Rémésy & Demigne, 1981). In a small passerine bird such as the horned lark, a 20 h fast reduces the liver glycogen content by $\approx$ 90%, but the blood glucose concentration changes insignificantly from 24.7 $\pm$ 1.84 mM to 22.9 $\pm$ 1.56 mM (Swain, 1992a).

Thus, normal blood glucose levels are high in birds compared with mammals and, in contrast with mammals, the steady-state levels are much less susceptible to change during starvation. The blood glucose level does not indicate its turnover rate, which depends on its rate of uptake and release. Three types of protein are mainly responsible for regulating blood glucose levels: glucose transporters, hexokinases and glucose 6-phosphatases. Glucose transporters are a family of transmembrane proteins that allow facilitated diffusion across the plasmalemma, i.e. movement down the concentration gradient. Avian blood glucose levels are higher than the intracellular glucose concentrations in the tissues, thus enabling facilitated diffusion of glucose into the tissues. The liver is an exception in exporting glucose, and the small intestine and the renal epithelial cells are involved in transepithelial glucose transport. Five different glucose

Table 3.1. *Blood glucose concentrations in different avian species*

Species	Order	Blood glucose concentration (mM)
Adélie penguin	Sphenisciformes	15.9 ± 1.4
Spoonbill	Ciconiiformes	14.7 ± 3.2
Night heron	Ciconiiformes	13.3 ± 2.1
Little egret	Ciconiiformes	16.6 ± 3.3
Mute swan	Anseriformes	6.9 ± 1.0
Marsh harrier	Falconiformes	18.1 ± 1.3
Black vulture	Falconiformes	7.9 ± 0.3
Bald eagle	Falconiformes	14.8 ± 0.3
Domestic fowl	Galliformes	11.1 ± 0.2
Quail	Galliformes	10.7 ± 0.4
Turkey	Galliformes	14.8 ± 0.2
Horned lark	Passeriformes	24.7 ± 1.8
American goldfinch	Passeriformes	22.5 ± 1.1
Rook	Passeriformes	13.1 ± 1.7
Mammals (Horse, cow, sheep, goat, pig, dog, cat, monkey, human)		2.5–8.3 (range)

Compiled from: Sarkar (1977); Veiga *et al.* (1978); Kaneko (1980); Didier *et al.* (1981); Marsh & Dawson (1982); McMurthy, Rosebrough & Steele (1987); Yorita *et al.* (1987); Donaldson *et al.* (1991); Minick & Duke (1991); Lavin *et al.* (1992); Swain (1992a,b); Miksik & Hodny (1992); Aguilera, Moreno & Ferrer (1993); Polo *et al.* (1994).

transporters have been found in mammalian tissues (GLUT 1–5). They have been characterised and their tissue distribution is known. Although there is evidence for glucose transporters in avian tissues (see Section 3.2.2) and it seems probable that a similar family exists in birds, so far they have not been studied in detail. Hexokinase and glucose 6-phosphatase also have an important role in controlling glucose transport by a substrate cycle (see Newsholme & Leech, 1983). They catalyse the phosphorylation and dephosphorylation of glucose, respectively. Phosphorylated sugars cannot generally cross the plasmalemma.

3.2.1 Hexokinase (EC.2.7.1.1.)

Hexokinase catalyses the following reaction:

$$glucose + ATP \rightleftharpoons glucose\text{-}6\text{-}phosphate + ADP + H^+$$
$$\Delta G' = 16.7 \text{ kJ/mol} \tag{3.1}$$

The large negative free energy change is indicative of an enzyme not operating near equilibrium, and, therefore, forming a rate-limiting step and an important potential control point. Hexokinase has been studied in a wide variety of mammalian tissues and has been found to exist as four isoenzymes. Hexokinase IV, generally known as *glucokinase*, has the most distinctive properties. It has a narrower substrate specificity, effectively restricted to glucose, it does not show product inhibition by glucose 6-phosphate and it becomes saturated at much higher glucose concentrations. The K_m for glucose of hexokinases I–III ranges from 0.02 to 0.13 mM, and that of glucokinase from 5 to 12 mM. Glucokinase is only found in the hepatocytes and the β-cells of the pancreas, both of which are served by the hepatic portal vein, which carries glucose absorbed by the small intestine. When the glucose concentration in the hepatic portal vein is high, liver glucokinase regulates the uptake of glucose into the liver where it is converted into glycogen for storage. With the high blood glucose concentrations found in birds, one might expect a homologous glucokinase to be present in avian hepatocytes. A number of studies were made between 1963 and 1978 (see Pearce, 1983) to detect glucokinase in the liver of domestic fowl, but reports appeared contradictory, some detecting its presence and others failing to do so. One of the difficulties lay in factors that interfere with the coupled spectrophotometric assay, which was generally used to detect it. This was overcome by using a radiochemical assay that measures directly the [U-^{14}C]glucose 6-phosphate formed, thus enabling glucokinase to be detected in the livers of the domestic fowl, the Mallard duck and pigeon (Stanley *et al.*, 1984). An additional reason why activity was not always detected was that in avian liver a large proportion of the activity is membrane bound, whereas in mammalian liver glucokinase activity is present almost entirely in the soluble fraction (Wals & Katz, 1981). It appears that the membrane-bound form is physiologically more important, as this is induced by glucose and insulin (Klandorf *et al.*, 1986). Upon fasting, the level of particulate glucokinase falls, and this may partially account for why avian blood glucose does not readily fall upon starvation, since high glucokinase activity might otherwise channel some of the blood glucose into the liver. Detailed structural studies have so far only been carried out on mammalian and yeast glucokinases.

Apart from the studies specifically on glucokinase (hexokinase IV) mentioned above, hexokinase activity has been measured in cultured chick embryo hepatocytes (Hamer & Dickson, 1990) and in liver homogenates from goose, duck, quail, lapwing, kelp gull, dove, parakeet, mocking bird, cow bird, blackbird and yellow finch (Ureta et al., 1973) and in the pectoral muscles of American goldfinch, pigeon and domestic fowl (Yacoe & Dawson, 1983; Blomstrand et al., 1983). The K_m values range from 0.036 to 0.216 mM, and in most cases two isoenzymes have been separated on DEAE–cellulose, but there was no evidence for a high K_m isoenzyme with glucokinase-like activities, although a radiochemical assay was used (Ureta et al., 1973). The tissue distribution of low-K_m hexokinases has not been studied in birds so far.

3.2.2 Glucose transporters

Glucokinase and hexokinase regulate the metabolism of glucose once it has entered a tissue and, therefore, exert an effect on its uptake, but before this occurs glucose must cross the plasma membrane by facilitated diffusion. The family of glucose transporters characterised in mammalian tissues have the following tissue distribution: GLUT 1 (erythrocyte and brain), GLUT 2 (liver and β-cells of the pancreas), GLUT 3 (foetal muscle), GLUT 4 (insulin-dependent carrier in adipocytes and muscle) and GLUT 5 (small intestine). It is difficult to purify these proteins because they are integral membrane proteins generally present in low densities in the cell membrane. The erythrocyte transporter is an exception and represents $\approx$ 5% of human erythrocyte membrane protein, and has been most studied (Silverman, 1991). Glucose transport in human erythrocytes is high, and the intracellular glucose concentration is close to that of the plasma. The situation appears rather different in birds. Shields, Herman & Herman (1964) observed that when [1-^{14}C]glucose was incubated with intact erythrocytes from domestic fowl, goose, duck and humans, very little [^{14}C]carbon dioxide was generated from the avian erythrocytes compared with human erythrocytes. If haemolysed avian erythrocytes were used, then the rate of [^{14}C]carbon dioxide generation was increased. This suggests that the avian erythrocyte membrane is much less permeable. More recent observations have confirmed and extended these findings. Negligible amounts of glucose are taken up by either mature erythrocytes or embryonic erythrocytes. Despite high circulating extracellular glucose concentrations in domestic fowl, little free intracellular glucose is detected (Mathew, Grdisa & Johnstone, 1993). Carrier-mediated uptake of glucose into erythrocytes from embryonic and adult domestic fowl has, however, been demonstrated (Ingermann et al., 1985). Erythrocytes from domestic fowl, when incubated with glucose, show low rates of uptake,

and intracellular concentrations of ATP are not maintained. (NB Glucose uptake is necessary to maintain intracellular ATP levels.) However, if incubated with adenosine alone, or with adenosine and glucose, then ATP concentrations are maintained. Adenosine appears to stimulate glycolysis, and when glucose is also added it has a synergistic effect (Espinet, Bartrons & Carreras, 1989). When a wider range of potential metabolites were tested for their ability to maintain ATP levels in domestic fowl erythrocytes, the most effective were found to be inosine, guanosine and glutamine. Mathew et al. (1993) suggest that the low rate of glucose transport in avian erythrocytes is compensated by a high capacity to transport and metabolise nucleosides and glutamine. The glutamine concentration in domestic fowl serum is $\approx$ 250 μM, and the combined inosine and guanosine $\approx$ 9 μM. Inosine is the metabolic precursor of uric acid, and its metabolism might be part of a salvage pathway in which the carbohydrate moiety is further utilised leaving the purine ring to form uric acid. Therefore, in the tissue in which glucose transport has so far been studied in avian species, namely erythrocytes, the pattern is quite different from that in mammals.

Using antibodies raised against rat GLUT-4, no immunostaining was detected in domestic fowl tissue (Duclos et al., 1993). However, when antibodies against GLUT-1 were used, immunostaining was detected in skeletal muscle, adipose tissue and brain, but not in erythrocytes. The protein recognised had a M_r value of 40 000, which agrees with that expected of GLUT-1 from other species. So far avian GLUT-1 has not been purified.

3.2.3 Glucose 6-phosphatase (EC.3.1.3.9)

Glucose 6-phosphatase catalyses the reaction:

$$glucose\ 6\text{-phosphate} + H_2O \rightarrow glucose + P_i$$
$$\Delta G^{o\prime} = -12.1\ kJ/mol \tag{3.2}$$

It enables the hexose moiety from phosphorylated glucose to be released into the bloodstream. It is present in the endoplasmic reticulum of liver and kidney, but activity in brain and skeletal muscle is generally low. Glucosyl units, after their removal from liver glycogen by phosphorylase, can also be released into the bloodstream. Glycogen, present in skeletal muscle, is used almost exclusively within muscle itself. Low glucose 6-phosphatase activities have been detected in the pectoral muscles of domestic fowl, pigeon and house sparrow (Lackner et al., 1984). This may be an indication that small amounts of glucose can be released from muscle for use in other tissues, or alternatively that glucose 6-phosphatase together with hexokinase present in skeletal muscle regulate glycolysis by a substrate cycle.

Higher glucose 6-phosphatase activities have been detected in the liver from domestic fowl, duck, turkey, pigeon, hummingbird and black vulture and also in domestic fowl kidney (Nordlie, 1974; Campbell & Langslow, 1978; Veiga, Roselino & Migliorini, 1978; Shen & Mistry, 1979; Donaldson & Christensen, 1992). The activity is higher in the liver of black vulture than the domestic fowl. The enzyme from domestic fowl liver has a K_m for glucose 6-phosphate of 3–8 mM and its activity in chick embryos increases to a maximum just before hatching (Campbell & Langslow, 1978). Glucose 6-phosphatase activity in turkey liver is related to the carbohydrate content of the diet, e.g. it is increased 4-fold on fasting compared with levels on a diet containing 55% carbohydrate (Donaldson & Christensen, 1991). Like hexokinase, the reaction it catalyses is not close to equilibrium under physiological conditions, and it is, therefore, a rate-limiting step in the release of glucose units from glycogen stored in the liver and kidney.

3.3 Glycogen and gluconeogenesis

Gluconeogenesis is the metabolic pathway by which glucose is synthesised from smaller metabolites. These generally include the intermediates of the tricarboxylic acid cycle, pyruvate, lactate, glycolytic intermediates, compounds such as glycerol that can be metabolised to glycolytic intermediates, and glucogenic amino acids. Glucogenic amino acids, e.g. aspartate, alanine, glutamate, serine and cysteine, are generally transaminated, or deaminated either directly or indirectly into pyruvate, oxaloacetate or oxoglutarate. For example, aspartate and alanine are transaminated to glutamate (reactions 3.3 and 3.4), and the latter is deaminated by glutamate dehydrogenase (reaction 3.5).

$$\text{aspartate} + \text{2-oxoglutarate} \rightleftharpoons \text{oxaloacetate} + \text{glutamate} \tag{3.3}$$

$$\text{alanine} + \text{2-oxoglutarate} \rightleftharpoons \text{pyruvate} + \text{glutamate} \tag{3.4}$$

$$\text{glutamate} + \text{NAD}^+ \rightleftharpoons \text{2-oxoglutarate} + \text{NADH} + \text{NH}_3 + \text{H}^+ \tag{3.5}$$

Several enzymes are common to both the glycolytic and gluconeogenic pathways, but four enzymes catalyse steps that only occur in gluconeogenesis: pyruvate carboxylase (reaction 3.6), phosphoenol-pyruvate carboxykinase (PEPCK) (reaction 3.7), fructose 1,6-bisphosphatase (reaction 3.8) and glucose 6-phosphatase (reaction 3.2).

$$\text{ATP} + \text{pyruvate} + \text{CO}_2 + \text{H}_2\text{O} \rightleftharpoons \text{ADP} + \text{P}_i + \text{oxaloacetate} \tag{3.6}$$

$$\text{GTP} + \text{oxaloacetate} \rightleftharpoons \text{GDP} + \text{PEP} + \text{CO}_2 \tag{3.7}$$

$$\text{fructose 1,6-bisphosphate} + \text{H}_2\text{O} \rightleftharpoons \text{fructose 6-phosphate} + \text{P}_i \tag{3.8}$$

They catalyse rate-limiting steps in the pathway and are important control points. Glucogenic intermediates of the tricarboxylic acid cycle and amino acids that are transaminated or deaminated to tricarboxylic acid intermediates do not require pyruvate carboxylase (Fig. 3.1). Pyruvate and metabolites such as lactate, alanine, serine, glycine, cysteine that are converted into pyruvate require all four enzymes (Fig. 3.1). Gluconeogenesis is often linked to glycogen synthesis catalysed by glycogen synthase (see Section 3.5).

The main glycogen reserves in birds are present in the liver, kidney and skeletal muscle. Glycogen, present in skeletal muscle is metabolised within muscle and is not generally released into the bloodstream as a source of glucose (see Section 3.2.3). However, it is metabolised anaerobically to lactate within muscle. Lactate is released into the bloodstream, where it is mainly taken up by the liver and used for gluconeogenesis. Skeletal muscle is unable to carry out gluconeogenesis, since it lacks PEPCK; its glycogen is largely derived from blood glucose. It does, however, have both glucose 6-phosphatase (Lackner et al., 1984; Asotra, 1986) and fructose 1,6-bisphosphatase. It is perhaps surprising that the breast muscle of domestic fowl, although it is unable to carry out gluconeogenesis, is a good source of fructose 1,6-bisphosphatase (Anjanayaki, Tsolas & Horecker, 1977). The enzyme is inhibited by AMP with $K_{i,\text{AMP}} = 0.04\ \mu\text{M}$. The intracellular concentration of AMP is about 3 mM, which would suggest that the enzyme is completely inhibited under physiological conditions. Its function is not yet clear. Glycogen, present in the liver and kidney, is converted to glucose and released into the bloodstream. Although avian liver contains much higher concentrations of glycogen than does kidney (Table 3.2), the latter is nevertheless an important source of blood glucose. There is a small glycogen body present in the lumbosacral region of the avian spinal cord, which is minute compared with the other three sources, and whose function is unknown. It is metabolically active in the synthesis and degradation of glycogen, which has led to the view that it may

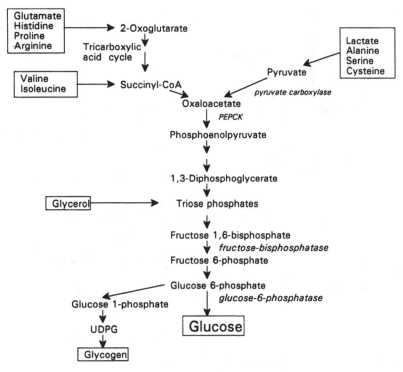

Fig. 3.1. Gluconeogenic pathway from lactate, glycerol and glucogenic amino acids.

Table 3.2. *Glycogen content of tissues of the domestic fowl*

Tissue	Glycogen content (mg/g wet weight)	Proportion of total glycogen (%)[a]	Comment
Liver	30.8	24	After 48 h starvation falls to 0.9 mg/g
Kidney	0.42	0.2	Maintained after 48 h starvation
Skeletal muscle	12	75	
Glycogen body	440 (at hatch)	0.01	Reaches maximum of 5–10% of total glycogen at E15–E21, but becomes a very small fraction in the adult

Source: From de Gennaro (1982); Asotra (1986); Yamano *et al.* (1988).
[a] Approximate values.

play an important role in the central nervous system of birds (de Gennaro, 1982). Both liver and kidney are capable of high rates of gluconeogenesis. In the domestic fowl, approximately 70% of gluconeogenesis is from the liver and the remainder from the kidney (Tinker, Brosnan & Herzberg, 1986). Birds appear capable of higher rates of gluconeogenesis than mammals, and carnivorous birds are generally able to carry out higher rates of gluconeogenesis than graminiferous birds. In a study comparing the black vulture and the domestic fowl when fasted, it was interesting to observe that the domestic fowl synthesised higher levels of PEPCK, glucose 6-phosphatase, alanine aminotransferase and aspartate aminotransferase in the liver, in order to increase the capacity for gluconeogenesis, whereas very little change occurred in the case of the black vulture, presumably because it already had sufficient capacity (Veiga *et al.*, 1978). Most of the detailed studies on gluconeogenesis in birds have been carried out using either the liver and kidney of the domestic fowl or pigeon, but the Japanese quail has also been studied

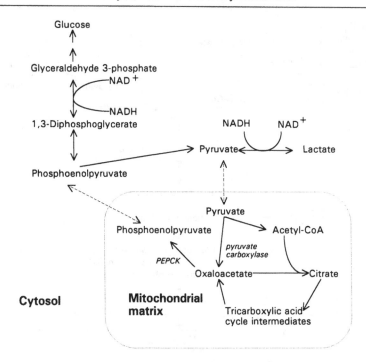

Fig. 3.2. Intracellular compartmentation of gluconeogenesis in avian liver.

although to a more limited extent (Golden, Riesenfeld & Katz, 1982). It is clear from these studies that although the pathways used are very similar to those used by mammals there are differences in their relative capacities and their control. In avian kidney, gluconeogenesis occurs predominantly in the cortex (Yorita *et al.*, 1987), and in liver it occurs in the parenchyma cells.

The metabolic steps in gluconeogenesis occur in two intracellular compartments (Fig. 3.2): the cytosol and the mitochondrial matrix. The enzymes of the tricarboxylic acid cycle reside in the mitochondrial matrix, apart from succinate dehydrogenase which is present in the inner mitochondrial membrane, whereas most of the enzymes of the gluconeogenic pathway are present in the cytosol. Transaminases, such as alanine aminotransferase and aspartate aminotransferase, are present both in mitochondria and cytosol of the domestic fowl liver (Sarkar, 1977). One of the control enzymes in gluconeogenesis, PEPCK, has a different intracellular distribution in avian liver compared with mammalian liver (Table 3.3). PEPCK in both pigeon and domestic fowl liver is present almost exclusively (> 99%) in mitochondria (Söling *et al.*, 1973), whereas in most mammals that have been studied, it is present mainly in the cytosol, and only present, if at all, in smaller amounts in

mitochondria. This accounts for a number of differences observed in gluconeogenesis in avian liver compared with mammalian liver.

A number of studies have been made on the efficacy of various metabolites as gluconeogenic precursors in avian liver (Söling *et al.*, 1973; Dickson & Langslow, 1978; Ochs & Harris, 1978; Brady, Romsos & Leveille, 1979; Sugano *et al.*, 1982a,b; Golden *et al.*, 1982). There is some variation in the rate of gluconeogenesis obtained, depending on the type of preparation used, but the order is typically: lactate ≈ glycerol > pyruvate > alanine > aspartate > serine (Langslow, 1978). In avian liver, lactate is the best substrate, but in mammalian liver pyruvate is better than lactate. Gluconeogenesis is a reductive process, requiring NADH to reduce 1,3-diphosphoglycerate to triose phosphates and, therefore, requires reducing potential. The difference between avian liver and mammalian liver arises from the requirement for reducing equivalents in the cytosol to promote gluconeogenesis. The probable mechanism for using lactate for gluconeogenesis is as follows. Lactate, present in the cytosol, is oxidised to pyruvate, generating NADH for the reduction of 1,3-diphosphoglycerate. Pyruvate enters the mitochondrial matrix, where both pyruvate carboxylase and PEPCK are present in avian liver (Fig. 3.2). The PEP formed is

Table 3.3. *Key enzymes in avian gluconeogenesis*

Enzyme	Tissue	Intracellular location	M_r and subunits	Cofactors and effectors	Comments
PEPCK	Liver, kidney (absent from muscle)	Mitochondria	75 kDa monomer	Mn^{2+} (cofactor)	Mainly lactate→glucose
PEPCK	Kidney	Cytosol	75 kDa monomer	Mn^{2+}	Mainly pyruvate and amino acids to glucose
Pyruvate carboxylase	Liver, kidney; other tissues in low activity	Mitochondria	Tetramer	K^+, Mg^{2+}, acetylCoA (allosteric activator: K_a 12 μM; n_h, 2.5)	
Fructose 1,6-bisphosphatase	Liver, kidney, skeletal muscle	Cytosol	144 kDa, tetramer	Mn^{2+} most effective cofactor; AMP, allosteric inhibitor (K_i, 10 μM)	Purified from turkey liver, domestic fowl muscle
Glucose 6-phosphatase	Liver, kidney, very low in muscle	Mainly endoplasmic reticulum, also nuclear membranes			Loses activity on removal of membrane

Source: From Han & Johnson (1982); Hebda & Nowak (1982); MacGregor *et al.* (1982).

transported out of the mitochondria in exchange for either malate or citrate (Söling & Kleinicke, 1976) and then converted to glucose in the cytosol. Pyruvate, or alanine which can be transaminated to pyruvate, could be converted into PEP by the same route but would not generate the reduced NADH required for gluconeogenesis.

Avian kidney has two distinct PEPCK isozymes, mitochondrial and cytosolic PEPCK, and so differs from avian liver which only has the mitochondrial PEPCK. Between 20 and 50% of the activity is present in the cytosol (Watford *et al.*, 1981). It is like mammalian liver in that pyruvate is a better gluconeogenic substrate than lactate. In avian kidney and rat liver, oxaloacetate is generated in the mitochondria, but the PEPCK that acts on it is present in the cytosol (Fig. 3.3). Oxaloacetate cannot cross the inner mitochondrial membrane. The oxaloacetate formed is reduced to malate within the mitochondria and then passes out in exchange for oxoglutarate, using the malate–aspartate shuttle (Fig. 3.4). On entering the cytosol, the malate becomes reoxidised to oxaloacetate and in so doing generates the necessary cytosolic reducing equivalents to reduce 1,3-diphosphoglycerate to triose phosphates.

The two avian isoenzymes of PEPCK have been purified and their cDNAs sequenced (Weldon *et al.*, 1990). There is 60% homology in their cDNAs. It is

clear that these are the products of distinct genes, although there are similarities particularly in their binding sites. Another important difference is that whereas the mitochondrial enzyme is constitutive the cytosol PEPCK is able to adapt in response to dietary or hormonal stimuli (Watford, 1985). The roles of the two enzymes differ. The mitochondrial enzyme is important for the conversion of lactate to glucose. Much of the lactate arises from anaerobic glycolysis in skeletal muscle. Lactate is released into the bloodstream, where it is largely taken up by the liver. The cytosol enzyme enables the avian kidney to adapt to dietary changes, thereby enabling pyruvate and amino acids to be converted to glucose. Tissues having little gluconeogenic capability, such as avian muscle, adipose tissue and gastrointestinal tract, have very low PEPCK activity, i.e. 1% or less of the activities present in liver and kidney (Wiese, Lambeth & Ray, 1991).

PEPCK requires, in addition to oxaloacetate, its second substrate GTP or ITP. If the enzyme is located in the mitochondria, then GTP will also be necessary in the mitochondria. GTP is generated in substrate level phosphorylation by the citric acid cycle enzyme succinyl-CoA synthetase (reaction 3.9), which is present in the mitochondrial matrix.

$$\text{succinyl-CoA} + \text{GDP} + \text{P}_i \rightleftharpoons \text{succinate} + \text{GTP} \quad (3.9)$$

There is also an additional enzyme present in the

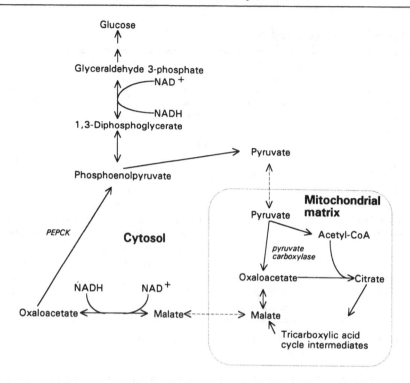

Fig. 3.3. Intracellular compartmentation of gluconeogenesis in avian kidney.

mitochondrial matrix, nucleotide diphosphate kinase (NDP kinase), which catalyses reaction 3.10.

$$GDP + ATP \rightleftharpoons GTP + ADP \qquad (3.10)$$

Söling (1982) has studied the intracellular distribution of NDP kinase in rat and pigeon livers. The NDP kinase activity in the rat liver mitochondrial matrix is so low that he concludes that all the GTP required in the matrix must come from succinyl-CoA synthetase. By contrast, pigeon liver mitochondria have significant NDP kinase activity, which may generate the GTP required for PEPCK.

Oxaloacetate is formed from pyruvate by pyruvate carboxylase, located in the mitochondria. Pyruvate carboxylase is a biotin-requiring enzyme; its kinetic mechanism has been studied in detail using the enzyme purified from domestic fowl liver (Attwood & Graneri, 1992). In tissues where PEPCK is located in the cytosol, its substrate oxaloacetate is required in the cytosol for the formation of PEP for gluconeogenesis. The malate–aspartate shuttle is required for gluconeogenesis in avian kidney, according to the scheme in Fig. 3.3, but whether or not it is also required for gluconeogenesis in avian liver is unresolved. There is evidence for the existence of a

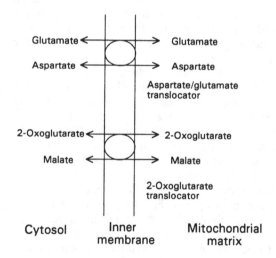

Fig. 3.4. The malate–aspartate shuttle.

shuttle in avian liver, since glutamate–oxaloacetate transaminases and malate dehydrogenases are present in both mitochondria and cytosol. Also, if mitochondria isolated from the liver of domestic fowl are incubated either with glutamate or with glutamate and malate, then aspartate is formed (Ochs & Harris, 1980).

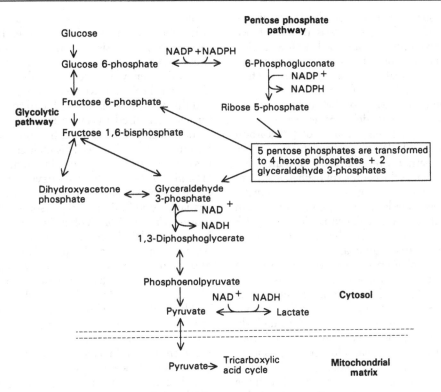

Fig. 3.5. Key intermediates in the glycolytic pathway and the pentose phosphate pathway.

Aminooxyacetate, an inhibitor of glutamate–oxalacetate transaminase, inhibits the formation of aspartate. Söling & Kleinicke (1976) observed that aminooxyacetate did not inhibit the formation of glucose from lactate and, therefore, concluded that the malate–aspartate shuttle was not essential for the lactate gluconeogenesis in avian liver. However, Ochs & Harris (1980) found that aminooxyacetate did block lactate gluconeogenesis when lower concentrations of pyruvate were used and incubation was for longer than 15 min. They concluded that the malate–aspartate shuttle was required.

Gluconeogenesis in the liver and kidney often results in the release of glucose into the bloodstream, but under certain conditions it may lead to the production of glycogen. For this, glucose 1-phosphate is converted to UDPG, and then glycogen synthase catalyses its conversion to glycogen (see Section 3.5).

3.4 Glycolysis, the pentose phosphate pathway and the tricarboxylic acid cycle

The breakdown of glucose occurs in all vertebrate tissues. Together with the breakdown of lipids (see Section 4.6) these processes generate the ATP and intermediary metabolites for a wide range of biosynthetic activity and for mechanical and other forms of work. The breakdown can be divided into two stages: that from glucose to pyruvate, and that from pyruvate to carbon dioxide and water. The first stage involves the glycolytic sequence or pentose phosphate shunt (Fig. 3.5). Glycolysis may occur under aerobic or anaerobic conditions; under anaerobic conditions the end-product is lactate, formed by reduction of pyruvate. This reduction enables the regeneration of NAD^+, which is required for one of the earlier steps in glycolysis catalysed by glyceraldehyde phosphate dehydrogenase.

3.4.1 Relative importance of the pentose phosphate pathway in avian tissues

The importance of the pentose phosphate shunt is that it enables the generation of NADPH and a number of pentose phosphates, but particularly ribose 5-phosphate. The NADPH is required for a number of biosynthetic processes, e.g. steroid and fatty acid biosynthesis, and ribose 5-phosphate for the synthesis of nucleotides and nucleic acids. The extent to which this pathway is used varies with the tissue type. In mammalian tissues, it is most used by those tissues that have high rates of lipid synthesis. Its importance in avian tissues has only been assessed in a few studies. Two methods can be used to assess the importance of the pentose phosphate pathway in tissues: the first is to measure the activities of the two rate-limiting enzymes, namely glucose 6-phosphate dehydrogenase and 6-phosphogluconate dehydrogenase, and the second is to compare the amounts of [^{14}C]carbon dioxide formed from [1-^{14}C]glucose and [6-^{14}C]glucose. The glycolytic pathway releases equal amounts from both sources, whereas the pentose phosphate pathway releases more from [1-^{14}C]glucose. No systematic studies have been made for a range of tissues from adult birds, but several have been made using chick embryos and growing chicks. The data obtained up to 1971 are summarised by Pearce & Brown (1971). The pentose phosphate pathway is clearly operative in a number of tissues, including liver, adipose tissue (Goodridge, 1968b), red blood cells (Shields et al., 1964), kidney (Shen & Mistry, 1979) and the nervous system (Larrabee, 1989). However, it seems unlikely that it is a major route in any of these tissues. Goodridge (1968b) observed that glucose 6-phosphate dehydrogenase and phosphogluconate dehydrogenase activities in liver do not increase in parallel with the increase in lipogenesis in young chicks and concluded that they do not play an important role in generating NAPDH for lipogenesis. Bannister (1984) has confirmed that the combined activities of glucose 6-phosphate dehydrogenase and 6-phosphogluconate dehydrogenase in adult liver from the domestic fowl are low, and he was unable to detect the presence of 6-phosphogluconate in liver. Shen & Mistry (1979) measured glycolytic and pentose phosphate pathway enzymes in the kidneys of developing chicks up to 58 days after hatching. The activities of glucose 6-phosphate dehydrogenase are generally an order of magnitude lower than those of 6-phosphofructokinase, again suggesting that the pentose phosphate pathway only plays a minor role. A correlation between chondrocyte proliferation within the epiphyseal growth plate in domestic fowl and glucose 6-phosphate dehydrogenase activity has been observed. Farquharson et al. (1992) suggest it may be related to the provision of ribose phosphate for nucleic acid synthesis.

3.4.2 Oxidative versus anaerobic metabolism

The tricarboxylic acid cycle only occurs under aerobic conditions, and much of the early work carried out by Krebs used pigeon breast muscle homogenates (Krebs & Lowenstein, 1960) because it is metabolically a very active tissue. The enzymes catalysing glycolysis and the pentose phosphate shunt occur in the cytosol, whereas the tricarboxylic acid cycle enzymes occur in the mitochondria. Glucose is broken down to pyruvate in the cytosol, and the pyruvate passes into the mitochondria to be further catabolised. Pyruvate dehydrogenase, present in the mitochondria, regulates the entry of acetyl-CoA into the citric acid cycle. This multienzyme complex has been purified from pigeon breast muscle (Furita & Hashimoto, 1982). The first enzyme of the multienzyme complex, pyruvate decarboxylase, is subject to covalent modification by phosphorylation, which causes its inactivation. A phosphoprotein phosphatase has been isolated from pigeon liver that is able to dephosphorylate the phosphorylated pyruvate dehydrogenase complex, thus reactivating it, but it is uncertain whether pyruvate dehydrogenase is its physiological substrate (Hinman et al., 1986).

Although glucose breakdown occurs in all tissues, the relative importance of glycolysis, the pentose phosphate pathway and the citric acid cycle varies between tissues. In skeletal muscle, glycolysis is of prime importance in the generation of ATP for muscle contraction, whereas in liver, where the rate of glycolysis is much lower, its role is primarily to provide intermediary metabolites for biosynthesis. The biosynthesis of triglycerides in liver is particularly important in birds, where adipose tissue has very limited capacity for de novo triglyceride biosynthesis (see Section 4.3). There are considerable differences in the importance of anaerobic and aerobic metabolism of glucose in different types of skeletal muscle. They can be divided into fast-twitch glycolytic muscle fibres and fast-twitch oxidative muscle fibres. The former are sometimes known as white fibres and the latter as red fibres, on account of the higher myoglobin and cytochrome content of the latter. Anatomic muscles may be made up of varying proportions of each. Two key enzymes, 6-phosphofructokinase and 2-oxoglutarate dehydrogenase, are useful indicators of the relative importance of glycolysis and the citric acid cycle, and hence the importance of aerobic and anaerobic metabolism. Citrate synthase is also a good marker enzyme for the citric acid cycle (Marsh, 1981). In glycolysis, the two rate-limiting steps between glucose 6-phosphate and pyruvate are 6-phosphofructokinase and pyruvate kinase. Both are non-equilibrium enzymes and control points in the pathways. 2-Oxoglutarate dehydrogenase is the enzyme that gives the best quantitative assessment of the capacity of muscle for aerobic metabolism.

Table 3.4. *Comparison of mammalian and avian pyruvate kinase isozymes*

	L isozyme	R isozyme	M_1 isozyme	M_2 isozyme
Mammalian tissue	Most abundant isozyme, present in liver, kidney, small intestine	Erythrocyte	Muscle, brain	Foetal tissues, most adult tissues
Avian tissue	Possibly present in liver in small amounts	Three R type isozymes in pigeon erythrocytes	Muscle	Most abundant form, present in liver, kidney, lung and other tissues
FBP activation	+	+		+
ATP inhibition	+	+	+	+
Alanine inhibition	+	+	−	+
Kinetics with PEP	Sigmoid	Sigmoid	Hyperbolic	Sigmoid

Blomstrand *et al.* (1983) have compared the activities of 6-phosphofructokinase and oxoglutarate dehydrogenase in pectoral and cardiac muscles from pigeon and domestic fowl as a measure of the capacities of the tissues for anaerobic versus oxidative metabolism. They found that cardiac muscle from both pigeon and domestic fowl were capable of generating approximately equal amounts of ATP from both glycolysis and the citric acid cycle. In contrast, the pectoral muscle of the domestic fowl, which is a good example of white fibres, had only a 1% capacity for aerobic metabolism compared with anaerobic metabolism, but the pigeon pectoral muscle, which has a high proportion of red fibres, had similar aerobic and anaerobic capacities. Higher citrate synthase activity is found in avian muscle than in any other vertebrate muscle, and this may be an indication of higher energy demands of flight muscle. The highest citrate synthase activity reported is in the pectoral muscles of the catbird, with similar activities in the house sparrow and American goldfinch. These have about twice the activity present in the pigeon pectoral muscle (Marsh, 1981). The capacity of pectoral muscles to carry out aerobic metabolism is related to the ability to perform sustained flight.

3.4.3 Pyruvate kinase isozymes

The second rate-limiting step in glycolysis is the conversion of PEP to pyruvate, catalysed by pyruvate kinase. There are a number of isoenzymes of pyruvate kinase, and they have been classified according to their properties and their tissue distribution in mammals. The avian enzymes have then been fitted to the mammalian classification where appropriate (Table 3.4). Those enzymes present in tissues that carry out gluconeogenesis generally have more

sophisticated control mechanisms than those in other tissues. All four isozymes are tetramers. The L isoenzyme is the main form present in mammalian liver (70–90% of the total), and is also present in small amounts in kidney and small intestine. High levels of the L form are induced in mammalian liver in response to high carbohydrate diets. The much smaller amounts of the M_2 isozyme present in mammalian liver are present in the non-parenchymal cells. Of the four groups listed, M_2 has the widest tissue distribution and controls the substrate cycle between pyruvate, PEP and oxaloacetate (Fig. 3.6). Pyruvate kinases have been less well studied in avian tissues, but it is clear that there are significant differences compared with mammalian tissues. The M_2 isozyme is also the most widely distributed, but, in contrast to mammals, it is the predominant form present in avian liver and is present in kidney and lung. It has been characterised and sequenced (Lonberg & Gilbert, 1983; 1985). It is inactivated by phosphorylation of a serine residue and is a tetramer of four identical subunits, M_r 58 000 (Fister *et al.*,1983). An L isozyme that occurs in much smaller amounts in the liver has been reported (Strandholm, Cardenas & Dyson, 1975; Eigenbrodt & Schoner, 1977); although more recently Abramova, Mil'man & Kuznetsov (1992) claim that no L isozyme is present in liver from domestic fowl throughout its development. They have evidence of two forms of the M_2 isozymes differing in isoelectric points (pI 6.0 and 6.8); the pI 6.0 form is converted to the pI 6.8 form on incubation with fructose 1,6-bisphosphatase. Unlike the pyruvate kinases from mammalian tissues, their activity appears unchanged in response to dibutyryl-cAMP (Ochs & Harris, 1978); although they are phosphorylated *in vitro* by a cAMP-independent mechanism (Engstrom *et al.*, 1987). Also the total

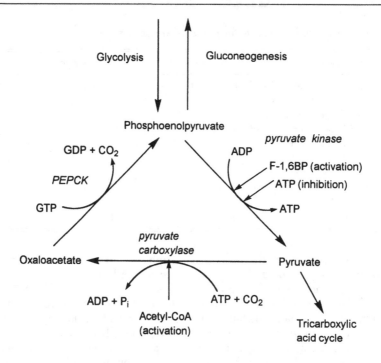

Fig. 3.6. Substrate cycle between pyruvate and phosphoenolpyruvate.

pyruvate kinase activity in avian liver changes little in response to an increase in the carbohydrate content of the diet, unlike the mammalian liver (where the L form is dominant) (Strandholm *et al.*, 1975). On the basis of amino acid composition, the M_1 and M_2 isozymes of pyruvate kinase from domestic fowl are closely related to the M_1 and M_2 isozymes from dog and rat (Becker *et al.*, 1986). Three isozymes of pyruvate kinase have been purified from pigeon erythrocytes, which have properties similar to those of the R-type mammalian enzymes (Calomenopoulou, Kaloyianni & Beis, 1989). The isozymes were activated by fructose 1,6-phosphate and inhibited by ATP, 2,3-diphosphoglycerate and inositol phosphates.

Modulation of the M_2 pyruvate kinase activity also occurs when chick embryo fibroblasts are transformed by Rous sarcoma virus. The pyruvate kinase shows a lower affinity for PEP. The oncoprotein from Rous sarcoma virus pp60[v-src] is able to phosphorylate the M_2 pyruvate kinase *in vitro* (Presek, Remacher & Eigenbrodt, 1988). Phosphorylation occurs on a tyrosine residue. Three other glycolytic enzymes have been found to be phosphorylated on tyrosine residues when chick embryo fibroblasts are transformed by Rous sarcoma virus. They are enolase,

phosphoglycerate mutase and lactate dehydrogenase (Cooper *et al.*, 1983). Cancerous cells are often found to have high rates of glycolysis. These phosphorylations may relate to the increased rate of glycolysis; however, of the four enzymes, only pyruvate kinase is rate limiting in glycolysis and is considered a control point.

3.5 Glycogenolysis and glycogen synthesis

Glycogen is formed from glucose 1-phosphate in two steps catalysed by UDPglucose pyrophosphorylase (reaction 3.11) and glycogen synthase (reaction 3.12), and it is degraded by phosphorolysis catalysed by phosphorylase (reaction 3.13).

$$\text{glucose 1-phosphate} + \text{UTP} \rightleftharpoons \text{UDPglucose} + \text{PP}_i \tag{3.11}$$

$$\text{UDPglucose} + (1,4\text{-}\alpha\text{-D-glycosyl})_n \rightleftharpoons \text{UDP} \\ + (1,4\text{-}\alpha\text{-D-glycosyl})_{n+1} \tag{3.12}$$

$$(1,4\text{-}\alpha\text{-D-glycosyl})_n + \text{P}_i \rightleftharpoons (1,4\text{-}\alpha\text{-D-glucosyl})_{n-1} \\ + \alpha\text{-D-glucose 1-phosphate} \tag{3.13}$$

These enzymes and their control mechanisms have been studied intensively in liver and skeletal muscle, since these are the main sites of glycogen

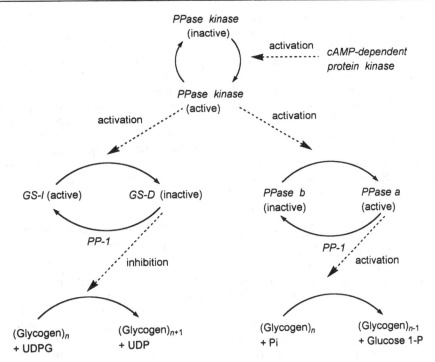

Fig. 3.7. Control of glycogen synthase and phosphorylase. PPase, phosphorylase; GS, glycogen synthase.

synthesis and breakdown. The synthesis and degradation of glycogen is regulated by a number of hormones, but mainly insulin, glucagon and adrenaline, and this aspect will be discussed in Sections 7.6.1 and 7.6.3. Both glycogen synthase and phosphorylase are controlled by covalent modification (phosphorylation/dephosphorylation) and by allosteric mechanisms; both mechanisms are well described in many standard biochemistry textbooks (see Price & Stevens, 1989; Mathews & van Holde, 1990; Hardie, 1991). Phosphorylation converts glycogen synthase from the active form (known as *a*, or *I* because its activity is independent of D-glucose 6-phosphate) to an inactive form (known as the *b* or *D*, because its activity is dependent on D-glucose 6-phosphate). Phosphorylation converts phosphorylase *b* to phosphorylase *a*. By contrast phosphorylase *b* is inactive except in the presence of the activator AMP, whereas phosphorylase *a* is active and does not require the presence of AMP. Phosphorylation is catalysed by a cAMP-dependent protein kinase, and dephosphorylation by protein phosphatase-1, as outlined in Fig. 3.7. The activities of glycogen synthase and phosphorylase have been measured in a number of different avian tissues, but their regulation has not generally been examined in detail. There is little reason to believe that it is

substantially different from that in mammals. Glycogen synthase is found in both the particulate fraction and the soluble fraction of liver from domestic fowl after homogenates are centrifuged at $50\,000 \times g$ for 60 min (Rosebrough, Mitchell & Steele, 1990). The *I* form comprises 50% to 80% of the total glycogen synthase in the liver of domestic fowl, as judged by its activity measured in the presence and absence of glucose 6-phosphate (Egana, Trueba & Sancho, 1986; Rosebrough *et al.*, 1990). This is a much higher proportion than is found in rat liver, where only $\approx 10\%$ is in the *I* form. Glycogen phosphorylase activity in domestic fowl liver is almost entirely in the *a* form and shows little increase in activity when assayed in the presence of AMP (Rosebrough *et al.*, 1990). Glycogen phosphorylase has been purified from skeletal muscle from domestic fowl. Its size (95 kDa), peptide mapping and antigenicity show it to be very similar to the rabbit muscle enzyme (Law & Titball, 1992). In skeletal muscle, much of the phosphorylase activity is associated with the glycogen particle, which also contains a number of glycolytic enzymes. The glycogen particle is thought to be a large multienzyme aggregate that facilitates the control of glycogen metabolism (see Price & Stevens, 1989). In dystrophic domestic fowl, there is a change in the intracellular distribution of phosphorylase in skeletal muscle, with a much larger proportion becoming membrane associated. Sokolove (1985) suggests that this change may relate to the underlying defect in dystrophic domestic fowl.

One avian tissue that has been studied and which has no mammalian equivalent is the gizzard. The gizzard from graminiferous birds is well developed and is a better source of smooth muscle than any from mammalian sources, and it has been a useful tissue in which to study the enzymes that control glycogen synthase and phosphorylase, although its glycogen content is lower than that of both striated muscle and the smooth muscle present in the uterus (Gröschel-Stewart & Zuber, 1990). Glycogen phosphorylase kinase has been purified from both skeletal muscle (Andreeva et al., 1986) and from the gizzard (Nikolaropoulis & Sotiroudis, 1985) of the domestic fowl. The enzyme from red and white skeletal muscle has a subunit structure similar to that from mammalian sources, namely $(\alpha,\beta,\gamma,\delta)_4$ in which γ is the catalytic subunit, and δ is calmodulin forming an integral part of the protein. Only one isozyme has been found. The enzyme partially purified from gizzard was the first to be purified from any smooth muscle. Unlike other phosphorylase kinases, it does not appear to be activated by cAMP-dependent protein kinase and does not seem to be regulated by phosphorylation (Fig. 3.7). It differs from other glycogen phosphorylase kinases in not having calmodulin as an integral part of its structure. It is, however, activated by exogenous calmodulin and thus modulated by intracellular Ca^{2+} like other phosphorylase kinases. The difference may simply be one of degree, in that it binds calmodulin less strongly than do other phosphorylase kinases. Phosphorylase kinase from the gizzard of domestic fowl has a particularly high affinity for phosphorylase b, approximately two orders of magnitude greater than that from mammalian sources. It is also not clear whether it contains a γ subunit, since no band was detected on polyacrylamide gels in the expected position (Nikolaropoulis & Sotiroudis, 1985). This might be because of proteolysis, or it might indicate a significantly different structure from other isozymes.

Protein phosphatase-1 has been purified from the myofibrils of the domestic fowl gizzard (Alessi et al., 1992). It comprises three subunits (130, 37 and 20 kDa) and is probably a heterotrimer. The 37 kDa subunit is the catalytic subunit, and the others are regulatory subunits. The regulatory subunits enhance the catalytic activity towards meromyosin but suppress activity against phosphorylase, phosphorylase kinase and glycogen synthase. Two protein phosphatases have been purified from turkey gizzard. Both are most active against light chain smooth muscle myosin and show some activity towards phosphorylase kinase (Pato & Adelstein, 1983a,b; Pato et al., 1983). The myosin light chain is phosphorylated to initiate contraction of smooth muscle, and it correlates with potentiation of the isometric twitch in striated muscle. Two protein phosphatases have also been isolated from hen oviduct, and although they will dephosphorylate phosphorylase kinase, it is more likely that their physiological role is in regulation of steroid hormone receptors (Kanyama et al., 1985).

3.6 Control of glycolysis and gluconeogenesis

The balance between the synthesis of glucose and its utilisation depends on its supply and demand, and it will vary between tissues. Gluconeogenesis is most active in the liver and kidney, whereas glycolysis is most active in skeletal and cardiac muscle. Both processes are catalysed by the same enzymes, except for the substrate-cycle enzymes 6-phosphofructokinase, fructose 1,6-bisphosphatase, pyruvate kinase, pyruvate carboxylase and PEPCK (Figs. 3.6 and 3.8). These are the rate-limiting steps in the pathways at which the flux is controlled. Homeostatic mechanisms exist within cells so that when particular metabolites build up they exert feedback inhibition on their formation, e.g. adenine nucleotides. In addition, regulation is imposed from without the cells by hormones interacting with receptors on the cell surface and causing the release of second messengers within the cells. This aspect is discussed in Chapter 7. The rate-limiting enzymes are generally non-equilibrium enzymes and often show the sigmoid kinetics characteristic of allosteric enzymes. Control of these pathways may be fine or coarse, or acute or chronic as it is sometimes described. The former involves the regulation of enzyme activity by effectors binding to regulatory sites and either increasing or decreasing the activity, or by reversible covalent modification, usually phosphorylation. The latter involves a change in the amount of a particular enzyme, either changing its rate of synthesis or degradation.

The first point of regulation is the interconversion of fructose 6-phosphate to fructose 1,6-bisphosphate (Fig. 3.8). 6-Phosphofructokinase has been purified from liver and from breast muscle of domestic fowl (Kono & Uyeda, 1973) and crystallised from the former. It is an oligomeric enzyme comprising identical subunits of $\approx$ 60 kDa. The important features

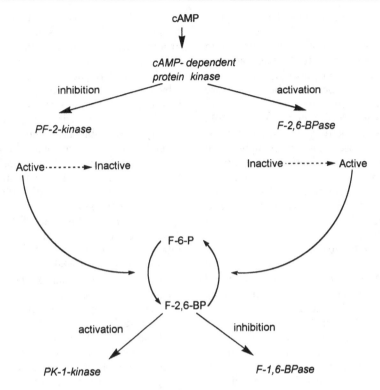

Fig. 3.8. Regulation of 6-phosphofructokinase and fructose bisphosphatase. PF-2-kinase, 6-phosphofructo-2-kinase; F-2,6-BPase, fructose 2,6-bisphosphatase; F-2,6-BP, fructose 2,6-bisphosphate; PK-1-kinase, 6-phosphofructo-1-kinase; F-1,6-BPase, fructose 1,6-bisphosphatase.

of all eukaryotic 6-phosphofructokinases are that they show cooperative binding of the substrate fructose 6-phosphate, inhibition by high concentrations of ATP and inhibition by citrate. The inhibition by ATP is relieved by P_i, AMP, cAMP, ADP, fructose 6-phosphate and fructose 1,6-bisphosphate. Fructose 1,6-bisphosphatase has been purified from chicken breast muscle (MacGregor *et al.*, 1982) and turkey liver (Han & Johnson, 1982). The enzyme shows hyperbolic kinetics with the substrate but is inhibited by AMP and high concentrations of the substrate fructose 1,6-bisphosphate.

The intracellular concentrations of ATP, ADP and AMP have an important influence on the balance between 6-phosphofructokinase and fructose 1,6-bisphosphatase activity. A high ratio of ATP:AMP will favour gluconeogenesis by inhibiting glycolysis, whereas a low ratio will favour glycolysis by inhibiting gluconeogenesis. However, more important still is the role of fructose 2,6-bisphosphate, which is both a very potent activator of 6-phosphofructokinase, having effects at 10^{-7} M and a very potent inhibitor

of fructose 1,6-bisphosphatase (Fig. 3.8). The synthesis and breakdown of fructose 2,6-bisphosphate is catalysed by the bifunctional enzyme 6-phosphofructo-2-kinase/fructose 2,6-bisphosphatase (reactions 3.14 and 3.15, respectively) which has been purified from both pigeon muscle and domestic fowl liver (van Schaftingen & Hers, 1986; Li, Pan, & Xu, 1991).

$$\text{fructose 6-phosphate} + \text{ATP} \rightleftharpoons \text{fructose}$$
$$\text{2,6-bisphosphate} + \text{ADP} \qquad (3.14)$$

$$\text{fructose 2,6-bisphosphate} + \text{H}_2\text{O} \rightleftharpoons \text{fructose}$$
$$\text{6-phosphate} + \text{P}_i \qquad (3.15)$$

6-Phosphofructo-2-kinase activity is inhibited by citrate and by PEP. A difference between this bifunctional enzyme in liver and in muscle is that the former is phosphorylated by cAMP-dependent protein kinase, causing inactivation of the kinase activity and activation of the phosphatase. The pigeon muscle enzyme, however, is not phosphorylated by cAMP-dependent protein kinase but, nevertheless, its activities are more like those of the phosphorylated enzyme, having a high ratio of fructose 1,6-bisphosphatase to 6-phosphofructo-2-kinase activity (van Schaftingen & Hers, 1986). 6-Phosphofructo-2-kinase activity and its product fructose 2,6-bisphosphate have been detected in chicken erythrocytes, where they are probably involved in the regulation of glycolysis *in vivo* (Espinet, Bartrons & Carreras, 1988; 1989). 6-Phosphofructo-2-kinase activity and the intracellular concentration of fructose 2,6-bisphosphate both increase when chick embryo fibroblasts are transformed by Rous sarcoma virus and when they are stimulated by the tumour promoter phorbol 12-myristate (Marchand *et al.*, 1992; Bosca, Rousseau & Hue, 1985). The mechanisms involved are still unclear but may involve protein kinase C.

The second substrate cycle includes the three enzymes pyruvate kinase, pyruvate carboxylase and PEPCK (Fig. 3.6). Whilst pyruvate carboxylase and PEPCK are probably the most important steps in the *in vivo* regulation of gluconeogenesis, pyruvate kinase (see Section 3.4.3) is probably less important than 6-phosphofructo-1-kinase in regulating glycolysis. Acetyl-CoA is a very potent activator of pyruvate carboxylase ($K_a = 2 \mu M$ in chicken liver) but is probably not involved in regulation under physiological conditions as it is present in saturating concentrations. The ATP:ADP ratio is likely to be more important physiologically, since ADP exerts an inhibitory effect on pyruvate carboxylase (Söling & Kleinicke, 1976). The amount of pyruvate carboxylase increases on starvation but PEPCK does not increase after 72 h starvation.

3.7 Changes in intermediary metabolism during development

A dramatic change in the source of fuel occurs when a chick hatches. The developing embryo depends on the egg-yolk, and to a much lesser extent the egg-white, as a source of nutrient during its embryonic development. Typically egg-yolk consists of approximately 55% protein, 44% lipid and 1% carbohydrate dry weight. The small amount of carbohydrate is used by E7 in the domestic fowl (Bate & Dickson, 1986). The main source of fuel throughout embryonic development is lipid. Upon hatching, the chick begins to feed. In the domestic fowl and in graminiferous birds in general, the diet is rich in carbohydrate. With carnivores, such as birds of prey, the diet is still predominantly lipid and protein. The changes in intermediary metabolism during development have only been investigated in any detail in the domestic fowl. On hatching, the fuel changes from almost exclusively lipid to one that is predominantly carbohydrate. The liver is the principal organ capable of interconverting fuels. In birds, the adipose tissue has very limited capacity for *de novo* lipid biosynthesis (see Section 4.4), and takes up lipid from the plasma, which is either dietary in origin or has been synthesised in the liver. During embryonic development, the liver has a high capacity for gluconeogenesis and is largely responsible for maintaining blood glucose levels in the later stages of incubation. Blood glucose rises during the last week of incubation from 3–5 mM to ≈ 10 mM, a level that is maintained in the adult (Langslow, 1978; Garcia *et al.*, 1986). Before hatching, there are increases in fructose 1,6-bisphosphatase, glucose 6-phosphatase and PEPCK (Shen & Mistry, 1979; McCaffrey & Hamilton, 1994), three of the rate-limiting enzymes in gluconeogenesis.

Glycogen is detectable in the liver and skeletal muscle from E11 (Garcia *et al.*, 1986). In both tissues, the glycogen concentration increases until just before hatching, when it falls, only to rise sharply again once feeding has started. Hamer & Dickson (1989) have examined the utilisation of liver glycogen in isolated hepatocytes. In hepatocytes from chick embryos, almost all the glycogen is used in glucose formation, whereas in the neonatal hepatocyte only 25–40% is used for glucose formation. Much of the rest of the glycogen is broken down by glycolysis and used for lipid synthesis. Increased 6-phosphofructo-1-kinase activity is detectable not only in the liver (Hamer & Dickson, 1987) but also in other tissues, e.g. brain, muscle and kidney (Newton & Hamer, 1989). There is no increase in cardiac muscle. Whereas the skeletal muscles are inactive *in ovo*, the same is not true of cardiac muscle. The concentration of fructose 2,6-bisphosphate in the liver increases in parallel with that of 6-phosphofructo-1-kinase after hatching, favouring glycolysis rather than gluconeogenesis. The kidney is also an important tissue capable of gluconeogenesis, but unlike the liver in which the gluconeogenic capacity is well developed by E13, in kidney it is only fully developed after hatching. In the late embryo where there is high provision of lipid, much is used for gluconeogenesis, but significant amounts are converted to ketone bodies (Bate & Dickson, 1986). Ketone bodies appear to be used as a fuel by embryonic heart, brain and kidney. All have substantial amounts of 3-hydroxybutyrate dehydrogenase (reaction 3.16) and 3-oxoacid-CoA transferase (reaction 3.17). These two enzymes are necessary for the formation of acetoacetyl-CoA, which can then be converted to acetyl-CoA and hence be metabolised via the tricarboxylic acid cycle.

$$\beta\text{-hydroxybutyrate} + NAD^+ \rightleftharpoons \text{acetoacetate} \\ + NADH + H^+ \qquad (3.16)$$

$$\text{acetoacetate} + \text{succinyl-CoA} \rightleftharpoons \text{acetoacetyl-} \\ \text{CoA} + \text{succinate} \qquad (3.17)$$

For the conversion of carbohydrate into lipid, the carbohydrate has to be broken down into glycerol 1-phosphate and acetyl-CoA, the precursors of triglycerides. Fatty acid synthesis also requires reducing equivalents in the form of NADPH. Goodridge (1968b) has shown that there are changes in the activities of glucose 6-phosphate dehydrogenase and 6-phosphogluconate dehydrogenase during development, but they do not coincide with the increase in lipid synthesis. These two enzymes of the pentose phosphate pathway are, in any case, only present in low activity in the liver of domestic fowl (Shen & Mistry, 1979). There are much more dramatic changes in the citrate cleavage enzyme and the malic enzyme that coincide with the increase in lipogenesis. The citrate cleavage enzyme increases 85-fold in the first 7 days after hatching, and the malic enzyme increases 15-fold in the first 5 days after hatching (Goodridge, 1968a). The citrate cleavage enzyme (reaction 3.18) could provide a source of acetyl-CoA and the malic enzyme (reaction 3.19) the reducing equivalents in the form of NADPH.

$$ATP + \text{citrate} + CoA \rightleftharpoons ADP + P_i + \text{acetyl-} \\ \text{CoA} + \text{oxaloacetate} \qquad (3.18)$$

$$\text{malate} + NADP^+ \rightleftharpoons \text{pyruvate} + CO_2 + NADPH \qquad (3.19)$$

4

Lipids and their metabolism

4.1 Introduction

Lipids include a range of chemical compounds that have the common physical properties of being insoluble in water but soluble in non-polar organic solvents, such as chloroform, hydrocarbons or alcohols. Their most common chemical feature is that of ester bonds, in which the acid component is one or more aliphatic carboxylic acid. They range in complexity from the simple triglycerides to lipids containing other groupings, such as phosphate in phospholipids, carbohydrates in glycolipids and gangliosides, and proteins in lipoproteins. A further group are the sterols, which may exist as free alcohols or in esterified forms. The molecular structures of the major lipid groups are shown in Fig. 4.1. The lipids present in largest amounts in vertebrates, including birds, are the triglycerides, where their principle role is as an energy reserve. Phospholipids have a largely structural role in the cell membranes. The sterols are also present as an integral part of the membranes, but certain sterols are also hormones, such as the sex hormones and the adrenocortical steroids (see Chapter 10). The glycolipids are particularly important as structural entities in the central and peripheral nervous system. Lipoproteins are important in the transport of lipids between tissues.

This chapter is primarily concerned with the structure and metabolism of the triglycerides and their transport between tissues in the form of lipoproteins. The constant structural feature of all triglycerides is the glycerol moiety esterified by three carboxylic acids. Triglycerides have a variety of carboxylates, varying in the length of the aliphatic chain, the degree of unsaturation and the position of the double bonds. Each of the three hydroxyl groups of glycerol may be esterified by the same or different fatty acids. The nature of the fatty acids alters the physical properties of the triglyceride, in particular its melting point.

4.2 Lipids present in avian tissues

Most information on avian triglycerides relates to the domestic fowl. Detailed studies have been made of their composition because of their importance as a source of dietary fat in humans. Research has centred on whether it is possible to select lean domestic fowl genotypes that have a low tendency to accumulate triglyceride (Leclercq, 1984), and to what extent the endogenous triglyceride can be modified by changes in dietary intake (Ajuyah et al., 1991). Triglyceride is the most abundant lipid present in birds; it is present throughout the tissues of the body, but the principal deposits are in the adipose tissues. Several cells of the body contain lipid droplets, but in all cells except the adipocyte, these droplets rarely exceed $2\,\mu m$ in diameter. By contrast in the adipocyte, the lipid droplet may be a large as $100\,\mu m$ in diameter, and the total triglyceride content may be as much as 920 g/kg tissue (Evans, 1977); this varies with the species and the particular type of adipose tissue. In the chick embryo at 19E, 80% of the lipid present in the liver is in the form of cholesterol ester (Noble & Connor, 1984), which is derived from yolk cholesterol (Shand et al., 1994). The abdominal adipose tissue is the most well-defined fatty tissue; together with subcutaneous and intramuscular adipose tissue it makes up the bulk of the lipid reserves (Cahaner, Nitsan & Nir, 1986). In the domestic fowl, the distribution of triglyceride between these tissues as a percentage of body weight, ranges between 0.6 and 2.5% in abdominal tissue, 0.9 and 2.6% in subcutaneous tissue and 0.4 and 0.6% in intramuscular tissues. The precise amount depends on whether the breed has been selected for high or low abdominal adipose tissue (Ricard, Leclerq & Touraille, 1983; Leclercq, 1984). The capacity of birds for storing lipids exceeds that of any other class of vertebrate.

The fatty acid components of triglycerides are predominantly C_{16} and C_{18} fatty acids, and a typical

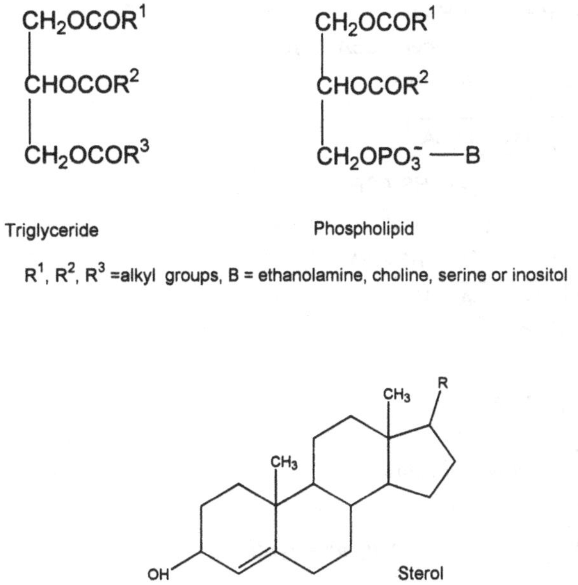

CH_2OCOR^1

$CHOCOR^2$

CH_2OCOR^3

Triglyceride

CH_2OCOR^1

$CHOCOR^2$

$CH_2OPO_3^- \text{—B}$

Phospholipid

R^1, R^2, R^3 = alkyl groups, B = ethanolamine, choline, serine or inositol

Sterol

Fig. 4.1. The structure of typical triglycerides, phopholipids and sterols.

Myristic acid	$CH_3(CH_2)_{12}COOH$
Palmitic acid	$CH_3(CH_2)_{14}COOH$
Stearic acid	$CH_3(CH_2)_{16}COOH$
Palmitoleic acid	$CH_3(CH_2)_5CH=CH(CH_2)_7COOH$
Oleic acid	$CH_3(CH_2)_7CH=CH(CH_2)_7COOH$
α-Linoleic acid	$CH_3(CH_2)_4(CH=CHCH_2)_2(CH_2)_6COOH$
Eicosatrienoic acid	$CH_3(CH_2)_7(CH=CHCH_2)_3(CH_2)_2COOH$
Arachidonic acid	$CH_3(CH_2)_4(CH=CHCH_2)_4(CH_2)_2COOH$

Fig. 4.2. The principal fatty acids found in avian triglycerides and phospholipids.

proportion of saturated:monounsaturated:polyunsaturated fatty acids in broilers is 0.31:0.45:0.23 (Yau *et al.*, 1991); a similar distribution is found in jungle fowl and dove (Koizumi, Suzuki & Kaneko, 1991). This is a significantly higher proportion of polyunsaturated fatty acids than occurs in mammalian tissues. Shifts in favour of polyunsaturated fatty acids can be achieved by feeding domestic fowl a high proportion of dietary polyunsaturated fats. In the quail, the proportion of saturated to unsaturated fatty acids is also influenced by environmental temperature (Christie & Moore, 1972; Durairaj & Martin, 1975). In certain species, the body fat may be as high as 50% of body weight, prior to migration. Following prolonged starvation, the total triglyceride may fall to a minimum of about 4% body weight; this amount is required to protect the integrity of tissues (Griminger, 1986). The most abundant fatty acids in the quail are palmitic, stearic, myristic, palmitoleic, oleic, linoleic, eicosatrienoic and arachidonic acids (Fig. 4.2), and the same is probably true for other domesticated birds (Griminger, 1986).

4.3 Lipid biosynthesis

Birds generally have a very high capacity for lipid biosynthesis. During growth between 1 and 6 weeks of age, the total body lipid in the domestic fowl doubles every 5.5 days (Scanes, 1987). During egg-laying from a normal commercial diet the fowl may receive 3 g of lipid per day, but the average egg contains 6 g lipid (Griminger, 1986). The rate of fatty acid synthesis in the quail is at its maximum in the liver at 7 weeks of age. In the female, this is just after the onset of laying (Furuse *et al.*, 1991). Unlike other vertebrates where the adipose tissue has a very high capacity for *de novo* lipid biosynthesis, in avian species the liver is the principal site of lipid synthesis; it has about 20 times the capacity for lipid biosynthesis compared with an equal weight of adipose tissue (Griminger, 1986). This has been demonstrated in domestic fowl, duck, turkey, pigeon, quail, common starling and white wagtail (Pearce, 1980). Adipose tissue in the domestic fowl has a very limited capacity for fatty acid synthesis (Goodridge, 1968b), most of its accumulating fat is of dietary origin (Griffin & Hermier, 1988). Bone marrow also has a high capacity for lipid synthesis (Nir & Lin, 1982).

The rate of *de novo* synthesis is reduced when the dietary lipid intake is increased, and also the specific activities of both the ATP citrate lyase and the malate enzyme in liver decrease (Griminger, 1986). Together they catalyse the production of acetyl-CoA and reducing equivalents needed for the synthesis of fatty acids.

ATP citrate (pro-S)-lyase (EC.4.1.3.8) catalyses the reaction

$$ATP + citrate + CoA \rightleftharpoons ADP + P_i + acetyl\text{-}CoA + oxaloacetate \qquad (4.1)$$

and malic enzyme (EC.1.1.1.40) catalyses the reaction

$$malate + NADP^+ \rightleftharpoons pyruvate + CO_2 + NADPH \qquad (4.2)$$

Malic enzyme is present in both mitochondria and cytosol, the former using NAD^+ and the latter

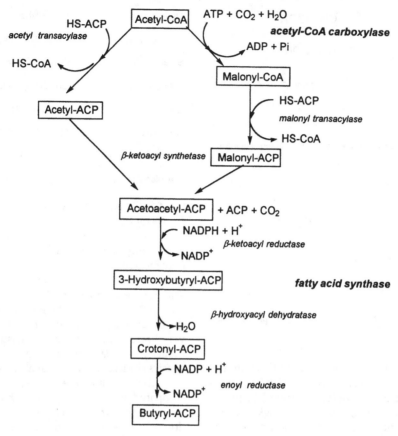

Fig. 4.3. The biosynthesis of fatty acids catalysed by acetyl-CoA carboxylase and fatty acid synthase. The steps from acetoacetyl-CoA to butyryl-ACP are repeated for each addition of two carbon atoms to the growing acyl-ACP.

$NADP^+$. The enzyme was first discovered in pigeon by Ochoa, Mehler & Kornberg (1947) and was later found to be widely distributed in nature. In the cytosol, it has an important role in catalysing the generation of NADPH for *de novo* long-chain fatty acid synthesis. The pigeon liver enzyme is a tetramer composed of four identical subunits (65 kDa). A detailed study of its kinetics have been made, and the cloning and expression of its cDNA carried out (Chou *et al.*, 1994). The mechanism of fatty acid biosynthesis and its control has been most studied in the liver of the domestic fowl, where the enzymes have been purified and their genes sequenced. A second tissue that is frequently studied, particularly in geese and ducks, is the uropygial gland. This gland is the main source of fatty oils used to cover the feathers in the process of preening. There are two main enzymes involved in fatty acid synthesis in birds, and in vertebrates in general, namely acetyl-CoA carboxylase and fatty acid synthase. The latter is a

multienzyme polypeptide, and in all vertebrates so far studied it contains the seven enzyme activities required for synthesis of long-chain fatty acids from malonyl-CoA. *Acetyl-CoA carboxylase* catalyses the first step (Fig. 4.3) in the biosynthesis from acetyl-CoA, and this is generally the rate-limiting step in the process and, therefore, a control point. The enzyme, which has been purified from chicken liver (Wakil, Stoops & Joshi, 1983), catalyses the carboxylation of acetyl-CoA to malonyl-CoA. It is a biotin-dependent enzyme, requiring ATP and having a regulatory site for the allosteric activator citrate. It is a large enzyme (4×10^3 to 8×10^3 kDa) made up of 265 kDa subunits, which have been sequenced as cDNA (Takai *et al.*, 1988). The enzyme has been detected in embryonic chick brain, where its activity increases to a peak at 19E, and this corresponds with a period of active myelination (Thampy & Koshy, 1991).

The malonyl-CoA formed by the action of acetyl-CoA carboxylase is then used by the fatty acid synthase. Each sequence of six steps results in the growing fatty acid chain increasing by two carbon atoms (Fig. 4.3). The final step is the release of the free fatty acid, which results from the seventh catalytic activity, i.e. the thioesterase. *Fatty acid synthase* has been purified from chicken liver and goose uropygial gland (Wakil *et al.*, 1983). The latter is a particularly good source, since between one third and one sixth of the total protein in that gland is fatty acid synthase (Buckner & Kolattukudy, 1976a). Enzyme from both sources comprises two identical subunits (α_2) M_r 250 000, on each of which six catalytic activities can be detected: acetyl transacylase, malonyl transacylase, β-ketoacyl reductase, β-hydroxyacyl dehydratase, enoyl reductase and thioesterase. The seventh activity, β-ketoacyl synthetase, is dependent on two thiol groups on opposite chains being juxtaposed to function (for details see Price & Stevens, 1989). The amino acid sequence of the chicken fatty acid synthase has been deduced from the cDNA sequence, and a partial sequence for the goose enzyme has been obtained (Kameda & Goodridge, 1991; Subrahmanyan *et al.*, 1989). The complete amino acid sequence of the fragment containing the thioesterase activity has been determined (Yang *et al.*, 1988). The mechanism of thioester hydrolysis is similar to that used by serine proteases, and the catalytic site has a similar charged relay involving serine, histidine and aspartate (Pazirandeh, Chirala & Wakil, 1991).

4.4 Control of fatty acid biosynthesis

There are a number of different of levels at which control of fatty acid biosynthesis is exerted, and these may involve regulation of the rate of synthesis and of the proportions of particular fatty acids synthesised. Liver, rather than adipose tissue, is the main site of lipid biosynthesis in birds; two thirds of the total lipid biosynthesis in the domestic fowl occurs in the liver (Saadoun & LeClerq, 1983). Diet has a greater effect on hepatic lipid biosynthesis than it has on that in other tissues of the body (Saadoun & Leclercq, 1987). Most studies on control of the rate of lipid biosynthesis have been based on the liver, but those on the control of the type of fatty acid synthesised have used the uropygial gland.

4.4.1 *Liver*

The main controls on the amounts of fatty acids synthesised are nutritional and hormonal (Wakil *et*

al., 1983). Long-term or coarse control is exerted at the transcriptional and translational level, as for example in the control by insulin. Short-term or fine control is generally by allosteric mechanisms, such as activation of acetyl-CoA carboxylases by citrate and fatty acyl-CoA, but also by covalent modification. The principal hormones that influence lipid biosynthesis in the liver are insulin, glucagon, triiodothyronine and glucocorticoids (Fischer & Goodridge, 1978; Vives, Sancho & Gomezcap, 1981). Both the thyroid hormones and insulin promote fatty acid biosynthesis, exerting their effects on gene transcription (Wilson *et al.*, 1986). The effect of triiodothyronine appears to be mediated through the promoter region of the fatty acid synthase gene (Kameda & Goodridge, 1991), in a similar way to that by which steroid hormones activate gene expression (see Section 10.1). Both glucocorticoids and glucagon have antilipogenic effects, glucagon inhibiting the accumulation of fatty acid synthase mRNA (Wilson *et al.*, 1986). Starvation causes a reduction in the rate of fatty acid synthesis, whereas refeeding generally has the reverse effect. When unfed 2-day-old goslings were fed for 24 h, the rate of synthesis of fatty acid synthase increased more than 42-fold. By using ^{32}P-labelled cDNA for the fatty acid synthase, Morris *et al.* (1982) showed that the level of mRNA for fatty acid synthase increased 70-fold, indicating that control probably occurs at the pre-translational level. Similar findings have been made in embryonic and neonatal chicks. Messenger RNA for fatty acid synthase was almost undetectable in embryonic liver at 16E but increased from then on until hatching. In neonatal chicks, feeding increased the abundance of mRNAs for fatty acid synthase and the malic enzyme, and starvation decreased their abundance. As a control, the concentration of mRNA for albumin was measured after feeding and starvation, and it was unaffected (Morris *et al.*, 1984). Similar effects were observed with neonatal ducklings (Goodridge *et al.*, 1984; Back *et al.*, 1986).

Fine control of fatty acid synthesis is by allosteric control and by covalent modification. Acetyl-CoA carboxylase, which is generally regarded as the rate-limiting step in biosynthesis, is allosterically activated by citrate and inhibited by long-chain fatty acyl CoA (Fig. 4.4). An increase in the level of intracellular cAMP, such as may be brought about by glucagon stimulation, causes a decrease in the activity of phosphofructokinase (see Fig. 3.8) and so decreases the rate of glycolysis, and hence the levels of pyruvate and citrate. The lower concentration of citrate decreases the activity of acetyl-CoA carboxylase

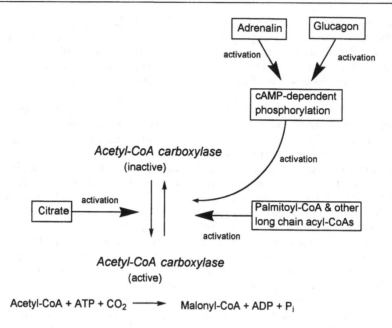

Fig. 4.4. The control of acetyl-CoA carboxylase activity by allosteric effectors and by phosphorylation.

(Fig. 4.4). Acetyl-CoA carboxylase exists as an inactive dimer ($\approx 500\,\text{kDa}$) and a polymeric active form ($\approx 20 \times 500\,\text{kDa}$). The allosteric effector palmitoyl-CoA competes with citrate and causes depolymerisation of acetyl-CoA carboxylase and inhibits its activity. Phosphorylation of acetyl-CoA carboxylase also causes its inactivation. The primary structure of domestic fowl liver acetyl-CoA carboxylase has been deduced from its cDNA sequence (Takai *et al.*, 1988).

Liver has the enzyme activity to catalyse the introduction of a single double bond into saturated fatty acids. *Stearyl-CoA desaturase* (EC.1.14.99.5; reaction 4.3), which catalyses the synthesis of oleic and palmitoleic acids, is present on the endoplasmic reticulum and is hormonally regulated (Joshi & Aranda, 1979).

$$\text{stearyl-CoA} + AH_2 + O_2 \rightarrow \text{oleoyl-CoA} + A + 2H_2O$$
$$(4.3)$$

Higher desaturase activity is found in the liver of genetically fat strains of domestic fowl compared with genetically lean strains (Legrand & Lemarchal, 1992). The synthesis of a higher proportion of monounsaturated acids in the genetically fat line is thought to facilitate their incorporation into very low density lipoproteins (VLDL), which leads to a higher plasma concentration of VLDL and, in turn, a higher uptake by adipose tissue. The presence of monounsaturated fatty acids in hepatic triglycerides modifies their physicochemical state so that they are more easily incorporated into VLDL. Turkeys have a lower proportion of body fat than domestic fowl; at 12 weeks typical abdominal fat contents in turkeys are 0.6–0.7% body weight compared with 2.0–3.5% in domestic fowl (Kouba, Catheline & Leclercq, 1992). This correlates with the lower activities of stearyl-CoA desaturase, acetyl-CoA carboxylase and malic enzyme in turkey liver (Kouba *et al.*,1992; Kouba, Bernard-Griffiths & Lemarchal, 1993). All three enzymes modulate the biosynthesis of mono-unsaturated fatty acids. Hepatic stearyl-CoA desaturase activity is also stimulated by oestradiol, which catalyses an increase in the proportion of oleic acid and palmitoleic acid in triglycerides (Pageaux *et al.*, 1992). During sexual maturation of female birds, there is intense proliferation of the cells of the oviduct, and this is accompanied by large increases in their triglyceride and phospholipid content. This has been extensively studied in female quail. Much of this increase in oviduct triglyceride and phospholipid is the result of increased hepatic biosynthesis and transport via the plasma to the oviduct. The oviduct is able to utilise certain plasma fatty acids to a greater extent than other fatty acids (Pageaux *et al.*, 1992).

4.4.2 Uropygial gland

Interest in fatty synthesis in the uropygial gland has centred on the types of fatty acid synthesised. These are unusual for vertebrates in being largely branched chain fatty acids. They differ from those present in other tissues in the body, and their synthesis is regulated by different control mechanisms. The gland is also of particular interest since it has a very high capacity for *de novo* fatty acid synthesis, the rate of which shows seasonal variation. The uropygial gland (sometimes known as the preen gland or oil gland) is a sebaceous gland lying dorsally and medially in the synsacrocaudal region. It secretes waxes and oils as a holocrine secretion. The gland contains a number of secretory tubules. Basal cells on the periphery of these tubules migrate to form a transition layer where intracytoplasmic lipid globules are produced. These cells eventually die and expel their contents into the lumen of the tubules. Autolysis of these cells is brought about by the combined action of lysosomal enzymes, and by oxidation of sulphide groups to disulphides (Suzuki, Maruyama & Morohashi, 1994).

Extensive comparative studies have been made of the uropygial glands, which vary considerable in size in different genera (see Jacob & Ziswiler, 1982). With a few exceptions, including ostrich, rheas, cassowarys, emu, some parrots and bustards (family *Otididae*), all birds possess a uropygial gland (Jacob, 1976). The largest recorded gland (relative to body weight) is in the Little grebe at 0.61% body weight, and the smallest is in the fruit pigeons (genera *Ducula* and *Ptilinopus*) at 0.02% body weight. The largest glands are generally found in birds that go in water, particularly ducks and geese, but they are also quite large in Galliformes. The secretions of the uropygial gland are important in coating feathers with wax to keep them supple, but they are not the only source of waxes and oils on the feathers, since other sebaceous glands throughout the skin also contribute. They also act as sex attractants (pheromones). The alkyl-substituted wax acids and alcohols have antimicrobial activity and may, therefore, have a further protective function. Daily secretions of waxes vary between species, from about 0.5 mg in Passeriformes, to 600 mg in Laridae (Jacob, 1976). Most biochemical studies on the uropygial gland have used the goose or ducks.

The main lipids secreted by the uropygial gland are monoester waxes, diester waxes, triester waxes, and glycerides (Fig. 4.5). More than 100 different species of *wax* may be secreted, and these have been

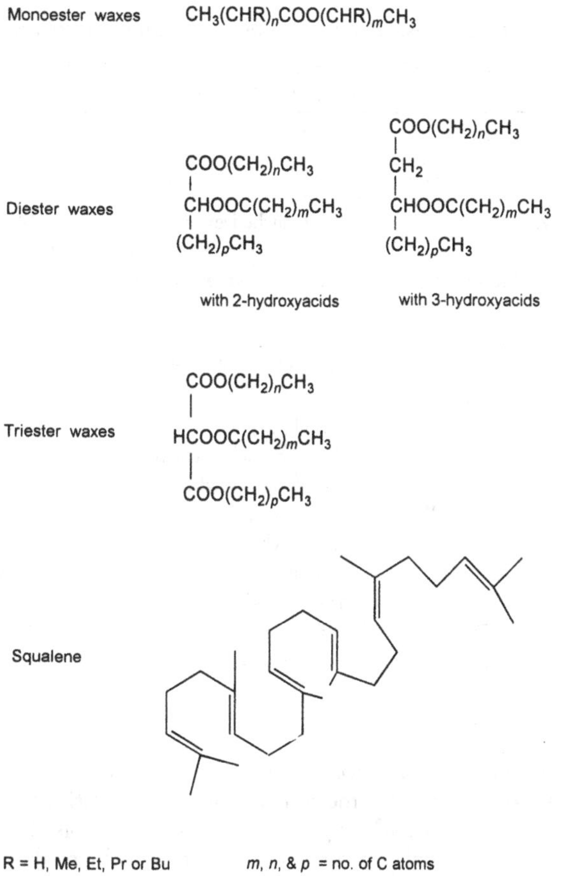

Monoester waxes: $CH_3(CHR)_n COO(CHR)_m CH_3$

Diester waxes:
$$\begin{array}{l} COO(CH_2)_n CH_3 \\ | \\ CHOOC(CH_2)_m CH_3 \\ | \\ (CH_2)_p CH_3 \end{array}$$ with 2-hydroxyacids

$$\begin{array}{l} COO(CH_2)_n CH_3 \\ | \\ CH_2 \\ | \\ CHOOC(CH_2)_m CH_3 \\ | \\ (CH_2)_p CH_3 \end{array}$$ with 3-hydroxyacids

Triester waxes:
$$\begin{array}{l} COO(CH_2)_n CH_3 \\ | \\ HCOOC(CH_2)_m CH_3 \\ | \\ COO(CH_2)_p CH_3 \end{array}$$

Squalene

R = H, Me, Et, Pr or Bu   *m, n, & p* = no. of C atoms

Fig. 4.5. Lipid components of uropygial gland secretions. Triglycerides are also secreted by uropygial glands of some species.

examined from several different orders of bird. There are different patterns of waxes produced by different taxonomic groups of birds (Table 4.1). In the Anseriformes, the monoester waxes form the bulk of the waxes, but in the Galliformes the diester waxes are also important. A novel triester wax in which alkylhydroxymalonic acids are esterified with *n*-alkanols and *n*-fatty acids has been found in unrelated orders such as Anseriformes, Piciformes, Turdidae and Corvidae. Two Anseriformes, Muscovy duck and Magpie goose, are unusual in secreting squalene (Fig. 4.5) in their uropygial gland (Jacob, 1976). Triglycerides, although often present in the wax secretions, are generally only minor constituents. The fatty acid components of the waxes are mainly methyl-substituted derivatives of carboxylic acids. In the goose, 2,4,6,8-tetramethyldecanoic acid and 2,4,6,8-tetramethylundecanoic acid are major components of the waxes, but unbranched chains are also

Table 4.1. *Distribution of uropygial gland wax esters among avian species*

Type	Fatty acid component	Alcohol component	Species distribution
Monoester	Straight chain	Straight chain	Grey heron, magpie goose, crested screamer, Atlantic puffin, but generally unusual
Monoester	Multibranched even numbered, e.g. 2,4,6-trimethyl, 2,4,6,8-tetramethyl	Straight chain	Common in ducks, geese and swans
Monoester	Multibranched, even numbered	Multibranched, even numbered	Falconiformes, Gruiformes, Charadriiformes, Lariformes, Strigiformes, Passeriformes
Monoester	3-methyl-branched chain	3-methyl-branched chain	Eurasian cuckoo, barn owl, green woodpecker, black woodpecker
Monoester	2-ethyl-, 2-propyl- and 2-butyl-branched-chain	Methyl-branched	Northern eagle owl, long-eared owl
Diester type I	α- or β-monohydroxy acids	Straight chain	White stork, brown kiwi
Diester type II	Straight chain	1,2- and 2,3-diols	Many Galliformes, brown kiwi
Triester	Straight chain	Glycerol	Some Ciconiiformes
Triester	Straight chain and alkylhydroxy malonic acid	Straight chain and alkylhydroxy malonic acid	Widely distributed

Source: Jacob (1976); Downing (1986).

present. The alcohol components of the monoester waxes are mainly methyl-substituted derivatives of the alkanols and long chain n-alkane-2,3-diols, but there is a smaller proportion of branched chains than with the fatty acids (for details, see Jacob & Ziswiler, 1982). Both the fatty acids and the alcohols are synthesised *de novo* in the uropygial gland using acetate and/or propionate as precursors, as shown by incorporation of [^{14}C]-labelled precursors.

4.4.3 The mechanism of lipid biosynthesis in the uropygial gland

Many animal tissues synthesise C_{16}- and C_{18}-unbranched fatty acids as the principal products of fatty acid synthase catalysis. By contrast, the uropygial gland from most avian species synthesises multibranched-chain fatty acids, which in the case of many waterfowl are also predominantly of shorter length (Jacob, 1976). Many have methyl branches on the even-numbered carbon atoms, e.g. 2-methyl, 2,4-dimethyl, 2,4,6-trimethyl, 2,4,6,8-tetramethyl. In some, the 4-position is unbranched, e.g. 2,6-dimethyl, 2,6,10-trimethyl in some Gruiformes and Charadriiformes, whilst in others, e.g. certain crows, the 4-position is favoured and 4-methyl, 4,8-dimethyl and 4,10-dimethyl branching is present. There are

also some where the branches are located only on the odd-numbered positions (Downing, 1986). What is the biosynthetic mechanism to account for these products? Experiments carried out *in vivo* showed that [^{14}C]acetate and [^{14}C]propionate were both incorporated into the branched-chain fatty acids present in the waxes of the uropygial gland. The labelling pattern for [^{14}C$_3$]propionate shows the methyl branches become labelled (Fig. 4.6). Experiments carried out *in vitro* using uropygial gland extracts show that both malonyl-CoA and methyl-malonyl-CoA are used in the biosynthesis of fatty acids. The properties of purified goose uropygial gland fatty acid synthase are very similar to those from rat liver and pigeon liver, although neither liver enzymes normally catalyse the synthesis of branched-chain fatty acids. It, therefore, seemed probable that different relative concentrations of fatty acid precursors in the uropygial gland might determine the nature of the fatty acid synthesised. Partially purified preparations of acyl-CoA carboxylase from goose uropygial gland have a similar value for K_m (1.5×10^{-5} M) and V_{max} (0.8 μmol/min per mg protein) for both propionyl-CoA and acetyl-CoA (Buckner & Kolattukudy, 1975). The branched-chain fatty acids are not generated as the result of carboxylation of propionyl-CoA to methylmalonyl-CoA. However,

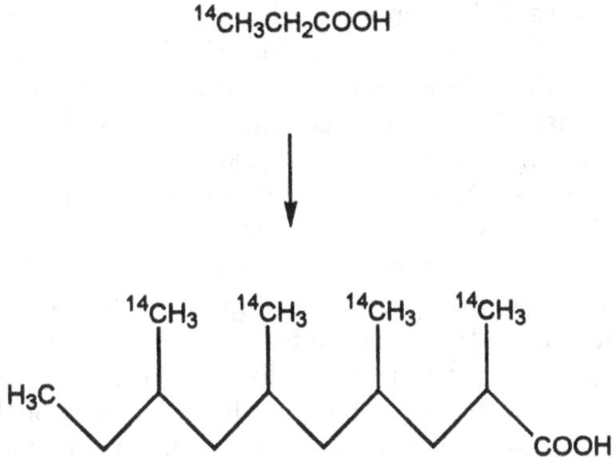

$^{14}CH_3CH_2COOH$

Fig. 4.6. The incorporation of [$^{14}C_3$]propionate into the branched chain fatty acid 2,4,6,8-tetramethyldecanoic acid.

the extracts were found also to contain a decarboxylase that could be separated from the fatty acid synthase and that decarboxylated malonyl-CoA at a high rate and methylmalonyl-CoA at a low rate. The rate at which the decarboxylase acted on malonyl-CoA is approximately 100-fold higher than the rate at which acetyl-CoA becomes carboxylated (Buckner & Kolattukudy, 1976a); therefore the concentration of malonyl-CoA is kept very low, enabling a higher proportion of methylmalonyl-CoA to be used for branched-chain fatty acid synthesis. How a 'skipped pattern', in which, for example, 4-position is unbranched, but positions 2, 6, and 8 are branched has not yet been established. This is also the case with the odd-carbon methyl branching pattern (Downing, 1986).

It is unclear how the synthesis of predominantly short-chain fatty acids occurs, especially as the fatty acid synthase from the uropygial gland appears to have the same characteristics as in other tissues. During the biosynthesis of fatty acids from malonyl-CoA and methylmalonyl-CoA by the fatty acid synthase, the growing chain is attached to the pantotheine group of the multienzyme polypeptide by a thioester link. One of the enzyme activities associated with a domain of the multienzyme polypeptide is the *thioesterase* (reaction 4.4), which has a specificity for cleaving thioesters of longer chain fatty acids esterified to the acyl carrier protein (ACP).

$$RCOS\text{-}ACP + H_2O \rightleftharpoons R.COO^- + HS\text{-}ACP \quad (4.4)$$

The uropygial gland also has a unique thioesterase (*S*-acyl fatty acid synthase thioesterase) separate from the multienzyme polypeptide, which can replace

the thioesterase domain and cause the release of short-chain fatty acids (Rogers, Kolattukudy & de Renobales, 1982). It also catalyses reaction 4.4 but with different fatty acid specificity. It binds the fatty acid synthase with K_a of $1\,\mu M$ (Foster *et al.*, 1985) and has a preference for C_{12} and C_{10} chains, degrading C_6, C_8, C_{14}, C_{16} and C_{18} at between 30–50% the rate (Rogers *et al.*, 1982).

The monoester and diester waxes are composed of fatty acids esterified to alkanols and alkane diols. These frequently have similar chain lengths to those of the fatty acids. The biosynthesis of the alcohols has been less extensively studied than that of the acids. The alkanols are formed by reduction of the corresponding acyl-CoA using NADPH as hydrogen donor. The reductases have an apparent preference for unbranched chains, and so there is a smaller proportion of branching in the alcohol moiety than the acid moiety of the ester. The reductases have been located both in the peroxisomes and in the endoplasmic reticulum (Bohnet *et al.*, 1991). The alkane diols are also formed from fatty acids, first by α-oxidation to the α-hydroxy fatty acid, followed by activation to the α-hydroxyacyl-CoA and then reduction to the diol (Buckner & Kolattukudy, 1976b). The formation of 3-hydroxy fatty acids from the corresponding saturated fatty acid has been demonstrated using cell-free extracts from Mallard uropygial gland. The reaction requires ATP, CoA and O_2, and it seems likely that it proceeds via dodec-2-enoyl-CoA by hydration to form 3-hydroxy-dodecanoyl-CoA, the major hydroxy acid present (Kolattukudy & Rogers, 1987). Ducks such as the Mallard produce diesters of 3-hydroxy fatty acids, which become esterified with one alcohol and one fatty acid. The enzymes catalysing these steps in diester wax synthesis are located in the peroxisomes, in contrast to those for monoester waxes which are formed on the endoplasmic reticulum (Bohnet *et al.*, 1991).

4.4.4 *Seasonal changes in the production of lipids by the uropygial gland*

There are interesting seasonal changes in the production of waxes and oils by the uropygial gland. These have been most extensively studied in male and female Mallards. For most of the year, the Mallard drake produces waxes in which over 80% have short-chain fatty acids, but during eclipse after the postnuptial moult, he produces waxes that have predominantly long-chain fatty acids, with about 20% short chain. This occurs for a two month period during June and July (Kolattukudy, Rogers & Flurkey, 1985). These changes are accompanied by a marked reduction in the S-acyl fatty acid synthase thioesterase, which catalyses the release of short-chain fatty acids during the eclipse period. This suppression of thioesterase activity is controlled at the transcriptional level, since there is a marked reduction in the mRNA. The eclipse period corresponds with the maximum level of circulating thyroxine and minimum level of testosterone, although there is no direct evidence so far for the molecular basis of this regulation. In the female Mallard, there is a different change in the secretions of the uropygial gland. During the breeding season, monoester wax produced throughout the rest of the year is replaced by diesters of 3-hydroxy fatty acids (Bohnet et al., 1991). The principal components of these waxes are fatty acids (abundance: $C_{10} > C_{12} > C_{14} \approx C_{16}$), 3-hydroxy acids (abundance: $C_{10} > C_{12} > C_8$), and fatty alcohols (principally C_{16} and C_{18}) (Bohnet et al., 1991). These diesters are believed to act as sex pheromones (Jacob, Balthazart & Schoffeniels, 1979).

The waxes have a distinctive species distribution (Table 4.1), with close similarity among related species. The diesters themselves are not very volatile, but there is evidence these waxes are partially cleaved to more volatile derivatives after distribution on the plumage. There is evidence that this change does occur on the plumage of the Marabou stork (Jacob et al., 1979). Plasma concentrations of oestradiol show seasonal changes in both female and male Mallard ducks. In the female, the normal level of 40–60 ng/l falls to about 10 ng/l in July after the postnuptial period. In the male, the normal level is much lower (2–3 ng/l) but rises in July during eclipse to about 10 ng/l (Humphries, 1973). The change in wax production is induced by oestradiol, which is highest in the circulation during the breeding season. Bohnet et al. (1991) have shown that a similar change can be induced in male Mallards by injection of oestradiol. There is a proliferation of the uropygial gland peroxisomes, in which 3-hydroxy fatty acids are produced, and this is accompanied by an increase in alcohol dehydrogenase activity, which may be concerned with the generation of fatty acids required for the formation of waxes from the corresponding alcohols, (Hiremath et al., 1992).

4.5 Lipid transport

The most abundant lipids in the body, the triglycerides, are the main energy reserves and are used by many tissues, e.g. skeletal and cardiac muscle. Cholesterol is an essential component of most cell membranes and is the precursor of the steroid hormones. Triglycerides, cholesterol and cholesteryl esters present in the tissues of the bird are either acquired through the diet or synthesised de novo. In the case of triglycerides, the bulk of de novo synthesis in avian species occurs in the liver. Unlike mammals, there is little de novo synthesis in the adipose tissue. Lipids have to be transported through the bloodstream from either the small intestine, in the case of exogenous lipid, or from the liver in the case of lipid synthesised de novo. The large lipid reserves of the adipose tissue are mobilised when dietary intake is low and energy output is high, as, for example, during migration. Lipid transport is then largely from adipose tissue to muscular tissue. In laying birds, the large quantities of lipids laid down in the egg yolk have to be transported from the liver to the developing oocyte. All these examples require the transport of lipids from one tissue to another. Most of the lipids are transported in the form of lipoproteins present in the plasma. Understanding of the avian lipid transport system has lagged behind our understanding of the mammalian system. The latter has been intensively studied in humans, particularly because of its relevance in coronary diseases.

4.5.1 The lipoproteins

Since triglycerides, cholesterol and cholesteryl esters are either insoluble, or only slightly soluble, in aqueous solutions, a vehicle is necessary for their transport. Lipoproteins are an association of lipid and protein of varying degrees of stability, involving secondary valence bonds between the two entities. The lipoproteins should be regarded as aggregates rather than distinct molecular species, as their composition is variable. The protein moiety serves two functions: (i) to solubilise the lipid, and (ii) to interact with the appropriate tissue receptors to allow the lipid to be taken up into specific tissues. Operationally, it is possible to separate the lipoproteins on the basis of their differences in buoyant densities, into three or four main fractions. Although the fractions are not homogeneous and there is some overlap of components, distinct functional roles have nevertheless been assigned to the types present in different fractions. The method most commonly used to fractionate plasma lipoproteins is by ultracentrifugation in a density gradient for up to 48 h (Hermier, Chapman & Leclercq, 1984; Hermier, Forgez & Chapman, 1985), but agarose gel electrophoresis has also been used (Alexander & Day,

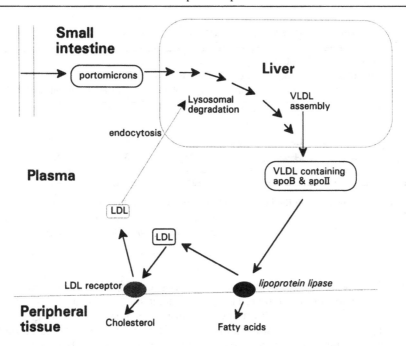

Fig. 4.7. The role of lipoproteins in lipid transport between avian tissues.

1973). Lipoproteins from the domestic fowl have been most studied amongst avian species, where interest has centred on the difference between the lipoprotein content of strains selected for high and low adiposity, since this is of importance in the production of broilers (Hermier, Salichon & Whitehead, 1991). The administration of oestrogens to domestic fowl stimulates production of VLDL by the liver and may cause marked hyperlipidaemia and atherosclerosis (Kudzma, Swaney & Ellis, 1979). This makes the fowl an important model for understanding atherosclerosis in humans and has also prompted studies on lipoproteins in the domestic fowl. There are also strains of quails and pigeons that are susceptible or resistant to atherosclerosis (Shih, Pullman & Kao, 1983; Hadjiisky et al., 1993) and these may also be useful. Separations of lipoproteins have also been carried out on other species, such as goose, quail, turkey, wood pigeon, European robin, garden warbler and pied flycatcher (Mills & Taylaur, 1971; Langelier, Connelly & Subbiah, 1975; Jenni-Eiermann & Jenni, 1992; Oku et al.,1993).

The three main fractions obtained by density gradient centrifugation of plasma from domestic fowl are VLDL, low-density lipoproteins (LDL) and high-density lipoprotein (HDL). Three equivalent fractions were obtained from quail serum (Oku et al.,

1993). Other classes sometimes included are intermediate density lipoprotein (IDL) and very-high-density lipoproteins (VHDL) (Schjeide & Schjeide, 1981; Hermier et al., 1984; 1985). There is some variation in the protocol for fractionation used by different research groups and also in the density range used for classification. All three main fractions contain characteristic proteins and have been assigned particular functions.

The role of VLDL is principally the transport of exogenous triglyceride from the small intestine, and endogenous triglyceride from the liver, to peripheral tissues (Fig. 4.7). With mammalian plasma lipoproteins, it is usual to distinguish the chylomicrons that transport exogenous triglyceride from the intestine from the VLDLs that transport endogenous triglyceride from the liver, the former being larger and having a higher lipid content. In avian species, the VLDL class includes lipoproteins containing lipid of both endogenous and exogenous origins. Birds do not have a well-developed lymphatic system, and triglycerides absorbed in the small intestine are transported directly via the portal system to the liver as lipoprotein particles referred to as *portomicrons*. The proportion of triglyceride of endogenous and exogenous origin in the VLDL fraction depends on the nutritional status of the bird (Hermier et al., 1984). LDLs are

Table 4.2. *Lipoproteins present in the plasma of domestic fowl*

Class	Buoyant density	Plasma concentration (mg/ml)	Protein (% of total)	Lipid (% of total)	Phospholipid (% of total)	Ch + ChE[a] (% of total)	Triglyceride (% of total)	Apoproteins present
VHDL	>1.25		80	20	14	2.4	3.6	Lipovitellin, phosvitin
HDL	1.063–1.25	4.6–5.2	42–48	52–58	22–30	16–27	1.6–7.4	ApoAI
LDL	1.006–1.063	1.2–1.5	24–29	71–76	17–23	33–42	7–22	ApoB
VLDL	<1.006	0.03–0.44	6–26	74–94	14–18	18–23	41–53	ApoB, ApoII

[a] Nutritional status has a marked influence on the lipid content, reflected in the range of values given. Cholesterol plus cholesterol esters.

Source: The data are compiled from Mills & Taylaur (1971); Kudzma *et al.* (1979); Schjeide & Schjeide (1981); Hermier *et al.*,(1984; 1985).

generally thought to be concerned with cholesterol transport to tissues, where cholesterol is required for membrane structure and as a steroid hormone precursor. They have been less intensively studied in avian species than in humans. LDLs arise from VLDLs after triglycerides have been lost, and as a result they become relatively richer in cholesterol and cholesteryl esters. HDLs are important in transport of cholesterol from peripheral tissues to the liver (referred to as reverse cholesterol transport), where they may be degraded. In humans, the HDL fraction is divided into two subfractions, HDL_2 and HDL_3, which can be distinguished by density gradient fractionation. The HDL_2 appears to have a stronger inverse relationship to the occurrence of cardiovascular disease in humans. In the domestic fowl, there is a single HDL class, which has a protein:lipid ratio closest to human HDL_2 but with other physical properties more closely resembling human HDL_3 (Kruski & Scanu, 1975).

4.5.2 Apoproteins

The composition of the various lipoprotein fractions are given in Table 4.2 and it can be seen that the proportion of lipid decreases from the VLDL to the HDL, but the total plasma concentrations of the three classes increases in the order VLDL (1%), LDL (16%) to HDL (80%). The LDL:HDL ratio of approximately 1:5 is much lower than in mammals, where it is around unity (Hermier *et al.*, 1984). The protein component of the lipoproteins, known as the apoproteins, are very important in the functioning of the lipoproteins. Apart from maintaining the lipid in a soluble form, the proteins have important binding properties and, in some cases, catalytic activity. The characteristics of the proteins associated with lipid transport in avian species are given in Table 4.3. The original nomenclature was based on the mammalian system, where more extensive work has been carried out, and the mammalian equivalent or near-equivalent is also shown in Table 4.3.

ApoA-I is the major protein present in HDL from domestic fowl. Its principal sites of synthesis are liver and intestine, but it is also expressed in a number of peripheral tissues (Bhattacharyya & Banerjee, 1993). This contrasts with mammals where synthesis appears restricted to liver and intestine. The complete cDNA sequence for apoA-I has been determined (Byrnes *et al.*, 1987; Rajavashisth *et al.*, 1987) and more recently the complete gene sequence (Bhattacharyya *et al.*, 1991; 1993). ApoA-I is the main protein component of domestic fowl HDL, making up 65% of the total protein. It has approximately 74% α-helical structure, which is significantly higher than the 60–70% found in mammalian apoA-I with which it has a fair degree of homology (Rajavashisth *et al.*, 1987). The sequence of 240 amino acid residues includes a number of 11 residue repeats. Unlike the mammalian apoA, the domestic fowl apoA-I is monomeric (28 kDa) when lipid free and some of its properties resemble those of mammalian apoE (Kiss *et al.*, 1993). It has a higher lipid-binding affinity than mammalian apoA and is expressed in liver, brain, intestine, adrenals, kidney and heart. Since there is only a single gene for apoA-I, there must be tissue-specific regulation to account for its presence in the range of peripheral tissues. The transcription of the apoA-I gene is regulated by a region 5′ upstream of the coding region. TATA and CACAAT boxes (see Section

Table 4.3. *Proteins involved in lipid transport*

Protein	M_r	Location	Possible function	Mammalian equivalent
Apoprotein A-I	28 000	Mainly HDL	LCAT activator and coat protein	ApoA-I and ApoE
Apoprotein B	500 000	Mainly VLDL and LDL	Receptor binding	ApoB
Apoprotein II	16 000(dimer)	Mainly VLDL and LDL	Lipoprotein lipase inhibitor (K_i, 40 mg/l)	ApoC
Lipoprotein lipase	56 000	Capillary cells	Hydrolyses TG	Lipoprotein lipase
Lipovitellin	200 000	Vitellogenin	Transports lipid to oocyte	–

8.5), and binding sites for transcriptional factors (see Section 8.5), have been identified in this region (Bhattacharyya & Banerjee, 1993; Bhattacharyya *et al.*, 1993). ApoA-I is synthesised on the rough endoplasmic reticulum in domestic fowl hepatocytes. In the early stages of its synthesis it is partially exposed on the cytosolic side of the endoplasmic reticulum, and like other secretory proteins, it is translocated during translation through to the luminal side before moving to the Golgi where it becomes conjugated with lipid prior to secretion (Dixon *et al.*, 1992). Triglyceride biosynthesis is catalysed by enzymes present in the cytosolic side of the endoplasmic reticulum. This raises the question of how they reach the apoproteins for lipoprotein assembly. The three principal apoproteins (apoA-I, apoB and apoII) cross the endoplasmic reticulum of domestic fowl hepatocytes at markedly different rates and, therefore, cannot be associated with each other at this stage. Measurements of the rate of movement of labelled triglycerides across the endoplasmic reticulum support the view that the lipoproteins assemble in the Golgi (Bamberger & Lane, 1990). Transport of triglycerides across the endoplasmic reticulum is an ATP-dependent process. Cholesterol esters are also synthesised in the rough endoplasmic reticulum, but phospholipid and triglycerides are synthesised on the smooth endoplasmic reticulum (Griffin, 1992).

A 26 kDa protein found in quail lipoproteins is very similar to apoA-I from domestic fowl. It is the same size, and the 36 N-terminal residues which have been sequenced show 92% homology with domestic fowl apoA-I (Oku *et al.*, 1993). It differs in being the predominant protein in HDL, LDL and VLDL, in contrast to domestic fowl where only traces are found in VLDL. ApoA-I has been isolated and sequenced from Pekin duck (Gu *et al.*, 1993). It shows a high degree of homology with that of domestic fowl, having only 7% of residues different. No apoE has been detected in avian species, and a number of pieces of evidence point to apoA-I serving an analogous role in avian species. Avian apoA-I is expressed in a similar range of tissues as that of mammalian apoE. A region of its structure is similar to the binding site for mammalian LDL receptors. An increase in the circulating level of apoA-I occurs upon hatching, and also after optic nerve ablation, the latter is followed by degradation of the myelin sheaf. In both of these, cholesterol metabolism is affected. ApoA-I functions both as an activator of the enzyme lecithin–cholesterol acyltransferase (EC.2.3.1.43; LCAT, reaction 4.5) thereby stimulating the efflux of cholesterol from peripheral tissues, as well as being a coat protein for HDL. LCAT is implicated in the transport of cholesterol from peripheral tissues.

$$\text{phosphatidylcholine} + \text{sterol} \rightleftharpoons$$
$$\text{1-acylglycerophosphocholine} + \text{sterol ester} \quad (4.5)$$

When the proteins are extracted from domestic fowl VLDL and special precautions are taken to avoid proteolysis, three main protein bands are seen on SDS–PAGE. These are apoB ($\approx$ 500 kDa) and the monomeric (9.5 kDa) and dimeric forms of apoII (Nimpf, Radosavljevic & Schneider, 1989a). Trace amounts of ApoA-I are also present. ApoB, which is also present in LDL, is involved receptor binding. ApoII is not involved in receptor binding, but is an inhibitor of lipoprotein lipase. ApoII comprises two identical polypeptide chains, each having 82 amino acid residues and joined by a single disulphide bridge (Williams, 1979). All of the apoproteins have some structural features in common, and it has been suggested that they evolved from a common ancestral gene (Boguski *et al.*, 1986).

4.5.3 *Lipoprotein lipase*

In avian species, the endogenous and exogenous triglycerides are transported via the VLDL to the peripheral tissues where fatty acids are released and taken up into the tissues (Fig. 4.7). In adipose tissue, uptake is primarily for storage, whereas in cardiac and skeletal muscle the fatty acids are used as a source of fuel. The release of fatty acids from the triglycerides present in VLDL is effected by the enzyme lipoprotein lipase. Lipoprotein lipase is synthesised in several tissues, but the largest amounts are associated with skeletal muscle and adipose tissue. It is secreted into the capillaries where it is bound by heparin sulphate on the luminal side of the endothelial cells. The gene for lipoprotein lipase has been cloned and sequenced from adipose tissue DNA (Cooper et al., 1989; 1992). Lipoprotein lipase from adipose tissue is a 56 kDa glycoprotein having 465 amino acid residues and showing 77% homology with mammalian lipoprotein lipases. Within its structure are a number of different binding sites which are important for its functioning. These include the catalytic site to which triglycerides are bound and hydrolysed, a heparin-binding site and an apoII-binding site. The cDNA was used in Northern blotting to probe for mRNA for lipoprotein lipase in heart and liver. It was present in heart but not liver. The levels of lipoprotein lipase and its mRNA in heart and adipose show reciprocal changes on starvation and refeeding. In adipose tissue, both enzyme and its mRNA fall on starvation and rise on refeeding, whereas the heart enzyme shows the reverse.

4.5.4 Lipid transport to peripheral tissues

The picture of lipid transport between tissues is still far from complete in birds, although it is clear that there are a number of differences from that of the mammalian system. Exogenous lipids (principally triglycerides, cholesterol and cholesteryl esters) are absorbed in the small intestine and transported as portomicrons directly via the portal system to the liver. The portomicrons undergo transformation in the liver and are released to form part of the VLDL fraction. VLDLs are heterogenous and include both modified portomicrons, which contain exogenous lipids, and also lipoproteins containing endogenous lipids. The latter are mainly synthesised in the liver. The proportions of endogenous and exogenous lipids will depend on the nutrition state of the bird. VLDLs, on reaching peripheral tissues, bind to lipoprotein lipase present on the capillary walls. The two main VLDL apoproteins regulate this process. ApoB contains a binding site for lipoprotein lipase, and ApoII is thought to inhibit lipid breakdown during transport. Once triglycerides have been hydrolysed by lipoprotein lipase, the fatty acids are taken up by the peripheral tissues. The VLDLs, having lost a portion of their lipid, have a higher density and become part of the LDLs of IDL. These LDLs return to the liver where they bind to cell surface receptors and become endocytosed, eventually breaking down in the lysosomes.

Cholesterol supplied to peripheral tissues may also be of endogenous or exogenous origin and is contained within VLDLs. When VLDLs reach the peripheral tissues and give up fatty acids, much of their cholesterol becomes esterified by association with HDLs, which contain the enzyme LCAT. The LDLs containing these cholesteryl esters are taken up by receptor-mediated endocytosis into the peripheral tissues. HDLs are important in maintaining cholesterol homeostasis and are involved in the reverse transport of cholesterol from peripheral tissues to the liver. Cholesterol returned to the liver may either become degraded or be stored in the form of cholesteryl esters. An enzyme, sterol O-acyltransferase (EC.2.3.1.26), present in the the endoplasmic reticulum of avian liver catalyses this transacylation (reaction 4.6) (Greer & Hargis, 1992):

$$\text{acyl-CoA} + \text{cholesterol} \rightleftharpoons \text{CoA} + \text{cholesterol ester}$$
(4.6)

4.5.5 Lipid transport to the developing oocyte

One form of lipid transport, which is exclusive to oviparous vertebrates, is the lipid synthesised in the liver and transported to the developing oocyte. Two lipoproteins are largely responsible for this process, VLDL and vitellogenin. These yolk precursors account for approximately 60% and 24%, respectively, of the dry mass of yolk (Burley, Evans & Pearson, 1993). At the onset of ovulation, there is a massive oestrogen-induced synthesis of VLDL and vitellogenin (Nimpf & Schneider, 1991). These lipoproteins circulate in the plasma and are then taken up by the developing oocyte. VLDLs present in the egg-yolk are similar to the VLDLs from plasma (Griffin, Perry & Gilbert, 1984). The protein and lipid composition of the plasma and yolk VLDL are very similar (Table 4.4), but there is a difference in the banding patterns of the apoproteins when separated by SDS–PAGE. ApoB is present in VLDL and is responsible for binding to the oocyte receptor. Evidence suggests that, although the triglyceride remains intact after receptor-mediated endocytosis into the developing oocyte, apoB is fragmented by proteolysis, by an enzyme having cathepsin D-type specificity (Nimpf, Radosavljevic

Table 4.4. *Comparison of plasma VLDL and yolk VLDL composition*

Component	VLDL	
	Plasma (%)	Yolk (%)
Protein	12.3	12.9
Total cholesterol (free + esterified)	4.6	4.3
Phospholipid	16.8	17.0
Triglyceride	66.3	65.8

Source: Data from Nimpf *et al.* (1989b).

& Schneider, 1989b). Cathepsin D is a lysosomal protease with an acid pH optimum and a preference for cleaving the carboxyl side of hydrophobic amino acid residues. The second apoprotein, which makes up approximately half of the total protein, is known as apovitellenine I. It appears identical with apoII found in plasma (Griffin *et al.*, 1984). ApoII, which is an inhibitor of lipoprotein lipase, is thought to ensure that triglyceride is transported intact in the VLDL (Schneider *et al.*, 1990). Oestrogen regulates the synthesis of ApoII in the liver. Using an ELISA (enzyme-linked immunosorbant assay) method to detect apoII, Pinchasov, Elmaliah & Bezdin (1994) have shown that it is undetectable in pullets up to about 12 weeks of age but increases slightly about 23 weeks and then increases sharply at the onset of lay. This occurs in parallel with the increased triglyceride synthesis in the liver.

Vitellogenin is also important in the transport of lipids to developing oocytes, although it contains a much lower proportion of lipid (15–20%). Vitellogenin is a dimer (500 kDa) of which 80% is polypeptide and the remainder lipid, covalently bound phosphate and carbohydrate (Banaszak, Sharrock & Timmins, 1991). The polypeptide chains show homologues with apoB and mammalian apoE. ApoB, apoE and vitellogenin are each able to bind mammalian LDL receptors. Vitellogenin interacts with the oocyte receptor and is endocytosed by the developing oocyte. Like VLDL, it also undergoes proteolytic cleavage into at least three fragments: lipovitellin-1 (120 kDa) arising from the N-terminus, phosvitin (35 kDa) from the central portion and lipovitellin-2 (30 kDa) from the C-terminus. The crystal structure of the lipovitellin complex from lamprey oocytes has been determined to a resolution of $\approx$ 28 nm, and it is probable that avian lipovitellin has a similar structure,

with a helical domain surrounding a cavity into which the phospholipid and triglyceride components can fit (Banaszak *et al.* 1991).

Both VLDL and vitellogenin bind the same site on the 95 kDa receptor localised on the plasma membrane of the oocyte (Nimpf & Schneider, 1991). They have similar binding affinities (VLDL K_d = 6.8 nM and vitellogenin K_d = 6.7 nM; Barber *et al.*, 1991). Somatic cells have a different plasmalemma receptor, which binds VLDL but not vitellogenin (Fig. 4.8). A total of four lipoprotein receptors have been detected in the tissues of the laying hen (Stifani *et al.*, 1991) The first is the 95 kDa receptor, which not only binds avian VLDL and vitellogenin but will also bind lipoproteins from other species that contain apoB, i.e. VLDL and LDL, but not HDL (George, Barber, & Schneider, 1987). The function of this receptor is most clearly demonstrated from the study of a genetic detect known as restricted ovulator. Hens with this defect are characterised by absence of egg-laying and by severe hyperlipidaemia. The lack of high-affinity binding of VLDL to oocyte membranes in restricted ovulator hens and the inability of antibody raised against the 95 kDa receptor to detect antigenic sites strongly suggest the absence of this receptor (Nimpf *et al.*, 1989b). It also demonstrates the role of the 95 kDa receptor in the uptake of VLDL and vitellogenin into the developing oocyte.

VLDLs synthesised in the liver by laying hens are smaller in diameter (30 $\pm$ 5 nm) than either VLDL synthesised by immature hens or portomicrons of dietary origin (Griffin, 1992). The basal lamina of the ovarian follicle acts as a filter preventing the larger portomicrons and VLDLs from passing through. Thus, only smaller VLDLs synthesised by the laying hen pass through the basal lamina and between the cells of the granulosa layer, where they bind to the receptors on the oocyte plasma membrane. Vitellogenin and the small VLDL bind the 95 kDa receptor and undergo receptor-mediated endocytosis (Griffin, 1992). The lipoproteins are released into the oocyte and the receptors recycle back onto the plasmalemma. ApoII is synthesised in response to circulating oestrogen. Griffin (1992) suggests it may act to control the size of the lipoprotein so that it can pass through the basal lamina.

The second receptor (130 kDa) has been detected in the fibroblast and granulosa cells and it is believed to be the counterpart of the mammalian LDL receptor. It is probably important in cholesterol homoeostasis in extraoocyte tissues (Hayashi, Nimpf & Schneider, 1989). The third and fourth receptors, which have been more recently discovered, are larger, but their

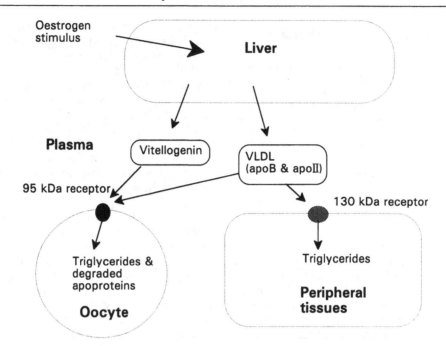

Fig. 4.8. Receptor-mediated uptake of lipoproteins by developing oocyte and by somatic cells (Nimpf & Schneider, 1991).

physiological role is not yet clear (Stifani *et al.*, 1991).

Within the developing embryo, lipoprotein lipase activity is high in both heart and adipose tissue at E14 but is absent from liver and brain. A big increase in activity occurs in adipose tissue between E12 and E16 and this coincides with the period of lipid uptake from the yolk and deposition in the adipocytes. More than 90% of the energy required by the developing embryo is obtained from oxidation of fatty acids present in yolk triglycerides. A further increase in lipoprotein lipase activity also occurs on hatching (Speake, Noble & McCartney, 1993).

4.5.6 *Plasma lipoprotein levels under different physiological conditions*

The concentrations of lipoproteins in the plasma depend on the rate of production of lipoproteins and on their rate of uptake and degradation by peripheral tissues. The main site of synthesis of lipoproteins is the liver, and this is greatly stimulated by oestrogens. Kudzma *et al.* (1979) found that administration of the oestrogen analogue diethylstilboestrol caused a 400-fold rise in VLDL, a 70-fold rise in LDL and a fall in HDL. The plasma lipoprotein concentrations also change, although less markedly, in response to fasting and refeeding. In the domestic fowl after fasting overnight, followed by feeding, the VLDL concentration rises by an order of magnitude, although the LDL and HDL concentrations change only slightly (Hermier *et al.*, 1984). There are also differences between strains of broilers that have been selected over a number of generations for high and for low abdominal fat.

During migratory flight, when there is no exogenous intake of triglyceride, a large amount of stored triglyceride is mobilised as fuel. The breakdown of triglycerides from adipose tissue provides fatty acids for skeletal muscle. This process has been studied in small passerine birds, where there are marked increases in the plasma levels of both VLDL and albumin-bound non-esterified fatty acid (Jenni-Eiermann & Jenni, 1992). These probably supply fatty acids as fuel to the flight muscles. VLDL would require the action of lipoprotein lipase in the capillary walls of the flight muscle to release fatty acids.

4.6 Lipolysis, fatty acid oxidation, ketone body formation and cholestrol metabolism

Triglycerides are the major form of fuel in birds, and the principal stores are in the adipose tissue and liver in adult birds, and in the yolk in developing embryos. Skeletal muscle and cardiac muscle are the tissues which use most lipid, but most tissues can use

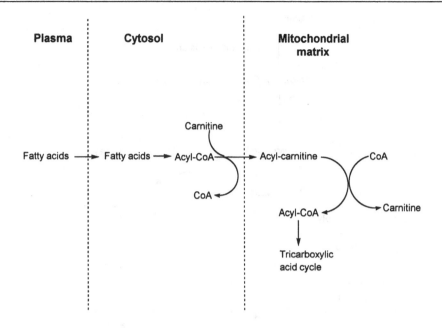

Fig. 4.9. Intracellular transport of long-chain fatty acids into mitochondria.

lipid-derived fuel in some form. The pathways of triglyceride catabolism in birds are basically the same as those in other vertebrates. Endogenous lipid is supplied to peripheral tissues from either the adipose tissue or the liver. The half-life of lipids from adipose tissue of domestic fowl varies with the age of the bird but is generally about 23 ± 3 days (Foglia *et al.*, 1994). Lipolysis occurs, releasing fatty acids from adipose tissue, and they are then transported in the plasma bound to the protein albumin and taken up by the peripheral tissue across the plasmalemma. Once in the cytosol, the fatty acid is converted to acyl-CoA and thence to acylcarnitine before entering the mitochondrial matrix, where it becomes oxidised via β-oxidation, eventually to acetyl-CoA (Fig. 4.9) and generating ATP as a primary source of energy. The fatty acids are able pass freely through the plasmalemma, but acyl-CoA derivatives are not able to cross the inner mitochondrial membrane. There is a shuttle mechanism by which they are converted to carnitine derivatives and are then able to enter the mitochondrial matrix. One of the steps regulating the oxidation of fatty acids by the mitochondrial enzymes is the transport of acyl groups across the inner mitochondrial membrane catalysed by one of the shuttle enzymes carnitine palmitoyltransferase I. Malonyl-CoA is an inhibitor of carnitine palmitoyltransferase in both mammals and birds (Griffin,

Windsor & Zammit, 1990). In mammalian tissues, an increase in the concentration of malonyl-CoA in the cytosol prevents acyl-CoAs from entering the mitochondria. This ensures that when fatty acid biosynthesis is active, β-oxidation is inhibited. However in the liver of domestic fowl, the concentration of malonyl-CoA does not increase sufficiently to inhibit the transferase, and so it appears that both lipogenesis and fatty acid oxidation can occur simultaneously in liver at a significant rate (Griffin *et al.*, 1990).

Much of the endogenous lipid that is eventually used by peripheral tissues is transported in the form of water-soluble *ketone bodies*, the two most important being β-hydroxybutyrate and acetoacetate. The metabolic pathway of ketone body formation and its relationship to cholesterol biosynthesis is shown in Fig. 4.10. Four enzymes are involved in the formation of ketone bodies, namely acetyl-CoA transferase (also known as thiolase), hydroxymethylglutaryl-CoA synthase (HMG-CoA synthase), hydroxymethylglutaryl-CoA lyase (HMG-CoA lyase) and β-hydroxybutyrate dehydrogenase. The last of these catalyses the interconversion of the two principal ketone bodies. All four enzymes are present in liver, the principal site of ketone body formation. Acyl-CoAs are unable to pass through the plasmalemma, and HMG-CoA lyase thus controls the release of ketone

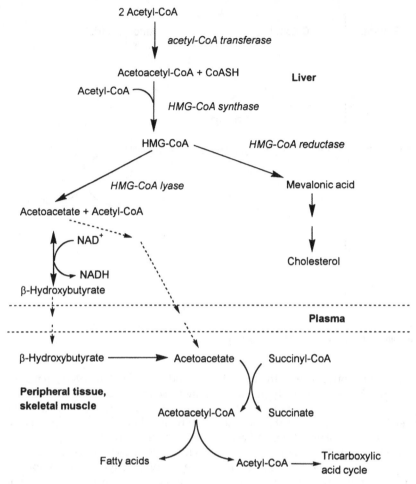

Fig. 4.10. Ketone body formation in the liver and its transport and utilisation in peripheral tissues, particularly skeletal muscle. HMG, hydroxymethylglutaryl.

bodies into the plasma. Coenzyme A is present in only small quantities in cells and, therefore, it is important that it is retained within the cells. The activity of avian HMG-CoA lyase, which has seven sulphydryls per polypeptide chain, is particularly sensitive to its state of reduction (Hruz & Miziorko, 1992). This may be a potential mechanism for regulation of its activity. The initial stages of cholesterol biosynthesis are shared with the pathway for ketone body formation (Fig. 4.10). The step catalysed by HMG-CoA reductase is the first in which metabolites are committed to the cholesterol biosynthetic pathway, and it is an important regulatory step.

Ketone bodies, once taken up by the peripheral tissues, are converted into acetoacetyl-CoA by the enzyme 3-ketoacyl-CoA transferase (Fig. 4.10). They can then be oxidatively metabolised. Although liver is the principal site of ketone body formation, its ability to utilise ketone bodies is very limited. The peripheral tissues differ in their preference for using ketone bodies and fatty acids as fuel. In the domestic fowl, β-hydroxybutyrate dehydrogenase activity is low in all tissues except the brain and kidney (Nehlig, Crone & Lehr, 1980). It is, therefore, probable that brain and kidney are the principal users of ketone bodies, whereas skeletal and cardiac muscle use fatty acids as their source of fuel. There is a high rate of uptake of β-hydroxybutyrate by chick brain, where it is a more efficient precursor of glutamate and related amino acids than glucose.

During embryonic development, the lipid of the yolk is the principal energy reserve, whereas after hatching there is a pronounced increase in carbohydrate

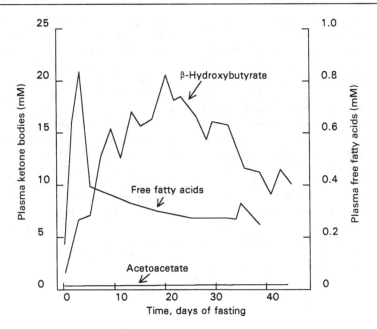

Fig. 4.11. Changing concentrations of plasma metabolites in fasting geese (Le Maho *et al.*, 1981).

utilisation (Nehlig *et al.*, 1980). In the chick embryo at 19E, about 80% of the lipid present in the liver is in the form of cholesteryl esters (Noble & Conner, 1984). From 14E to 21E, there is a steady increase in the concentration of cholesteryl esters and this is followed by a sharp fall for the first 7 days after hatching, during which the cholesteryl esters are partially replaced by triglycerides. Until hatching, the yolk lipid reserves form the main source of energy, whereas after hatching domestic fowl chicks generally use a high proportion of exogenous carbohydrate as the source of energy. The plasma concentrations of β-hydroxybutyrate are fairly steady from E16 to E20, when there is a sharp rise, followed by a fall after hatching (Linares *et al.*, 1993). The rise is preceded by an increase in β-oxidation in embryonic liver, which generates acetyl-CoA, a precursor of β-hydroxybutyrate. The synthesis of both amino acids and lipids from ketone bodies increases after hatching. The extent of this varies between tissues, but kidney shows the highest capacity to synthesise amino acids from ketone bodies. HMG-CoA reductase activity increases during the first four weeks after hatching (Iñarrea *et al.*, 1992), and this is probably to allow a faster rate of cholesterol biosynthesis. During this period, the peripheral tissues will have a high demand for cholesterol, needed for membrane synthesis, particularly for the myelin sheaths in nervous tissue.

The level of HMG-CoA reductase is regulated by the concentration of plasma cholesterol. High concentrations of plasma cholesterol inhibit HMG-CoA reductase. White Carneau pigeons develop atherosclerotic plaques chemically similar to those in humans and this has made the White Carneau pigeon a model for the study of atherosclerosis (Hadjiisky *et al.*, 1993). Cholesterol feeding induces hypercholesterolaemia and accelerates the formation of atherosclerotic lesions. Pravastatin, an inhibitor of HMG-CoA reductase, is able to reduce the hypercholesterolaemia and also the macroscopically visible aortic lipidosis.

A number of species of birds undergo a period of fasting during their normal life cycle, and this is associated with breeding, migration and moulting. The goose *Anser anser* fasts for periods of up to three months whilst breeding, and this is preceded by the build up of large fat reserves. The chicks of the king penguin may be abandoned by their parents during the whole of the subantarctic winter, at a time when there is little food in the sea. They may then survive a fast for up to 6 months, during which they lose upto 70% of their body mass (Cherel, Robin & Le Maho, 1988). These two species are both good subjects in which to study lipid catabolism, the goose being the more accessible. The adipose tissue of such species makes up a substantial part of the body weight, and it comprises over 90% triglyceride. At the onset of the period of starvation that accompanies breeding, the triglycerides are mobilised. Free fatty acids are released from the stored triglycerides, either directly into the plasma in the case of triglycerides from adipose tissue, or after oxidation to β-hydroxybutyrate in the case of the liver. This is illustrated in fasting in the goose in Fig. 4.11. It can be seen that there is a rapid increase in both β-hydroxybutyrate and free fatty acids within the first days of fasting, but no change in acetoacetate concentration. The falls in concentrations after about 40 days indicate the triglyceride reserves becoming depleted. The plasma concentrations of β-hydroxybutyrate may rise to 20 mM after days of fasting (Le Maho et al., 1981) and the free fatty acids to 1 mM (Cherel et al., 1988), but the acetoacetate levels are less than 0.1 mM. In other avian species, β-hydroxybutyrate concentrations in the plasma are at least an order of magnitude higher than those of acetoacetate. In humans during starvation, the total ketone body concentration may rise to 6 mM, of which about 80% is β-hydroxybutyrate and 20% acetoacetate, and free fatty acids to 1 mM. In adult pigeons, the plasma β-hydroxybutyrate concentration may rise 6-fold, whereas the acetoacetate concentration only doubles (Bailey & Horne, 1972). The smaller, or undetectable, increases in acetoacetate concentrations, when compared with those observed in humans, have been attributed to the low concentrations of β-hydroxybutyrate dehydrogenase present in avian tissues (Nehlig et al., 1980). Plasma concentrations of β-hydroxybutyrate in fully fed birds are high, which suggests that a substantial proportion of the lipid is oxidised in the liver (Griffin et al., 1992). There is also an increase in β-hydroxybutyryl-CoA dehydrogenase activity in mitochondria during premigratory fattening, in anticipation of the increased lipolysis about to occur (Marsh, 1981).

5

Protein and amino acid metabolism

5.1 Introduction

The principal nitrogenous compounds present in cells are proteins and their amino acid precursors, nucleic acids and their nucleotide precursors, peptides such as glutathione, compounds derived from amino acids, such histamine, choline and creatine, and excretory compounds such as urea and uric acid. The dietary nitrogen requirements for birds can be met by the essential amino acids (Section 2.2) together with the nitrogen-containing B vitamins. A typical avian diet will generally contain a wider range of nitrogenous compounds, including the non-essential amino acids. Birds are able to synthesise all their body proteins, nucleic acids and other nitrogenous compounds from the minimum dietary requirements. The mechanisms of protein and nucleic acid synthesis are essentially the same in all vertebrates, are well described in many textbooks (Mathews & van Holde, 1990; Zubay, 1993) and are, therefore, not discussed in this chapter. The hormonal control of protein synthesis in the hen oviduct has been extensively studied, and this is discussed in Section 10.3.

The relative importance of anabolic and catabolic pathways varies between tissues and between growing and mature birds and is considered under the general heading of protein turnover (Section 5.2). The mechanism of protein degradation is much less well understood than that of protein synthesis, although much progress has been made recently on this front (see Section 5.3). The final stages of protein degradation leading to nitrogen excretion are quite distinct in birds compared with those in ureotelic vertebrates and are considered in Section 5.4. Another aspect of excretion is the ability to eliminate foreign compounds (xenobiotics) from the body. Many xenobiotics become chemically modified before their elimination from the body, often by conjugation with amino

acids or glutathione. This general mechanism is used by a wide range of vertebrates. Perhaps because birds use glycine for uric acid biosynthesis, they use a different amino acid from mammals, namely ornithine, for conjugation with many xenobiotics prior to their excretion (Section 5.5).

5.2 Protein turnover

Since the late 1930s when isotopic tracers began to become available, and Schoenheimer and Rittenburg carried out their classic studies, it became evident that intracellular proteins are being continually synthesised and degraded throughout the life of a cell (see Price & Stevens, 1989). Since the 1970s, the rates of synthesis and degradation have been measured for some individual proteins in a variety of different tissues. Both the synthesis and the degradation of proteins involve a number of sequential reactions, but in each case there is a rate-limiting step. In the case of protein synthesis, the kinetics of the overall process are zero order and can be defined by a zero order rate constant, k_s, and the rate of degradation shows first-order kinetics defined by the rate constant, k_d (for details, see Price & Stevens, 1989). When referring to the rates of protein synthesis in whole tissues, k_s and k_d are often expressed as a fractional synthesis rate and fractional degradation rate, and these are simply a weighted average for the rates for all the proteins in that tissue. There is a wide range of rates of synthesis and degradation of individual proteins in a tissue. Various methods are available for measurement of k_s and k_d. A summary of these are given by Sugden & Fuller (1991), and some are discussed below.

Most avian studies of protein turnover have been carried out using the domestic fowl, and the tissue most studied has been skeletal muscle. Growing birds (domestic fowl, ducks and turkeys) accumulate

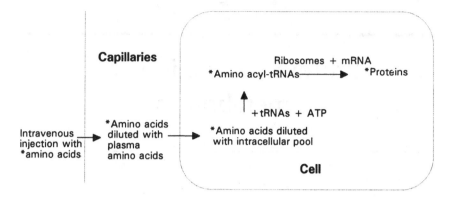

Fig. 5.1. The pathway of incorporation of radioactively labelled (*) amino acids into tissue proteins.

body protein at a rate of about 0.6% body weight/day, whereas laying hens accumulate protein at a rate of about of 0.3% body weight/day (Fisher, 1980). In the latter, protein accumulates mainly in the oviduct and liver, whereas in the former it is accumulating in most tissues. Skeletal muscles grow faster than overall body tissue, e.g. they comprise $\approx$ 6% body weight at one week, rising to 9% body weight at two weeks (Kang, Sunde & Swick, 1985). The rate of protein synthesis varies between tissues. The rates of incorporation of ^{14}C-labelled amino acids are fastest in intestine, intermediate in liver and kidney and slowest in skeletal muscle (Saunderson & Whitehead, 1987). The rate of protein synthesis is much higher than the rate of protein accumulation because of extensive turnover. Protein turnover may be about 5-fold higher than the dietary nitrogen intake, since $\approx$ 80% of the amino acids arising from turnover are reutilised (Swick, 1982). The rates of protein turnover measured in different tissues of quail decrease in the order: liver > heart $\approx$ brain > pectoral muscle (Park, Shin & Marquardt, 1991). The free amino acid pool in tissues is about 0.5% total tissue protein.

A protein that is synthesised rapidly during the process of moulting is keratin. It is the major protein present in feathers, claws and beak. Unlike the intracellular proteins, it is secreted and once formed is inert for the life of the feather, claw, scale or beak. Therefore, any amino acids incorporated into keratin will not be returned to the intracellular pool and there is no intracellular turnover. The sulphur-containing amino acids make up about 10% of the amino acids present in keratin. Murphy & King (1985) have suggested that the tripeptide glutathione (glutamylcysteinylglycine) acts as an intracellular buffer supplying free cysteine when demand is high, such as during moulting. In the white crowned sparrow, intracellular glutathione pools vary between the pre- and post-moult stages, and also show diurnal variations; the range in liver is between 2.5 and 5.0 mM, and in pectoral muscle between 0.5 and 1.0 mM.

In order to determine protein turnover, both k_s and k_d must be measured. There are many problems associated with their determination (Price & Stevens, 1989; Sugden & Fuller, 1991). Almost all the measurements of turnover in birds have been concerned with skeletal muscle. The rate of synthesis, k_s has been determined by measuring the rate of incorporation of ^{14}C-labelled amino acids into the tissue protein. For an accurate assessment, the dilution of the isotopically labelled amino acid with the endogenous amino acid pool has to be taken into account, preferably by measuring the specific radio-activity of the aminoacyl-tRNAs (Fig. 5.1). The rate of degradation can be determined by measuring the rate of loss of radioactivity from the tissue protein (or from the specific protein under investigation). This method assumes there is no reincorporation of the released isotope. Whilst this is a reasonable approximation for tissues with rapid protein turnover, it is very inaccurate for those in which the turnover is slow, as in skeletal muscle. A further disadvantage with the use of radioactively labelled amino acids is that they require the sacrifice of one bird for each time point, making the process very costly. An alternative method, which is easier to perform and which can be used without the sacrifice of the bird, is to measure the excretion of 3-methylhistidine. *3-Methylhistidine* occurs in certain proteins as a post-translational modification of a small proportion of the histidine residues. 3-Methylhistidine only occurs in significant quantities in muscle proteins, particularly actin. In myosin, 3-methylhistidine occurs

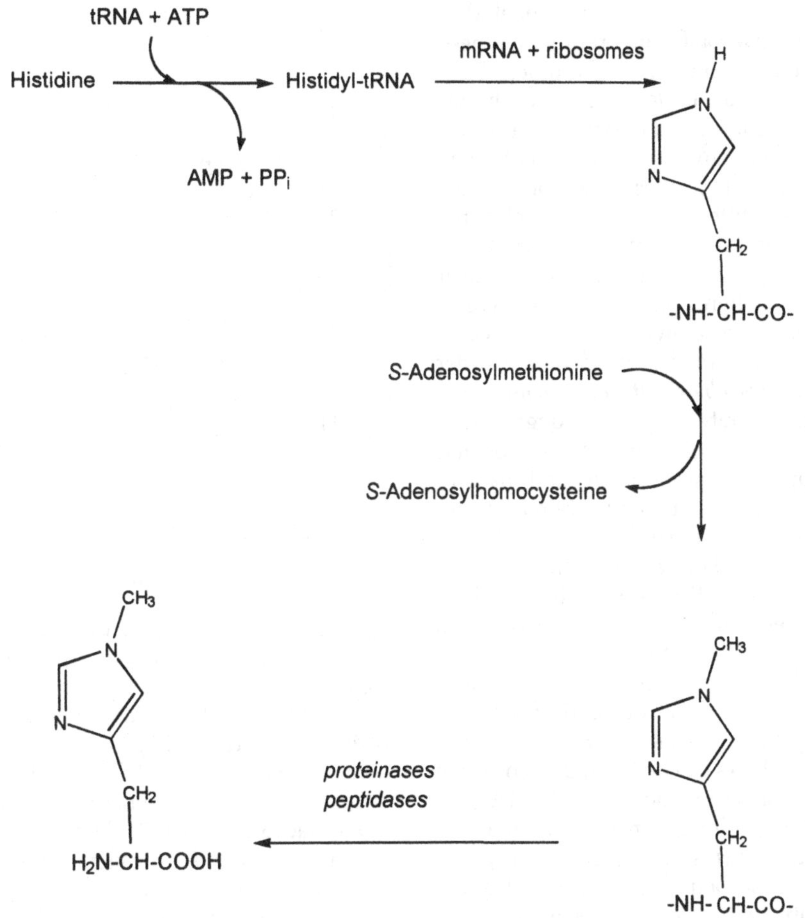

Fig. 5.2. Post-translational modification of histidine residues in skeletal muscle proteins, and the release of 3-methyl histidine after proteolysis.

only in fast-twitch (white) muscle and not in slow-twitch (red) muscle (Tinker, Brosnan & Herzberg, 1986). When proteins containing 3-methylhistidine are degraded, the 3-methylhistidine is excreted unchanged in the urine (Fig. 5.2). Since N-methylation of histidine is a post-translational modification, 3-methylhistidine is not reincorporated into protein. In order to use the rate of 3-methylhistidine excretion to measure k_d in muscle, two criteria must be met. First, it is necessary to demonstrate that 3-methylhistidine is excreted unchanged and not metabolised. Second, most of the 3-methylhistidine excreted must arise from skeletal muscle. These can be tested, first by injecting 3-[^{14}CH$_3$]methylhistidine and checking its recovery in urine; and second by measuring the amount of protein-bound 3-methylhistidine in tissues throughout the body. In domestic fowl and Japanese quail, there was a quantitative recovery of injected

3-[^{14}CH$_3$]methylhistidine in the urine, although this is not the case in turkey poults (Saunderson & Leslie, 1983; Maeda et al., 1986). The distribution of 3-methylhistidine in tissue proteins was very largely in skeletal muscle: 87% in the case of quail (Maeda et al,. 1986), and about 76% in the domestic fowl (Hayashi et al., 1985). It is, therefore, a valid method to study the rates of protein degradation in skeletal muscle of quail and domestic fowl, where most measurements of protein turnover have so far been made.

A further method that has been used to study the rate of protein degradation in vitro is to measure the release of a particular amino acid from the tissue after protein synthesis has been inhibited. Klasing & Jarrell (1985) have used this method to measure the rates of protein degradation in different skeletal muscles of the domestic fowl. They measured the release of tyrosine when isolated muscles were incubated in Krebs–Ringer bicarbonate buffer after protein synthesis had been inhibited using cycloheximide. They previously showed that under these conditions tyrosine was not reincorporated into protein nor was it further metabolised to any significant extent. They compared the rates of growth and of protein degradation in three different skeletal muscles, the *extensor digiti communis*, the *extensor digitorum longus* and the *ulnaris lateralis*. All three have similar proportions of aerobic fibres. They found an inverse correlation between the growth rate and protein turnover. The growth rate decreased in the order *extensor digitorum longus, ulnaris lateralis, extensor digiti communis*, whereas the rates of protein degradation followed the reverse order. A similar situation has been found to apply to muscle in general, namely that during the rapid growth phase protein turnover is lowest. Protein turnover has been studied both using radioactively labelled amino acids and by measuring the rate of excretion of 3-methylhistidine. Amino acids labelled with ^{14}C may be given as a constant infusion, and the rate of labelling of the protein measured, but it is found to be more satisfactory in the case of the domestic fowl (and probably in many birds) to take advantage of the presence of a crop. Chicks can be trained to consume a meal containing amino acids, including a ^{14}C-labelled amino acid, fat, carbohydrate, minerals and vitamins mixed with agar gel. This is taken into the crop and then is gradually absorbed through the intestinal wall. It is comparable with a continuous infusion into the bloodstream (Maruyama, Sunde & Swick, 1978). During the first seven weeks after hatching, the rates of protein synthesis and degradation in skeletal muscle decline from an initial high value (Fig 5.3). Skeletal muscles are fully differentiated in newly hatched birds, i.e. the ratio of sarcoplasmic proteins to myofibrillar proteins is the same as in adults. During this early period, the growth rate of breast muscle is greater than that of leg muscle, and this is reflected in the faster rate of protein synthesis in the former at one week, but not at two weeks (Maruyama *et al.*, 1978). Muscle protein turnover declines from 30% per day in chicks to 5% per day in adults (Swick, 1982). Under conditions of dietary limitation, for example when an essential amino acid is deficient, the rate of protein synthesis in skeletal muscle changes very little, but the rate of degradation increases markedly (Sunde, Swick & Kang, 1984). This generates an amino acid pool from endogenous protein, enabling synthesis to continue for a range of different proteins.

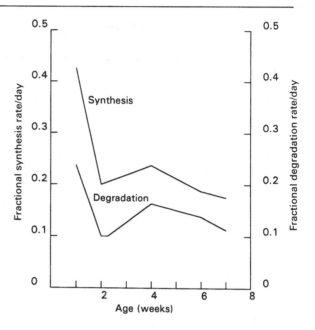

Fig. 5.3. Rates of protein synthesis and degradation in chick skeletal muscle during during growth (McDonald & Swick, 1981).

A number of studies use domestic fowl that have been selected for rapid growth or for high or low fat contents (Saunderson & Whitehead, 1987; Saunderson & Leslie, 1988) and in genetic strains with small body size (Maeda *et al*, 1987). The lower rate of protein degradation found in broilers compared with layers appears to account for the more rapid increase in total protein and body weight of the former (Saunderson & Leslie, 1988). There is a commercial interest in the domestic fowl with a dwarfing genotype. Because this gene (*dw*) is a sex-linked recessive, it is possible to use small hens ($Z^{dw}W$) as parents of broilers ($Z^{Dw}Z^{dw}$ cockerels, and $Z^{Dw}W$ pullets) in which the latter show normal grow rates. (In birds, the male is the homogametic sex ZZ and the female the heterogametic sex ZW.)

Parents: $Z^{dw}W$ (♀) × $Z^{Dw}Z^{Dw}$ (♂)

F$_1$ progeny: $Z^{Dw}W$ (♀) and $Z^{Dw}Z^{dw}$ (♂)

This enables a food saving when rearing the dwarf hens. Fowl possessing the dwarfing gene and having a lower body weight than the normal have a lower rate of protein synthesis for about the first two weeks after hatching (Maeda *et al.*, 1987). Similar studies have been carried out using quail that had been selected for about 60 generations for large and for small body size. Those selected for large body size show both higher rates of protein synthesis and lower rates of protein degradation than the small body size strains (Maeda *et al.*, 1986). It has also been possible to compare the laboratory bred quail with wild captured quail by measuring 3-methylhistidine excretion (Maeda, Okamoto & Hashiguchi, 1992). The wild quail had slightly lower body weight and a significantly higher rate of protein degradation: k_d, 4.1% per day compared with 2.5% per day. In avian muscular dystrophy, higher rates of 3-methylhistidine excretion are found, indicative of higher protein degradation (Hillgartner *et al.*, 1981).

In addition to measuring protein turnover in skeletal muscle in general, two specific enzymes have been studied, namely glycogen phosphorylase (Flannery, Easterby & Beynon, 1992) and ornithine decarboxylase (Bullfield, Isaacson & Middleton, 1988). Glycogen phosphorylase is one of the most abundant proteins in skeletal muscle, comprising 4% of the total. It is important for the utilisation of glycogen reserves in muscle (see Section 3.5). Glycogen phosphorylase has the cofactor pyridoxal phosphate strongly bound to it, and this can be utilised to measure its degradation. Glycogen phosphorylase can be radioactively labelled *in vivo* by injecting [G-^{3}H]pyridoxine into growing chicks. The release of the label into low M_r fraction can be used to measure glycogen phosphorylase degradation. The results of comparing glycogen phosphorylase degradation in broilers and layers, are similar to those found for total muscle protein. Both strains show similar rates of glycogen phosphorylase synthesis, about 15% per day, but the rate of degradation is 5% per day higher in the layers.

A more dramatic difference is seen with ornithine decarboxylase (EC 4.1.1.17). This enzyme catalyses the rate-limiting step in polyamine synthesis, and its activity in many tissues and organisms correlates well with the rate of DNA synthesis and cell proliferation. Its turnover is one of the most rapid of all enzymes, generally having a half-life of less than 20 min. Bullfield *et al.* (1988) have found a 20-fold higher activity in the skeletal muscle from the broiler strain compared with that in a layer strain of domestic fowl at one week of age. This increased activity is almost certainly achieved by an increase in k_s with little change in k_d.

Erythrocytes are unusual compared with other tissues in that there is little turnover during the lifetime of the cell, and the proteins become degraded when the cell itself is degraded. In humans, erythrocytes have an average life of 120 days, whereas avian erythrocytes have an average life of 35 days. The principal protein in erythrocytes is haemoglobin, which is degraded by proteolysis, and its prosthetic group, haem, is oxidised by haem oxygenase to biliverdin. Haem oxygenase, present in domestic fowl hepatocytes, catalyses the degradation of both haem released from senescent erythrocytes and from cytochrome present in the liver (Evans *et al.*, 1991).

5.3 Protein degradation

Protein turnover is important since enables a change in the protein composition of different cells and tissues to occur. Protein degradation is often more important in regulating protein turnover than protein synthesis. The mechanism of protein synthesis is well understood, but that of protein degradation is much less well understood. Proteins become degraded to amino acids, where they can act as precursors for the synthesis of new proteins. Different proteins are degraded at widely differing rates. The major problem in understanding protein degradation is how different proteins are selected for degradation. A number of the enzymes which make up the proteolytic system have now been elucidated, and so the later stages are better understood. Protein degradation has been studied in a number of higher eukaryotes, and it seems likely that most of the features are common to a wide spectrum of organisms. Proteins can be conveniently divided into short lived and long lived (Hershko & Ciechanover, 1982). Long-lived proteins are taken up into the lysosomes and degraded by a group of proteolytic enzymes known as cathepsins. Proteins that come into this category are 'housekeeping enzymes' and proteins that are not tissue specific but are present in a wide range of tissues. Short-lived proteins are, by contrast, broken down by a non-lysosomal pathway and include a number of pacemaker enzymes and regulatory proteins, e.g. ornithine decarboxylase is the pacemaker enzyme for polyamine biosynthesis. Short-lived proteins are generally degraded by an energy-dependent pathway, i.e. ATP is required for proteolysis. Operationally it is possible to distinguish the two. Inhibitors of ATP synthesis or uncouplers, such as dinitrophenol, inhibit the non-lysosomal pathway. Substances, such as chloroquine, that are able to collapse the lysosomal pH gradient inhibit the lysosomal pathway.

5.3.1 *Short-lived proteins and energy-dependent degradation*

Much effort has been spent trying unravel the mechanism by which short-lived proteins are selected for degradation by an ATP-dependent mechanism.

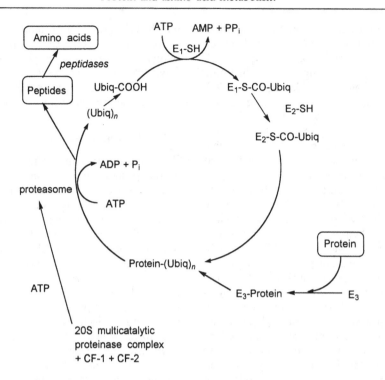

Fig. 5.4 The ubiquitin pathway for protein degradation (Hershko & Ciechanover, 1992). Ubiq, ubiquitin; E_1, ubiquitin-activating enzyme; E_2, ubiquitin-carrier protein; E_3, ubiquitin-protein ligase; CF-1 and CF-2, conjugate-degrading factors.

The process often involves modification by the protein ubiquitin. In skeletal muscle from the domestic fowl, which is the avian tissue most studied to date, there are a number of different pathways of proteolysis. These include lysosomal and non-lysosomal routes, some of which require ATP and ubiquitin (Fagan *et al.*, 1992). *Ubiquitin* is a widely occurring protein in eukaryotes, and it becomes covalently attached to amino groups on proteins, which are then selected for degradation. The detailed steps involved in ubiquitin conjugation have been worked out in rabbit reticulocytes (Fig. 5.4).

Ubiquitin is a small highly conserved protein, having 76 amino acids residues. There are multiple copies of ubiquitin genes in a number of eukaryotes. In the domestic fowl, there are two polyubiquitin genes *UbI* and *UbII*, both of which have been sequenced (Bond & Schlesinger, 1986; Mezquita *et al.*, 1993). *UbI* contains four copies of the protein-coding sequence in tandem, and *UbII* contains three in tandem. They are flanked by a cluster of tRNA genes (Mezquita & Mezquita, 1992). The organisation of the ubiquitin genes is unusual in that it contains no intervening sequences (see also Section 8.4). The cAMP-response element at the 5′ end enables it to be hormonally controlled (Mezquita & Mezquita, 1991). *UbII* is expressed at higher levels in testis than in somatic tissues (Mezquita & Mezquita, 1991). Ubiquitin makes up 0.05–0.23% of the total protein in domestic fowl tissues (Agell & Mezquita, 1988). Ubiquitin conjugation has been demonstrated *in vitro* in the skeletal muscles of domestic fowl where the activating enzymes are present (Arnold & Gevers, 1990).

Once a protein has been selected for degradation and conjugated with ubiquitins (Fig. 5.4), it is then believed to be degraded by a large *multicatalytic proteinase complex* (Orlowski, 1990; Rivett, 1993). This 700 kDa multicatalytic proteinase complex makes up 0.5–1.0% of the tissue protein. It has a number of endopeptidase activities and is made up of 15–20 subunits ranging in size between 21 and 32 kDa. The multicatalytic proteinase complex has been studied in erythrocytes and liver from the domestic fowl (Strack *et al.*, 1992; Sato & Shiratsuchi, 1990) and in duck erythroblasts (Coux *et al.*, 1992a,b). Little is known about the role of the individual subunits, although a 21 kDa protein isolated from a complex particle in duck erythroblasts is homologous to apoferritin (Coux *et al.*, 1992a). The smallest subunit in the multicatalytic proteinase complex from the liver of domestic fowl has chymotrypsin-like activity (Sato & Shiratsuchi, 1990). Coux *et al.* (1992b) have shown that the complex is arranged in the form of a hollow cylinder comprising a ring of six subunits. The multicatalytic proteinase complex has been purified from both nuclei and cytoplasm of erythrocytes from domestic fowl (Strack *et al.*, 1991; 1992).

After the discovery of this multicatalytic proteinase complex, which has a sedimentation coefficient of 20 S, it was realised that it was very similar to another particle that had been isolated independently. The latter, known as a *prosome*, is associated with mRNA that is repressed from translation. Generally a cell has two pools of mRNA, one that is being actively translated and the other that is inhibited from translation. Prosome-like particles have been detected in duck erythroblasts where they inhibit translation of the portion of the mRNA that is bound to the erythroblasts (Akhayat *et al.*, 1987; Nothwang *et al.*, 1992). The multicatalytic proteinase (20S) associates with two factors (CF1 and CF2) to form a 26S complex referred to as a *proteasome* (Fig. 5.4). The proteasome (26S) and the multicatalytic proteinase complex (20S) are thought to participate in ubiquitin-dependent and ubiquitin-independent proteolysis, respectively.

ATP-dependent proteolysis has been studied during embryonic muscular development (Ahn *et al.*, 1991), and during spermatogenesis (Agell & Mezquita, 1988) in the domestic fowl. In both these tissues, changes in the content of different proteins would be expected to occur during development. During embryonic muscle development between E8 and E20, the endopeptidase activities against certain peptide substrates decrease, whilst the proteinases activities, measured using casein as substrate, increase. During that period, the subunit composition of the proteasome changes, and this appears to relate to the change in enzyme activities. These changes suggest that the proteasome plays an important role in avian muscle development. In order to understand the situation further, it will be necessary to identify the natural substrates of the proteasome. Ubiquitin and ubiquitin conjugates are present in developing spermatids from domestic fowl in higher levels than in somatic cells, and the same was found to be the case when comparing *UbII* mRNA in testis with that in somatic cells (Mezquita & Mezquita, 1991). The maximum concentrations were found in the round spermatids (stage II), and from that stage until the fully developed spermatozoa the concentrations declined until hardly detectable (Agell & Mezquita, 1988). Ten stages are recognised in the development of spermatids. Stage II is an early stage in which the tail filament is just beginning to develop (Gunawardana, 1977). The high level of ubiquitin conjugates suggests that protein degradation is important in the restructuring that occurs during development of the spermatid. The level of ubiquitin–protein conjugates increases sharply (i) in the notochord when it becomes vacuolated, (ii) in the myotome when it is first formed by migration of the cells from the epithelial dermamyotome, and (iii) in the development of the lens as the cells elongate (Scotting, McDermott & Meyer, 1991). These are also consistent with ubiquitin-dependent proteolysis playing an important role in restructuring during development.

In addition to ubiquitin-dependent proteolysis, two other proteolytic activities have been identified in skeletal muscle of domestic fowl. The first is an energy-dependent protease that requires ATP ($K_m = 0.027$ mM) and shows the characteristics of a serine protease (Fagan & Waxman, 1989). It is a very large protein (600 kDa) and has a high specific activity when lysozyme or globins are used as substrates. The second proteolytic activity has been detected in ATP-depleted muscles and is stimulated by Ca^{2+}. This protease activity is accounted for by a group of Ca^{2+}-activated cysteine proteases known as calpains. Fagan *et al.* (1992) suggest that the calpains are primarily involved in the turnover of myofibrillar proteins, whereas the substrates of the ATP-dependent proteases are non-myofibrillar proteins of skeletal muscle.

5.3.2 Calpains

The Ca^{2+}-requiring proteinases known as calpains have been isolated from the skeletal muscle of a number of mammalian species and they are divided into two groups depending on their sensitivities to Ca^{2+}: μCANP (Ca^{2+}-activated neutral protease or calpain I) is activated by micromolar concentrations of Ca^{2+} and mCANP (calpain II) by millimolar concentrations. Both calpain I and II comprise two subunits. The smaller one (30 kDa) is identical in both, but the larger one (80 kDa) is distinct. The large subunit contains the active site and also a C-terminal calmodulin-like binding domain. Early work on skeletal muscle from domestic fowl suggested that only calpain II, which has $K_a = 0.15$ mM, was present (Kawashima *et al.*, 1984). Hathaway, Werth & Haeberle (1982) have isolated a similar calpain II from the smooth muscle of domestic fowl gizzard, where it is present in large amounts (15 mg/kg muscle). The intracellular concentrations of free

Ca^{2+} are in the micromolar range, but it has been suggested that ions other than Ca^{2+} may activate or have a synergistic effect, e.g. Mn^{2+} (Suzuki & Tsuji, 1982). The first reported purification of a calpain II in domestic fowl skeletal muscle suggested that it had a single polypeptide chain (80 kDa) and low sensitivity to Ca^{2+} ($K_a = 1.8$ mM; Ishiura et al., 1978). It has been suggested that the smaller subunits might be degraded by proteolysis. Calpain II undergoes autolysis in the presence of Ca^{2+} and the modified calpain formed has a much greater sensitivity to Ca^{2+}. In the case of calpain II from the domestic fowl gizzard, K_a for Ca^{2+} activation changes from 150 μM to 5 μM (Hathaway et al., 1982). This may be part of the normal physiological activation mechanism. The probable sequence of calpain II activation (Fig. 5.5) has been worked out by Crawford, Willis & Gagnon (1987). More recent work on domestic fowl skeletal muscle calpains suggests at least two and possibly three distinct enzymes do exist (Wolfe et al., 1989; Birkhold & Samms, 1994). Two that have been purified using anion exchange chromatography require 100 μM and 500 μM Ca^{2+} for half-maximal activity. The former fits the activation range for a calpain I (Birkhold & Samms, 1994). The principal calpain present in gizzard smooth muscle has wide substrate specificity, but a second calpain is also present in smaller amounts and is specific for the protein desmin, present in intermediate filaments (Johnson, Parkes & Barrett, 1984). An endogenous protease inhibitor (68 kDa), has also been isolated from skeletal muscle of domestic fowl (Ishiura et al., 1982). It is present at a lower concentration in skeletal muscle than the calpain and could only inhibit about one sixth of the calpain activity. The range of physiological substrates for calpains is beginning to emerge. They degrade cytoskeletal proteins and some have been located attached to the Z-line of the myofibril (Ishiura et al., 1982).

5.3.3 Long-lived proteins and the lysosome pathway

The other group of proteinases, responsible for degrading the long-lived proteins (see Hershko & Ciechanover, 1982), are those present in the lysosomes. The long-lived proteins become sequestered into the lysosomes before being degraded. This generally entails lysosomes fusing with autophagic vesicles present in the cytoplasm and then digesting their contents, eventually releasing low M_r substances such as amino acids. The lysosomes contain a range of proteinases, known as *cathepsins*, that have pH optima roughly corresponding with that of the

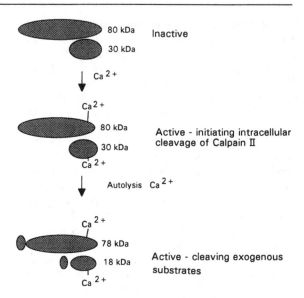

Fig. 5.5. The probable sequence of activation by Ca^{2+} of domestic fowl two subunit calpain II.

lysosomal milieu, i.e. about pH 5. They are relatively inactive at the cytosolic pH. The cathepsins B, H and L are cysteine proteinases and account for most of the lysosomal protein degradation. Inhibitors of cysteine proteinases such as leupeptin or E-64 inhibit the bulk of lysosomal proteolysis (Katanuma, 1989). The evidence so far suggests that the same range of cathepsins are present in avian tissues as are present in mammalian tissues. Avian cathepsins A, B, D and L have been purified and characterised (Iodice, Leong & Weinstock, 1966; Wada & Tanabe, 1988; Dufour et al., 1987; Wada, Takai & Tanabe, 1987).

5.3.4 Peptidases

The proteinases mentioned in Sections 5.3.2 and 5.3.3 generally degrade the protein substrates to small peptides. A further group of enzymes then degrades them to amino acids. These peptidases are subdivided into aminopeptidases and carboxypeptidases, depending on whether they attack the peptide bond at the N- or C-terminus. They are also categorised as endopeptidases and exopeptidases depending on whether the initial point of attack is at the end of the peptide chain or internally. Neutral aminopeptidases appear to be the principal peptidases responsible for degradation to free amino acids in skeletal muscle (Rhyu et al., 1992). A number of different neutral aminopeptidases have been detected in mammalian tissues, differing in their specificity; so far only aminopeptidases H and C have been characterised from skeletal muscle of domestic fowl (Nishimura *et al.*, 1991; Rhyu *et al.*, 1992). This is most probably because these enzymes have been less-thoroughly investigated in the domestic fowl, rather than because there are fewer of them. Aminopeptidase H shows higher activity against tetrapeptides than against di- and tripeptides; it is least active against dipeptides having hydrophobic or acidic amino acids or proline at their N-terminus.

5.4 Nitrogen excretion and the formation of uric acid

The amino acids present in different tissues are either of endogenous or exogenous origin. The former arise from the intracellular degradation of proteins and other nitrogenous substances, whereas the latter are of dietary origin. Most tissues maintain relatively small intracellular concentrations of free amino acids. Unlike fats and carbohydrates, there are no specific storage protein molecules, although skeletal muscle protein is the largest potential reservoir of amino acids. When dietary protein is either scarce or absent, endogenous protein, particularly from skeletal muscle, is degraded to generate a source of amino acids. When there is an excess of dietary protein, it cannot be stored as such and becomes degraded and deaminated, providing carbon skeletons for bio-synthesis of fats and carbohydrates. The surplus nitrogen is excreted. Unlike in mammals, the principal form of nitrogen excreted by birds is uric acid. As long ago as 1815, Proust discovered that the excrement of 'birds and serpents' contained large amounts of uric acid. In 1877 von Knieren showed that birds' urine contained 20–60 times more weight of uric acid than urea. Coindet was able to show as early as 1825 that eagles' urine contains approximately equimolar amounts of uric acid and ammonia. In 1914, well before the urea cycle was discovered by Krebs & Henseleit in 1932, Clementi had pointed out that arginase is present in the livers of *ureotelic* animals, but absent from *uricotelic* animals such as birds, although he was able to detect arginase in avian kidneys (for the early references, see Baldwin (1967) and Brown (1970)).

Subsequent studies suggested that avian liver might contain small amounts of arginase. Brown (1966) made a comprehensive study of arginase activity in avian livers (Table 5.1) and was able to detect it in all species studied. He attributes Clementi's failure to detect activity to the lower sensitivity of the methods then available and to suboptimal

Table 5.1. *Arginase activities in avian livers*

Species	Arginase activity ([a])	
	Units/g liver[b]	Units/mg protein
Uricotelic species		
Royal tern	140–200	0.685–1.17
Great blue heron	189 ± 28	
Little green heron	18.8 ± 2.5	0.136 ± 0.018
Belted kingfisher	220 ± 13	1.15 ± 0.07
Laughing gull	89.2 ± 10.1	0.462 ± 0.048
Herring gull	46.2 ± 0.15	
American crow	60.7 ± 7.3	0.357 ± 0.043
Sanderling	53.0 ± 3.0	
Boat-tailed grackle	34.5 ± 1.8	0.139 ± 0.008
American robin	22.9 ± 4.4	
Black-footed albatross	6.25 ± 0.07	
Goose	25.3 ± 1.7	0.151 ± 0.010
Domestic fowl	4.00 ± 0.42	
Pigeon	0–0.262	0–0.0016
Ureotelic species		
Frog	1000 ± 96	7.13 ± 0.70
Lion	658 ± 31	5.13 ± 0.24
Cattle	485	2.7
Rat	388 ± 8.3	2.38 ± 0.05

Source: Data from Brown (1966).
[a] Arginase activities were measured in liver homogenates at pH 9.5, the optimum pH *in vitro*. The incubation mixture contained 0.085 M arginine and 0.5 mM $MnCl_2$.
[b] Liver weight as wet weight.

conditions of assay. The activity in the domestic fowl is one of the lowest studied (Table 5.1). Generally higher arginase activities are detected in avian kidneys than in livers. This raises the question of whether there is a functional urea cycle in avian liver. Although arginase is detectable, the $K_{m,Arg}$ is in the region of 100 mM, whereas the K_m for mammalian liver is about 5 mM (Reddy & Campbell, 1970). The high K_m of the avian enzyme suggests that its activity is likely to be very low under physiological conditions. This is further borne out by more recent work on arginase activity in mammalian liver.

Of the five enzymes necessary for the functioning of the urea cycle, arginase is usually found to have the highest activity *in vitro*. It requires Mn^{2+} for activity and is generally assayed using optimal concentrations of both Mn^{2+} and arginine. However, the physiological concentrations of Mn^{2+} are less than 1.0 μM (Maggini *et al.*, 1992). When measured using physiological concentrations of Mn^{2+} and arginine, the activity is much lower, such that the enzyme may be the rate-limiting step in urea synthesis (Maggini et al., 1992). Avian liver arginase activity in vivo is also likely to be significantly lower than that measured in vitro by Brown (1966), for the same reasons. Arginase activity is about 10-fold higher in the kidney of domestic fowl than in liver (Robbins & Baker, 1981), and this explains why Clementi was able to detect arginase in avian kidneys but not liver. Also its activity is increased in response to either high-protein diets or a deficiency in an essential amino acid. In both circumstances, there would be an increased need to remove unwanted or unusable nitrogen. Ornithine carbamoyltransferase (EC 2.1.3.3), which catalyses the conversion of ornithine into citrulline, has been purified from the kidney of domestic fowl (Tsuji, 1983), although its role in the probable absence of a functioning urea cycle is not clear.

Although uric acid is the most important nitrogenous excretion product in urine, other nitrogenous compounds are present in significant amounts (Table 5.2). The percentage of uric acid in carnivorous birds may be higher, e.g. in the turkey vulture it ranges from about 75 to 85% of the total nitrogen (McNabb et al., 1980). The differences among higher vertebrates have been shown to be correlated with the mode of reproduction. Ureotelic metabolism is associated with viviparity, uricotelic metabolism with the development within a cleidoic egg. Cleidoic eggs develop under conditions in which there is a strict limitation of water supply. The metabolism that occurs in the developing embryo includes deamination reactions, and the resulting ammonia has to be disposed of. Ammonia is highly toxic, and instead of being converted to urea, which is highly soluble in water, it is converted to uric acid. Uric acid has very low solubility in water ($\approx 1\,mM$) and crystals become deposited in the allantois during the development of the embryo. When developing eggs are incubated under different conditions of humidity, those in which the rate of water loss is greatest show the highest rates of uric acid accumulation (Bradfield & Baggot, 1993). In adult birds, the concentration of uric acid in urine is in the region of 0.1–1.0 M. This greatly exceeds its solubility and most is present as a colloidal suspension (Skadhauge, 1983).

5.4.1 The biosynthesis of uric acid

The incorporation of ammonia into urea or into uric acid requires both energy and building blocks. The synthesis of uric acid is more costly in both ATP and

Table 5.2. *Nitrogenous constituents of hen ureteral urine*

Component	High-protein diet	Low-protein diet
Total N (mg/kg per h)	4.9–9.8	13.6–17.6
Uric acid/urates (% total N excreted)	54.7–58.7	72.1–79.2
Ammonia (% total N excreted)	17.3–29.4	10.8–14.9
Urea (% total N excreted)	3.4–7.7	1.2–9.7
Creatine/creatinine (% total N excreted)	6.8	0.9
Other N (% total N excreted)	1.7–20.3	3.8–7.4

Source: Skadhauge (1983).

organic carbon (Table 5.3), but the 'trade off' is that water is conserved by the developing embryo. The immediate biosynthetic precursors of uric acid are those required for purine biosynthesis in general, namely glycine, glutamine, aspartate, bicarbonate and formyltetrahydrofolate (see Fig. 2.1, p. 11 and Fig. 5.6). This was first demonstrated by administering isotopically labelled ammonia, glycine, formate and carbon dioxide to birds, followed by systematic chemical degradation of the uric acid excreted in order to identify the precursors of the purine ring (Baldwin, 1967). The three amino acids, glycine, glutamine and aspartate, may arise directly from proteolysis. Glutamine may arise from glutamate via glutamine synthetase. Glutamate itself and aspartate may arise from transamination of other amino acids, for example:

$$alanine + 2\text{-}oxoglutarate \rightleftharpoons pyruvate + glutamate$$

In this way, the nitrogen from several amino acids can be transferred to aspartate, glutamate or glutamine. The ketoacid acceptors may originate from glucose or gluconeogenic precursors. The C-1 atom on formyltetrahydrofolate may arise from serine or methionine. In addition HCO_3^- is required for the carboxylation of 5′-phosphoribosyl-5-aminoimidazole.

The importance of the avian liver in the synthesis of uric acid became clear as the result of two early experiments (see Baldwin, 1967). When geese were hepatectomised, there was a rapid rise in the concentration of ammonia in the blood and when goose liver was perfused with ammonia, the formation of uric acid could be detected. The biosynthetic pathway leading to uric acid formation shares many of the enzymes required for the purine precursors of the purine nucleotides. The three steps that are distinct for uric acid formation are the conversion of

Table 5.3. *Energy balance sheet for the urea and uric acid cycles*

Urea cycle

Overall equation:

$3ATP + 2H_2O + 2NH_3 + CO_2 \rightarrow 2ADP + 2P_i + AMP + PP_i + $ urea

The two reactions requiring ATP are those catalysed by carbamoyl phosphate synthetase ($2ATP \rightarrow 2ADP + 2P_i$) and argininosuccinate synthetase ($ATP \rightarrow AMP + PP_i$)

ATP requirement:

2ATP/N atom

Uric acid cycle

Overall equation:

PRPP + glycine + 2 glutamine + aspartate + $4ATP + CO_2 + 2MeTHFA \rightarrow$ ribose + 2

glutamate + $4ADP + 5P_i + PP_i + 2THFA + $ fumarate + uric acid

Regeneration of glutamine:

2 glutamate + $2ATP + 2NH_3 \rightarrow$ 2 glutamine + $2ADP + 2P_i$

Overall equation plus regeneration reaction:

PRPP + glycine + $2NH_3$ + aspartate + $6ATP + 2MeTHFA + CO_2 \rightarrow$ ribose + $6ADP + 7P_i + PP_i + 2THFA + $ fumarate + uric acid

Changes in ATP:

The formation of glycine is equivalent to 9ATP, making the total 15ATP. Fumarate is converted back to aspartate via the citric acid cycle and transamination and does not require any ATP equivalents

The precise energy requirement depends on how PRPP (phosphoribose pyrophosphate) is regenerated; if, as shown above, it is regenerated from ribose, then a further 3ATPs are required, but if the PRPP is split from inosinic acid by pyrophosphorolysis, then no further ATP is required

Reactions requiring ATP:

Phosphoribosyl-glycineamide synthetase ($ATP \rightarrow ADP + P_i$)

Phosphoribosyl-formylglycineamidine synthetase ($ATP \rightarrow ADP + P_i$)

Phosphoribosyl-aminoimidazole synthetase ($ATP \rightarrow ADP + P_i$)

Phosphoribosyl-aminoimidazole succinocarboximide synthetase ($ATP \rightarrow ADP + P_i$)

ATP requirement:

15ATP/4N atoms, i.e. 3.75 ATP/N assuming pyrophosphorolysis, otherwise the ATP/N becomes 4.5

Source: Mapes & Krebs (1978).

inosinic acid (IMP) to hypoxanthine, which is then oxidised via xanthine to uric acid. Mapes & Krebs (1978) have suggested two alternatives for the first step. Either a nucleotidase-catalysed step is followed by one catalysed by nucleosidase (reactions 5.1 and 5.2, respectively) or the conversion is catalysed by pyrophosphorylase (reaction 5.3).

$$IMP + H_2O \rightarrow \text{inosine} + P_i \quad (5.1)$$

$$\text{inosine} + H_2O \rightarrow \text{hypoxanthine} + \text{ribose} \quad (5.2)$$

$$IMP + PP_i \rightleftharpoons \text{hypoxanthine} + \text{phosphoribosylpyrophosphate} \quad (5.3)$$

More energy is expended in route 1 (reactions 5.1 and 5.2) because ATP would be required to regenerate 5-phosphoribosepyrophosphate (PRPP) from ribose (reactions 5.4 and 5.5). PRPP is necessary for the biosynthesis of inosine in the uric acid cycle (Fig. 5.6).

$$\text{ribose} + ATP \rightleftharpoons \text{ribose-5-phosphate} + ADP \quad (5.4)$$

$$\text{ribose-5-phosphate} + ATP \rightleftharpoons \text{5-phosphoribosyl-}\alpha\text{-pyrophosphate} + AMP \quad (5.5)$$

The oxidation steps from hypoxanthine to xanthine and then to uric acid are catalysed by a single enzyme, which in the case of birds is xanthine dehydrogenase (reaction 5.6).

$$\text{hypoxanthine} + NAD^+ + H_2O \rightleftharpoons \text{xanthine} + NADH + H^+ \quad (5.6)$$

$$\text{xanthine} + NAD^+ + H_2O \rightleftharpoons \text{uric acid} + NADH + H^+$$

Both ureotelic and uricotelic animals form uric acid, but in the former it is only derived from purine catabolism. In uricotelic animals, such as birds, uric acid formation assumes much greater importance,

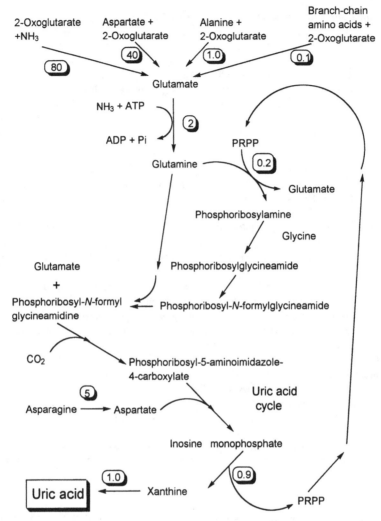

Fig. 5.6. The deamination of amino acids and the formation of uric acid in avian liver. The relative activities of key enzymes in domestic fowl liver are indicated in boxes.

since it is the end-product of both purine metabolism and the deamination that occurs in protein catabolism. Since much of the pathway leading to uric acid synthesis is common to adenine and guanine nucleotide biosynthesis, the regulatory mechanisms might be expected to be responsive to the requirements of the two processes. Uric acid biosynthesis has been most thoroughly investigated in the liver from the domestic fowl. In the uric acid cycle proposed by Mapes & Krebs (1978), PRPP is regenerated at each complete turn of the cycle, rather like that of oxaloacetate in the tricarboxylic acid cycle (Fig. 5.6). Ammonia or amino groups formed from amino acids are channelled into purine nucleotides and eventually uric acid. The

enzymes involved are transaminases, glutamate dehydrogenase, glutamine synthetase and phosphoribosylaminoimidazolesuccinocarboxamide synthase (EC.6.3.2.6) (Fig. 5.6). Several transaminases have been detected in avian tissues. Aspartate aminotransferase is present in both the mitochondria and cytosol from domestic fowl liver (Bertland & Kaplan, 1970). The two isozymes are the products of different genes, but show 46% amino acid sequence homology and are believed to have diverged about 10^9 years ago (Jaussi et al., 1987). Branched-chain amino acids (leucine, isoleucine and valine) are transaminated by branch-chain amino acid aminotransferase (EC 2.6.1.42) in reaction 5.7.

Table 5.4. *Comparison of the rates of nitrogen excretion in domestic fowl and rat*

Diet	Rate of uric acid excretion in domestic fowl [a]	Rate of excretion of urea nitrogen by rat[b]
Standard diet	≈ 1	10
High protein diet	8	12

Source: Wiggins, Lund & Krebs (1982).
[a] Uric acid nitrogen excretion, μmol/min per g liver wet weight.
[b] Urea nitrogen excretion, μmol/min per g liver wet weight.

branched chain amino acid + 2-oxoglutarate $\rightleftharpoons$ branched
$$\text{chain oxo-acid} + \text{glutamate} \qquad (5.7)$$

This enzyme shows highest activity in avian kidney and is also present in liver and skeletal muscle of domestic fowl and Japanese quail (Featherston & Horn, 1973; Mason & Ward, 1981; Sakhri, Jeacock & Shepherd, 1992). Glycine contributes directly both carbon and nitrogen atoms in the reaction catalysed by phosphoribosylamine-glycine ligase (EC.6.3.4.13) (reaction 5.8):

$$\text{ATP} + \text{5-phosphoribosylamine} + \text{glycine} \rightleftharpoons \text{ADP}$$
$$+ P_i + \text{5'-phosphoribosylglycinamide} \qquad (5.8)$$

5.4.2 Regulation of uric acid formation

Figure 5.6 shows the activities of some of the enzymes involved in uric acid formation when the fowl is maintained on a standard diet. Amidophosphoribosyltransferase (EC.2.4.2.14) activity is approximately the same as the overall rate of uric acid synthesis and is the rate-limiting step in the sequence (Wiggins, Lund & Krebs, 1982). It has been purified from pigeon liver (Wingaarden & Ashton, 1959). Its activity increases in domestic fowl liver by about 4-fold after transfer to a high-protein diet, and there is a similar increase in the rate of uric acid formation. The flux through this cycle depends on the enzyme activity of the rate-limiting enzyme, amidophosphoribosyltransferase, and also on the availability of one of its substrates. Lipstein, Boer & Sperling (1978) found the intracellular concentration of PRPP is about 200-fold higher in the domestic fowl compared with that in rat. The total rate of purine biosynthesis (uric acid and purine nucleotides together) is about 15-fold higher in the domestic fowl than in the rat. It is interesting to compare the rates of disposal of surplus nitrogen in ureotelic and uricotelic animals. On a standard diet, rats excrete more urea nitrogen than domestic fowl excrete uric acid nitrogen, but on a high-protein diet the adaptive response of the

domestic fowl is much greater (Table 5.4). This is also the case with the turkey vulture, where the rate of excretion of uric acid increases by about 5-fold after feeding on carrion (McNabb *et al.*, 1980).

The two tissues in which most uric acid formation occurs are the liver and kidney. Using intact cockerels, Tinker *et al.* (1986) have measured the arteriovenous difference in metabolite concentrations across whole organs or tissues. From the data obtained, it is possible to determine whether particular metabolites have been taken up or secreted by given tissues. In fed birds, the amino acids that were most readily taken up across the liver were arginine, isoleucine, leucine, phenylalanine, glutamine, glutamate, aspartate and proline. The removal of glutamine and asparagine was consistent with the observed production of urate by the liver, since glutamine provides two of the nitrogen atoms and aspartate one nitrogen for uric acid biosynthesis. Aspartate is readily formed from asparagine. Uric acid was also released from the muscle and adipose tissues. This was somewhat surprising, since it seemed unlikely that skeletal muscle and adipose tissue were actively synthesising urate. The most likely explanation seems to be that the liver and kidney release xanthine and hypoxanthine and these were converted to uric acid in the skeletal muscle and adipose tissue (Fig. 5.7). Hypoxanthine released from the liver may also be converted to uric acid in the kidney. Within a single organ, there may be differences between cell types in the ability to form uric acid. For example, epithelial cells present in the bursa of Fabricius readily form uric acid from hypoxanthine, whereas lymphocytes salvage any hypoxanthine and convert it to inosinic acid (Senesi *et al.*, 1992).

A surprising difference between the domestic fowl and the pigeon is the lack of detectable xanthine dehydrogenase in pigeon hepatocytes. This was first noted by Krebs and co-workers (Edson, Krebs & Model, 1936) and has subsequently been confirmed (Landon & Carter, 1960; Nagahara *et al.*, 1987). It

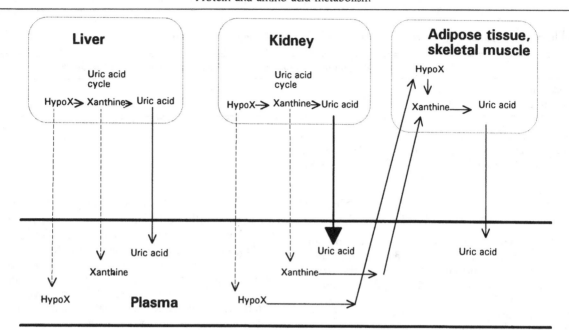

Fig. 5.7. Formation of hypoxanthine, xanthine and uric acid and their release from different avian tissues. The thicknesses of the arrows are indicative of the rates of release and uptake.

seems that in the pigeon hypoxanthine is produced in the liver and transported to the kidney, where it is oxidised by xanthine dehydrogenase to uric acid and excreted (Landon & Carter, 1960; Tinker *et al.*, 1986). Approximately 20% or less of the uric acid excreted by the domestic fowl is synthesised in the kidney (Chim & Quebbeman, 1978), with the bulk synthesised in the liver. Xanthine dehydrogenase has been purified from turkey liver and detailed kinetic studies have been made (Cleer & Coughlan, 1975). Xanthine dehydrogenase appears at different stages of embryonic development in liver and kidney. In the former it is not detectable until hatching, whereas in the latter it is detectable about 6 days earlier (Strittmatter, 1965). The serum concentration of uric acid is an index of the rate of protein catabolism. In migratory geese, where hyperphagy occurs before migration, the serum concentration of both protein and uric acid rises (Mori & George, 1978). The latter is indicative of higher rates of protein catabolism.

5.4.3 Is there a functional urea cycle in avian liver?

Although uric acid is the principal nitrogenous constituent of urine, there is some urea also present (Table 5.2). There are also significant concentrations of urea detectable in the plasma of many species, higher than those of uric acid (Table 5.5), but the uric acid concentrations are close to its solubility limit.

This raises some unresolved questions about whether there is a functioning urea cycle in avian liver. Arginase is detectable in avian liver extracts (Table 5.1), albeit in small amounts, and its activity *in vivo* may be much lower than *in vitro*. Another important enzyme necessary for a functioning urea cycle is carbamoylphosphate synthetase I:

$$2ATP + NH_3 + CO_2 + H_2O \rightleftharpoons 2ADP + P_i + \text{carbamoyl phosphate}$$

In mammalian tissues, two enzymes of the urea cycle are present in mitochondria (carbamoylphosphate synthetase I and ornithine carbamoyltransferase) whereas the remaining three (arginase, argininosuccinate lyase and argininosuccinate synthetase) are present in the cytosol. Carbamoylphosphate synthetase is one of the most abundant mammalian hepatic mitochondrial proteins comprising up to 26% of the matrix protein (Watford, 1991). Is it present in avian liver? Tamir & Ratner (1963) were unable to detect carbamoyl phosphate synthetase in liver or kidney of the domestic fowl, and Emmanuel & Gilanpour (1978) conclude that chick embryo lacks a complete urea cycle. This poses another problem, since carbamoyl phosphate synthetase activity is necessary for pyrimidine biosynthesis. In most higher organisms, this is a separate enzyme, carbamoyl phosphate synthetase II, which is part of a multienzyme complex catalysing pyrimidine biosynthesis and is present in

Table 5.5. *Plasma concentrations of urea and uric acid in different species*

Species	Urea (mM)	Uric acid (mM)	Urea-N (mM)	Uric acid-N (mM)
Duck		0.39		1.56
Golden eagle	2.33		4.66	
Imperial eagle	1.92		3.84	
Eurasian buzzard	1.17		2.34	
White stork (adult)	1.9	0.15	3.8	0.6
White stork (chicks)	1.39	0.85	2.78	3.4
Black stork (chicks)	1.82	0.75	3.62	3.0
Western marsh harrier	1.07	0.43	2.14	1.72
Domestic fowl	0.94	0.47	1.88	1.88
Griffon vulture	2.48		4.96	
Black kite	1.86		3.72	
Great bustard	2.8	1.3	5.6	5.2
Eastern white pelican	1.25	0.33	2.5	1.32
King vulture	1.93		3.86	
Cormorant	1.76		3.52	
Andean condor	2.3		4.6	

Source: Brown (1970); Balash *et al.* (1976); Sturkie (1986); Puerta *et al.* (1989; 1991); Alonso *et al.* (1990; 1991); Lavin *et al.* (1992).

the cytosol (see Price and Stevens, 1989). Carbamoyl phosphate formed by carbamoyl phosphate synthetase II probably remains bound to the multienzyme complex, and is, therefore, not available for arginine biosynthesis. Maresh *et al.* (1969) detected carbamoyl phosphate synthetase in chick embryo liver. The enzyme showed the properties of carbamoyl synthetase II, was detected in the cytosol and was assumed to be involved in pyrimidine biosynthesis. Its activity sharply declined from a maximum at E7 to very low activity at hatching. A more recent search to detect carbamoylphosphate synthetase in avian tissues does not appear to have been carried out, and it is assumed to be absent from avian liver (Baker, 1991).

Urea is found in avian excreta, but this is largely of faecal origin, arising from bacterial degradation. It is, however, also detectable in ureteral urine (Skadhauge, 1983) and also in the plasma (Table 5.5). The tissues from which it originated have not been determined.

5.5 Metabolism of xenobiotics using amino acids or glutathione

Uric acid, ammonia and to a lesser extent urea are the principal forms in which surplus nitrogen is excreted in birds. In addition to these substances, nitrogen from certain amino acids and glutathione are con-

jugated to foreign compounds (xenobiotics) prior to their excretion. This process is sometimes referred to as *detoxification*, although this is not strictly accurate. In some cases, the xenobiotics are not particularly toxic, and in others the modified compound may even be more toxic. Even where metabolism of the xenobiotic leads to the formation of less-toxic products, the products can cause considerable harm. The classic example is DDT (1,1,1-trichloro-2,2, bis(*p*-chlorophenyl)ethane, where its main degradation product DDE (1,1,-dichloro-2,2-bis(*p*-chlorophenyl)ethylene) is generally less toxic, although it often has a longer retention time within the organism. The problem of persistent pesticides was highlighted in 1961 when DDT metabolites were discovered in the eggs of peregrine falcons. DDT has been incriminated as the chief aetiological agent of various factors which together contribute to population declines in avian species (for references see van Wyk *et al.*, 1993). Living organisms are capable of chemically modifying a wide range of foreign organic compounds, including carcinogens, drugs and herbicides. Chemical modification includes oxidation, reduction, hydroxylation, hydrolysis, acylation, dealkylation, desulphuration and conjugation. Only conjugation is considered here, since it is the only mechanism involving endogenous nitrogenous compounds. However, oxidation may precede conjugation, e.g. for aromatic

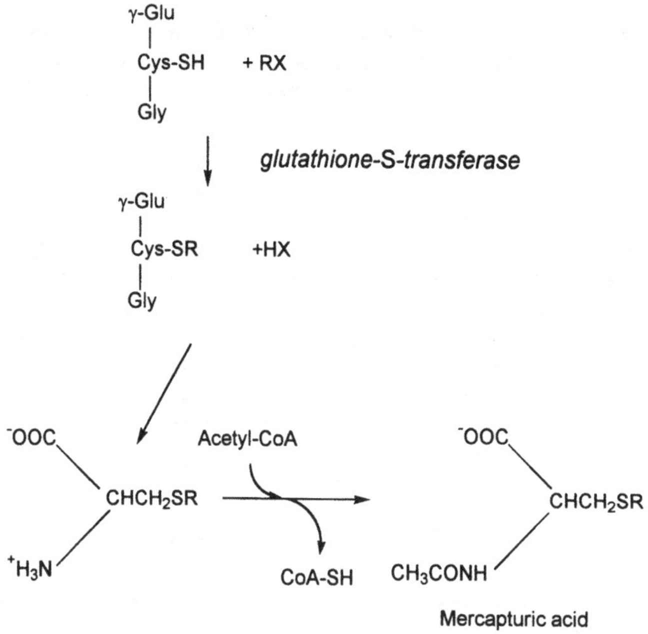

Fig. 5.8. Glutathione-S-transferase and its role in mercapturic acid formation.

hydrocarbons. The toxic substance often induces the range of enzymes necessary for its detoxification, e.g. β-naphthoflavone induces higher activities of cytochrome P-450, O-demethylase, aniline hydroxylase and glutathione-S-transferase in domestic fowl liver (Ehrich & Larsen, 1983). Cytochrome P-450, arachidonic acid epoxygenase and aryl hydrocarbon hydroxylase are induced in chick embryos at E16 by polychlorinated aromatic hydrocarbons (Rifkind et al., 1994). In birds, the oxidation occurs almost entirely in the liver using mixed function oxidases (Smith, 1968). Dealkylation has been shown to occur in endoplasmic reticulum of the grey partridge (Riviere, 1980). The two methods of conjugation involve either amino acids or the tripeptide glutathione.

Amino acids are used in the conjugation of aromatic acids, forming a peptide bond between the amino group of the amino acid and the carboxyl group of the aromatic acid. The most commonly used amino acid in ureotelic vertebrates is glycine. However uricotelic vertebrates, such as birds and reptiles, more commonly use ornithine, excreting ornithuric acids. The conjugation process occurs only in the liver of the domestic fowl, although it occurs in both kidney and liver of other species. The ornithuric acid formed by birds is N^2,N^5-diaroylornithine. It appears that the N^5

position is conjugated first. N^2- and N^5-monoaroylornithines have been detected in excreta, but they appear to be artifacts of subsequent bacterial degradation. In the chick embryo, ornithine conjugation first occurs at E6, approximately a day after uric acid synthesis has begun (Smith, 1968). This coincides with establishment and functioning of the mesonephros.

Evidence suggests that ornithine is most frequently, though not exclusively, used for conjugation in birds, e.g. the turkey, goose and duck use ornithine for conjugation of benzoic acid, the domestic fowl uses mainly ornithine together with a little glycine, pigeon and dove use glycine, but the crow and parrot appear to be unable to conjugate benzoic acid (Smith, 1968).

The second method of conjugation uses glutathione, which is present in millimolar concentrations in the liver. There are generally three steps in this process; the first is the formation of the glutathione conjugate with the xenobiotic, the second is the removal of the glycine and glutamic acid residues, and the third the N-acetylation of the conjugated cysteine to form a mercapturic acid (Fig. 5.8). This form of conjugation occurs with a range of aromatic hydrocarbons and substituted hydrocarbons, including chlorinated insecticides. Aflatoxin B_1, which is both a hepatotoxin and carcinogen, is believed to be oxidised in the domestic fowl by cytochrome P-450 monooxygenase and the epoxide formed conjugated to glutathione via glutathione-S-transferase (Beers, Nejad & Bottje, 1992). Hepatic glutathione levels increase in response to a single dose of aflatoxin B_1. There are a number of isozymes of glutathione-S-transferase, and they have been classified on the basis of their substrate specificity using a range of substituted aromatic and steroid substrates (Mannervik, 1985). The enzymes all exist as homodimers or heterodimers and have been most extensively studied in rat and human livers. In domestic fowl, glutathione-conjugating activity is detectable in the early embryo, e.g. bromosulphalein is conjugated at E9 (Smith, 1968). Glutathione-S-transferase (EC 2.5.1.18) has been purified from the liver of domestic fowl (Yeung & Gidari, 1980; Chang et al., 1990), where it is present on the endoplasmic reticulum (Morgenstern et al., 1984). It has also been detected in wild pigeons (*Columba livia*), although the activity is lower than that in rodents (Husain, Kumar & Mukhtar, 1984).

Five different subunits (CL1–CL5) of glutathione-S-transferase, each 27 kDa, have been separated from chick liver cytosol (Chang et al., 1990). Of these, CL2, CL3 and CL5 can form homodimers, whereas CL1 and CL4 exist only as the heterodimers CL1–2 and CL3–4. One of these has been purified and has a K_m for glutathione of 0.35 mM. Since the likely intracellular glutathione concentration is $\approx$ 2.5–5.0 mM, it suggests that glutathione will not be rate limiting in

well-nourished birds. The enzyme has been detected in a number of different tissues, including liver, kidney, duodenum, brain and testes, and in a number of different species including domestic fowl, Bobwhite quail, pigeon, kite, vulture and crow (Maurice et al., 1991; Das et al., 1981). Chemical modification studies so far carried out on the chick liver CL3–3 isozyme have shown that the single cysteine residue is not essential for activity (Chang, Wang & Tam, 1991). The nucleotide sequence of the cDNAs for a number of these glutathione-S-transferases have been determined (Liu, Wu & Tam, 1993).

The enzyme rhodanese (thiosulphate sulphurtransferase, EC.2.8.1.1) has a wide distribution among vertebrate tissues, and catalyses the conversion of cyanide into the less toxic thiocyanate:

$$\text{thiosulphate} + \text{cyanide} \rightleftharpoons \text{sulphite} + \text{thiocyanate}$$

Cyanide occurs in the environment, but perhaps more important are the cyanogenic glycosides present in a wide range of plants. Rhodanese has been detected in a number of tissues of the domestic fowl, with highest amounts in the liver and proventriculus, but it is also present in other parts of the gut and in the heart and kidneys. Aminlari & Shahbazi (1994) suggest that much of the cyanide becomes detoxified in the gut, where cyanogenic glycosides may be hydrolysed to release cyanide, the remainder being acted on by the liver enzyme.

6

Metabolic adaptation in avian species

6.1 Introduction

The ability of most birds to fly is the clearest behavioural feature distinguishing birds from most other vertebrates. Although not all species of birds are able to fly, e.g. ratites and penguins have lost the ability to fly, even the flightless birds have wings, although not so well developed. The ratites are believed to have diverged from the main stream of evolving birds some 100 million years ago (Cracraft, 1974). There is a wide range of flying ability amongst birds, from those that spend most of their time on the ground, e.g. Galliformes, to those that spend most of their time in flight, catching and eating their food on the wing, e.g. Apodiformes. Flight gives increased mobility, enabling birds to take advantage of seasonally available food resources in different geographical locations. Many migratory birds are able to fly long distances using only endogenous energy reserves. The ability to fly also has a constraining influence on anatomical evolution. Birds are shaped on aerodynamic lines, and their power to weight ratio has to be high enough if they are to become airborne. Flying requires a higher rate of energy production than other methods of locomotion.

The first section of this chapter considers metabolic adaptation required for flight. This includes ATP generation necessary for muscle contraction and the supply of nutrients and oxygen to muscle. Long-distance migration without refuelling requires pre-migratory storage of a large fat reserve. Amongst the 30 orders of birds, seven orders include a majority of strong swimmers and divers. The divers are capable of sustaining anaerobic glycolysis for long periods. Birds are found in all regions of the globe, including the polar regions. Since they are homeotherms, this means that, particularly in the polar regions, they maintain a body temperature that is much higher than the environmental temperature,

and this requires the expenditure of energy. Feathers serve both for flight and for insulation. The production of feathers requires the synthesis of a large amount of protein, which is effectively lost to the body, i.e. it is not turned over like other proteins (see Section 5.2). The principal protein present in feathers is keratin, which has a high cysteine content. Both when the initial feathering occurs in the growing chick and during the regrowth of feathers after moulting, metabolism has to adapt to these demands.

Birds and reptiles are oviparous, and the cleidoic eggs that they produce contain all the nutrient required until hatching. This nutrient, which is mainly protein and lipoprotein, is synthesised in the liver and oviduct prior to oviposition. Lipoproteins are discussed in Section 4.5, and the control of egg protein synthesis in Section 10.3. Birds excrete a semi-solid urine, and this requires a lower water intake than is possible in ureotelic animals. The metabolic adaption that allows this to occur is the excretion of nitrogen principally in the form of uric acid. Uric acid is sparingly soluble in water and is present in avian ureters largely as a colloidal suspension. This is discussed in Section 5.4.

This chapter is mainly devoted to energy provision and utilisation and is considered under the following headings: (i) the transport of oxygen to muscles, (ii) differences in energy metabolism in different types of muscle fibre, and their relative distribution in different species, (iii) energy reserves in migratory birds, (iv) adaptation to cold environments, and (v) adaptation for diving. It is followed by sections on metabolic requirements associated with feather growth and in egg-laying.

6.2 Metabolic demands of flight

The process of flying places considerable metabolic demands on an organism, although when the minimal

Table 6.1. *Types of muscle fibre present in skeletal muscle*

Type	Characteristics
Fast twitch, glycolytic (FG)	White muscles, rapid contraction, high ATPase activity, few mitochondria, anaerobic glycolysis, fatigue rapidly
Fast twitch, oxidative, glycolytic (FOG)	Reddish muscles, rapid contraction, aerobic and anaerobic glycolysis, fatigue slowly
Slow tonic, oxidative (SO)	Red muscles, slow contraction, concerned with posture maintenance, rich in mitochondria and myoglobin, fatigue very slowly

cost of transport is compared for different methods of locomotion, then running and crawling can be more expensive than flying (Hill & Wyse, 1989). Nevertheless, birds that are active show some of the highest rates of metabolism measured. Flight strategies vary considerably in their energetic demands, e.g. gliding and soaring use relatively little energy, whereas hovering is the most energy-intensive form of flight (Rayner, 1982). Flapping flight increases the oxygen consumption 5–14 times above the resting level (Saunders and Klemm, 1994). Many birds, e.g. raptors and petrels, whilst often appearing stationary during hovering are often flying slowly into the wind, or on an updraught. The humming birds are amongst the few birds capable of sustained true hovering, and they show some of the highest rates of metabolism measured. The energy required for flight is that produced by the hydrolysis of ATP, mainly in the pectoral and supracoracoideus muscles. The pectoralis muscle is the one that brings about the downward stroke of the wing and is, thus, the one providing most of the power, whilst the supra-coracoideus lifts the wings on the recovery stroke. Skeletal muscle constitutes 40–45% body mass, and between 40–60% of skeletal muscle comprises the two flight muscles: the pectoralis and the supra-coracoideus (see Swain, 1992a). The pectoralis is generally much bigger than the supracoracoideus. In most birds, the supracoracoideus is most important at take-off but is much less important in sustaining flight. However, in birds in which the upstroke of the wing is a propelling power stroke rather than simply a recovery stroke, the supracoracoideus is much larger. In humming birds it is about half the size of the pectoralis.

Individual muscles are made up of different types of fibre, the proportions of which determine the contractile properties of the muscle. Muscles fibres have been classified in a number of different ways. The original division into red and white muscle is an oversimplification, since there is much variation within these two types. It is clear that the fibres do not fit rigidly into discreet types; nevertheless some form of classification is useful. Complex classifications of avian muscle fibres involving five different classes have been proposed (see Harvey & Marshall, 1986) and there is an extensive discussion of classification by Rosser & George (1986). The division into three classes (Table 6.1) is a useful compromise. The muscle fibres can be classified on the basis of their rates of contraction and on the basis of their metabolism. The maximum rate of contraction depends to a large extent on the rate of hydrolysis of ATP by myosin, which is required for the cross-bridge cycle. Sustaining contractions depend on the rate of regeneration of ATP, by breakdown of secondary fuel sources. This may be either by anaerobic glycolysis, or aerobically involving glycolysis and the citric acid cycle. A useful index of aerobic or oxidative metabolism is the level of succinate dehydrogenase in the muscle. The three types of muscle fibre are sometimes referred to as (i) fast twitch, glycolytic (FG); (ii) fast twitch, oxidative, glycolytic (FOG); and (iii) slow tonic, oxidative (SO). The FG fibres are capable of rapid contraction (fast twitch) since they have the highest myosin ATPase activities. Their ATP is generated by anaerobic glycolysis. They have relatively low concentrations of myoglobin and low mitochondrial densities. For this reason, they have a whitish appearance and are sometimes referred to as white muscles. Although glycolysis generates ATP faster than oxidative phosphorylation, it depletes the substrate much more rapidly, and the muscles fatigue rapidly. The FOG represent an intermediate category of muscle fibres, capable of generating ATP both anaerobically and oxidatively, are capable of rapid contraction and do not fatigue readily. The SO fibres are able to sustain contractions for long periods without fatigue. They have a rich supply of myoglobin and abundant mitochondria. They also have a much higher capillary density. The SO tonic fibres' prime function is posture. The myoglobin, and to a lesser extent the cytochromes, present in mitochondria together with the capillaries give the FOG and SO muscles a red colour, and they are often termed red fibres.

Rosser & George (1986) have studied avian pectoralis muscles in a wide range of species, from which a number of generalisations can be made. The vast majority of pectoralis muscles are made up largely of FG and FOG fibres. SO fibres were only found in a minority of birds, particularly flightless birds such as the ostrich and emu. In the majority where both FG and FOG are present, the FG fibres tend to be on the surface of muscles, whereas the FOG are present in the deeper layers. The pectoralis muscle of the domestic fowl is largely white muscle (FG), with a deep red strip (FOG and SO) making < 1% of the total fibres. Domestication has not significantly affected the proportions of fibres present. When the pectoralis from the domestic fowl is compared with that of the jungle fowl, and the Pekin and Muscovy ducks with their wild counterparts, the proportions of the different fibres are similar. Some of the most highly aerobic fast twitch muscles studied in vertebrates are the pectoralis muscle of the pigeon and the Rufous humming bird. In the pigeon, these are predominantly the FOG, and in humming birds they are entirely that type. These muscles have a high capillary density: it is estimated in the humming bird to be $\approx$ 8000 capillaries/mm^2 of fibre cross-sectional area, five to six times greater than in mammalian soleus muscle (Mathieu-Costello, 1993). In the pectoralis muscle, the capillaries form manifolds around the muscle fibres, thus ensuring a good supply of oxygen. The muscle fibres have a high density of mitochondria, reckoned to occupy about 35% of the total fibre volume. Most of the mitochondria are concentrated near the fibre membrane. The capillary density is so high that there are about 26 km of capillaries in close proximity with each millilitre of muscle mitochondria (Mathieu-Costello, 1993)

6.3 Transport of oxygen and carbon dioxide

Oxygen enters the body at the lungs, which have an enlarged surface area brought about by the development of alveoli. The oxygen passes into the bloodstream where it becomes bound to haemoglobin, present in erythrocytes, and is pumped around the body's circulatory system. All tissues receive oxygen through a network of capillaries. Oxygen enters the tissues, where it is used for the oxidation of metabolites coupled to the generation of ATP. The driving forces that enable oxygen to pass from the lungs to the

tissues are the continuous pumping of the heart and the fall in partial pressure of oxygen from the lungs to the tissues. The partial pressure of oxygen in the lungs is about 13 kPa, and this drops to between 2.6 kPa and 5.3 kPa in the capillaries, eventually dropping to about 0.27 kPa in the cytoplasm of tissues such as skeletal muscle (Wittenberg & Wittenberg, 1989). The partial pressure of oxygen in the lungs varies under different physiological conditions. The transport of oxygen is dependent on two important carrier proteins, haemoglobin and myoglobin.

6.3.1 Haemoglobin

In birds and other vertebrates, haemoglobin is contained within erythrocytes. Avian erythrocytes differ from those in mammals in that they are nucleated and are larger in size. In birds they are oblate ellipsoid in shape, ranging in size from 5.0 μm $\times$ 9.7 μm to 5.0 μm $\times$ 15.9 μm (King & McLelland, 1984). The smaller sized ones generally occur in birds having higher metabolic rates, e.g. passerines and humming birds, since these have a higher surface:volume ratio. Mammalian erythrocytes are biconcave, with typical dimensions of 7 μm $\times$ 2.5 μm. The enucleation of mammalian erythrocytes has been claimed to make them more efficient oxygen carriers. However, some birds are capable of higher metabolic rates than mammals. The presence of a nucleus, ribosomes, Golgi, and a few mitochondria occupies some of the intracellular volume, but haemoglobin is present in both the cytosol and the nucleoplasm, and its subunit concentration is estimated to be $\approx$ 25 mM in birds compared with $\approx$ 20 mM in mammals (Lapennas & Reeves, 1983). Although avian erythrocytes have nuclei, they are also terminally differentiated like mammalian erythrocytes, since their chromatin is condensed and there is little DNA or RNA metabolism (Harris, 1983 and Section 9.6). They have an average lifespan of 20 to 35 days compared with 120 days in humans (King & McLelland, 1984).

Haemoglobin is a protein of M_r 68 000 made up of four subunits, each having one haem prosthetic group and hence one oxygen-binding site. As with most other vertebrates, there are two distinct globin chains, giving the general structure A_2B_2. An unusual feature of avian haemoglobin is the presence of two different definitive (adult) haemoglobins, both present in substantial proportions, e.g. in the case of the domestic fowl 70% haemoglobin A ($\alpha_2^A\beta_2$) and 30% haemoglobin D ($\alpha_2^D\beta_2$) (Baumann et al., 1984). The

Table 6.2. *Haemoglobin A and haemoglobin D in avian blood*

Bird	Altitude (m)	% HbD	O_2 affinities of Hb
Bar-headed goose	>8848	10	HbD>HbA
Andean goose	6000	4–40	HbD and HbA both high affinity
Rüffell's griffon	11278	16	HbD/D'>HbA'>HbA
White-headed vulture	4500	30	HbD>HbA
Northern goshawk	4000	15	HbD>HbA
Golden eagle	7500	35	HbD>HbA
Andean condor	6035	17	HbD>HbA
Black vulture	4500	19	HbD>HbA
Black-headed gull		15	HbD>HbA
Eurasian tree sparrow		15	
Pigeon		0	
Parakeet		0	
Penguin		0	
White stork		0	
Grey heron		0	
Macaw		0	
Duck			HbA>HbD
Pheasant			HbA>HbD
Turkey			HbA>HbD
Domestic fowl		30	

Source: Boyer *et al.* (1971); Schneeganss *et al.* (1985); Hiebl, Braunitzer & Schneeganss (1987); Hiebl *et al.* (1987; 1988; 1989); Godovac-Zimmermann *et al.* (1988).

proportions vary between species (see Table 6.2) and haemoglobin D is absent from some species, e.g. pigeons, parakeets or penguins (see Cobb *et al.*, 1992). There is a very high concentration of haemoglobin in avian erythrocytes ($\approx$ 7–8 mmol/kg cells). The solubility of deoxyhaemoglobin D in water is lower than the concentration found in avian erythrocytes. Its solubility is increased by the presence of ATP and inositol hexa*kis*phosphate (Fig. 6.1). Inositol hexa*kis*phosphate is particularly important in keeping haemoglobin in solution in avian erythrocytes (Baumann *et al.*, 1984).

The binding of haemoglobin to oxygen and other ligands is one of the most studied protein–ligand interactions (see Zubay, 1993) and was used by Monod, Wyman & Changeux (1965) as a model allosteric protein. The binding to oxygen shows positive cooperativity, i.e. after one molecule of oxygen has bound it facilitates the binding of subsequent oxygen molecules. This explains the sigmoid binding curve (Fig. 6.2), which can be described by the Hill equation (for details, see Dickerson & Geiss, 1983; Price & Stevens, 1989).

$$Y = \frac{[A]^h}{K + [A]^h}$$

(Y is the fractional saturation, $[A]$ is the ligand concentration, K is dissociation constant for oxygen binding to haemoglobin, and h is the interaction coefficient or Hill coefficient). The Hill coefficient is a useful indicator of the strength of cooperativity of ligand binding. A value close to 1.0 indicates weak cooperativity, or no cooperativity, whereas a value approaching the number of binding sites (four in the case of haemoglobin) indicates strong cooperativity, as can be seen in Fig. 6.2. A second feature of the oxygen-binding curve is that its sigmoidicity may be influenced by a number of small ions or molecules, in particular carbon dioxide, H^+, Cl^-, and organic phosphates; the effect of protons and carbon dioxide is known as the Bohr effect (for a detailed discussion, see Dickerson & Geiss, 1983).

Avian haemoglobins show a number of noteworthy differences from those of other vertebrates. Their interaction with oxygen is generally more cooperative than in other vertebrates. A typical Hill

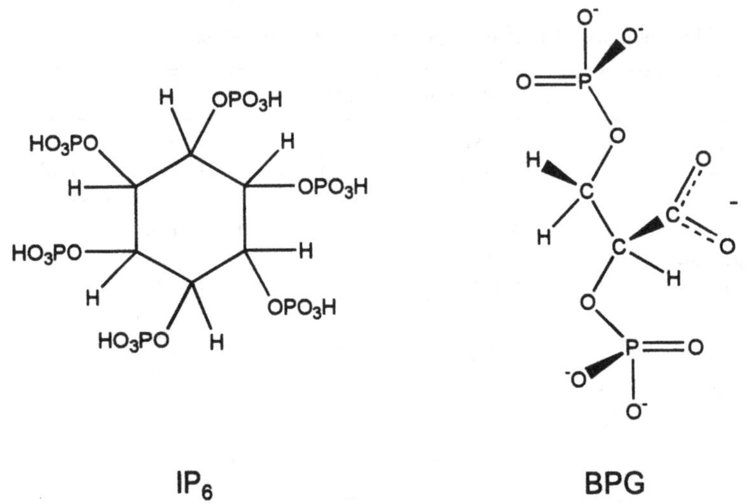

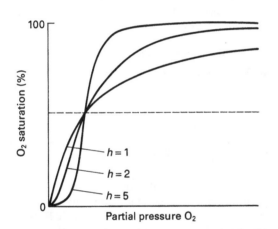

Fig. 6.1. Structures of inositol hexaphosphate (IP$_6$) and bisphosphoglycerate (BPG).

Fig. 6.2. Haemoglobin-oxygen binding curve. Note the increase in sigmoidicity as the Hill coefficient (h) increases from 1 to 5.

coefficient (h) for mammals is 2.8, whereas that for certain fishes may be as low as 1.1 (Brittain, 1991); by contrast, that for avian haemoglobin may be as high as 5–7 (Cobb et al., 1992). The physiological advantage of this high cooperativity is that the delivery of oxygen to the tissues is increased; in the case of humming birds this increase may be as much as 30% (Johansen et al., 1987). A value of $h > 4$ for haemoglobin is puzzling. For a tetrameric protein with four binding sites, the theoretical maximum would be $h = 4$. It is known that haemoglobin D has a relatively low solubility and a tendency to

aggregate. Cobb et al. (1992) have examined the sedimentation behaviour of both haemoglobins A and D. The latter does have a tendency to form dimers of the form $(\alpha_2{}^D\beta_2)_2$, although haemoglobin A does not. The dimer could have a theoretical $h \leq 8$. This can only be a partial explanation, since some birds only have HbA and yet have $h > 4$.

The other differences concern the importance of carbon dioxide and organic phosphates. The *Bohr effect*, originally describing the effect of carbon dioxide on the oxygen-binding curve (see Dickerson & Geiss, 1983) can be accounted for by two processes. First, carbon dioxide forms carbonic acid, and the protons have a greater affinity for oxyhaemoglobin than deoxyhaemoglobin, causing the release of oxygen

$$HbO_2 + H^+ \rightleftharpoons HbH^+ + O_2$$

(there is not a stoichiometric relation between H^+ binding and O_2 release).

Second, carbon dioxide combines directly with amino groups near the N-terminus forming carbamates more readily with deoxyhaemoglobin than oxyhaemoglobin:

$$globin\text{-}NH_2 + CO_2 \rightarrow globin\text{-}NH\text{-}COO^- + H^+$$

Although both these processes occur in mammalian erythrocytes, only the first effect occurs in avian erythrocytes. This is apparent because the magnitude of the effect of carbon dioxide on avian haemoglobin is the same as that with any other acid and is thus through $[H^+]$ alone (Maginniss, 1985). This difference

Table 6.3. *Concentrations of ATP, BPG and IP$_5$ in avian erythrocytes*

Species	ATP (mM)	BPG (mM)	IP$_5$ (mM)
Domestic fowl (adult)	4.5–10.6	0	8.5–21.7
Domestic fowl (embryo)	7.6	6.5	0.9
Pheasant	1.67 ± 0.09	0	3.34 ± 0.3
Quail	2.97	0	3.32
Common peafowl	1.32 ± 0.6	0	4.31 ± 0.93
Guinea fowl	3.68 ± 0.3	0	3.11 ± 1.23
Duck (adult)	7.3	0	13.0
Duck (embryo)	4.2	7.0	1.9
Emu (adult)	14.9–23.4	0	5.9–9.3
Emu (embryo)	13.8–46.4	0	0–3.8
Rhea (adult)	3.9–8.4	0	5.5–10.7
Rhea (embryo)	7–29.8	0	0–4.7
Ostrich (adult)	12.48 ± 0.1	0	6.66 ± 0.09 $(+ 13.13 \pm 0.96$ IP$_4$)
Ostrich (embryo)	5.0	5.7	1.0
Sparrow	7.71 ± 0.23	0.18 ± 0.05	12.3–21.7
Bar-headed goose	2.0	0.3	4.5
Greylag goose	2.6	0.4	4.3
Canada goose	3.0	0.3	5.2

Source: Bartlett & Borgese (1976); Isaacks *et al.* (1976; 1977; 1983); Petschow *et al.* (1977); Bartlett (1982); Maginniss (1985).

has been attributed to the strong binding to haemoglobin of inositol pentaphosphate present in avian erythrocytes. This leads on to another difference in behaviour of avian haemoglobin *in vivo*. Various organic phosphates act as modulators by decreasing the affinity of oxygen for haemoglobin. They play an important regulatory role in oxygenation and deoxygenation of haemoglobin. In mammals, bisphosphoglycerate (BPG, or DPG in earlier references) is the principal organic phosphate modulator (Fig. 6.1). In avian erythrocytes, the most important are inositol phosphates (Table 6.3). Inositol pentaphosphate binds more strongly to avian haemoglobin than to mammalian haemoglobin, because of the six basic residues at the phosphate-binding site of avian haemoglobin (Brygier *et al.*, 1975; Rollema & Bauer, 1979). It binds both the oxygenated and deoxygenated forms (avian oxyhaemoglobin, $K_a = 3 \times 10^5$, cf human oxyhaemoglobin $K_a = 2.5 \times 10^3$; and avian deoxyhaemoglobin, $K_a = 4 \times 10^9$, cf human deoxyhaemoglobin $K_a = 1 \times 10^6$). The principal inositol phosphates found in avian erythrocytes are inositol pentaphosphate and inositol tetraphosphate (Bartlett & Borgese, 1976; Bartlett, 1982; Maginniss, 1985; Brittain, 1991). It is interesting to note that BPG is transiently present in some avian embryos (turkey,

pheasant, guinea-fowl, duck, domestic fowl, ostrich) with the peak concentration occurring shortly before hatching (Bartlett & Borgese, 1976; Isaacks *et al.*, 1976; Bartlett, 1982), although it has not been detected in the embryos of emus or rheas at any stage (Bartlett, 1982). The inositol phosphates have a greater effect on the affinity of oxygen for HbA than for HbD in some species (Brittain, 1991). However, both HbD and HbA from the white-headed vulture and the ostrich are equally sensitive to inositol hexa*kis*phosphate (Isaacks *et al.*, 1977; Hiebl *et al.*, 1989). They can increase the oxygen affinity for haemoglobin by 10-fold (Hiebl, Schneeganss & Braunitzer, 1986). The effects of organic phosphates on P_{50} (the partial pressure of oxygen giving 50% saturation) of ostrich haemoglobin decrease in the order IP$_6$ > IP$_5$ > IP$_4$ > BPG > ATP (Isaacks *et al.*, 1977). The adult ostrich appears unusual in having a higher concentration of IP$_4$ than IP$_5$ in erythrocytes. IP$_4$ comprises 35% of the total phosphates.

6.3.2 Myoglobin

Myoglobin is a small protein present in all higher animals but is particularly abundant in cardiac muscle and red skeletal muscle. The myoglobin content of most muscles is closely related to their cytochrome oxidase content, and hence the number of mitochondria. It is richest in the FOG, and SO fibres. Avian gizzard muscle is unusual for smooth muscle in being able to sustain hard work; it also contains a high concentration of myoglobin.

Myoglobin is present in the sarcoplasm. It is a spherical molecule with a hydrophilic surface and this is believed to assist its free movement (sliding past other myoglobin molecules) within the sarcoplasm, where it diffuses between the plasmalemma and the mitochondria. It has a single binding site for oxygen, shows a hyperbolic binding curve for oxygen (P_{50} 0.27 kPa). Like haemoglobin it is also able to bind carbon monoxide, which is of similar size to oxygen. The binding does not involve oxidation of Fe^{2+}, since the Fe^{2+} that is coordinated in the porphyrin ring lies in a hydrophobic pocket of the globin chain. Unlike haemoglobin, myoglobin binding to oxygen is independent of pH and of effectors. Myoglobin diffuses within the sarcoplasm at about 5% the rate of diffusion of molecular oxygen. There has been much discussion of the importance of myoglobin in intrafibrillar oxygen transport, and as an oxygen reserve in aerobic muscles. The amount of myoglobin in muscle varies both with the species and with the particular muscle (Table 6.4). Muscles that sustain aerobic metabolism generally have higher concentrations of myoglobin. If myoglobin has a role as an oxygen reserve, it represents a fairly small reservoir. James (1972) has calculated the length of time that the oxygen bound to myoglobin could sustain the aerobic pectoral muscle of the pigeon. Based on the rate of oxygen uptake in respiring muscle, and the concentration of myoglobin in pigeon pectoral muscle, the oxygen stored in myoglobin would last 120 ms. It seems most likely that its principal role is to transport oxygen from the capillaries to the mitochondria. It plays an important role in transport of oxygen within the myofibrils. It has been estimated that, in a typical aerobic muscle fibre, about 30 times more oxygen is bound to myoglobin than is present as free oxygen in solution. This is based on the solubility of oxygen in water (3.2 μM at 37 °C), and a myoglobin concentration in muscle of about 200 μM, assuming about 50% saturation at the partial pressure of oxygen in the muscle fibres (Wittenberg & Wittenberg, 1989). Myoglobin-bound oxygen diffuses at about 5% the rate of molecular oxygen in aqueous solution.

6.3.3 Transport of oxygen to the tissues

Oxygen is taken up into the erythrocytes and bound to haemoglobin in the lungs where the alveolar air is at a P_{oxygen} of about 13.3 kPa. The tissues that consume the highest proportion of oxygen are muscles. Red muscle fibres (FOG and SO) operate at P_{oxygen} about 0.4 kPa or less, and about 85% of the oxygen taken up is used in mitochondrial oxidations. The P_{oxygen} at the capillary–muscle fibre interface is $\approx$ 2.6–3.3 kPa. The steepest P_{oxygen} gradient is between the capillary–fibre interface, with a much shallower drop between the sarcoplasm and the inner mitochondrial membrane (Wittenberg & Wittenberg, 1989). The much lower P_{oxygen} at the capillary–fibre interface causes haemoglobin to discharge a large proportion of the oxygen that has been transported to the tissues. Myoglobin has the important role of transporting oxygen from the fibre–capillary interface to the mitochondria.

6.4 Adaptation for high-altitude respiration

Birds are generally well adapted for respiration at low partial pressures of oxygen (hypoxia), such as encountered at high altitudes. This not only applies to those birds whose normal habitat is at high altitude or those that fly at high altitudes during migration, but also to those that are resident at altitudes near sea levels. House sparrows do not show any adverse effects when moved to altitudes of 6100 m (Faraci, 1991), whereas humans require a substantial period of acclimatisation to cope with large changes in altitude, such as climbing Mount Everest. A change in altitude from sea level to 7000 m entails a reduction of barometric pressure to $\approx$ 300 mmHg, and a $P_{oxygen} \approx$ 8.6 kPa. This would result in a significant reduction in uptake of oxygen in the lungs. High altitudes are often accompanied by a drop in temperature. To cope with these changes involves anatomical, physiological and biochemical adaptations (Faraci, 1991). Extensive biochemical studies have been made of avian haemoglobins in relation to high-altitude respiration (see Hiebl et al., 1989; Jessen et al., 1991). Two different types of adaptation to hypoxia have to be considered. The first is where a bird is exposed to transitory hypoxic stress, when a bird normally living at low altitudes flies at high altitudes from one location to another. An extreme example is the bar-headed goose living in the plains of northwest India but migrating over the Himalayas to the Tibetan lakes (altitude 4000–6000 m) to hatch their young. Such birds may begin

Table 6.4. *Myoglobin concentrations in the muscles of various avian species*

Species	Cardiac muscle (mg/g wet weight)	Pectoral muscle (mg/g wet weight)
Low-altitude, weak-flying birds		
Domestic fowl	1.3–3.6	0.3–2.0
Domestic quail	3.1–6.8	0.5–2.8
Ruffed grouse	–	0.2
Spruce grouse	–	2.5
Low-altitude, strong-flying birds		
Pigeon	3.0–4.3	2.2–4.7
European starling	2.0	2.0
Starling (arrival from migration)	7.2	9.5
Starling (resting on winter ground)	2.1	2.1
Black-headed gull	6.4	5.5
Hummingbird	2.8 ± 1.1[a]	3.4 ± 0.7[a]
Tufted duck	7.0	4.7
High-altitude, strong-flying birds		
Grey heron	–	6.3
Ural owl	–	5.9
Andean cock-of-the-rock	–	8.5
Bar-headed goose	6.0	6.4
Diving birds		
Adele penguin	–	29.2
Gentoo penguin	–	46.2
Bearded penguin	–	32.0
Rock hopper penguin	–	14.7
King penguin	–	35.9
Pigeon guillemot	16.1	21.6

[a] The average value obtained from 11 species of hummingbirds (Johansen *et al.*, 1987).
Source: Catlett, Walters & Dutro (1978); Pages & Planas (1983); Johansen *et al.* (1987); Saunders & Fedde (1991).

migration near sea level and reach altitudes near 9000 m in less than 1 day. The second are birds that live permanently at high altitudes, such as the Andean goose whose usual habitat is high in the Andes at ≈ 6000 m. The first of these requires a wider than normal range of oxygen affinities for haemoglobin to cope with both high and low altitude, whereas the second simply requires higher oxygen affinity for haemoglobin than normal.

The amino acid sequences of both α- and β-chains of haemoglobin have been determined for a number of birds, including geese, vultures, gulls, emu, rhea, ostrich and sparrow. From the sequence information, the importance of certain regions of the protein in determining the oxygen-binding affinity is apparent. Many species of birds have two types of haemoglobin (Table 6.2). HbA is usually the major constituent and in many species has the lower affinity for oxygen.

The presence of two haemoglobins, each having a different affinity for oxygen, gives rise to a wider range of P_{oxygen} over which the erythrocytes can bind and release oxygen. The high-affinity component ensures sufficient oxygen uptake, and the low-affinity form allows sufficient oxygen release at the tissues. This arrangement has an advantage for those species of birds that have to cope with the largest variations in P_{oxygen} throughout the year. There is some evidence that the proportions of HbA/HbD vary in an age-dependent manner (Hiebl, Braunitzer & Schneeganss, 1987).

When the P_{50} for whole blood from a range of avian species is measured, the values obtained relate to the degree of hypoxia the birds normally encounter. Birds whose normal habitats are at high altitudes have the lowest P_{50} (Table 6.5), for example (i) Rüppell's griffon, living in Central Africa is known to

Table 6.5. *Oxygen-binding parameters for whole blood from various avian species*

Species	P_{50} (mmHg)	Hill coefficient (h)
Ostrich	24.9	
Rhea	20.7	3.1
Domestic fowl	35.7	3.4
Turkey	33.4	3.2
Quail	42.4	2.4
Common pheasant	34.7	
Pekin duck	42.6	
Mallard	35.2	
House sparrow	41.3	
Pigeon	29.5	
Greylag goose	39.5	
Canada goose	42.0	
Andean goose	33.9	
Bar-headed goose	27.2	2.6–2.8
European black vulture	21.3	2.6–2.8
Rüffell's griffon	16.4	2.6–2.8

Source: Lutz, Longmiuir & Schmidt-Nielsen (1974); Isaacks *et al.* (1977); Petschow *et al.* (1977); Faraci (1991); Jessen *et al.* (1991).

fly as high as 11000 m; (ii) the European black vulture inhabits Mediterranean countries, Western and Central Asia and is the characteristic bird of Mongolic and Tibetan prairies where it lives at an altitude of more than 4500 m; (iii) the bar-headed goose can range from sea level to > 8000 m. Rüppell's griffon has four types of haemoglobin present in the following proportions, and with the P_{50} in parentheses: HbA, 48% (4.6 kPa); HbA', 24% (3.1 kPa); HbD 16% (1.93 kPa) and HbD' 9% (1.93 kPa) (Hiebl *et al.*, 1988). Each has the same β-chain, but they differ in their α-chains. These haemoglobins enable it to cope with a wide range of P_{oxygen}.

The explanation for the different sigmoidicities and hence P_{50} for the oxygen-binding curves lies in the conformation of the α- and β-chains in haemoglobin. According to the Monod–Wyman–Changeux model (see Price & Stevens, 1989) haemoglobin exists in two conformations, T (tense) and R (relaxed), which exist in equilibrium (Fig. 6.3). The interactions between the four globin subunits are stronger in the T conformation than in the R conformation. Oxygen has a greater affinity for subunits in the R conformation. When deoxyhaemoglobin is exposed to increasing concentrations of oxygen, binding occurs preferentially to the R conformation, causing more haemoglobin molecules to change to the R conformation in order

to maintain equilibrium. Conformational changes in a single subunit are relayed to adjacent subunits through $\alpha\alpha$-, $\alpha\beta$- and $\beta\beta$-subunits contacts. The extent of cooperativity of oxygen binding, therefore, depends on subunit contacts. The differences in binding affinities of haemoglobins from avian species adapted to high altitudes, compared with those living near sea level, can be partly be explained by differences in amino acid sequences at the position of subunit contacts. Some of the subunits contacts that appear to be important in this respect are 119α to 55β, and 34α to 124, 125 and 128β.

Proline at 119α, present in the greylag goose, is replaced by alanine in the bar-headed goose. The replacement of proline by alanine leaves a two carbon gap, which weakens the contact between the α- and β-chains, relaxing the tension of the T state and increasing the oxygen affinity. Similarly leucine at 55β in the greylag goose is replaced in the Andean goose by serine, again leaving a two carbon gap and relaxing the tension of the T state (Godovac-Zimmermann *et al.*, 1988; Jessen *et al.*, 1991). To test the importance of the residues at α119 and β55 in accounting for high-oxygen affinity, Jessen *et al.* (1991) performed site-directed mutagenesis on human globin chains synthesised in *Escherichia coli*. In the two human haemoglobin mutants in which proline at 119α is replaced by alanine and methionine at 55β is replaced by serine, the oxygen-binding affinities are greatly increased.

In Rüppell's griffon, the difference between haemoglobin A and A' lies in 34α, where threonine is replaced by isoleucine. The isoleucine in haemoglobin A' interrupts the $\alpha_1\beta_1$ interface. The difference between haemoglobin D and D' is at 38α, where glutamine is replaced by threonine and this affects the $\alpha_1\beta_2$ interface (Hiebl *et al.*, 1988). The differences in oxygen-binding affinities between haemoglobins A and A', and D and D' can, therefore, be accounted for by changes in the subunit interactions. Velarde *et al.* (1991) have suggested that the domestic fowl is able to adapt to high altitudes in a period of less than 500 years. They studied a population of domestic fowl from the Peruvian Andes (4000 m) which has a high haemoglobin–oxygen affinity. The domestic fowl is believed to have been introduced into Peru during the Spanish conquest of South America less than 500 years ago. The oxygen dissociation curve for this Peruvian population shows a P_{50} of 3.84 kPa compared with P_{50} of 6.77 kPa for domestic fowl living at sea level. When the birds adapted to high altitude were bred at sea level for three generations, they still showed the high oxygen affinity. Further

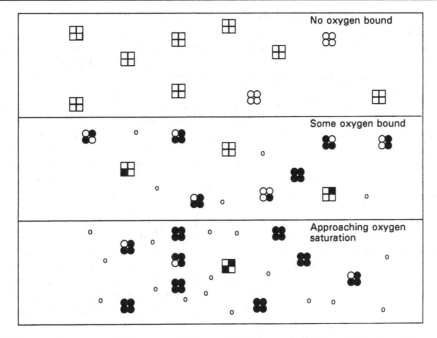

Fig. 6.3. Monod–Wyman–Changeux model for haemoglobin as an allosteric protein. R (relaxed) subunits are shown as circles, and the T (tense) subunits as squares. The filled circles and squares indicate oxygen binding. With increased oxygen saturation, the proportion of R subunits increases

work is necessary to understand the biochemical basis of this adaptation.

The differences in globin sequence influence the haemoglobin–oxygen binding curve, and hence the range of partial pressures of oxygen for which it is best adapted. The concentrations of organic phosphates, particularly inositol phosphates, within erythrocytes have an important modulating effect on the curve. Small changes have a pronounced effect on the P_{50}. The concentrations of organic phosphates in erythrocytes of the bar-headed goose (7.2 mM), the greylag goose (7.9 mM) and the Canada goose (9.1 mM) are similar, but differences in intrinsic affinities of oxygen for haemoglobin arise from the different affinities each haemoglobin has for inositol phosphates (Petschow *et al.*, 1977). Similar concentrations of IP_6 interact less strongly with haemoglobin from the bar-headed goose than from the greylag or Canada goose, and so they depress the haemoglobin–oxygen binding to a lesser extent. It is assumed that IP_5, the naturally occurring inositol phosphate in goose erythrocytes, has a similar effect to that of IP_6. The difference in intrinsic affinities of haemoglobin from the greylag goose and the bar-headed goose for oxygen (measured as P_{50}) is only 0.17 kPa, but the presence of IP_5 in whole blood is sufficient to raise

the difference 10-fold to 1.73 kPa (Rollema & Bauer, 1979).

6.5 Adaptation for diving

Adaptation for diving also requires energy, and hence ATP generation when the supply of oxygen is restricted. In contrast to high-altitude flight, where low partial pressures of oxygen are experienced for long periods, diving entails zero oxygen intake for much shorter periods, e.g. guillemots dive for up to 2 min, and penguins for up to 9 min (Haggblom, Terwilliger & Terwilliger, 1988). It can be seen from Table 6.4 that birds which can carry out prolonged dives have the highest myoglobin content in their muscles. However, even with this elevated myoglobin content, the oxygen stored would only sustain aerobic metabolism for a very short period, as can be seen from the calculation above (see Section 6.3). The haemoglobin concentrations in the blood of diving birds are similar to non-diving birds, but the pectoral muscle myoglobin content is substantially higher in diving birds (Table 6.4).

The haemoglobins of two diving birds, the black-necked grebe and great cormorant have been compared with those from domestic fowl, duck and turkey (Giardina *et al.*, 1985). All showed similar P_{50} values and sensitivities to IP_6. The higher myoglobin content of the muscles probably ensures that transport to the mitochondria of the limited oxygen available is highly efficient, rather than providing a significant oxygen reserve. It is possible that for the period of the dive, a high proportion of the ATP is generated by anaerobic glycolysis. Haggblom *et al.* (1988) have studied both myoglobin content and lactate dehydrogenase activities in the heart and pectoral muscles of the pigeon guillemot during development from chick to adult. They found a significant increase in myoglobin content of both cardiac and pectoral muscles, corresponding approximately with the time when the developing fledgling starts diving activity. During this period, the lactate dehydrogenase activity of cardiac muscle hardly changed, but its activity in pectoral muscle increased about 3-fold. It is difficult to assess glycolytic activity in these tissues from lactate dehydrogenase activities alone. In many mammalian tissues, lactate dehydrogenase is not a rate-limiting enzyme but is present in sufficient concentrations for the reactants and products to be present at equilibrium concentrations (see Price & Stevens, 1989). It, therefore, is not a good indicator of the rate of glycolysis, although the amounts present in some tissues do seen to relate to their rates of glycolysis. The rate-limiting step catalysed by phosphofructokinase, used by Marsh (1981) to assess glycolytic activity in the pectoral muscle of the catbird, would be a more suitable choice.

6.6 Metabolic adaptation associated with migration

Migration is an important feature in the life history of many birds. Many migrants exploit the seasonally abundant resources of the temperate North and Arctic and then return south in the summer. For example, over 180 species leave Europe and Asia for Africa, and over 200 species leave North America for the New World tropics (Gill, 1989). The distance of migration varies very considerably between species, from local migrations of a few thousand metres to movements of thousands of kilometres, e.g. the Arctic tern. At the lower end of the range, it is sometimes difficult to distinguish migratory from sedentary species, e.g. the American goldfinch. In some cases, not all birds of one species migrate. Metabolic adaptation for migration is most pronounced in those that travel furthest, but it also depends in whether this involves a non-stop flight or the use of a number of staging posts. The Western Atlantic

Ocean, the Gulf of Mexico, the Sahara Desert and the Mediterranean Sea present the largest geographical barriers that birds encounter during migration, where non-stop flights of 1000 to 4000 km for periods of 20–80 h are necessary. The metabolic rate during such flights may be five to ten times the basal metabolic rate (Dawson, Marsh & Yacoe, 1983). Metabolic adaptation required for the most strenuous migrations has been most extensively investigated and will be focused on here. The metabolic aspects of migration can be considered as occurring in two phases: the pre-migratory phase, during which the food reserves are built up, and the migratory phase, which involves the large energy output associated with long-distance flight, often in the absence of feeding. The first of these is more accessible to biochemical study. Relatively little is known about metabolism during the migratory flight, other than what can be deduced from comparing birds before and after migration and a limited number of measurements made on birds caught during migratory flight.

6.6.1 Pre-migratory metabolic changes

The fuel used on long migratory flights is almost entirely fat, principally triglycerides. Fat is in a more highly reduced state than either carbohydrates or proteins, and so more energy can be generated per gram than with either the other two (fat 37 kJ/g, carbohydrate 16 kJ/g, and protein 17 kJ/g). In addition, triglycerides can be stored in the adipose tissue and other tissues in the form of droplets, separate from the aqueous phase, thus providing a much more concentrated reserve. During the pre-migratory phase, large amounts of lipids are stored as the result of hyperphagy. The extent of fat deposition correlates positively with the distance of migration. In most cases this is estimated as the mass gain in the period prior to migration. It is assumed that most of the increase in weight arises from fat deposition. These can be very impressive: for passerines crossing the Sahara Desert and Mediterranean Sea, the Gulf of Mexico or the Atlantic Ocean 40 to 70% and, in a few cases, approaching 90% lean body mass, and for waders crossing the Atlantic including the Greenland ice cap or the Pacific Ocean 50 to 90% with a maximum in a few of 100% relative to lean body mass, i.e. a doubling of body weight in the premigratory period (Alerstam & Lindström, 1990). Where there are possibilities for stopovers during long migratory flights, e.g. crossing the mountains and deserts of Central Asia, the strategy adopted by passerines is to rely on effective refuelling stops rather than on maximal fuel storage before the start of a migratory flight (Dolnik, 1988). Whilst foraging at refuelling stops, a migrant passerine may accumulate enough fat to support 1.14 h of migratory flight in spring and 0.58 h in autumn. The average duration of an overnight flight is about 4.5 h in both spring and autumn (Dolnik, 1988). The measurement of fat reserves of Steppe buzzards (*Buteo buteo vulpinus*) trapped during migration from Southern Africa to Eastern Europe suggested that they carried insufficient lipid fuel to complete migration without replenishing their energy reserves (Gorney & Yom-Tov, 1994).

Most of the fat is stored in the adipose tissue. Avian adipose tissue has limited ability for *de novo* triglyceride synthesis (see Section 4.3), and most of the lipid accumulated is either of dietary origin or as the result of *de novo* synthesis in the liver.

The principal metabolic changes that occur during the pre-migratory period are the build-up of fat reserves and the adaptation of the flight muscles, principally pectoralis muscle, for sustained flight. In the wood thrush, there is evidence for a change in the fatty acid composition of lipids stored before the autumn migration, compared with the lipids present during the breeding season. The bulk of the fatty acids are C_{16} or C_{18} in both breeding and migrating birds; however, in breeding birds, oleic acid accounts for about one third of the total, whereas in migrating birds it accounts for nearly two thirds (Conway, Eddleman & Simpson, 1994). In some species, there is evidence for the build-up of protein reserves prior to migration. The black-headed gull, which is generally a short-distance migrant, increases the pectoral muscle protein content by almost 40% at the beginning of its spring migration (Cantos, Alonso-Gómez & Delgado, 1994). The changes are frequently assessed by measurement of enzymes catalysing rate-limiting steps in the pathways. The rate-limiting steps in fatty acid synthesis are catalysed by acetyl-CoA carboxylase and fatty acid synthetase. There is as yet no evidence to indicate a big increase in the activities of either of these enzymes in the livers of free-living pre-migratory birds (Ramenofsky, 1990). However, in captive dark-eyed juncos, there is a vernal increase in the activity of hepatic fatty acid synthetase prior to the onset of pre-migratory hyperphagia (Ramenofsky, 1990). For uptake of dietary lipid into adipose tissue, lipoprotein lipase is the key enzyme (see Section 4.5.3) and this increases in the autumnal premigratory period. It is probable that the increased lipogenesis observed at this stage may result from increased levels of precursors and by allosteric activation.

Detailed studies have been made on the adaptation of flight muscle (Marsh, 1981; Lundgren & Kiessling, 1985; 1986; Lundgren, 1988) where a number of marker enzymes have been used. The questions generally asked are, which is the predominant fuel used and in the case of carbohydrate is it degraded anaerobically or aerobically. 6-Phosphofructo-1-kinase is used to assess glycolysis, citrate synthase the citric acid cycle, cytochrome oxidase the respiratory chain, and 3-hydroxyacyl-CoA dehydrogenase (HOAD, EC.1.1.1.35; reaction 6.1) β-oxidation.

$$3\text{-hydroxyacyl-CoA} + NAD^+ \rightleftharpoons 3\text{-oxoacyl-CoA} + NADH + H^+ \qquad (6.1)$$

Lactate dehydrogenase is sometimes used to assess anaerobic glycolytic activity, but it is less useful as it does not catalyse a rate-limiting step. In comparisons that have been made between (i) breeding and migrating passerines (Lundgren & Kiessling, 1985), (ii) pre-migratory and migratory reed warblers, and (iii) the migratory and stationary individuals in the partially migratory species (goldcrest, the great tit and yellowhammer), the following pattern emerges. In all species, there is an increase in the oxidative capacities during the pre-migratory phase. This can be seen by comparing the enzyme activities in breeding or non-migrating groups with activities in migrating groups (Table 6.6). In most cases, there is an increase in the HOAD activity and a decrease in glycolytic capacity. These results suggest an adaptation towards a higher use of fatty acids and decreased use of carbohydrates. The change is similar to that which occurs in human athletes during the training period. In the catbird, where a portion of the population migrates across the Gulf of Mexico, the HOAD increases substantially during the pre-migratory period and the glycolytic activity, as determined by PFK, remains unchanged (Marsh, 1981). The specific activity of citrate synthetase remains unchanged, but since there is muscle hypertrophy during this period, the total activity increases by about 35%. In the catbird, the pectoralis has a high aerobic capacity at all times and possesses the highest reported activity in any skeletal muscle. The most pronounced changes occur in the pectoral muscle, but changes also occur in the supracoracoideus muscle.

6.7 Seasonal acclimatisation and adjustment to cold

The average body temperature of birds is between 38 and 42 °C (Whittow, 1986) and this is maintained throughout the seasons of the year, where temperature may vary by as much as 50 °C. As the external

Table 6.6. *Enzyme activities in the avian pectoral muscles*

Species	6-Phosphofructokinase	Citrate synthase	β-Hydroxybutyrate dehydrogenase	Hexokinase	Cytochrome oxidase
Goldcrest					
migrant	40	200	90		150
non-migrant	55	170	70		80
Great tit					
migrant	55	110	50		110
non-migrant	60	100	40		65
Yellowhammer					
migrant	85	170	70		100
non-migrant	80	160	55		70
House finch	23	100–119	12–19	1.1	
American goldfinch					
summer	50	210	44	0.7	
winter	62	203	80	0.4	
spring	41.5	319	54	0.7	
Catbird	40–69	131–292	21.8–103		
Reed bunting					
migrating	55	150	55		110
breeding	55	180	50		65
Sedge warbler					
migrating	30–60	200	70		80–100
breeding	55	170	55		70
European robin					
migrating	45	180–200	60		85–105
breeding	60	130	35		60
Blackbird					
migrating	60–80	140–160	50–70		55
breeding	100	100	40		40
Reed warbler					
migrating	30–50	220–240	70–80		90–100
breeding	50	160	60		70

Enzyme units as μmoles/min per g wet weight.

Source: Marsh (1981); Marsh & Dawson (1982); Yacoe & Dawson (1983); Lundgren & Kiessling (1985); Carey *et al.* (1988); Lundgren (1988).

temperature is lowered, a larger amount of energy has to be expended in maintaining body temperature. Birds that are cold acclimatised in the laboratory for one to four weeks increase their metabolic rate by up to 85% (Dawson & Marsh, 1988). Shivering appears to be the main way in which birds effect thermogenesis, and the flight muscles (pectoralis and supracoracoideus), which comprise 15–30% of body mass, are most involved in this activity. Although there is limited evidence for non-shivering thermogenesis in birds, only a few studies have so far been carried out and its general importance is not yet clear (see Connolly, Nedergaard & Cannon, 1988). In mammals, the principal tissue for thermoregulatory heat production through non-shivering thermogenesis is brown adipose tissue. Heat is generated through uncoupling of mitochondrial phosphorylation. In mammals, this is regulated by an uncoupling protein, and the presence of this protein can be used to distinguish brown adipose tissue capable of non-shivering thermogenesis from white adipose tissue (Brigham & Trayhurn, 1994). No evidence has been found for the presence of the uncoupling protein in adipose tissue from winter-acclimated common pheasants, quail, feral rock pigeons, house sparrows or great tits (Saarela et al., 1991), nor in birds that hibernate and enter deep torpor (body temperature < 5.0 °C) such as common poorwills. The search for the uncoupling protein in avian adipose tissue has been carried out using antisera raised against the mammalian uncoupling protein. Brigham & Trayhurn (1994) conclude that, unless an avian uncoupling protein exists which is sufficiently different from the mammalian protein so as not to cross-react with the antisera, it seems unlikely that brown adipose tissue is present in birds. Non-shivering thermogenesis has been demonstrated in the king penguin, although brown adipose tissue is lacking (Duchamp et al., 1991).

Seasonal adjustment is most important in small birds since they have less insulation and have to use a higher proportion of their energy reserves to maintain body temperature. The metabolic changes during acclimatisation have been studied in American goldfinches and house finches (Dawson & Marsh, 1988). Both species spend the winters in areas of the United States where cold winters are common and so have to adapt to similar cold conditions. However, under laboratory conditions their capacities to withstand cold differ. American goldfinches adapted to winter conditions remain homoeothermic for more than 6–8 h below − 60 °C, during which the metabolic rate increases 5.5-fold, whereas winter-adapted house finches can tolerate similar temperatures for only 90 min with a 6.4-fold increase in metabolic rate (Carey et al., 1988). There are also differences in their metabolic adaptation. The American goldfinch uses primarily fatty acid as fuel for thermogenesis and only a little carbohydrate, whereas the house finches show only a modest increase in lipid reserves from 6.5% to 7.5% body weight during winter acclimatisation and depend more on carbohydrate fuel (Carey et al., 1988).

The adaptive changes shown by the American goldfinch are those that one might expect with the increased dependence on lipid fuel. There is a significant increase in the triglyceride reserves in the adipose tissue during winter acclimatisation. Glycogen contributes as a fuel to only a minor extent. This is evident from the changes in enzyme activity in pectoralis muscles. Of the rate-limiting enzymes measured, the largest increase is in HOAD activity (50–100%), smaller increases occur in glycogen phosphorylase and phosphofructokinase (14–28%), and no change occurs in hexokinase, citrate synthase and succinate dehydrogenase (Yacoe & Dawson, 1983). The increased lipolysis in the adipose tissue is probably controlled by glucagon secretion. Fatty acids are released into the plasma, where the concentrations are in the region of 1 to 2 mM, which is appreciably higher than in mammals (0.2–0.4 mM). The fatty acids are taken up into the pectoral muscles, stimulating an increase in the rate of β-oxidation and, probably, inhibiting pyruvate dehydrogenase. The latter decreases the flow of metabolites into the citric acid cycle (Yacoe & Dawson, 1983). Glucose turnover in the plasma of winter-acclimatised birds is also decreased. Measurements of enzyme activities in vitro do not necessarily correlate with activities and fluxes in vivo. This is particularly so with allosteric enzymes such as phosphofructokinase, which is activated by AMP and inhibited by ATP and citrate. Although phosphofructokinase activity increases by 14% on cold acclimatisation when measured in vitro, the changes in the concentrations of glycolytic metabolites suggest that it is inhibited in vivo (Marsh et al., 1990).

Pyruvate kinase is another potential site for regulation of glycolysis, but the changes in concentrations of metabolites do not identify it unequivocally as a regulatory step. They do suggest that the higher concentrations of pyruvate as well as fatty acids in the winter-acclimatised birds may cause inhibition of pyruvate dehydrogenase, and hence metabolism through the citric acid cycle. This is some of the first work studying the underlying control mechanisms of cold adaptation in birds, and further work will be necessary, such as examining the fluxes of adenine nucleotides and fructose 2,6-bisphosphate, in order to get a full understanding of its regulation.

King penguin chicks are able to survive subantarctic winters for as long as 4–6 months with very little food, and during this period their body mass may decrease by 70% (Cherel, Robin, & Le Maho, 1988). When captive king penguin chicks were cold acclimatised and compared with chicks reared at thermoneutrality, the former group showed non-shivering thermogenesis, although not possessing brown adipose tissue. The non-shivering thermogenesis is brought about by an increased oxidative capacity of skeletal muscle, shown by increased cytochrome oxidase. The muscle mitochondria show increased oxidative capacity, but this is not accompanied by a corresponding increase in ADP phosphorylation, suggesting an uncoupling that results in heat generation. The morphological changes that occur in the skeletal muscle on acclimatisation suggest that subsarcolemmal mitochondria are responsible (Duchamp *et al.*, 1991).

Cold acclimatisation has also been studied in ducklings (Barré, Nedergaard & Cannon, 1986). The basal respiration of intermyofibrillar mitochondria was increased, whereas that of liver mitochondria was unchanged. In mammalian tissues, fatty acids have been implicated as regulators of the uncoupling of oxidative phosphorylation associated with thermogenesis. The oxidative metabolism in intermyofibrillar mitochondria from cold-acclimatised ducklings showed increased sensitivity to free fatty acids. Although there is no direct evidence that free fatty acids have a regulatory role in ducklings, exposure to cold is known to increase the plasma concentration of free fatty acids, and it, therefore, seems a plausible mechanism (Barré *et al.*, 1986).

6.8 Metabolic adaptation associated with fasting

There are times when birds in the wild do not eat in spite of having food readily available, e.g. during moulting, breeding and egg incubation. These are thought to be times when feeding would compete with otherwise more important activities. When incubating eggs, birds often fast or eat little and lose considerable amounts of weight. This kind of fasting is particularly pronounced in geese and penguins, where the associated metabolic changes have been studied. The domestic goose may be anorexic for 2.5 months during the breeding season, whereas the king penguin may last for 4–6 months without food. Prior to fasting, birds generally build up their endogenous food reserves, principally in the form of lipid. The size of the reserve is related to the natural length of the fast, as is evident in comparing nocturnal fasts, breeding fasts and migratory fasts (Table 6.7). Some smaller birds reduce the rate at which the reserves are used by going into a state of torpor. For example, some hummingbirds decrease their resting metabolic rate by as much as 27% by entering nocturnal torpor. During torpor, body temperature may be reduced by as much as 20 °C. Larger birds, such as geese are able to survive long periods without food, because of their large lipid reserves. Many small birds do not have large lipid reserves and have to use endogenous protein as a source of energy, particularly during brood rearing. For example, the horned lark has sparse lipid reserves

Table 6.7. *Fat storage in various avian species before natural periods of fasting*

Status	Species	Fatness as % body weight
Nocturnal fast	Yellow-vented bulbul	5
	Bullfinch	6
Breeding fast	Common eider, ♀	16
	Emperor penguin, ♂	27
Migratory fast	Bay-breasted warbler	31
	Red-necked phalarope	40

Source: Cherel *et al.* (1988).

and although it forages and feeds for 10–12 h per day during brood rearing, it has to degrade endogenous protein overnight as a source of energy and metabolic precursors. Protein is lost from flight muscles, postural muscles and liver, but to a lesser extent from leg muscles (Swain, 1992b). Sarcoplasmic proteins are degraded to a greater extent than myofibrillar proteins, and an increase in cathepsin D activity in flight muscles can be detected after an overnight fast. By contrast, Jones (1991) found myofibrillar proteins were degraded in preference to sarcoplasmic proteins in a similar situation in the laying house sparrow.

6.8.1 Phases of adaptation

The changes in metabolism in the goose during fasting have been studied in detail (Le Maho *et al.*, 1981) and can be divided into three phases. The corresponding changes in metabolite concentrations are given in Table 6.8. The same three phases have also been identified in the king penguin (Cherel *et al.*, 1988). During phase I, the body weight drops sharply, and adaptation for using lipid as the principal energy source takes place. During the second phase, lipid is used as the principal source of fuel, the resting metabolic rate decreases (reduced by 30–50% after 30 days fasting) and the rate of nitrogen excretion decreases. It has been calculated that during starvation in geese, 93% of the energy source is triglyceride and 7% protein (Le Maho *et al.*, 1981). Phase II corresponds to the increased level of fatty acids and the ketone body β-hydroxybutyrate in the plasma (Table 6.8). The plasma concentration of the other ketone body, acetoacetate, remains low. This is because avian livers have very low concentrations of β-hydroxybutyrate dehydrogenase (reaction 6.2), unlike mammalian livers.

Table 6.8. *Biochemical changes during long-term fasting in geese*

Measurement	Phase I (0–3 days fasting)	Phase II (3–30 days) fasting	Phase III (>30 days fasting)
Body weight loss during phase (%)	5.7–8.8	24–28	>28
Plasma glucose (mM)	8–10	8–10	8–10
Plasma free fatty acids (mM)	Peaking at day 2 at 0.8	Falling steadily to 0.2	≈0.2
Plasma alanine (mM)	0.5	0.5 falling steadily to 0.4	Rising sharply to 0.7
Plasma β-hydroxy-butyrate (mM)	Rising steeply 0 to 7	Rising to peak 22 at 15–20 days, then falling	Continuing to fall to 5–10
Principal changes	Adaptation to lipid utilisation	Adaptation complete, reduced resting metabolic rate	Increased use of protein as lipid becomes depleted

Source: Le Maho *et al.* (1981).

$$\beta\text{-hydroxybutyrate} + NAD^+ \rightleftharpoons \text{acetoacetate} + NADH + H^+ \qquad (6.2)$$

Nitrogen excretion increases in phase III, because of endogenous protein breakdown (Fig. 6.4). In the goose this increase in nitrogen excretion occurs before the lipid reserves are exhausted (Le Maho *et al.*, 1981). In the domestic fowl, phase III is reached after fasting for over 20 days and the daily nitrogen excretion may increase by as much as 5-fold (see Cherel *et al.*, 1988). The plasma concentrations of glucose and alanine in the goose remain fairly constant through phases I and II. The former is maintained by reduced utilisation rather than by increased glucose turnover (Groscolas & Rodriguez, 1981). This is because the lipid fuels exert a sparing effect. The increase in plasma alanine after about 40 days fasting is the result of protein mobilisation. By phase III when endogenous protein is being more extensively catabolised, it is important for penguins' survival that refeeding begins. At this stage, they leave the colony and walk to the sea, in order to start refeeding (Cherel *et al.*, 1988).

6.8.2 Metabolic adaptation during moulting

During feather regrowth after moulting, birds are often in negative nitrogen balance. The integument, which includes the feathers, the feather sheath and other epidermal structures in the mass ratio of approximately of 20:4:1, consists of ≈95% protein. Pigments account for ≈3–5% of feather mass and they also are synthesised from amino acid precursors. The main amino acids required for pigments are tyrosine and cysteine. Both exogenous and en-dogenous sources of amino acids are used for feather growth. The endogenous sources are generally cell proteins, and proteolysis is required to mobilise amino acids. In the Canada goose, breast muscle protein is catabolised in order to provide the materials for feather growth (Hanson, 1962). Piersma (1988) studied moulting in the great-crested grebe to differentiate between the 'nutritional stress' hypothesis (Hanson, 1962) and the 'use–disuse' hypothesis. The former places the emphasis on nutritional deficiency, whereas the latter explains the loss of flight muscle as the result of atrophy. Since the birds are flightless after complete moult, they do not use the pectoral muscles to the same degree until feather regrowth has occurred. During moulting, the great-crested grebe loses about 22% of flight muscle mass, but there is very little change in leg muscle. Since the timing of muscle atrophy more closely follows the onset of flightlessness than the onset of moulting, Piersma (1988) favours the 'use–disuse' hypothesis. Biochemical studies, in which radioactively labelled amino acid precursors are used to label muscle proteins and then follow their subsequent release and reincorporation into feather protein, are needed to clarify the dynamics of the process.

The blue-winged teal is flightless for about 30 days during the moulting period. During this period, there is atrophy of the flight muscles and hypertrophy of the leg muscles. There is also a change in the aerobic and anaerobic capacities of the muscles. Using citrate synthase and lactate dehydrogenase as indicators of citric acid cycle and glycolytic capacities (see Section 3.4.2), Saunders & Klemm (1994) have shown that moulting is accompanied by a decrease in the aerobic capacities of the flight muscles. This

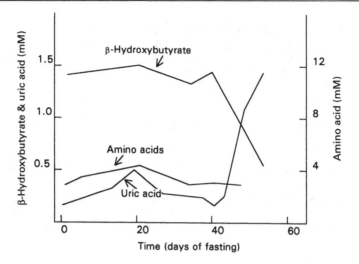

Fig. 6.4. Changes in fuel utilisation in geese after prolonged fasting, reflected in the plasma concentrations of β-hydroxybutyrate and uric acid (Le Maho *et al.*, 1981).

capacity is rapidly restored just prior to regaining flight.

The difference between fasting associated with breeding and fasting associated with moulting has been most clearly differentiated in the emperor and king penguins. During fasting associated with breeding, there is considerable sparing of protein and a high dependence on fat as fuel. In the emperor penguin during the breeding fast, it is estimated that 96% of the energy expenditure is from lipid and 4% from protein, whereas in the fast associated with moulting the corresponding figures for energy expenditure are 80% fat and 20% protein (Groscolas & Cherel, 1992). During the moulting fast, endogenous protein has to meet the amino acid requirements for keratin biosynthesis. The changes in plasma concentration (Fig. 6.4) of amino acids, uric acid and β-hydroxybutyrate are indicative of the endogenous reserves being mobilised. For the first 40 days, β-hydroxybutyrate derived from β-oxidation of fatty acids provides much of the energy. Endogenous protein is also being mobilised, both as a source of amino acids for keratin biosynthesis and also for deamination to provide carbon skeletons that can be catabolised as a source of energy. Beyond 42 days, the available lipid becomes sparse, and protein is increasingly being used as a source of energy. Hence the dramatic increase in uric acid concentration. Plasma thyroxine concentration increases at the onset of feather synthesis and may be involved in its regulation. Plasma corticosterone and glucagon increase after 25–30 days and may regulate the increased use of protein.

Corticosterone has a marked effect on protein catabolism, and glucagon promotes gluconeogenesis from amino acids and may regulate the increased use of protein after 30–40 days of fasting.

It is clear that endogenous protein provides the precursors for feather protein synthesis, and so it is important to consider how well matched the amino acid pool generated by endogenous proteolysis is to the requirements for feather protein synthesis. In most penguins, there is an increase in body mass in anticipation of moulting. This involves not only an increase in body fat but also in protein. Pre-moulting king penguins have about 52% dry mass as protein compared with 48% in pre-breeding penguins, the increase being largely in skeletal muscle (Cherel, Charrassin & Handrich, 1993). During feather synthesis, large amounts of endogenous amino acids are used, but in spite of this, there is also high nitrogen excretion. This is assumed to be because of the mismatch between amino acids required for keratin biosynthesis and those released from muscle protein (Cherel & Freby, 1994). Between 2.9 and 3.6 g tissue protein is required to produce 1 g feather protein. The excess mismatched amino acids are deaminated, the carbon skeletons can then act as gluconeogenic precursors and the nitrogen is excreted as uric acid. This has been studied in the white-crowned sparrow, which has the advantages of being one of the most rapid moulters and, therefore, has one of the highest rates of integument production. Plumage regeneration probably occurs at the expense of endogenous protein only during overnight fast. At peak moult,

Table 6.9. *Egg-yolk proteins from domestic fowl and their origin*

Protein	Site of synthesis	Dry matter (% yolk)	Dry matter (% total protein)
VLDL (apoB-48)	Enterocyte	16	22 (combined VLDL)
VLDL (apoB-100 and II)	Hepatocyte	50	
Vitellogenin	Hepatocyte	20	48
Plasma albumin	Hepatocyte	2	30 (combined albumin, transferrin and IgG)
Transferrin	Hepatocyte	5	
Immunoglobulin G	B lymphocyte	3	

Source: Moran (1987).

the white-crowned sparrow uses approximately one third of its daily intake of essential amino acids for integument replacement, but for individual amino acids the extremes are 17% for histidine to 75% for sulphur-containing amino acids. The latter is high because of the large amounts of sulphur in keratin.

6.9 Metabolic adaptation for egg-laying

Egg-laying generally requires a more substantial increase in protein synthesis than for feather regrowth, and it also requires lipid biosynthesis. By contrast, during moulting much of the energy is provided by lipid catabolism. There is considerable species variation in the average clutch size, in the size of the egg in relation to the bird and, therefore, in the energy and amino acid demands placed on the bird. The emphasis on the use of endogenous versus exogenous reserves varies between species. In some species, there is a marked decline in body weight, and hence nutrient reserves, whilst in others the body weight is maintained and an increased food intake through hyperphagia balances the requirements for egg-laying. The proportion of yolk:white varies between species from yolk constituting 15% egg volume up to 60%. Egg-yolk typically comprises about 60% dry mass as lipid (mainly triglyceride) and the rest mainly protein. Almost all the dry mass of egg-white is protein. The energy cost of egg production has been studied most extensively in the domestic fowl (Scanes, Campbell & Griminger, 1987). The costs are both in the nutrients transferred to the egg and in the energy

required for synthesising the egg. In the domestic fowl, these amount to about 700 kJ/day. Egg-yolk proteins are synthesised principally in the liver (Table 6.9) whereas the egg-white proteins are synthesised in the tubular cells of the oviduct. Their synthesis is under hormonal control and involves activation of transcription and translation of genes encoding egg proteins (see Sections 10.3 and 10.4). In birds, the liver is also the main site of triglyceride synthesis (see Section 4.3). In the domestic fowl feeding *ad libitum*, the daily intake increases when hens come into lay, and less food is consumed on days when either ovulation or oviposition is missed (Scanes *et al.*, 1987). Under these conditions, there is no depletion of protein or lipid reserves except whilst fasting during the hours of darkness. In many species, the protein content of the body declines during egg laying, e.g. Adélie penguin, American coot, lesser black-backed gull, common eider, red-billed quelea and grey-backed camaroptera (Blem, 1990), although Krementz & Ankney (1988) found no increase in protein and lipid reserves of house sparrows before ovulation and no decline during ovulation. Several migratory waterfowl build up extensive lipid and protein reserves to be used subsequently for egg-laying (Blem, 1990). In all species, egg-laying involves increased protein and lipid synthesis in the liver, and increased protein synthesis in the tubular gland of the magnum, but the extent of the changes and their timing varies between species, depending on their habitat, source of nutrient and egg size in relation to body size.

7

Avian hormones and the control of metabolism

7.1 Introduction

Hormones are generally defined as substances released into the bloodstream by specialised tissues, known as endocrine glands, and which interact with distant target tissues where they exert their effect. The general awareness of hormones as regulatory molecules dates from the early part of this century. More recently other types of regulatory molecule have been discovered that are released from one type of cell to diffuse through the intercellular fluid and exert their effects on other types of cell. Because of their different transport route these are referred to as paracrine secretions. A third group of substances are released by one type of cell and exert their effects on the same cell type. These are referred to as autocrine secretions. Endocrine secretions are, therefore, just one type of a more general intercellular messenger. This chapter is largely restricted to the role of endocrine secretions, as defined above, in controlling metabolism (topics covered in Chapters 2 to 6). Paracrine secretions are discussed in connection with morphogenesis in Chapter 12, paracrine and autocrine secretions in connection with oncogenesis in Chapter 11, and steroid hormones in Chapter 10.

Current understanding of the mechanism of action of hormones suggests that there are two general mechanisms operative. In the first, the hormone or first messenger (the intercellular messenger) interacts with receptors on the cell surface of the target tissue, activating the release of the second messenger (intracellular mesenger), which in turn interacts with various intracellular enzymes to modulate intracellular reactions. In the second type, the hormone enters the target cells, crossing the plasmalemma, and interacts with nuclear receptors to bring about a change in gene expression. The hormones described in this chapter are of the first type. The most extensively investigated examples of the second type are the sex hormones and adrenal steroids (see Chapter 10). The most important hormones controlling metabolism are those secreted by the endocrine pancreas and the adrenal glands and these will be the principal subject of this chapter.

7.2 The endocrine pancreas

A brief description of the endocrine pancreas is given here, but for the detailed anatomy and embryology see Hazelwood (1986a) and Cramb & Langslow (1984). The anatomy and histology of the pancreas from a number of avian species has been described (Guha & Ghosh, 1978), but most detailed studies have been made of domesticated species, particularly the domestic fowl. The avian pancreas is approximately 0.1% of the total body weight, but only 1–2% of pancreatic tissue is the endocrine pancreas, the other 98% being the exocrine pancreas, which secretes large amounts of digestive enzymes into the duodenum. The endocrine and exocrine pancreas appear to have evolved independently. Comparative studies suggest that the most primitive association of the two glands occurs in the lamprey (Epple & Brinn, 1987). Avian endocrine tissue is concentrated in clusters, referred to as the *islets of Langerhans*, after their discoverer. The avian pancreas is a tongue-shaped structure generally comprising three morphologically distinct lobes; splenic, dorsal and ventral. In the domestic fowl, the ventral lobe has a longitudinal cleft running throughout its length giving the appearance of two separate lobes. The central portion of this, closest to the dorsal lobe, is sometimes confusingly referred to as the 'third' lobe. This has also been found in other galliform birds but is absent from other orders, e.g. Passeriformes and Columbiformes (Guha & Ghosh, 1978).

Within the islets, distinct cell types can be distinguished. The principal types are A cells, B cells,

D cells and PP cells, capable of secreting glucagon, insulin, somatostatin and avian pancreatic polypeptide (APP), respectively. The avian pancreatic polypeptide was first discovered in the domestic fowl by Kimmell, Pollock & Hazelwood (1968) and was subsequently found also in mammals. Birds are unusual in having two distinct types of islet. The islets, which are made up of different proportions of these cell types, are classified as dark or light islets. The predominant cell type in the darker islets is A (glucagon secreting), but with some B, D, and PP cells, whereas the lighter islets have B (insulin) cells as the predominant type. PP cells are fairly uniformly distributed throughout the lobes. In galliform birds, the dark islets are found almost exclusively in the splenic and 'third' lobes, whereas the light islets are distributed throughout the pancreas (Epple & Brinn, 1987). In other orders that have been studied, e.g. Passeriformes and Columbiformes, the dark islets are more evenly distributed throughout all of the lobes (Guha & Ghosh, 1978). In the domestic fowl, the B cells in the 'light' islets of the splenic lobe are very large compared with the B cells located elsewhere, and thus the concentration of insulin in the splenic lobe is 10–40 times greater than in the other lobes (Hazelwood, 1984). Foltzer, Harvey & Mialhe (1987) estimate that 80% of the glucagon and 63% of the somatostatin are concentrated in the splenic lobes of the duck, and 53% of the insulin is concentrated in the dorsal lobe. The higher concentration of glucagon in the splenic lobe of the ostrich pancreas suggests the cell distribution in this species is similar to that of the domestic fowl and quail (Ferreira et al., 1991). Type A cells are generally more abundant in avian species than in other vertebrates (Guha & Ghosh, 1978).

7.3 Structure and biosynthesis of avian pancreatic hormones

The biosynthesis of avian endocrine pancreatic hormones has only been studied in detail for insulin and glucagon in the domestic fowl, where it follows the pattern that has been established for secretory proteins. The mRNA transcript is synthesised in the nucleus, initially as a much larger precursor RNA. During its maturation it becomes spliced to remove the transcribed introns. The mature mRNA then exits the nucleus. Synthesis of the precursor proteins containing the signal peptide occurs on the rough endoplasmic reticulum. The signal peptide is cleaved off the protein as it is released into the lumenal side of the rough endoplasmic reticulum. The main peptide,

or propeptide, is then transported to the Golgi apparatus.

7.3.1 Insulin

Insulin is synthesised by this general method. Within the Golgi apparatus, at least 95% of the proinsulin is cleaved and the insulin produced is stored in secretory granules. Membrane-bound vesicles transport the secretory granules to the plasma membrane where their contents are secreted into the extracellular space. Insulin circulates in the bloodstream, probably as a dimer. The domestic fowl pre-proinsulin gene is a single-copy gene with a structure similar to that of mammalian insulin genes (Perler et al., 1980). Transcription of the insulin gene has been studied during embryogenesis in the domestic fowl (Serrano et al., 1989). The mRNA transcript is ≈ 600 nucleotides(nt) and its concentration increases about 30-fold from E10 to three weeks post-hatching. The coding region of the insulin gene (324 nt) together with the putative untranslated 5′ and 3′ regions (134 nt) adds up to 458 nt, leaving ≈ 150 nt poly(A) tail (Perler et al., 1980). Although there is a major increase in insulin mRNA at E16, there is not a corresponding increase in the plasma concentration of insulin, which remains low until hatching. It is assumed that there is some form of post-transcriptional control, and that the mRNA is stored in the B cells (Serrano et al., 1989). Insulin is synthesised as a single polypeptide precursor *preproinsulin* (103 amino acid residues in the domestic fowl) (Fig. 7.1). The signal peptide (24 amino acid residues) is cleaved by an endopeptidase. The single chain proinsulin then folds and the disulphide bridges form, and the C-peptide (28 amino acid residues in the domestic fowl) is cleaved leaving the mature insulin, which is initially stored in secretory granules.

The amino acid sequences of four avian insulins have been determined, namely domestic fowl, turkey, duck and goose. Those of domestic fowl and turkey are identical but differ from that of the duck and goose at three positions (Fig. 7.2). There are also differences in the size of the C-peptide, which is 28 residues in the domestic fowl and 26 in the duck. Another feature that is evident from the proinsulin sequence (Fig. 7.2) is that the C-peptide is less conserved than the A- and B-peptides. The C-peptide simply acts as a link between the A- and B-peptides to enable the disulphide bridges to form. Domestic fowl insulin is more potent than bovine insulin in elliciting changes in metabolism in the domestic fowl. This has been attributed to the six differences in the amino acid sequences of the two proteins.

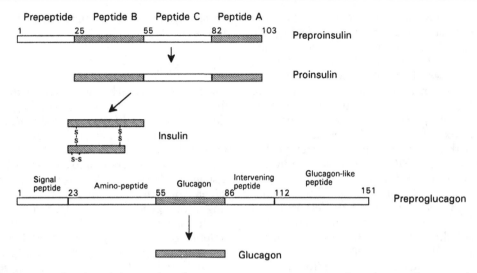

Fig. 7.1. The formation of insulin and glucagon from their respective precursors, pre-proinsulin and pre-proglucagon in domestic fowl.

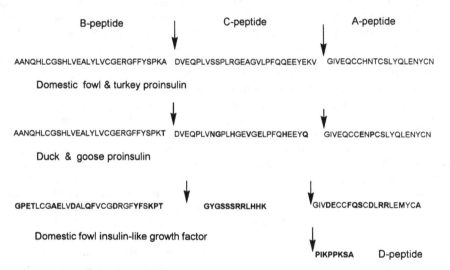

Fig. 7.2. Sequences of avian proinsulins together with domestic fowl IGF-1. Sequence differences compared with domestic fowl proinsulin are shown in bold. Note IGF-1 has an additional D-peptide.

7.3.2 Glucagon

Glucagon is a small polypeptide of 29 residues and is synthesised as part of a larger precursor, *preproglucagon*, which in the case of the domestic fowl is 151 amino acid residues. The structure and nucleotide sequence of the glucagon precursor from domestic fowl have been determined (Hasegawa *et al.*, 1990). In addition to containing the 29 residue sequence for glucagon, it also contains a signal peptide (22 residues), an N-terminal peptide (30 residues), an intervening peptide (24 residues) and a C-terminal peptide (37

residues) (Fig. 7.1). The C-terminal peptide is referred to as glucagon-like peptide I (GLP-I), because in the case of the domestic fowl it has 41% sequence identity. The signal peptide is cotranslationally removed leaving 129 amino acid residues. Further post-translational processing involving proteolysis adjacent to certain basic amino acid residues cleaves it into four polypeptides (Fig. 7.1). Domestic fowl pre-proglucagon is smaller than those from mammalian sources since it lacks GLP-II. The functions of GLP-I and GLP-II are not known. What appears to be a much larger glucagon precursor has been reported to be present in pigeon islets, having an M_r value of 69 000 (Tung, 1973), but in view of the more recent work using other species, this needs confirming. The mechanism of glucagon secretion is similar to that of insulin. Glucagon is a highly conserved polypeptide showing few differences from those in mammals and reptiles, but more from fish. It has been sequenced from four species of birds. The sequences in domestic fowl and turkey are identical, and there is a single replacement of Ser-17 by Thr-17 in duck and ostrich (Ferreira *et al.*, 1991).

7.3.3 Somatostatin

The biosynthesis of *somatostatin* (a tetradecapeptide, S14) and avian pancreatic polypeptide (APP) have not been studied in avian species. Higher M_r forms of somatostatin, including the 28 amino acid residue form (S28) have been detected in the plasma of neonatal ducks (Foltzer *et al.*, 1987). S28 exerts more lasting and profound physiological effects than S14 (Epple & Brinn, 1987). S14 has identical sequences in pigeon, pig, sheep, rat and anglerfish (see Cramb & Langslow, 1984; Hazelwood, 1986b). A large somatostatin-like peptide, M_r 11 000 to 12 500, has been detected from pigeon pancreas (Speiss *et al.*, 1979). It dissociates in the presence of urea and thioglycol, but not in the presence of urea alone, into fragments the size of S14. This suggests that the larger molecule is held together by disulphide bridges. S14 has a single disulphide bridge, making it a cyclic peptide. The avian pancreas is reported to contain from 14 (domestic fowl) to 180 (pigeon) times the amount found in rat or humans (Hazelwood, 1984).

7.3.4 Avian pancreatic polypeptide (APP)

APP was first detected as a contaminating protein during the isolation of insulin from the domestic fowl (Kimmel *et al.*, 1968). APP from duck, turkey and domestic fowl have identical 36 amino acid residue sequences, but there are four amino acids different in goose (Hazelwood, 1986b; Glover, Barlow & Pitts, 1985). The 36 amino acid residue polypeptide has a tightly folded structure in which the N- and C-terminus are close together, and this is an important feature for receptor recognition (Hazelwood, 1993a). This can be seen from the turkey APP, on which X-ray crystallographic studies have been performed (Glover *et al.*, 1985). APP is one of a family of related polypeptides; two others with closely related structures and functions are polypeptide-YY (PYY) released from the lower intestinal tract, and neuropeptide-Y (NPY) present in the brain and autonomic nervous

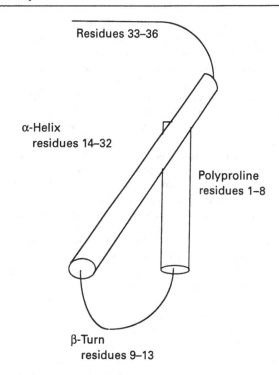

Fig. 7.3. The PP fold protein structure found in avian pancreatic polypeptide (APP) and related peptides.

system. Both have tyrosine residues (Y) at the N- and C-terminus, the latter being amidated. APP has tyrosine only at the C-terminus. All three have a high degree of structural similarity, more evident in the conformation than in the primary sequence. Many homologous peptides have been found in mammalian species. They are sometimes referred to collectively as PP-fold proteins, on account of their structure (Fig. 7.3), which arises from having proline residues in positons 2, 5 and 8, closely packed against an amphiphilic α-helical region of residues 14 to 32 (Hazelwood, 1993a). APP is synthesised as a prohormone (Hazelwood, 1984). It has diverse effects on the brain and gastrointestinal tract but also has metabolic effects, causing the mobilisation of liver glycogen and free fatty acids and elevating plasma triglycerides. Glucagon, somatostatin and APP first become detectable in the domestic fowl embryo at E16 (Martinez, Lopez & Sesma, 1993).

7.4 Control of secretion of avian pancreatic hormones

The newly synthesised pancreatic hormones are stored in granules in the cytoplasm of the islet cells.

They are subsequently released by a process of exocytosis. There are three main types of agent that regulate their release: (i) metabolites such as glucose and amino acids, (ii) other hormones, particularly pancreatic hormones, and (iii) signals from the autonomic nervous system. Much of the current understanding results from studies of insulin and glucagon using isolated perfused pancreas from the domesticated fowl and other domesticated species (Cramb & Langslow, 1984; Simon, 1989). Insulin and glucagon tend to act antagonistically: insulin promotes glycogen synthesis, fatty acid synthesis and protein synthesis, whereas glucagon promotes gluconeogenesis, glycogenolysis and lipolysis. Their roles are similar throughout vertebrates, although the effectiveness of the two is different in birds from in mammals. Glucagon appears to play a more dominant role in birds and the circulating glucagon:insulin ratio is higher in birds than in mammals (Sitbon et al., 1980). The dominant effects of glucagon are evident after pancreatectomy, which in goose and domestic fowl does not cause hyperglycaemia, in contrast to mammals. The early investigations into the effects of pancreatectomy overlooked the splenic lobe, which has an abundant supply of A cells, but more recent investigations in which the splenic lobe was removed, confirm the dominance of glucagon (Epple & Brinn, 1987). Pancreatectomy causes a decrease in the plasma concentration of free fatty acids (Sitbon et al., 1980); these are principally mobilised from adipose tissue after stimulation by glucagon. Pancreatectomy does not induce hypoglycaemia in all birds, e.g. in raptors. The normal plasma concentration of glucose in birds is higher than that in mammals (see Table 3.1, p. 30). In the domestic fowl, it is about 10–12 mM compared with about 5–6 nM in mammals (Simon, 1989).

High blood-glucose concentrations promote the release of insulin, but in the domestic fowl this does not occur until the concentration is 15–30 mM, which is higher than the normal physiological concentration (Simon, 1989). Insulin is also released in response to increased intracellular cAMP. Cyclic AMP has a rapid turnover and so elevated levels can be induced by inhibiting its breakdown with theophylline, an inhibitor of phosphodiesterase.

$$cAMP + H_2O \rightleftharpoons 5'\text{-}AMP$$

Cyclic AMP or acetyl choline potentiate the release of insulin from the B cells induced by elevated glucose concentrations in the plasma (Rideau & Simon, 1993). Amino acids such as arginine also exert a strong synergistic effect on the glucose-stimulated release of insulin, whereas fatty acids exert a weak synergistic effect (Simon, 1989). Metabolites of glucose and of amino acids also exert a synergistic effect, and Rideau & Simon (1992) suggest that it is through their effects on intracellular metabolism that they promote insulin release. The release of glucagon is promoted by low plasma glucose concentrations, e.g. 3–4 mM, and inhibited by high plasma glucose. Fasting, which results in a decrease in the circulating plasma glucose concentrations, causes a 2- to 3-fold increase in the circulating glucagon concentration in domestic fowl, ducks, geese and pigeons (Hazelwood, 1986b). The plasma glucagon concentrations in fasting eagles is much lower than in domesticated birds (Minick & Duke, 1991), but there is also very little decrease in plasma glucose concentrations in carnivorous birds on fasting, probably on account of their higher gluconeogenic capacity (Veiga, Roselino & Migliorini, 1978). Prolonged fasting in penguins causes a decrease in plasma insulin concentrations (Table 7.1).

There is much evidence to suggest that glucagon plays a more dominant role than insulin in controlling carbohydrate metabolism in avian species than in vertebrates in general. There are many reported cases of spontaneous diabetes mellitus in mammals, e.g. horse, cow, pig, sheep, dog, cat and monkey, whereas it has only been reported in parakeets, a pigeon and the red-tailed hawk among birds (Altman & Kirmayer, 1976; Guha & Ghosh, 1992; Wallner-Pendleton, Rogers & Epple, 1993). In the last of these, the blood glucose level was over 50 mM. It is difficult to induce experimental diabetes in many species of bird, either by pancreatectomy or by using cytotoxic agents such as alloxan or streptozotocin. The former may be partly because of the difficulty in completely removing the pancreas (see below). The normal ratio of plasma glucagon:insulin is high in birds. The avian pancreas has a large number of somatostatin-secreting cells, and in the pigeon, for example, the pancreatic somatostatin level is over 150 times that present in humans. The plasma circulating levels of somatostatin in the domestic fowl are 3- to 5-fold higher than in mammals (Hazelwood, 1984). Somatostatin release is promoted by glucagon, insulin, acetyl choline, glucose and amino acids and is suppressed by hypoglycaemia. High circulating levels of somatostatin inhibit its further release. Somatostatin is also a strong suppressor of insulin release but a weak suppressor of both glucagon and APP release (Hazelwood, 1984). It may act as a paracrine secretion in this respect (Hazelwood, 1984). The high concentration of

Table 7.1. *Pancreatic hormones in avian blood*

Hormone	Species	Plasma concentration (nM)	Reference
Insulin	Domestic fowl:	0.1–2	c, h, j, m
	overnight fasting	0.04–0.2	j
	lean strain, after meal	0.71	d
	fat strain, after meal	1.6	d
	Muscovy duck	0.26	h
	Pekin duck	0.16	h
	Goose:		
	overnight fasting	0.096	k
	1 min after feeding	0.26	k
	Quail	0.04–0.19	n
	Bald eagle	0.01–0.04	i
	Pigeon	0.049	f
	Penguin:		
	24 h fast	0.28	a
	12 day fast	0.07	a
Glucagon	Domestic fowl	0.4	e
	Bald eagle	0.07–0.36	i
APP	Domestic fowl	0.48–4.0	b, l
	Pigeon	0.14	f
IGF-I	Domestic fowl	2.0–7.0	g

Source: (a) Chieri *et al.* (1972); (b) Langslow, Kimmel & Pollock (1973); (c) Cramb, Langslow & Philips (1982); (d) Simon & LeClercq (1982); (e) Cramb & Langslow (1984); (f) John *et al.* (1989); (g) Ballard *et al.* (1990); (h) Constans *et al.* (1991); (i) Minick & Duke (1991); (j) Simon *et al.* (1991); (k) Karman *et al.* (1992); (l) Hazelwood (1993b); (m) Taouis *et al.* (1993); (n) Tedford & Meier (1993).

somatostatin may, therefore, have a role in preventing diabetogenesis in birds (Guha & Ghosh, 1992).

APP is released from PP cells adjacent to the A cells of the pancreas. It circulates as a dimer in concentrations up to 4 nM in birds (Table 7.1), which is 30–50 times that of the mammalian counterpart (Hazelwood, 1993b). It is released in response to elevated levels of amino acids or fatty acids, hypoglycaemia or nerve stimulation (Hazelwood, 1984). Its effects are diverse and include inhibition of exocrine pancreatic and intestinal secretion, and inhibition of intestinal motility. APP and the two related hormones, PYY from the intestine and NPY from the brain, have overlapping biological actions concerned with feeding behaviour and selection of nutrients.

Both glucagon and insulin respond primarily to the nutritional status of the bird. There is considerable variation between species in the nutritional stress likely to be encountered. For example, domestic fowl eat regularly during daylight hours and store food in their crops, enabling a fairly constant rate of digestion to occur. This may occur in graminiferous birds generally, and is in contrast to birds of prey that eat more spasmodically and will be subjected to high nutritional loads after catching prey. Langslow, Kimmel & Pollock (1973) have shown that the great horned owl has at least four times the pancreatic insulin content of the domestic fowl. Carnivorous birds have a high protein:carbohydrate ratio diet and also have a higher capacity for gluconeogenesis than graminiferous birds (Veiga *et al.*, 1978). Pancreatectomies carried out on carnivorous birds cause hyperglycaemia and glucosuria, in contrast to most other species of bird (Cramb & Langslow, 1984). Complete surgical pancreatectomy is difficult to achieve in many birds because of the inaccessibility of the splenic lobe isthmus (Hazelwood, 1986a). When successfully achieved in a duck it led to fatal hypoglycaemia (Hazelwood, 1986a). The balance of evidence, therefore, suggests that in the domesticated graminiferous species which have been most extensively studied glucagon has the dominant role in regulating metabolism under different nutritional conditions (Hazelwood, 1986b), whereas in carnivorous birds insulin has a quantitatively more important role.

The second factor that is important in controlling the release of pancreatic hormones is the interaction between the hormones. This may result from transport of a pancreatic hormone from one cell type to another, either by endocrine or paracrine secretions. Insulin release from the B cells is influenced by other pancreatic hormones. In domestic fowl, duck and penguin, glucagon induces insulin secretion, whereas somatostatin inhibits insulin release (Simon, 1989). Conversely insulin decreases the release of glucagon in the short term but increases it in the longer term. Somatostatin supresses glucagon secretion more strongly than insulin secretion and so may have a role in controlling the insulin:glucagon ratio. This ratio is highest when birds are in 'anabolic' mode, particularly after feeding, and low in 'catabolic' mode, when the bird is exercising or migrating and is consuming energy reserves. Extrapancreatic hormones also influence insulin and glucagon release, but to a lesser extent.

Secretion of insulin and glucagon is also under nervous control. A clear example of this is the early insulin response to food intake. Within 1 min of oral food intake by the domestic geese, there is a rapid increase in insulin secretion, and plasma levels increase by 2.5-fold (Karmann et al., 1992). This is a much more rapid increase than any hyperglycaemic effect and is independent of the nature of the food ingested. There are also indications that circadian neural control centres regulate the release of insulin in quail (Tedford & Meier, 1993).

7.5 Tissue uptake of hormones: the role of hormone receptors

Once the pancreatic hormones are released into the bloodstream, they circulate and become bound to receptors on the plasmalemma of cells of the target tissues. The tissues most sensitive to changes in the concentrations of circulating insulin and glucagon are muscle and adipose tissue, and most research has focused on these tissues. Circulating hormone concentrations in different species are illustrated in Table 7.1. In order for the hormone receptors to bind hormones at these plasma concentrations, the receptor must have K_d for the hormone of a similar order of magnitude (Table 7.2) to that of the circulating hormone concentration. The half-life of circulating insulin is estimated at $\approx$ 5 min in the

domestic fowl, and $\approx$ 7 min in the turkey (Simon, 1989), but the factors controlling its clearance in avian plasma are not well understood. The half-life of circulating APP is 6–7 min (Hazelwood, 1993a). The circulating hormones also bind to the receptors on the target tissue. Both the insulin receptor and the glucagon receptor have been isolated from a few avian tissues, and specific binding of the hormones has been detected in many more. One of the earliest to be studied was the insulin receptor on turkey erythrocytes (Ginsberg, Kahn & Roth, 1976). Erythrocytes make good starting material, since they are readily collected in reasonable amount, no tissue homogenisation is required, they are readily lysed releasing their intracellular contents, mainly haemoglobin, leaving a membrane preparation from which the receptor can be isolated. Insulin receptors have also been isolated from domestic fowl brain, liver and skeletal muscle (Simon & Leroith, 1986; Adamo et al., 1987), and from duck liver (Constans et al., 1991); they have also been detected in avian cardiac muscle, thymocytes, chondrocytes and myoblasts (Simon & Taouis, 1993). It is perhaps surprising that the receptors that have high-affinity binding sites for insulin are found in tissues such as brain and erythrocytes, which are not generally regarded as major targets of insulin. The reason for this is not clear. In the brain, insulin receptors are present on neurons. They have a distribution that is distinct from that of IGF-1 (insulin growth factor-1) receptors. Their function does not appear to be associated with glucose homeostasis but rather, in growing birds, as a growth promoting or neurotropic factor (Heidenreich, 1991). In adult brain, the function is unclear.

The insulin receptor is a transmembrane protein comprising four polypeptide chains, two α and two β, linked by disulpide bridges (Fig. 7.4). The subunit composition of the insulin receptor and its phosphorylation sites have been determined after crosslinking the receptor–[^{131}I]-labelled insulin complex, using disuccinimidylsuberate. The α- and β-chains are of similar size in all insulin receptors studied. The β-chain is 97–99 kDa. In domestic fowl liver, the α-chain is 139 $\pm$ 2 kDa, whereas the brain α-chain is 127 $\pm$ 2 kDa (Simon & Leroith, 1986). Both receptors are glycosylated, but it seems the difference in size may be the result of additional sialic acid residues on the liver receptor. [^{125}I]-labelled insulin is generally used to detect receptors and to measure the strength of binding, since low concentrations (10^{-9} to 10^{-11} M) must be used to detect the highly specific, high-affinity binding, and, therefore, only a hormone analogue

Table 7.2. *Pancreatic hormone receptors*

Hormone	Species	Tissue	K_d (nM)[a]	No. binding sites
Insulin	Domestic fowl	Liver	≈ 0.3 and 20	2200
Insulin	Domestic fowl	Erythrocyte	–	180
Insulin	Domestic fowl	Brain	–	
Insulin	Domestic fowl	Skeletal muscle	0.1	
Insulin	Quail	Oviduct	0.35 and 8.23	3500 (high affinity)
Glucagon	Domestic fowl	Liver	≈ 9.4 and 400	12 900
APP	Domestic fowl	Liver	0.72 and 200	300 000 (high affinity)
APP	Domestic fowl	Brain	135	

[a] Where two values are given at least two types of binding site exist.

Source: Simon, Freychet & Rosselin (1977); Cramb *et al.* (1982); Simon & Leroith (1986); Ukhanova & Leibush (1988); Inui *et al.* (1990); Kato *et al.* (1990).

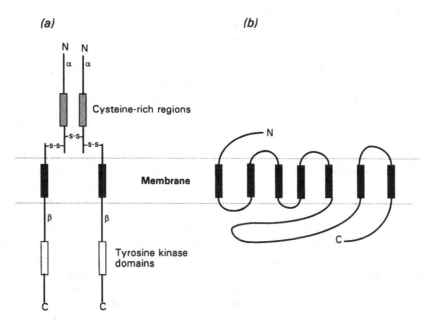

Fig. 7.4. Structure of (*a*) insulin receptor and (*b*) β-adrenergic receptor.

having high specific radioactivity would be detected. The receptor shows higher affinity for avian insulin than porcine or guinea-pig insulin, and this is because of the histidine at position 8 in the A chain in domestic fowl insulin. The avian insulin receptor from liver, muscle and brain will also bind IGF-I and IGF-II but with low affinity (Simon & Taouis, 1993). However, it is clear that there are separate receptors for insulin and for IGF, and these are found in different locations in the brain. The number of insulin receptors per cell is generally lower in birds than in mammals. This may be a partial explanation for the lower sensitivity of birds to insulin. There is evidence for a correlation between the concentration of insulin receptors and insulin sensitivity (see also Section 7.6). When domestic fowl were given daily injections of corticosterone, they showed both a reduced response to insulin and a reduced number of insulin receptors (Taouis *et al.*, 1993). The number of insulin receptors is also regulated by the circulating insulin concentration. The number of insulin receptors detectable on hepatocyte membranes may decrease by over 50% in response to high circulating concentrations of insulin. This decrease is the result of internalisation rather than degradation, since the total number of receptors detectable after cell soublisation remains constant. Fasting has the effect of increasing the number of insulin receptors detectable.

There are two direct consequences of insulin binding to the receptor. First the tyrosine kinase activity on the β-subunit is activated. This causes autophosphorylation and also phosphorylation of intracellular proteins. Although a number of proteins have been phosphorylated *in vitro*, it is not yet clear what is the normal physiological substrate or substrates. In mammals, several cellular protein substrates have been proposed, but the precise role of any of the proteins is unclear. A 70 kDa protein has been proposed as a substrate in cultured foetal chick neurons, and a 72 kDa phosphoprotein in domestic fowl hepatoma, but the functions of these polypeptides are unclear (Simon & Taouis, 1993). Second, binding causes internalisation of the receptor, after which the insulin is removed and degraded in the lysosomes and the receptor recycled back to the plasmalemma. In domestic fowl hepatocytes, the half-life of the receptor is about 10 h (Simon, 1989).

7.6 Effects of pancreatic hormones on metabolism

7.6.1 Insulin

Insulin has a number of effects on both carbohydrate and lipid metabolism, generally stimulating their uptake and storage (Table 7.3). The principal site of triglyceride biosynthesis in birds is the liver (see Section 4.3). A number of *in vitro* studies have examined the effects of pancreatic hormones on lipid metabolism. Rosebrough & McMurtry (1992) have used liver explants to study the short-term effects of insulin and glucagon. With their preparations, insulin did not affect the rate of lipogenesis, whereas glucagon caused a decrease. Laurin & Cartwright (1993), however, have shown that physiological concentrations of insulin, when incubated in a suitable medium with hepatocytes from domestic fowl, stimulate triglyceride synthesis and also increase the activity of the malic enzyme, an enzyme which generates NADPH required for fatty acid synthesis (reaction 4.2, p. 47). This effect occurred in livers from 4-day-old birds; the effect decreased dramatically when they used older birds. Insulin does not exert an antilipolytic effect on adipose tissue in birds, by

comparison with mammals, where insulin antagonises the lipolytic effects of glucagon.

Avian liver is much less sensitive to the lipogenic effects of insulin than mammalian liver, and this may be the result of the lower concentrations (2–5-fold) of insulin receptors in the avian liver (Simon & Taouis, 1993). However, not all the evidence supports a direct relationship between insulin sensitivity and the density of insulin receptors in a tissue. Domestic fowl containing the sex-linked dwarfing gene (*dw*) show a greater sensitivity to insulin than normal birds and have higher concentrations of the insulin receptors in their livers (Simon & Taouis, 1993). However, genetically fat domestic fowl are more sensitive to exogenous insulin than their lean counterparts, although no differences in the number of insulin receptors or their affinity have been found (Simon et al., 1991).

Insulin has a major effect on carbohydrate metabolism. This has been studied in cultures of primary hepatocytes from chick embryos shortly before hatching (Hamer & Dickson, 1990). At this stage there is an increase in glycolysis and decrease in gluconeogenesis (see Section 3.7). When the hepatocytes are incubated with insulin, there is a dramatic stimulation of 6-phosphofructo-1-kinase and 6-phosphofructo-2-kinase activities, an increase in the concentration of fructose 2,6-bisphosphate concentration and increased glycolytic flux. The changes in 6-phosphofructo-2-kinase and fructose 2,6-bisphosphate occur before that of 6-phosphofructo-1-kinase. 6-Phosphofructo-1-kinase catalyses a rate-limiting step in glycolysis. The mechanism by which insulin stimulates glycolysis involves a complex network of components. These may involve both control of enzyme amount and allosteric regulation.

Insulin also has effects on protein metabolism, but these may be secondary effects. It stimulates the activity of a number of enzymes involved in nitrogen metabolism, e.g. ornithine decarboxylase, arginase, tyrosine aminotransferase, xanthine oxidase and purine nucleoside phosphorylase (see Simon, 1989). There are both examples where insulin increases and decreases gene transcription. Expression of VLD apolipoprotein II (apoVLDLII) in liver is dependent on oestrogen (see Section 10.4.2). Physiological concentrations of insulin strongly inhibit expression of apoVLDLII mRNA (Berkowitz, Chu & Evans, 1993). There is still much uncertainty as to how the effects of insulin binding to the receptor are relayed to sites within the target cells. Insulin stimulation does result in changes in the activities of specific protein kinases and phosphatases (Cohen, 1992).

Table 7.3. *The effects of avian pancreatic hormones on metabolism*

Hormone	Tissue	Effect on metabolism
Insulin	Liver	Increased glucokinase: increased glucose uptake
		Increased glycogen synthase activity: glycogen deposition
		Inhibited gluconeogenesis
		Increased malic enzyme, acetyl-CoA carboxylase, fatty acid synthetase and stearoylCoA desaturase: increased lipogenesis
		Release of VLDL from hepatocytes
Insulin	Adipocytes, fibroblasts, skeletal and cardiac muscle	Increased glucose transport across the plasmalemma
Glucagon	Liver	Increased cAMP, decreased fructose 2,6-bisphosphate, increased phosphorylase activity, glycogen mobilisation
		Inhibited accumulation of mRNA for malic enzyme and other lipogenic enzymes
Glucagon	Adipocytes	Stimulated lipolysis
APP	Liver	Increased glycogenolysis, but probably the result of increasing glucagon secretion
APP	Exocrine pancreas	Inhibited secretion
APP	Intestine	Inhibited secretion
APP	Adipocytes	Inhibited lipolysis

These have not been studied specifically in avian species, but it is likely that the underlying mechanism is the same as that which occurs in mammals.

7.6.2 Insulin-like growth factor

Polypeptides structurally related to insulin are IGF-I and IGF-II, which are synthesised in the liver, brain, skeletal and cardiac muscle (Kajimoto & Rotwein, 1989). Both insulin and IGFs appear to have evolved from a common ancestral molecule, since they have about 60% of amino acids identical (Epple & Brinn, 1987). IGFs are single chain polypeptides of 70 amino acid residues, with three disulphide bridges as in proinsulin (Fig. 7.2). IGF-I is synthesised as a 153 residue primary translation product (Kajimoto & Rotwein, 1989). A 48 amino acid residue peptide is cleaved from the N-terminus, and a 35 amino acid residue peptide from the C-terminus. The complete amino acid sequence of IGF-I derived from the nucleotide sequence of cDNA has been determined (Fawcett & Bulfield, 1990). The amino acid sequence has also been determined directly (Ballard et al., 1990). The proteins are highly conserved; IGF-I from domestic fowl having 88.5% identity with that from human. Insulin and IGFs have overlapping physiological effects, affecting both metabolism and growth. The predominant effects of insulin are on metabolism, whereas those of IGF are predominantly growth

promoting (Epple & Brinn, 1987). IGF-1 stimulates protein synthesis in chick embryo fibroblasts at much lower concentrations than does insulin (Jeffrey & Pain, 1993). It also stimulates glucose uptake in domestic fowl myotubes, with a potency equal to that of insulin (Duclos et al., 1993). The plasma concentrations of both IGF-I and IGF-II increase in the first few weeks after hatching and later fall (Ballard et al., 1990; Decuypere et al., 1993). The tissue distribution of IGF receptors has been studied in both domestic fowl (Canfield & Kornfeld, 1989) and turkey (McFarland et al., 1992). Unlike in mammals where two types of receptor exist (type-I and cation-independent mannose 6-phosphate/type II), there is only evidence for type I in birds, which binds both IGF-I and IGF-II. This receptor has wide distribution, being present in most tissues of the body (McFarland et al., 1992).

7.6.3 Glucagon

In many species of birds, glucagon appears to have a more dominant role than insulin (see Section 7.4). In domestic fowl liver, there are more glucagon receptors than insulin receptors although the latter have higher sensitivity (Table 7.2). The mechanism of action of glucagon is fairly clear. Binding of glucagon to the receptor causes the activation of adenylyl cyclase, and thus an increase in the intracellular concentration of cAMP. Acting through cAMP-dependent protein kinases, this stimulates glycogenolysis, gluconeogenesis and inhibits lipogenesis in avian liver and promotes lipolysis in adipose tissue. Avian liver is more sensitive to stimulation by glucagon than by adrenaline. Many of the effects of glucagon release are the converse of those of insulin, and include an increase in lipolysis (Simon, 1989), glycogen mobilisation and stimulation of gluconeogenesis from amino acids and from intermediates of the glycolytic pathway (Table 7.3). The gluconeogenic action of glucagon involves the inhibition of 6-phosphofructo-1-kinase, which catalyses a rate-limiting step in glycolysis, thereby shifting the flux in favour of gluconeogenesis (Sugano et al., 1982b). Glucagon stimulates lipolysis in adipose tissue by activation of cAMP-dependent lipoprotein lipase, and this results in the release of long-chain fatty acids into the plasma. The increased lipolysis has been demonstrated using adipocytes prepared from broilers and maintained in primary culture (Oscar, 1991). A 2–3-fold stimulation of glycerol release from adipocytes was observed using 1.4 nM glucagon, which is close to normal physiological concentrations of plasma glucagon (Table 7.1). There are, however, some differences between birds and mammals. Whereas glucagon causes a decrease in plasma fatty acid levels in rat plasma, elevated levels are observed in birds such as domestic fowl, ducks and pigeons (see Ballantyne, John & George, 1988). The difference can be accounted for by uptake and increased rate of mitochondrial oxidation of fatty acids in the form of acyl carnitines in mammals, and by decreased rate of oxidation of long-chain fatty acyl carnitines by pigeon liver mitochondria (Ballantyne et al., 1988).

When broiler chickens are selected over a number of generations for high body weight ('fat' line) and for food efficiency ('lean' line), they show differences in sensitivities to glucagon. Glucagon stimulated release of glycerol from adipose tissue in the 'lean' line, but not in the 'fat' line (Buyse et al., 1992). When adipose tissue from both lines was incubated in vitro with dibutyryl-cAMP (a cAMP analogue resistant to hydrolysis by phosphodiesterase), lipolysis was stimulated in both tissues to an equal extent. Buyse et al. (1992) conclude that the 'fat' line must have a lower density of glucagon receptors than the 'lean' line. There is also evidence that growth hormone affects lipolysis in adipose tissue from domestic fowl (Campbell & Scanes, 1987). When explants of adipose tissue are incubated with glucagon, there is a rapid but short-lived lipolytic response, but when incubated with both glucagon and growth hormone the lipolytic response is initially less pronounced but more prolonged. This would enable a more prolonged release of glycerol and fatty acids, important during chronic dietary stress.

Lipid catabolism is often associated in mammals with an elevation of ketone bodies in the plasma. In birds, the only ketone body detectable in plasma is β-hydroxybutyrate, and although its concentration increases in fasting as fat stores are mobilised, glucagon infusions do not promote a detectable increase in the plasma level (Epple & Brinn, 1987).

7.6.4 Avian pancreatic polypeptide

The tissues having APP receptors have been studied using ^{125}I-labelled APP, either in in vivo perfusion studies, or in vitro binding studies using membrane preparations (Adamo & Hazelwood, 1992). APP can be labelled either in Gly-1 (N-terminus) or in Tyr-36 (C-terminus). APP has high-affinity binding sites on adipocytes from domestic fowl ($K_d = 5.6 \times 10^{-9}$ M) where it has been shown to inhibit glucagon-stimulated lipolysis (Adamo & Hazelwood, 1990). In vivo perfusion studies using C-terminal ^{125}I-labelled APP failed to reveal binding, although N-terminally labelled APP became bound to some tissues (Adamo & Hazelwood, 1992). In vitro binding studies showed little difference between N- and C-terminal labelling of APP. Cerebellar and spleen membranes exhibited most specific binding, whereas no binding was detectable with membranes from pancreas, duodenum, proventriculus or cerebral cortex. Liver membranes bound C-terminally, but not N-terminally labelled APP, whereas brain and spleen membranes of domestic fowl bound APP whether it was labelled at the N- or C-terminus. Adamo & Hazelwood (1992) suggest that binding may involve primarily the N-terminal portion of APP, so that the introduction of the bulky iodine atom may have prevented binding in the case of liver. Cross-linking studies indicate that APP binds a 67 kDa receptor protein in brain membranes (Hazelwood, 1993a). There are few studies on the role of APP, but the evidence so far suggests that it may have an antilipolytic effect (John et al., 1989). APP injections cause an increase in hepatic glycogenolysis, but this is probably the indirect effect of stimulating glucagon secretion (Cramb & Langslow, 1984). Insulin does not exert an antilipolytic effect in birds, as it does in mammals, and Hazelwood (1993b) has suggested that APP, by exerting an antilipolytic effect, influences the distribution of lipids in birds rather as insulin does in mammals, although further work is required to clarify this. In addition a number of physiological effects have been ascribed to APP (see Hazelwood, 1993a)

7.7 The adrenal hormones

The second group of hormones that have important effects on the control of carbohydrate and lipid metabolism are those secreted by the adrenal gland. There are two distinct types of hormone produced, the catecholamines, which are synthesised from the amino acid tyrosine, and the corticosteroids, synthesised from cholesterol. Within the adrenal gland, there are two distinct types of tissue: *chromaffin*, which is neural tissue homologous to that of sympathetic ganglia, and *steroidogenic tissue*, which is of mesodermal origin (Bentley, 1982). In avian adrenal glands, the chromaffin tissue, which synthesises catecholamines, occurs as small islands of cells comprising 15–25% of the total tissue of the gland. The chromaffin tissue is distributed throughout the steroidogenic tissue, unlike the gland in the eutherian mammal, where there is a distinct medulla and cortex, in which catecholamines are synthesised in the former, and corticosteroids in the latter (Holmes *et al.*, 1991). Adrenaline and noradrenaline are synthesised in two distinct cell types (Unsicker, 1973). The proportions of adrenaline to noradrenaline vary considerably between species. At one extreme the Javavese cormorant has only noradrenaline-secreting cells, whereas the house sparrow has 95% adrenaline-synthesising cells. The domestic fowl, the pigeon and the golden oriole have 80%, 57% and 21% noradrenaline-synthesising cells, respectively (Gapp, 1987). It has been postulated that the more primitive species have less efficient methylation systems and, hence, a higher proportion of noradrenaline. The conversion of noradrenaline to adrenaline is catalysed by noradrenaline *N*-methyltransferase (EC.2.1.1.28; reaction 7.1). This enzyme increases in activity at E13 to E15 in domestic fowl embryo, and this corresponds with a rise adrenaline concentration (see Freeman, 1983).

$$S\text{-adenosyl-L-methionine} + \text{noradrenaline} \rightarrow S\text{-adenosyl-L-homocysteine} + \text{adrenaline} \quad (7.1)$$

The four corticosteroids, corticosterone, cortisol, cortisone and aldosterone, are structurally and biosynthetically related (see Fig. 10.1, p. 161). The principal ones are corticosterone and aldosterone. The former is present in the highest concentrations in plasma. The ratio corticosterone:aldosterone in plasma is approximately 14:1 in domestic fowl, and 63:1 in the duck (Harvey, Scanes & Brown, 1986). The enzymes together with P-450 cytochrome, which catalyse the biosynthesis of corticosteroids, are present on the endoplasmic reticulum.

Unlike mammals, where aldosterone has a mineralocorticosteroid function and the other three have glucocorticoid functions, the avian corticosteroids all have overlapping activities. Therefore, all corticosteroids and catecholamines have some role in the regulation of carbohydrate and lipid metabolism. Exposure to stress stimulates the release of catecholamines and also corticosteroids. The adrenal glands are larger in birds of marine habitats. The glaucous-winged gull reared with only salt water to drink has much larger adrenals than those given fresh water (Holmes, Butler & Phillips, 1961). This is indicative of the importance of its mineralocorticoid function.

7.8 Adrenaline and noradrenaline receptors

Adrenaline and noradrenaline interact with receptors present on the plasmalemma of target tissues and cells. The receptors are referred to as α- and β-receptors, and they are classified according to the potency of various agonists (stimulating) and antagonists (inhibiting) that elicit responses. More recently, the two main classes have been subdivided, largely on the basis of mammalian studies, into at least four classes (Table 7.4). Both β_1- and β_2-receptors are linked to the effector adenylyl cyclase, via the *transducer G_s-protein* and their stimulation thus increases intracellular levels of cAMP.

$$\text{ATP} \rightarrow \text{cAMP} + \text{PP}_i$$

The α_1-receptors activate phospholipase C (1-phosphatidylinositol-4,5-bisphosphate, EC.3.1.4.11), through an as yet unknown transducer. *Phospholipase C* acts on phosphoinositides to catalyse the formation of diacylglycerol and inositol triphosphate.

$$\text{1-phosphatidylinositol}$$
$$\text{4,5-bisphosphate} + H_2O \rightarrow \text{1,2-diacylglycerol} + \text{inositol}$$
$$\text{1,4,5-triphosphate phosphate}$$

Diacylglycerol, in turn, activates protein kinase C, which catalyses the phosphorylation of a number of intracellular proteins:

$$\text{protein} + \text{ATP} \rightarrow \text{phosphoprotein} + \text{ADP}$$

Inositol triphosphate promotes the release of intracellular Ca^{2+}, principally from the endoplasmic reticulum.

The α_2-receptors inhibit adenylyl cyclase, acting via the protein transducer G_i. They also stimulate the Na^+/H^+ antiporter, but the transducer for this system is unknown (Lefkowitz & Caron, 1988). Less is known about avian α- and β-receptors than

Table 7.4. *Classification of adrenergic receptors*

Type	Agonists (order of potency)	Other agonists	Antagonists	Probable mode of action
α_1	Adr $\geq$ Nor $>$ Iso	Phenylephrine	Prazosin	Activation of phospholipase C, transducer unknown
α_2	Adr $\geq$ Nor $>$ Iso	Clonidine	Yohimbine	Inhibition of adenylyl cyclase, transducer G_i-protein
β_1	Iso $>$ Nor $\approx$ Adr		Atenolol, practolol	Activation of adenylyl cyclase, tranducer G_s-protein
β_2	Iso $>$ Adr $\gg$ Nor		Butoxamine	Activation of adenylyl cyclase, transducer G_s-protein

Adr, adrenaline; Nor, noradrenaline; Iso, isoproterenol.
Source: Lefkowitz & Caron (1988).

mammalian receptors, although the β-receptor on the turkey erythrocyte membranes has been used extensively as a model system for studying signal transduction (Yarden *et al.*, 1985). It is easy to isolate erythrocyte membranes, and preparations from turkey remain very responsive to activators of the membrane transducer G-proteins. The turkey erythrocyte β-receptor is one of the family of transmembrane proteins that has seven transmembrane domains (Fig. 7.4). This family also includes rhodopsin and the receptors for acetyl choline, thyrotropin and angiotensin. The turkey erythrocyte β-receptor was one of the first seven-span transmembrane proteins to be sequenced. It is a 50 kDa glycoprotein. The N-terminus is on the extracellular side and contains the catecholamine-binding site. [^{125}I]Iodocyanopindolol has been used as an affinity label to identify the receptor-binding site (Eshdat, Chapot & Strosberg, 1989). Glutamate or aspartate residues are present at the binding site, and the asparagine-linked oligosaccharides are also close to the binding site (Cervantes-Oliver *et al.*, 1985). The C-terminal domain, which is on the intracellular side is unusually long ($\sim$ 139 residues) compared with other homologous seven-span transmembrane proteins, e.g. 25 amino acid residues in muscarinic receptors (Luxembourg, Hekman & Ross, 1991; Parker & Ross, 1991).

Turkey erythrocytes also possess a *purinergic receptor*, which is linked through a G-protein to a phosphoinositide phospholipase C (Ruiz-Larrea & Berrie, 1993). Purinergic receptors are activated by extracellular ATP, ADP and AMP, which occur in low concentration outside cells and have physiological effects. They are known to be released by exocytosis from adrenal chromaffin granules.

Phospholipase C has been purified and is a 150 kDa protein. The β-receptor is directly coupled to the guanine nucleotide-dependent phospholipase C. Thus, the β-adrenergic receptor can couple independently to both adenylyl cyclase and to phospholipase C (Rooney, Hager & Thomas, 1991). Isoproterenol, an adrenaline agonist, stimulates both adenylyl cyclase and phospholipase C. Both purine-receptors and β-receptors may activate phospholipase C by similar mechanisms. The turkey erythrocyte β-receptor is often described as β_1 type, but it shows significant differences both in its mechanism of desensitisation and in its agonist sensitivity from the mammalian receptors. Incubation of turkey erythrocyte preparations with catecholamines causes desensitisation of the β-receptor, and this is accompanied by phosphorylation of serine and threonine residues (Parker & Ross, 1991).

Adrenaline and noradrenaline function both as endocrine secretions and as neurotransmitters. Most of the adrenergic receptor sites that have been studied in birds, other than the β-receptor on avian erythrocytes, have been neurotransmitter receptors. Beta-adrenergic receptor sites have been mapped in the spinal cord and cerebellum of the domestic fowl using a fluorescent β-adrenergic antagonist 9-amino-acridin-propanolol (Bondok, Botros & El-Mohandes, 1988). An α_2-agonist, [^{125}I]clonidine, has been used to detect α_2-receptors in the hypothalamus of the Pekin duck (Müller & Gerstberger, 1992). Receptors were detected in a number of regions of the hypothalamus, including the the antidiuretic hormone-synthesising nucleus paraventricularis. This supports the physiological observation that release of anti-diuretic hormone is modulated by stimulation with α_2-adrenergic agonists. Quantitative autoradiography, using the α_2-adrenergic agonist p-[^{3}H]-amino-clonidine

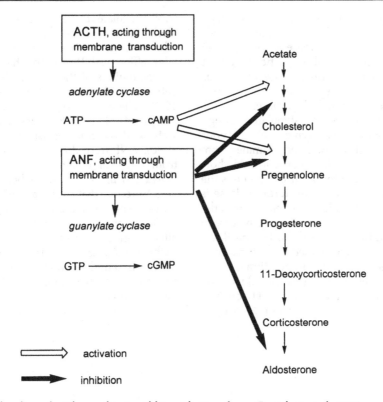

Fig. 7.5. The effect of cyclic nucleotides on the steroid biosynthetic pathway (Rosenburg *et al.*, 1989).

has been used to detect α_2-receptors in the midbrain of the quail (Ball & Balthazart, 1990). Regions of the midbrain contain a high density of α_2-receptors, and there is a sex difference, the males having the higher density. The number of receptors in the male is reduced after castration and restored by administering testosterone.

7.9 Effects of catecholamines on metabolism

The principal effects of adrenaline and noradrenaline secretion on metabolism are stimulation of glycogenolysis and lipolysis, and the inhibition of fatty acid synthesis and glycogen synthesis. The tissues most affected are liver and skeletal muscle, and adipose tissue to a lesser extent. Many of the effects of adrenaline are similar to those of glucagon, but in birds adrenaline is a less potent agent. Adrenaline and noradrenaline are released in response to a 'fright and flight' situation, whereas glucagon is released in response to nutritional stress. In each case, the effects are mediated through the intracellular concentration of cAMP, which stimulates phos-

phorylation of the key enzymes by activation of cAMP-dependent protein kinase. Corticosterone, the principal corticosteroid, has a slower acting mechanism, affecting the induction or repression of protein synthesis, rather than the covalent modification of existing enzymes. Corticosterone generally has anabolic effects on liver and kidney and catabolic effects on other tissues. It promotes protein catabolism in skeletal muscle and increased lipogenesis and gluconeogenesis in liver. Plasma corticosterone concentrations rise in winter populations of Harris' sparrows, and in dark-eyed juncos when there is a sharp drop in temperature. This is associated with increased feeding and lipogenesis. Corticosterone may mediate these responses (Rogers *et al.*, 1993). When turkeys were given an infusion of physiological concentrations of corticosterone for 5 h, there was no significant increase in the plasma glucose. This is attributed to the slower acting nature of the response, requiring protein synthesis and eventually producing a modest increase in gluconeogenesis (Thurston, Bryant & Korn, 1993). By contrast, an infusion of physiological concentrations of adrenaline, or equivalent concentrations of isoproterenol (β-agonist), cause a marked increase in plasma glucose levels within 1 h. In this case the increase is the result of increased hepatic glycogenolysis. The α-agonist phenylephrine did not affect the glucose levels, thus confirming that it is the β-receptors on the hepatocyte membrane that are the stimulatory route.

Secretion of adrenal hormones is stimulated through a number of mechanisms. The sympathetic nervous system triggers the release of catecholamines from the chromaffin granules. Adrenocorticotrophic hormone (ACTH) from the anterior pituitary causes the release of corticosterone, and aldosterone release is controlled by renin and angiotensin release from the kidney. An acute dose of ACTH causes an increase in plasma glucose and free fatty acids both in quail and domestic fowl. This is accompanied by an increase in circulating corticosterone and glucagon (Bray, 1993). It is probably the latter that has greatest influence on plasma glucose and free fatty acids. ACTH, together with another hormone atrial natriuretic peptide (ANP), regulates the release of aldosterone in domestic fowl. ANP is a 29 amino acid residue peptide released from the atrium of the heart (Inagami, 1989). High-affinity ANP receptors have been detected in cardiac regions of the quail with K_d values ranging between 0.03–0.23 nM (Cerra, Canonaco & Tota, 1993). There are also specific ANP receptors in the salt glands of the Pekin duck, where ANP controls water and salt homeostasis (Schütz & Gerstberger, 1990). ANP from domestic fowl differs in sequence from those of mammalian origin at 10 positions (Miyata et al., 1988). Rosenberg, Pines & Hurwitz (1989) propose that ACTH and ANP act via membrane transducers to activate adenylyl cyclase and guanylyl cyclase, respectively. The cyclic nucleotides, in turn, affect various stages in the steroid biosynthetic pathway (Fig. 7.5). The adrenal hormones, in general, are released under conditions of stress, and it can be seen that they receive signals both from the nervous system directly and also from other hormones released from the pituitary, kidneys and heart.

Part 2

The avian genome and its expression

8

The avian genome

8.1 Introduction

The genome comprises all the genetic material present in the haploid cell and, therefore, includes all the nuclear and mitochondrial genes. Two basic approaches that have been used to study genome structure and organisation can be described as cytogenetic and molecular genetic. The cytogenetic approach began with the differential staining of the giant polytene chromosomes from dipteran salivary glands in the 1930s. However, it was not until the late 1960s that differential staining methods were developed that could distinguish banding patterns in normal mitotic chromosomes (Caspersson *et al.*, 1968). More selective staining methods have allowed more detailed resolution of banding patterns; and these have made considerable impact on understanding the organisation of genomes at a microscopic level.

The second approach has been to study the genome at a molecular level. This can be done directly by analysing the DNA, or indirectly by studying the products of transcription and translation, i.e. mRNAs and proteins. Nucleic acid hybridisation techniques, developed since the late 1960s, and DNA sequencing, developed in the mid-1970s, have been central to the molecular approach. Studying DNA sequences provides detailed information about the structure and organisation of genes, whereas studying the number of different mRNAs and proteins synthesised in a cell helps to estimate the number of genes being expressed in a particular cell or tissue.

These different approaches, the cytogenetic and the molecular, complement one another; by using the former the whole chromosomes can be examined *in situ*, but to a lower resolution than by adopting a molecular approach. The latter enables individual genes to be examined in much more detail. So far the cytogenetics of many avian species have been studied, e.g. the karyotype of approximately 700 of

the 8000 species has been determined, although detailed banding patterns have been studied in far fewer species (Christidis, 1990). By contrast, DNA sequencing of genes has been largely restricted to a few domesticated species. In the domestic fowl, which has been most intensively studied, over 400 genes have been sequenced, or partly sequenced.

8.2 Genome size

All cells from a typical eukaryote have the same amount of DNA per haploid set of chromosomes. With the exception of germ cells, most cells are diploid and contain twice the haploid amount of DNA, but there are a few cells with more than twice the haploid amount, e.g. the liver contains a number of polyploid cells. The amount of DNA in all the diploid cells of an individual organism is constant, and there are only small variations in the amount of DNA per cell within a given species. One might expect that, because the genetic information is carried on DNA, the amount of DNA per cell would increase directly in relation to the complexity of the organism. This is generally the case in prokaryotes. Mycoplasma, which are intracellular parasites, have the genetic information for only about 350 proteins, close to the minimum to support autonomous life, whereas more complex bacteria may have up to 8000 genes. The amount of DNA in prokaryotes is directly related to the number of genes that the DNA encodes.

Eukaryotes are less straightforward. In eukaryotes, there is an 80 000-fold range in genome size, but probably only about a 50-fold difference in the number of protein-encoding genes. The genome size is often referred to as the *C value* (C standing for constant or characteristic), and this is defined as the amount of DNA in the haploid genome. The C value is usually expressed in picograms DNA per haploid cell, or kilobases (kb) DNA per haploid cell. There appears to be no simple relationship between C value and organismic complexity (often referred to as the C value paradox), e.g. some unicellular protozoans have higher C values than some mammals. Also, there are sometimes large differences in the C value between closely related species. The range of C values in vertebrates is illustrated in Table 8.1. In some vertebrate groups the range is very wide, but an even wider range is found within different species of plants, i.e. 6000-fold. The variation in the C value within mammals, reptiles and birds is much narrower, i.e 4-fold at most and only 1.9-fold in birds. In many species, only a small proportion of the total DNA codes for protein, and this in part accounts for the paradox.

C values have been determined in over 130 species of birds by four research groups (Mirsky & Ris, 1951; Bachmann, Harrington & Craig, 1972; Venturini, D'Ambrogi & Capanna, 1986; Tiersch & Wachtel, 1991). The C values are compared with domestic fowl as standard, using the value of 2.5 pg DNA/diploid nucleus from domestic fowl; a range from 2.0 to 3.8 pg, and a mean $\pm$ standard deviation of 2.82 $\pm$ 0.33 pg DNA/diploid cell was obtained for 135 avian species from 17 different orders of birds (Tiersch & Wachtel, 1991). The range and mean value is smaller than for any other vertebrate group (Table 8.1). There appears to be no correlation of the DNA content of different avian species with their evolutionary hierarchy.

If the sizes of different eukaryote genomes are compared with the estimated number of proteins produced in a species, it is apparent that only a small percentage of the total genome (2–3%) is required to code for proteins. The large size of the genome, and the size variation between closely related species (C value paradox) is greatly affected by the amount of non-coding DNA. DNA in eukaryote cells is usually divided into three types: (i) unique sequences, of which there are only single copies, (ii) moderately repetitive sequences (up to about 1000 copies), and (iii) highly repetitive sequences (10^3 to 10^5 copies). There is some variation in the copy number used to define the moderate and highly repetitive groups by different workers. The proportions of these three classes are usually estimated by measuring the rates of renaturation of denatured genomic DNA (see Lewin, 1994). The proportion of each of the three types of sequence in a given taxon has an important bearing on the C value.

Several determinations of the amounts of the different types of DNA have been made in avian species; most work has been carried out in the domestic fowl (Schultz & Church, 1972; Epplen et al., 1978; Arthur & Straus, 1983; Venturini, Capanna & Fontana, 1987). Although there is some variation in the results from different workers, the range for unique sequence DNA from most recent studies lies between 60 and 75%, for moderately repetitive DNA between 10 and 20% and for highly repetitive DNA between 10 and 20%. Avian species have a smaller proportion of repeat sequences compared with other vertebrates (Table 8.1). It is possible to estimate the proportion of the unique sequence DNA needed to code for all gene functions, but this requires certain assumptions to be made and can only be approximate. The total number of different mRNAs produced in a eukaryote cell is of the order of 50 000. An average size protein contains about 300 amino acid residues; therefore, it requires 1800 bp of double-stranded DNA, corresponding to 300 codons. A typical eukaryote gene also contains introns, a promoter and leader sequence, and various control regions (see Fig. 8.1 and Section 8.5). This might increase the amount of DNA per gene by about 3-fold. The number of base pairs accounted for on this basis would be $1800 \times 3 \times 50\,000 = 27 \times 10^7$ which is equivalent to 0.275 pg DNA ($1\,bp = 1.02 \times 10^{-9}\,pg$). The average haploid DNA content of the avian genome is 1.41 pg. Assuming 73% of this is unique sequence DNA (Table 8.1), only about a quarter of the unique sequences are thus accounted for.

Many active genes may exist as multiple copies of the same or similar genes, e.g. multigene families (see Chapter 9). These, along with pseudogenes, i.e genes that are not expressed, could account for much of the rest of the single-sequence DNA. Multiple copies of certain genes are well documented, e.g. in the domestic fowl there are between 200 and 290 copies of the ribosomal RNA genes, 10 to 20 copies of the histone genes and four copies of the ubiquitin genes (see Stevens, 1986; Sharp & Li, 1987). The organisation of the gene sequences including the intervening sequences and control elements is discussed in Section 8.5, and the organisation of repetitive sequences in Section 8.4.

8.3 The microscopic organisation of the avian genome

The organisation of the genome can be studied at a number of different levels; in this section we examine the organisation that can be discerned using light or fluorescence microscopy. The avian genome is unusual

Table 8.1. *DNA content and types of sequence in vertebrates*

Taxon	C value (pg/haploid cell)	Ratio of sizes (highest /lowest)	Unique sequence (%)	Repeat sequence (%)
Cartilagenous fish	1.5–16.1	10.7	41	59
Bony fish	0.39–142	364	40	60
Amphibia	0.95–86.0	90.5	23	77
Reptiles	1.26–5.45	4.3	59	41
Birds	1.0–1.9	1.9	73	27
Mammals	1.45–5.8	4	61	39

Source: Olmo, Capriglione & Odierna (1989); Li & Graur (1991); Tiersch & Wachtel (1991).

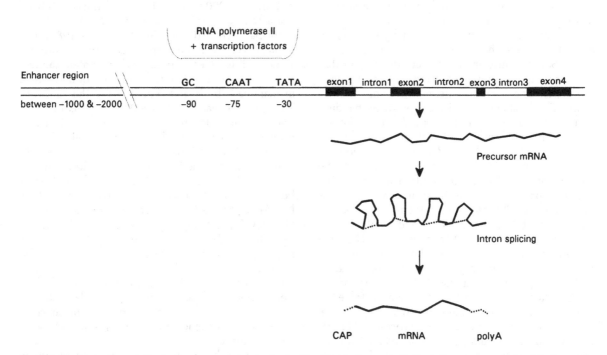

Fig. 8.1. Diagram of a typical eukaryote gene that is transcribed by RNA polymerase II. Control regions upstream of the promoter are involved in the regulation of transcription. The primary transcript is processed to remove the introns. The 5′ cap and 3′ polyA tail are added before the mature mRNA is transported from the nucleus.

in that it is dispersed over a large number of chromosomes that can be subdivided into *macrochromosomes* and *microchromosomes*. The diploid number of chromosomes ranges from 40 in the stone curlew to 126 in the hoopoe, but most avian species have between 76 and 82 (Christidis, 1990). A typical avian species has from 14 to 16 macrochromosomes and 60 to 64 microchromosomes. In the domestic fowl, approximately 70% of the DNA is distributed amongst the macrochromosomes (pairs 1 to 10) and 30%

amongst the microchromosomes (pairs 11 to 39) (Bloom, 1981).

8.3.1 Methods used to obtain chromosome banding patterns and their significance

A number of methods have been developed to stain the chromosomes differentially. Many of these methods, developed in the late 1960s, were initially used to determine the karyotype of various eukaryotes, particularly humans. The *karyotype* is the somatic chromosomal complement of an organism and is usually displayed as a photomicrograph of the metaphase chromosomes arranged in descending order of size, or in diagrammatic form as an idiogram. To recognise each of the different chromosomes from the full complement (e.g. 23 pairs in the case of humans), distinguishing features of each must be found. Initially, the size and position of the centromere constriction can be used, but when two chromosomes of similar size and with similar centromere positions are found, further methods become necessary. It was this that gave the initial impetus to develop differential staining methods. Chromosomes are generally examined after arrest in the metaphase of mitosis, since this is when they appear most distinct.

The first method to be developed, by Caspersson *et al.* (1968), used the fluorescent compound chloroquine. Subsequently methods were developed that did not require fluorescence microscopy and gave a more permanent record, generally using Giemsa stain. These staining methods produce banding patterns that are characteristic of individual chromosomes. This not only enables chromosomes to be identified but also allows detection of deletions, inversions and translocations of regions of the chromosomes. Although the molecular basis of the banding patterns is not fully understood, it is becoming possible to relate them to particular features of DNA composition and conformation (Bickmore & Sumner, 1989; Sumner, 1990). The most important staining methods are those which give rise to C-banding, G-banding, R-banding and Q-banding and also $AgNO_3$–AS staining. A summary of the regions stained by different stains is given in Table 8.2.

Chromatin can be divided into two categories on the basis of the various staining procedures (Table 8.2): the densely staining *heterochromatin*, and the much more diffusely staining regions known as *euchromatin*. Euchromatin corresponds to regions of the chromosomes that are transcriptionally active, whereas the more densely coiled heterochromatin is transcriptionally inactive. Regions of the chromosomes that are transcriptionally inactive can be divided into three categories: (i) chromatin that is composed of non-coding sequences which are not transcribed, e.g. highly repetitive sequences; (ii) tissue-specific condensed chromatin, e.g. chromatin coding for genes that are not expressed in a particular tissue; and (iii) chromatin that is inactivated during early development and which remains so for many cell generations, e.g. one of the X chromosomes in mammals where the female is XX. The first of these categories is referred to as *constitutive heterochromatin*. Some authors refer to both the second and third categories as *facultative heterochromatin*, but others prefer to use the term tissue-specific condensed euchromatin for the second and to restrict the term facultative heterochromatin to the third category (John, 1988).

The main regions of constitutive heterochromatin occur at the centromeres and at most telomeres. The *centromere* is important for the segregation of the chromatids and is the point at which the spindle fibres attach during mitosis. Centromeres have been studied in most detail in *Saccharomyces*. They contain a sequence of 220–250 bp within which there is an AT-rich repeat region. A highly repetitive sequence comprising 8000 units of 187 bp each has also been characterised from the centromere of chromosome 2 of the Sarus crane (Chen, Lin & Hodgetts, 1989). The *telomeres* are the end regions of each of the chromosomes. All telomeres so far studied contain tandem repeat sequences of five to eight nucleotides, which in mammals, birds and reptiles have the sequence TTAGGG repeated 250–1000 times. It is believed that the role of the repetitive sequences is to allow proper completion of chromosome replication, to prevent 'stickiness' of the ends of chromosomes (which might otherwise cause translocations to occur) and to prevent the ends of chromosomes being degraded by exonucleases (Blackburn, 1991). Nanda & Schmid (1994) studied the distribution of $(TTAGGG)_n$ in metaphase chromosomes of domestic fowl. In addition to being present in the telomeric regions, they were also detected in centromeric and interstitial regions of both macrochromosomes and microchromosomes. The significance of this distribution is not yet clear, but it may be related to the evolution of the avian karyotype. The presence of $(TTAGGG)_n$ sequences may be the result of chromosome fusion or fission.

Facultative heterochromatin, such as that found in the second of the X chromosomes of female mammals (XX), does not appear to be present in birds. In mammals, a dosage compensation mechanism operates, which prevents the expression of one of the X chromosomes in the female; otherwise females would express two copies of any sex-linked genes. In a normal female mammal (XX constitution), each somatic cell has a single *Barr body* composed of facultative heterochromatin. The Barr body is a partially or completely condensed X chromosome that is transcriptionally inactive. There is no evidence for an analogous dosage compensation mechanism in birds. No Barr body has yet been identified in a normal male (ZZ) bird, and evidence from a study of the

Table 8.2. *Principal methods for selectively staining regions of chromosomes*

Method	Reagents used	Microscopic detection	Regions of the chromosome stained
C-banding	Alkali, followed by Giemsa	Light	Heterochromatin; mainly centromeres, also telomeres
G-banding	Saline or trypsin, followed by Giemsa	Light	Mainly AT-rich regions; repetitive DNA, some tissue-specific genes, long interspersed repeated sequences
R-banding	Heat, followed by Giemsa	Light	Mainly GC-rich regions; 'housekeeping' genes, short interspersed repeated sequences
Fluorescence Q-banding	Quinacrine, Dapi or Hoescht 33258	Fluoresence	AT-rich regions, similar to G-banding
Fluorescence R-banding	Chromomycin A_3 or mithramycin	Fluorescence	GC-rich regions, similar to R-banding obtained with Giemsa after heat treatment
Ag–AS staining	$AgNO_3$, followed by ammoniacal $AgNO_3$ and formalin	Light	Nucleolar organiser region; rRNA synthesis

sex-linked gene for aconitase shows that male birds have higher aconitase levels than females (Baverstock *et al.*, 1982), suggesting that no dosage compensation mechanism operates in birds.

8.3.2 Chromosome banding patterns in different avian species

C-banding patterns have been studied in over 100 different avian species (Christidis, 1990). There is significant variation between species, and some variation has been demonstrated within species (de la Sena, Fechheimer & Nestor, 1991). Nevertheless, some important generalisations can be made. In birds, the banding patterns of the macrochromosomes are easier to discern than those of the microchromosomes, and often only the macrochromosomes have been studied in detail. The most conspicuous C-band in all the larger chromosomes is that of the centromere, but other interstitial bands are present in many chromosomes. Telomeric bands are also visible on most chromosomes. An example of an idiogram, that of the C-banded chromosomes in quail, is shown in Fig. 8.2.

The microchromosomes in many avian species show predominantly heterochromatic staining. For example, those from the goose, turkey, quail and domestic fowl have GC-rich heterochromatin, whereas those from pigeon have AT-rich heterochromatin (Fritschi & Stranzinger, 1985; Auer *et al.*, 1987; Mayr, Lambrou & Schleger, 1989; Mayr *et al.*, 1990). The substantial proportion of heterochromatin in microchromosomes led to the suggestion that micro-

chromosomes might differ fundamentally from macrochromosomes and were perhaps a DNA reserve that was not transcribed. However, since then a number of important genes have been located on microchromosomes, e.g. the major histocompatibility complex, the nucleolar organiser region, several oncogenes, thymidine kinase and the gene for dystrophin (Somes, 1988; Bitgood & Somes, 1990; Dominguez-Steglich *et al.*, 1990). It is now generally assumed that the microchromosomes have normal genetic functions.

In birds, the female possesses the heteromorphic pair of chromosomes, having the Z and W sex chromosomes, whereas the male has a pair of Z chromosomes. The W chromosome is much smaller than the Z chromosome, being the size of a microchromosome. In most species examined, the W chromosome also has a high proportion of heterochromatin, usually GC-rich (Fritschi & Stranzinger, 1985; Auer *et al.*, 1987; Mayr *et al.*, 1989; 1990). Chandra (1994) suggests that the W sex chromosome shows the characteristics of an inactive chromosome in somatic cells and could be regarded as facultative heterochromatin.

Idiograms of chromosomes 1–4 and the sex chromosomes from the domestic fowl after chromomycin staining (GC-rich) and after Dapi staining (AT-rich) illustrate the reciprocal staining pattern obtained (Fig. 8.3).

Some of the early studies suggested that there were several nucleolar organiser regions (NORs), but the results obtained for most species using the $AgNO_3$–AS method suggest that there are two

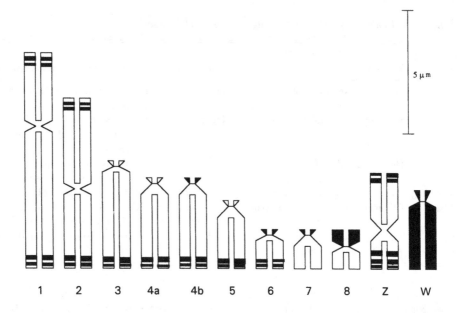

Fig. 8.2. Idiogram of C-banded macrochromosomes of the Japanese quail. 4a and 4b are variants of chromosome 4 (de la Sena *et al.*, 1991).

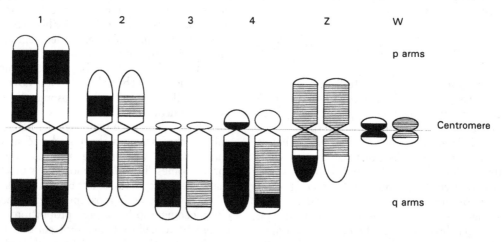

Fig. 8.3. Idiogram of chromosomes 1–4 and the sex chromosomes Z and W from the domestic fowl. Chromosomes on the left of each pair are stained with chromomycin, which shows a preference for GC-rich sequences, whereas those on the right are treated with 4,6-diamidino-2-phenylindol·2HCl (Dapi), which preferentially stains AT-rich sequences (Fritschi & Stranzinger, 1985).

nucleolar organiser regions per diploid cell and that these are located on a pair of microchromosomes (Christidis, 1990). In the domestic fowl, where most detailed studies have been made, the nucleolar organiser region is located on chromosome 17 (one of the microchromosomes) (Auer *et al.*, 1987). In almost all species where the location of the NOR has been studied, it has been found on one of the microchromosomes. An exception is that of the osprey where it is present on chromosome 2 (Kohler, Schaadt & Vekemans, 1989). This is perhaps surprising since in all other species of the order Falconiformes so far studied the NOR is on one of the micro-chromosomes. It may have arisen by a translocation from a microchromosome.

Many of the studies of chromosome-banding patterns in birds have been aimed at understanding the evolution of the karyotype and the taxonomic relationships between species, (see, for example, Stock & Bunch, 1982; Ansari, Takagi & Sasaki, 1986; 1988; Christidis, Shaw & Schodde, 1991). However, they also provide a framework on which to place the results obtained from the molecular biological studies described in the next section.

8.4 Molecular organisation of the avian genome

Cytological studies of chromosomes after specific staining procedures are valuable in giving an overall picture of the banding pattern of each of the chromosomes, although the precise nature of the banding material is not so clear. By contrast, molecular methods have much higher resolving power, but it is sometimes more difficult to locate the chromosomal origin of the material once it has been extracted from the cells or tissues, e.g. the location of a DNA fragment to a particular chromosome.

The genome of the domestic fowl comprises about 70% unique sequence DNA and 30% moderate and highly repetitive sequences (see Section 8.2); the question is how are these sequences organised on the 39 pairs of chromosomes. Various methods have been useful in probing the organisation of the DNA on the chromosomes. In eukaryotes, there is not very much variation in the total GC content of the DNA from different species. It generally ranges between 40 and 45%, by contrast with prokaryotes where the range is from 25 to 75% in different organisms. There are, however, local differences in the distribution of the bases in different regions of the eukaryote genome, which give rise to GC-rich regions. These differences in base composition have been used to fractionate different regions of the genome.

The most widely used method for separating DNAs having different base compositions is by *density gradient centrifugation* using either CsCl or Cs_2SO_4 solutions. Heavy metal ions, e.g. Ag^+ or Hg^{2+}, are sometimes complexed to the DNA preparations (Bernardi, 1993). These have the effect of enhancing the differences in buoyant density of DNA fragments because of their selective binding to certain regions. The buoyant density of fragments of uncomplexed DNA is directly related to their GC content. When DNA is fragmented and centrifuged in a CsCl gradient, the fragments are resolved into bands of different buoyant density.

Bernardi *et al.* (1985) have made extensive studies of the genomes of a wide range of eukaryotes using this method. Before carrying out the CsCl density gradient centrifugation they reduced the average size of the DNA by controlled fragmentation to between 30 kb and 100 kb. This is about an order of magnitude smaller than a typical chromosome band (obtained by using one of the staining methods given in Table 8.2), which may be around 1250 kb. When genomic DNA is fractionated in this way, the pattern of bands obtained (known as *isochores*) is found to differ between organisms (Fig. 8.4), in particular, those from homoiothermic animals, e.g. mouse, human and domestic fowl, and those from poikilothermic animals, e.g. carp and *Xenopus*. The high density bands (H) represent a higher proportion of the total in the homoiothermic animals.

Bernardi *et al.* (1985) interpreted these patterns as suggesting that nuclear DNA from homoiothermic animals consists of a mosaic of alternating light (L) and heavy bands (H), which become sheared as shown in Fig. 8.5. This is consistent with the G- and R-banding patterns obtained from homoiothermic animals, and the paucity or absence of Q-banding found in the chromosomes from poikilothermic animals (Bickmore & Sumner, 1989). The densely stained G-bands seen under the light microscope (AT-rich regions) probably give rise to the L bands.

Having separated the nuclear DNA from domestic fowl into the L_1, L_2, H_1, H_2 and H_3 fractions, it is possible using various probes, such as mRNAs, to locate particular genes. This is done by mixing ^{32}P-labelled mRNAs with the appropriate H or L fractions under conditions in which complementary regions will hybridise to one another (for details, see Adams, Knowler & Leader, 1992). Protein-coding genes are found to be non-randomly distributed throughout the fractions. The highest number of genes are located on the smallest band, H_3, which represents only 3–5% of the total genome. In general, housekeeping genes (including oncogenes) are more abundant in the H isochores, whereas tissue-specific genes are more abundant in the L isochores (Bernardi, 1993). Housekeeping genes encode proteins that are constitutively expressed in all cell types, e.g. citric acid cycle and glycolytic enzymes. The genes for α^A-, α^D-, β-, and ρ-globin from the domestic fowl are present on H_2, whereas those for conalbumin, ovalbumin Y and X (ovalbumin pseudogenes) and vitellogenin are on L_2 (Bernardi *et al.*, 1985). The ovalbumin multigene family is known to be clustered on a DNA fragment of about 40 kb (Royal *et al.*, 1979) which is of a size that would fit on single fragment generated by shearing.

The repeated DNA sequences may be arranged either in clusters or in a dispersed fashion throughout the genome. The percentage of GC present in

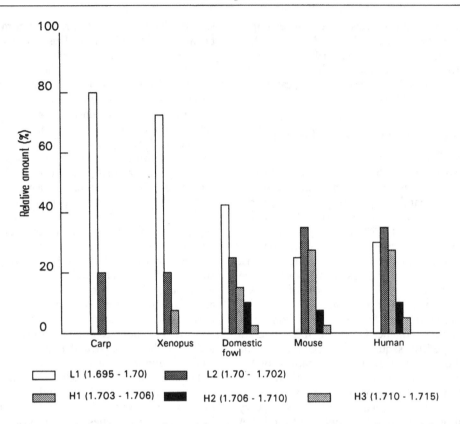

Fig. 8.4. Vertebrate DNA isochores. Histograms showing the relative amounts (%) and buoyant densities of the major isochores, designated H (heavy) or L (light), isolated from carp, *Xenopus*, domestic fowl, mouse and man (Bernardi *et al.*, 1985). Buoyant densities are shown in parentheses.

clustered or tandemly repeated sequences may differ sufficiently from the other DNA fragments so as to form a *satellite band* of DNA in density gradient centrifugation. The roles of the repetitive sequences present in the centromere and telomere are becoming clear (see Section 8.3.1), but that of much of the rest of the satellite DNA is still unknown. The dispersed repeated sequences are found in introns, regions flanking genes and intergenic regions. These sequences are often subdivided into two types, based on the length of their repeat. Short interspersed repeated sequences (*SINES*) are typically less than 500 bp and may be present as 10^5 or more copies. They are generally transcribed by RNA polymerase III and most are retrosequences. Long interspersed repeated sequences (*LINES*) are typically longer than 5 kb and are present as 10^4 or more copies. The most well known of the SINE family is the *Alu* family of the human genome. This is a family of very similar sequences of about 300 bp in length, of which there are about 300 000 copies throughout the genome.

This repeated element was called *Alu* because it contains the sequence AGCT that is recognised by the restriction enzyme *AluI*. The function of the sequence is not clear.

In the domestic fowl there is a family of SINES homologous to the *AluI* family, known as CR1. There are approximately 7000 copies of CR1 per haploid genome (Olofsson & Bernardi, 1983), i.e. CR1 only constitutes about 0.1% of the genome. CR1 sequences have also been detected in duck and common peafowl (see Shapira, Yarus & Fainsod, 1991). They are concentrated almost exclusively in the heaviest of the isochores, H₃. Sequences of the CR1 family have been found flanking a number of avian genes, including ovomucoid, vitellogenin and homeobox genes (Stumph *et al.*, 1981; Haché & Deeley, 1988; Broders, Zahraoui & Scherrer, 1990; Shapira *et al.*, 1991). Baniahmad *et al.* (1987) has shown that CR1 repeats can function as silencer elements (see Section 8.5).

Other short repeat sequences have been isolated

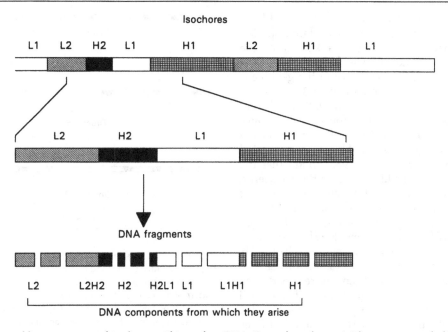

Fig. 8.5. The possible arrangement of isochores within nuclear DNA (Bernardi *et al.*, 1985). The segments with different degrees of shading correspond to stretches of sequences differing in GC content. After random breakage of the DNA, fragments can be separated by density gradient centrifugation. This separates the fragments according to buoyant density which is determined by the GC content.

from various avian species, but their functions are not known. A repeat sequence has been isolated from the nuclear envelopes of erythrocytes of the domestic fowl (Matzke *et al.*, 1990). It consists of tandem repeats of 41–42 bp and accounts for about 10% of the genome. It appears to be present mainly on the microchromosomes and is distinct from, although related to, a 21 bp repeat sequence found in the W chromosome (see Section 8.7); both have AT-rich regions. Tandem repeats of 42 bp have been isolated from DNA from goose, swan and domestic fowl. All three have completely different sequences, but that from domestic fowl has 52% homology with the W-specific repeat (Zhu *et al.*, 1992). The sequences from swan and goose both have binding sites for the transcriptional factor AP2. The presence of certain tandemly repeated DNA sequences forms the basis of multilocus probes used in DNA fingerprinting (see Section 8.9.). A number of microsatellite repeats (sequences of $\approx$ 20 nucleotides, having mono-, di-, tri- or tetranucleotide repeats) have been identified in over 60 different genes from the GenBank database for the domestic fowl (Moran, 1993). These will be useful markers for future gene mapping both in domestic fowl and other avian species.

An alternative method used to study the organisa-

tion of DNA sequences is to fragment DNA and then study the renaturation profiles of the fragments using $C_o t$ plots (for details see Lewin, 1994). Arthur & Straus (1983) used a modification of the $C_o t$ plot method to distinguish parts of the genome that contain unique sequences adjacent to repeat sequences from long tracts of unique sequences that are devoid of repeated sequences. If DNA is cut into sufficiently small fragments so that the repeated sequences are completely cleaved from the unique sequences, then the renaturation profiles will distinguish the unique and repeated fragments. From such a profile, the repeat sequences can be separated by chromatography on a hydroxyapatite column. If, however, a more limited fragmentation occurs, larger fragments containing repeated sequences linked to adjacent unique sequences will be present. Arthur & Straus (1983) used the isolated repeat sequences as a means of distinguishing fragments containing unique sequences interspersed with repeated sequences from fragments containing long tracts of unique sequences with no adjacent repeat sequences. The isolated repeat sequences would readily hybridise with the former, but not with the latter, and so the two fractions could be separated on a hydroxyapatite column.

Prior to this study (Arthur & Straus, 1983), two

kinds of pattern of organisation of DNA had been found when DNA was isolated from different species. The more widely occurring was known as the 'Xenopus pattern', since it was first discovered in Xenopus. In this arrangement most of the genome is composed of SINES of approximately 300 bp interspersed with unique sequences about 0.7–1.1 kb long (Fig. 8.6). The other type, first found in Drosophila, contains much larger units in the interspersion pattern, unique sequences of 13 kb or more, separated by LINES. Arthur & Straus (1983) found that the domestic fowl genome fits neither of these patterns. A fraction making up 34% of the genome consists of unique sequences of about 4 kb long interspersed with LINES, closer to the 'Drosophila pattern', but another 38% of the genome consists of long stretches of unique sequences having a minimum length of 22 kb.

Arthur & Straus (1983) also examined these two fractions to see how structural genes were apportioned between the two. To do this they took the total mRNA from E10 chick embryos and hybridised it separately with the two fractions (for details, see Adams et al., 1992). They found that it hybridised to roughly an equal extent of both fractions, indicating that structural genes are equally distributed between both types.

How do these results relate to those of Bernardi et al. (1985)? Analysis using buoyant density (Bernardi et al., 1985), separates DNA on the basis of base composition, but not on sequence, whereas hybridisation studies (including $C_o t$ plots) depend on sequences (Arthur & Straus, 1983) rather than base composition. The fragment size used by Arthur & Straus (1983) is smaller (0.5–22 kb) than the isochores (30–100 kb) described by Bernardi et al. (1985). The interspersion pattern described by Arthur & Straus (1983) could, therefore, occur within some or all of the isochores described by Bernardi et al. (1985). With the increasing availability of genes cloned from the domestic fowl, it will be possible to design a range of specific hybridisation probes to locate a number of genes both on the isochores (Bernardi et al., 1985) and on the two classes of unique sequence separated by Arthur & Straus (1983).

Apart from the general organisation of unique sequence and repeated sequence DNA within the domestic fowl genome, progress is being made on mapping genes on specific chromosomes. The methods used include fluorescent in situ hybridisation, radioactive in situ hybridisation, chromosome fractionation (Bumstead & Palyga, 1992; de Leon & Burt, 1993; Tixier-Bouchard, 1993). A summary of some of the assignments is given in Table 8.3.

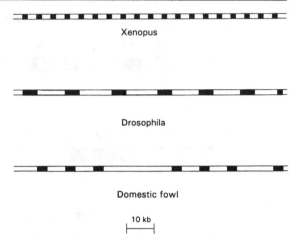

Xenopus

Drosophila

Domestic fowl

10 kb

Fig. 8.6. The possible interspersion pattern of repetitive sequence (shaded blocks) and non-repetitive DNA (open blocks) as found in Xenopus, Drosophila and the domestic fowl. The pattern shown for the domestic fowl only represents about 34% of the genome; another 38% consists of tracts of unique sequences that are not interspersed, and a further fraction of the genome consists of foldback elements and their adjacent sequences (Arthur & Straus, 1983).

In the next sections we examine, first, an example of a structural gene with its associated control regions, and then two examples of repeat-sequence DNA, namely the nucleolar organiser region and the W chromosome. After these, the contrasting organisation of the mitochondrial genes, with what appears to be a much more economical use of DNA, is discussed. Finally, an important technique with potential applications to a wide area of avian biology and which arises from the organisation of the genome, namely DNA fingerprinting, is described.

8.5 Structural genes and their control regions

The length of DNA that constitutes a structural gene is generally much greater than the number of codons needed to specify the amino acid sequence of the encoded protein (see Fig. 8.1). There are two main reasons for this; the first is the presence of non-coding intervening sequences and the second is the requirement for control modules.

Most proteins synthesised in eukaryotes contain intervening sequences, known as introns. These are non-coding sequences that interupt the coding sequence. A large number of genes have been cloned from the domestic fowl, and so far only those coding for rRNA (Muscarella, Vogt & Bloom, 1985), histones

Table 8.3. *Gene assignments to chromosomes in the domestic fowl*

Chromosome	Genes
1	Cytochrome P-450 (androgen aromatisation), progesterone receptor, growth hormone, histone H5, glyceraldehyde 3-phosphate dehydrogenase, β-globin (chromosome 1 or 2), histone multigene family, heat shock protein *HSP108*
2	Ovalbumin, c-*erbB*, c-*myc*, β-actin
3	c-*myb*
4	Phosphoglycerate kinase
5	Tyrosine hydroxylase, transforming growth factor β-3
6	Phosphoglucomutase, phosphoribosylpyrophosphate amidotransferase, serum albumin, vitamin D-binding protein
Z	Ornithine carbamyl transferase, iron responsive element binding protein
Micro	Dystrophin, ovotransferrin, ovomucoid, G-protein (β-subunit), rRNA, myosin heavy chain, adenylyl kinase, β-nerve growth factor, $β_2$-microglobulin, thymidine kinase, hypoxanthine phosphoribosyltransferase, c-*mil*, c-*fes*, c-*ets*, c-*src*, neural cell adhesion molecule, histocompatability antigens, α-globin

Source: Stevens (1991a); de Leon & Burt (1993); Rauen *et al.* (1994); van Hest *et al.* (1994).

(D'Andrea *et al.*, 1985), protamines (Oliva & Dixon, 1989) and ubiquitin (Sharp & Li, 1987) lack introns. On the basis of genes studied in other species, one might also expect the genes for heat shock proteins to lack introns (Lindquist, 1986). For the bulk of genes that do contain introns, the initial transcript is much longer than the final mRNA that becomes translated. Following transcription, introns are excised before translation occurs. For many genes the total length of the introns exceeds that of the exons.

The presence of *control modules* is a second reason for the additional DNA sequences. In order that a gene may be transcribed, the RNA polymerase has to bind to the DNA close to the start of the coding sequence. With eukaryotes there are three different RNA polymerases: RNA polymerase II is the enzyme responsible for transcription of mRNA coding for proteins, whereas RNA polymerases I and III transcribe the genes encoding rRNAs and tRNAs. The control element directing the RNA polymerase II to the coding region is known as the upstream *promoter*, as it usually lies within 100 to 200 bp upstream from the coding sequence (Fig. 8.1 and Table 8.4).

By sequencing a number of normal and defective promoters for different genes, it is possible to determine a consensus sequence for the promoter. Three short sequences constitute the most important elements of the promoter and are present in most, but not all promoters; they are recognised by a group of proteins known as transcription factors. These sequences lie approximately 30, 75 and 90 bp upstream of the initiation site and are referred to as the *TATA*

box, the *CAAT box* and the *GC box*, respectively.

The promoter region is necessary to ensure that transcription of a particular gene occurs; a second group of elements, the *enhancer* and *silencer elements*, regulate the rate at which transcription of a particular gene occurs. Enhancer elements act positively, and silencer elements act negatively. A remarkable feature of both the enhancer and silencer elements is the distance they may be from the structural gene; this can be between 1000 and 20 000 bp. Their functioning is independent of their distance from, and orientation with respect to, the gene. A particular structural gene has the same set of control modules in every cell, although not necessarily in the same control mode (active or inactive). This explains how tissue-specific proteins may arise.

A particular group of enhancers are those that are controlled by hormones and are known as *hormone response elements*. The hormones that directly affect gene expression are those which enter the target tissues, such as steroid hormones, thyroxine and triiodothyronine, and retinoic acid and are discussed further in Chapter 10.

Avian lysozyme is a particularly good example with which to illustrate how such a control system works. The lysozyme gene is expressed in different ways in different tissues of the domestic fowl. In the tubular gland cells of the hen oviduct, expression is induced by steroid hormones (Palmiter, 1972), but the same gene is expressed constitutively in macrophages using the same promoter (Hauser *et al.*, 1981). Its expression is also developmentally regulated,

Table 8.4 *The genetic domain of lysozyme from the domestic fowl and the differences in DNAase sensitivity of the domain in different tissues where the gene exists in differing functional states*

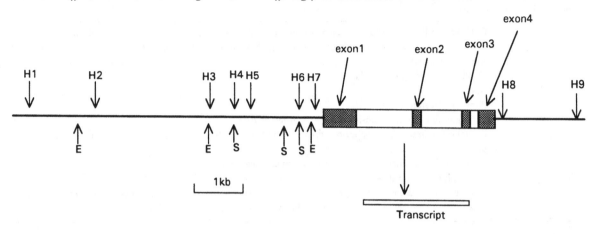

Tissue or cell type	Hypersensitive site number									Functional state of gene	Transcription
	1	2	3	4	5	6	7	8	9		
Immature oviduct +DES	+	+		+	+		+	+	+	Induced	Yes
Immature oviduct −DES	+	+		+			+	+	+	Deinduced	No
Mature oviduct	+	+		+	+		+	+	+	Active	Yes
Macrophage	+	+	+			+	+	+		Constitutive	Yes
Liver, kidney, brain, embryo	+			+				+	+	Inactive	No
Erythrocyte						+				Dormant	No

E, enhancer; S, Silencer; H, hypersensitive site; DES, diethylstilboestrol (an oestrogen analogue); ↓, DNAase hypersensitive site. The closed blocks are the exons and the open blocks the introns. Plus signs in the table means that the site is present.
Source: Gross & Garrard (1988).

serving as a marker for myeloid differentiation in the haemopoietic system and is progressively and selectively activated in the late stages of macrophage differentiation (Cross *et al.*, 1988; Bonifer *et al.*, 1990).

One method which has been particularly useful in helping to identify control modules is to identify the *DNAase-sensitive regions* of chromatin (Gross & Garrard, 1988). Eukaryote chromosomes participate in transcription, replication, meiotic and mitotic condensation, pairing, recombination and segregation. These processes require interactions between nuclear proteins and particular regions of the chromosome. The organisation of the chromosomes is such that certain regions are more exposed, enabling particular proteins to gain access. These regions are much more sensitive to DNAases than the bulk of the chromosomes and may be nucleosome free. They are referred to as

hypersensitive sites and are generally two orders of magnitude more sensitive to DNAase than the rest of the chromatin. They represent a minor fraction (*c.* 1%) of the total DNA. Study of the hypersensitive regions has been very useful in identifying changes in chromatin structure associated with the changed expression of genes, particularly after hormone induction as in the case of avian lysozyme and apoVLDLII (Kok, Snippe & Gruber, 1985).

Using this technique, it has been found that most of the control elements regulating lysozyme expression lie within a region of about 20 kb that includes the gene itself. It can be seen from Table 8.4 that the gene, including exons and introns, occupies about 4 kb. There are nine hypersensitive sites in this 20 kb region and some of these have been identified with enhancer or silencer elements. Those tissues in which the gene is not expressed show fewer hypersensitive sites. Macrophage-specific enhancers have been identified at E-0.7 kb, E-2.7 kb, E-6.1 kb and a silencer at S-2.4 kb (Baniahmad *et al.*, 1987; Grewal *et al.*, 1992). (The convention is to indicate the position in relation to the 5′ end of the transcriptional unit, e.g. E-0.7 kb is an enhancer 0.7 kb upstream of the first exon of the lysozyme gene). The enhancers are generally active in promacrophages and mature macrophages, whereas the silencer is inactive in mature macrophages only. The mechanism by which enhancers increase the rate of transcription is not well understood. As a result of studies on the S-2.4 kb silencer element, Baniahmad *et al.* (1990) suggest that the repression may be mediated by different factors in different tissues.

A second lysozyme gene exists in the domestic fowl, referred to as the goose-type lysozyme since its structure is homologous to that expressed in goose egg-white. It is substantially larger than the chicken-type gene. In the domestic fowl, the goose-type lysozyme gene is expressed in bone marrow and lung but is not detectable in other organs. It has a quite distinct genomic organisation (Nakano & Graf, 1991) with different exon–intron boundaries from those in the chicken-type gene. A TATA-like sequence and two CAAT sequences have been found at -19 to -24, -52 to -55 and -58 to -61, respectively. No steroid hormone receptor-binding sites have been identified within 1.2 kb upstream from the start site.

The lysozyme control elements are among the most studied. Those concerned with the haemoglobin, ovalbumin, and vitellogenin families of genes are described in Sections 9.6, 10.3.1 and 10.4.1.

8.6 The nucleolar organiser region

One group of genes for which there are multiple copies are the rRNA genes. This group can be readily identified as secondary constrictions on metaphase chromosomes after staining with $AgNO_3$–AS stain (Table 8.2): the NOR. The genes are transcribed during interphase of the cell cycle by RNA polymerase I to produce 45S RNA, which is then cleaved to form equimolar amounts of 5.8S, 18S and 28S rRNA (Sommerville, 1986). However, during mitosis, this group of genes exists in an inactive state, appearing as NOR. After mitosis, transcription is resumed and the genes reorganise to form bodies known as nucleoli. Cells may contain one or more nucleoli. In the domestic fowl, it has been shown by hybridisation analysis of erythroid DNA, that there are between 200 and 290 copies of the rRNA genes per diploid

cell, arranged in tandem (Muscarella *et al.*, 1985). In triploid and tetraploid cells there are proportionately more copies. Multiple copies of the rRNA genes are assumed to be required to maintain a sufficiently high rate of ribosome biogenesis to maintain protein synthesis. Chromosome 17, which in the domestic fowl contains the NOR (Auer *et al.*, 1987), also contains the major histocompatibility complex genes, and together they occupy opposite ends of the chromosome (Bloom & Bacon, 1985). In many animals there are several NORs, e.g. six in mice, 10 in humans, located on more than one chromosome, whereas in the domestic fowl there is only a single pair located on homologous chromosomes.

8.7 The W chromosome

The W chromosome is the smaller of the sex chromosomes and is only present in the female. It is the equivalent of the Y chromosome in most other vertebrates, where the male is the heterogametic sex. In most avian species it is the size of a typical microchromosome, although in some species it may be as large as the macrochromosomes. The W chromosome has not been identified in all the avian species studied, e.g. it has not been identified in certain ratites, but there is reason to believe that the ZZ/ZW system of sex determination operates in all birds (Christidis, 1990). The smaller size of the W chromosome as compared with the Z chromosome means that male birds (ZZ) generally have a larger amount of DNA per cell than female birds (ZW). This has been used by Nakamura *et al.* (1990) as a means of sexing birds. They measured the amounts of DNA in propidium iodide-stained nuclei from erythrocytes of male and female birds from 30 different species using a flow cytometer. The differences in the amounts of DNA between males and females varied with the species and was closely related to the differences in the sizes of the Z and W chromosomes. The largest differences were in the Sphenisciformes (3.5% difference), and the smallest in the Casuariformes (0.6% difference).

In the domestic fowl, the W chromosome comprises 1.4% of the DNA present in the nucleus, compared with $\approx$ 10% for the Z chromosome (Tone *et al.*, 1982). Being small, and only present in the one sex, its molecular structure has been easier to study than that of any of the other chromosomes, although at present no genetic loci have been definitely assigned to it. There have been suggestions that a sex-determining antigen, HW, may be associated with the W chromosome (McCarrey *et al.*, 1981), but this has not been confirmed. Two open reading frames have been identified (Kodama *et al.*, 1987) but it is doubtful if they are transcribed. A distinctive feature of the W chromosome, which it shares with the Y chromosome in mammals, is that it is largely although not entirely composed of heterochromatin. The evidence for this is that in some avian species, differential C- and G-banding is observed (Christidis, 1990). G-banding is a general indicator of heterochromatin, whereas C-banding tends to be found mainly in the region of the centromeres and telomeres. Heterochromatin is generally indicative of repetitive DNA, and this is so for the W chromosome from the domestic fowl, which has been studied in detail (Tone *et al.*, 1982; 1984; Kodama *et al.*, 1987; Saitoh, Harata & Mizuno, 1989; Saitoh *et al.*, 1991; Saitoh & Mizuno, 1992).

Two families of repetitive elements were identified when DNA from White Leghorn hens was digested with the restriction endonuclease *Xho*I. Two repeat sequence fragments of 0.7 kb and 1.1 kb were isolated. These were inserted into an *E. coli* plasmid vector and used as probes for *in situ* hybridisation with metaphase chromosomes from White Leghorn hen embryos. They were localised in the W chromosome (Tone *et al.*, 1984). The smaller of these two repeat sequences was found to be 717 bp in length and is repeated 14 000 times throughout the chromosome. The larger repeat is 1.1 kb in length and is repeated 6000 times. Together these make up about half of the total DNA of the W chromosome. Subsequently a third family, consisting of an element 1211 bp in length, was found by using *Eco*RI digestion in place of *Xho*I digestion. This *Eco*RI family together with the two *Xho*I families makes up between 70–90% of the total DNA of the W chromosome (Saitoh *et al.*, 1991).

The smaller (717 bp) family has been studied most. It comprises a sequence of 14 tandem repeats 21 bp in length. Not all the repeats are identical (as can be seen from Fig. 8.7), but a consensus sequence containing a run of $(A)_{3-5}$ and $(T)_{3-5}$ is present in most members of the family. These $(A)_{3-5}$ and $(T)_{3-5}$ have a bearing on the DNA conformation. Early information from X-ray diffraction studies on fibre preparations of DNA suggested that it had a fairly uniform structure of a double helix in which each base pair is spaced at intervals of 0.34 nm along the helical axis (Watson & Crick, 1953). More recent X-ray diffraction studies using oligonucleotide crystals have shown that there are small sequence-dependent differences in the spacing of bases pairs, which in some cases can result in curved DNA. *Curved DNA* is the term used to describe a non-linear helical axis

Position	Sequence
6	GAAATACCACTTTTCTCTC
25	GAAAATCATGCATTTTCATCC
46	AAAAATACCACCTGTCTCCC
66	AAAAATTCTGCACTTCCTTCCC
88	GAAAATACCACTTTTGCGTGG
109	GAAATAACACATTTCTACCC
129	CAAATATAACACGCTTCACTCA
151	CAAAGCACGCATTTTCACCCC
172	GAAAGTACCACTTTTCAGCC
192	GAAAATTACGCTTTTTCCTCCA
214	GAAAATACCACTTCTCAAACA
235	GAAATATCACGTTTCGCCAA
255	GAAAATAGCACCATTCACCC
275	CAAAATCACGCGTTTTCTCTCCA
298	GAACTACCACTTTTCTCAC
317	GGAAATCACACATTTTCTTCCC
339	GAAAGTACCACCTTGCACAC
359	GAAAATCACGCATTTTCTGCGC
381	GAAACAACCCCATTTCACCC
401	CAAAATCAGTCTTTTTCTCCG
423	GAAAATACAACTTTTCTAAC
443	GAAATCCATGCGTTATCACTCT
465	GAAAATACACGTTTTTGCCC
485	GAAAATCACGCATTTTGCCTTC
507	GTAAATTGCCCATGTCGCCA
527	GAAAATAATTCATTTCCTTACC
549	GTAAATGCCCCTTTTCACC
568	CAAAAATCACGATTTCCCCCCG
591	GAAAAACGACCTCTGTCC
610	AAACATCACGCATTTTCTACCC
632	GAAAATACCACTTTTGGCTGG
653	GAAATAACACATTTCTCCCC
673	CAAATATACCACCTTTCACCC
694	CAAAATCACGCATTTTCTCTC

GAAAATAC/ACACNTTTTCTCCC

Consensus sequence

Fig. 8.7. The sequence of the *Xho*I family from the W chromosome of the domestic fowl (Kodama *et al.*, 1987). There are 34 tandem repeat sequences. Each sequence is between 19–22 nucleotides long, giving a total of 715 nucleotides. The consensus sequence, consisting of clusters of $(A)_n$ and $(T)_n$, is shown at the bottom.

arising from intrinsic factors, in contrast to *bent DNA* which is used to describe a non-linear axis arising from extrinsic factors, e.g. DNA-binding proteins. The runs of $(A)_{3-5}$ and $(T)_{3-5}$ found in the consensus sequence cause DNA curvature, and this can be detected by the slower mobility of the DNA fragments on gels (Hagerman, 1990). The other *Xho*I family and the *Eco*RI family also contain 21 bp repeats with a similar arrangement of As and Ts. Elements of the 1.1 kb *Xho*I family will cross-hybridise with those of the 717 bp family, also suggesting structural similarities.

The repeat sequences account for the heterochromatic staining. This seems to be brought about by the curvature of DNA having this repeat sequence, together with a non-histone nuclear protein (W protein), which is able to bind to the repeat sequences with high affinity (Harata *et al.*, 1988). The minimum length of repeated sequence DNA required to bind the W protein is about 300 bp, and this is the length necessary for a complete turn around the protein. DNA footprint analysis has been used to determine the region of the sequence which binds to the W protein. This entailed binding the W protein to the DNA fragment and then subjecting it to digestion with DNAase I. The results showed the periodic appearance of protected sites along the sequence (footprints), suggesting that the DNA wraps itself around the W protein, with the runs of As and Ts facing the protein (Fig. 8.8). There are similarities between this structure and the wrapping of DNA around the histone octamer in the nucleosome structure. This arrangement is further supported by the observation that distamycin, an antibiotic which binds preferentially to AT-rich regions of DNA, inhibits the binding of the W protein (Harata *et al.*, 1988). The location of these repeat sequences on the W chromosome has been studied by *in situ* hybridisation using fluorescence labelled probes of the repeat unit (Saitoh & Mizuno, 1992) and a possible distribution is shown on Fig. 8.9.

Although most work on the W chromosome has concentrated on the domestic fowl, comparisons have been made with pheasants and other domesticated species. When Southern blot hybridisation was performed with restriction enzyme digests made using *Hpa*II or *Msp*I from DNA from the domestic fowl, red jungle fowl and grey jungle fowl, similar patterns were obtained, but a different pattern was obtained with the green jungle fowl, probably indicating a different methylation pattern at the sites around the repeating units (Tone *et al.*, 1984). Similar 0.7 kb sequences were detected in pheasant and turkey by hybridisation at low stringency; the W protein isolated from domestic fowl will bind heterologous repeat units from these species (Saitoh *et al.*, 1989).

During meiosis, the short arms of the W and Z chromosomes pair, and this suggests that they share some homologous sequences (Chandra, 1994). These may be comparable with the pseudoautosomal regions found in X and Y chromosomes of mammals, in which crossing over occurs. The *Xho*I and *Eco*I families have not so far been detected on the Z chromosome, so it may be other regions in which the pairing occurs during meiosis.

8.8   The avian mitochondrial genome

Apart from the bulk of the genes located on the chromosomes within the nucleus, there are also

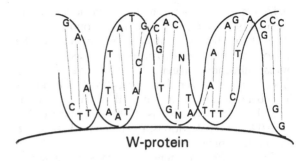

Fig. 8.8. A proposed arrangement for the binding of W-protein to the curved region of the *Xho*I sequence from the W-chromosome (Harata et al., 1988).

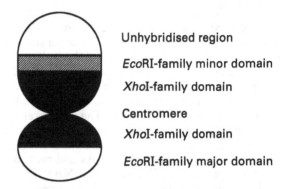

Fig. 8.9. A diagram of the W-chromosome from the domestic fowl showing the possible positions of the *Xho*I and *Eco*RI sequences (Saitoh & Mizuno, 1992).

extranuclear genes within the mitochondria. The latter are able to replicate throughout the cell cycle, unlike those of the nucleus which are confined to the mitotic phase. The mitochondrial genome is very small compared with that of the nucleus, containing approximately 0.3% the amount of DNA of an average-sized microchromosome and 0.001% that of the nuclear genome in the domestic fowl. However, the amount of mitochondrial DNA as a fraction of the total is much larger, because individual mitochondria contain multiple copies of the genome, and there are several mitochondria per cell. Mitochondrial DNA can be separated from the nuclear genome by density gradient centrifugation, because it has a significantly higher GC content (49%) than the nuclear DNA (43%). All mitochondrial genomes so far studied are double stranded, and most are circular. The number of mitochondria per cell varies with cell type, being highest in the tissues most active in oxidative metabolism, such as the heart. Once

DNA-sequencing methods had been developed in the mid-1970s, the small size of mitochondrial DNAs made them attractive subjects for sequencing. The first complete sequence to be resolved was that of the human mitochondrial genome (Anderson *et al.*, 1981). Since then several vertebrate mitochondrial genomes have been sequenced, including that of the domestic fowl (Desjardins & Morais, 1990) and much of the quail (Desjardins & Morais, 1991) and duck (Ramirez, Savoie & Morais, 1993).

The mitochondrial genomes so far studied mostly code for the same genes. The difference between the larger and smaller genomes is mainly in the amount of non-coding DNA present. The vertebrate mitochondrial genomes, which are amongst the smallest, contain only a very small proportion of non-coding sequences. The human mitochondrial genome contains 16 569 bp, of which 15 368 bp code for RNA or proteins. Of the remaining 1201 bp, a substantial proportion is the specific sequences of the replication origin, and only about 90 bp is apparently genetically unimportant material. The genetic material is used very economically, containing no introns, leader sequences or intergenic sequences. The more recently sequenced domestic fowl mitochondrial genome is of very similar size, 16 775 bp (Desjardins & Morais, 1990), and codes for the same RNAs and proteins, namely 13 proteins, 2 rRNAs and 22 tRNAs. Only a small proportion of the mitochondrial proteins are encoded by the mitochondrial genome. The majority are transcribed in the nucleus, translated in the cytosol and imported across the mitochondrial membranes. It is estimated that 69 polypeptides are required for oxidative phosphorylation. By contrast, all the RNAs used by the mitochondria are synthesised *in situ*, since RNA cannot pass across the inner mitochondrial membrane. The 13 proteins encoded on the mitochondrial genome are cytochrome b, two of the subunits of ATP synthetase, three subunits of cytochrome c oxidase, six subunits of NADH dehydrogenase and there is one unassigned reading frame (URF).

In order to synthesise proteins, mitochondria require ribosomes, together with a full complement of tRNAs. To use all of the codons, making allowance for wobble in the third position of the codon, it would be expected that 32 different tRNAs would be required. Mitochondrial translation apparatus has been found not only to have slightly different codon usage (see Table 8.5) but also to use a more extensive wobble ('superwobble') than the cytoplasmic system, and this enables them to incorporate all the amino acids with only 22 tRNAs.

The locations of all the coding positions in the domestic fowl mitochondrial genome is shown in Fig 8.10. The arrangement is very similar to that of human and *Xenopus*, except for the different positions of the genes for tRNA^{Glu}, NADH dehydrogenase subunits 5 and 6 and cytochrome b (Desjardins & Morais, 1990). These changes in the domestic fowl have not been found in any other vertebrate, but they are present in other Galliformes, e.g. quail, guinea-fowl, pheasant, turkey and duck (Desjardins & Morais, 1990; 1991; Ramirez *et al.*, 1993). This suggests that the transposition occurred before the divergence of Galliformes, but more information is required to establish whether this also preceded the separation of birds and reptiles.

Another feature of the mitochondrial DNA that has been extensively studied is its evolution. Brown, George & Wilson (1979) first reported that mammalian mitochondrial DNA undergoes sequence divergence at a rate 5–10-fold higher than that of single-copy nuclear DNA. At the time this was a surprising discovery because small genes having essential functions have been found to be amongst the most highly conserved (for a full discussion of this, see Gray, 1989). For some time, there had been a debate as to whether avian species had a lower rate of evolution than other vertebrates, or whether the period in which they had evolved was shorter than originally supposed (Wyles, Kunkel & Wilson, 1983). Had they evolved from fossil birds with teeth (like *Archaeopterix*) dating from 135 to 165 million years ago, or was there a second round of evolution after a catastrophe about 65 million years ago? Evidence from mitochondrial DNA isolated from different species of geese indicates that it is evolving at approximately the same rate as that from mammals, namely 2% divergence per million years (Shields & Wilson, 1987), and this is also the case with pheasants (Helm-Bychowski & Wilson, 1986), thus favouring evolutionary divergence 65 million years ago. The rapid rate of evolution of mitochondrial DNA, together with the maternal inheritance of mitochondria, has made it particularly useful for the study of evolution of closely related species, where no differences might be expected in the more slowly evolving genes. (NB On fertilisation the sperm transfers only the nuclear genes to the ovum, and thus all the mitochondrial genes are maternally inherited.) It has been possible to confirm evolutionary relationships and generate dendrograms for closely related species and subspecies of ducks (Avise, Ankney & Nelson, 1990), geese (Shields & Wilson, 1987), sparrows (Avise & Nelson, 1989; Zink, 1994),

Table 8.5. *The genetic code of the mitochondrial genome*

First position	Second position				Third position
	U	C	A	G	
U	Phe (AAG)		Tyr (AUG)	Cys (ACG)	U
		Ser (AGU)			C
	Leu (AAU)		*stop*	**Trp (ACU)**	A
				Trp(ACC)	G
			His (GUG)		U
C	Leu (GAU)	Pro (GGU)		Arg (GCA)	C
			Gln (GUU)		A
					G
	Ile (UAG)		Asn (UUG)	Ser (UCG)	U
A		Thr (UGU)			C
	Met (UAU)		Lys (UUU)	*stop*	A
	Met (UAC)				G
			Asp (CUG)	Gly (CCU)	U
G	Val (CAU)	Ala (CGU)			C
			Glu (CUU)		A
					G

The anticodons of the corresponding 22 mitochondrial tRNAs are given in brackets beside the names of the amino acids. The codons that differ from those of the nuclear genome are shown in bold.

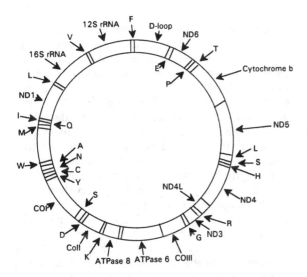

Fig. 8.10. The organisation of the mitochondrial genome from the domestic fowl (Desjardins & Morais, 1990). The genes for the tRNAs for each amino acid are indicated by the single amino acid code. Genes for proteins are indicated thus: Cyt b, cytochrome b; ND, NADH dehydrogenase; CO, cytochrome oxidase; ATPase, adenosine triphosphate synthetase. In each, the subunit number is given as the suffix. The two concentric circles represent the L- (light) and H- (heavy) strands.

tits (Mack *et al.*, 1986), white eyes (Degnan & Moritz, 1992), dunlin (Wenink, Baker & Tilanus, 1993) and auks (Moun *et al.*, 1994) on the basis of their mitochondrial DNA sequences.

8.9 DNA fingerprinting applications

An important analytical technique that has found increasing application in a number of genetic, behaviourial and ecological problems in birds is DNA fingerprinting. (For details of the methods, see Avise 1994.) When restriction fragments of human DNA are separated by agarose gel electrophoresis and hybridised with ^{32}P-labelled *minisatellite DNA probes*, an individual-specific banding pattern results (Jeffreys, Wilson & Thein, 1985), hence the term fingerprint. The principle underlying DNA finger-printing arises from the occurrence of tandemly repetitive DNA sequences dispersed throughout the genome, similar to those described in Section 8.4. Many of the tandem repeats show individual variation, but with a common consensus sequence. The function of these tandemly repeat sequences is unclear, although they are similar to the generalised recombination signal (χ) of *E. coli*. They may serve as recombination signals causing unequal crossing-over during meiosis,

Table 8.6. *Examples of the application of DNA fingerprinting in birds*

Species	Probe[a]	Application	Reference
House sparrow	33.6	Demographic population studies	Wetton et al. (1987)
Willow warbler	L13, L18	Organisation of minisatellite	Gyllensten et al. (1989)
Tits	33.15	Paternity testing	Gullberg, Tegelstrom & Gelter (1992)
Swallow	(TG)$_n$	Paternity testing	Ellegren (1991)
Domestic fowl	M13mp9	Correlation with inbreeding coefficients	Kuhnlein et al. (1990)
	M13	Correlation with susceptibility to Marek's disease	Kuhnlein & Zadworny (1990)
	33.6 & R18.1	Selection for body weight and antibody response	Dunnington et al. (1991)
	33.6	Genetic distances	Haberfield, Dunnington & Siegel (1992)
Common peafowl	cPcr	Paternity analysis and sexual selection	Hanotte et al. (1991)
Hispaniolan amazon	33.15	Estimate degree of inbreeding	Brock & White (1991)
Swan spp.	33.6	Population genetics, evolution	Meng, Carter & Parkin (1990)
Kestrel spp.	M13mp8	Sex-linkage, kinship within captive populations	Longmire et al. (1991)
Various birds of prey	(GGAT)$_4$, (GAGA)$_4$, (CAC)$_5$, (CA)$_8$	Paternity testing for forensic work	Wolfes, Mathe & Seitz (1991)
Antarctic skua	pV47-2	Sex determination	Millar et al. (1992)

[a] The probes listed are multilocus probes, except for cPcr, L13 and L18.

and thereby generating both multiple and variable numbers of copies (Jeffreys et al., 1985).

After the initial discovery using human DNA, the method was tested on a number of other vertebrates using the same two probes, designated 33.6 and 33.15, derived from human DNA. The validity of the method was assessed in birds (Burke & Bruford, 1987) and applied in a demographic study of a wild house sparrow population (Wetton et al., 1987). It has since been applied in studies of parentage and kinship, sex determination, measurement of genetic distance and gene mapping. Examples of these are given in Table 8.6.

The 33.6 and 33.15 probes are referred to as *multilocus* probes, since they hybridise to a number of loci dispersed throughout the genome, although the precise loci have not been identified. For many applications of DNA fingerprinting, the critical factor is the probability of two individuals sharing the same banding pattern. This depends on the number of bands resolved in the gel and how many of these arise from independent loci. If more than one band arises from a single locus, then the discriminatory power is reduced. When the method is first tested with a new species or taxon using a particular probe, the number

of loci corresponding with the bands should be determined in order to evaluate the resolving power for the particular application under study (Burke et al., 1991). For example, in the house sparrow all the loci that hybridised with the 33.15 probe are dispersed throughout the genome (Burke & Bruford, 1987), and hence each band corresponds to one individual locus, whereas the DNA fingerprints from the red grouse contained 16 bands, 12 of which correspond to a single locus (Burke et al., 1991). This latter finding significantly increases the chances of two unrelated individuals having the same banding pattern.

DNA fingerprinting using multilocus probes cannot be used to estimate relatedness of two particular individuals with very great precision, nor can it be easily used in gene mapping. For these single locus-specific probes are needed. A combination of about five different locus-specific probes would provide more precise information, and locus-specific probes can be used to map genes by determining the recombinant frequency between the gene in question and the presence of particular bands. At present multilocus probes are more readily available, but the situation will change as locus-specific probes are developed.

9

Avian multidomain genes and multigene families: their evolution and function

9.1 Introduction

The existence of a large number of multigene families has become evident since the 1970s, largely as a result of the enormous amount of DNA sequence data available. Hood, Campbell & Elgin (1975) describe multigene families as groups of tandemly repeated genes, characterised by their multiplicity, linkage in clusters, sequence homology and related or overlapping phenotypic function. Searches for members of a multigene family can be made using DNA probes to search for related sequences within the genomic DNA, or, alternatively, the Gene Bank data base for that particular species can be searched to see whether related sequences have already been discovered. Multigene families are generally considered to have arisen by processes such as gene duplication and gene conversion. Two terms are used to describe homologies between genes and gene products. Genes are said to be *orthologous* if their sequences have diverged as the result of a speciation event, but they are *paralogous* if they have diverged as the result of gene duplication or some related process. For example, α-globin from domestic fowl and dog are orthologous proteins, whereas α-globin and β-globin both from domestic fowl are paralogous proteins. In this chapter we are mainly concerned with paralogous proteins.

Dayhoff (1978) has suggested that the proteins ($\approx$ 500 000) expressed by the human genome can be grouped into about 500 superfamilies, each containing about 100 sequences. The term 'superfamily' is used here to distinguish distantly related proteins from closely related proteins. Proteins showing greater than 50% similarity at the amino acid level are considered members of a family, whereas those with only $\approx$ 50% similarity are regarded as members of a superfamily (Dayhoff, 1978). The average avian genome is about one third the size of the average

mammalian genome, although the former has a slightly higher proportion of unique sequence DNA (Table 8.1, p. 119). The number of 'superfamilies' might be slightly less in avian genomes, but nevertheless of the same order of magnitude.

The importance of gene duplication in evolution was first postulated by Haldane (1932) long before any proteins sequences were known, but the first evidence from amino acid sequences for gene duplication of a specific group of proteins was the globin family (Ingram, 1961). The importance of gene duplication and gene evolution is that once a duplicate copy of the gene has been incorporated into the genome mutations can occur that either modify the function or cause loss in function. These are then no longer subjected to the same selective pressure as when only a single gene copy exists. There are varying degrees of duplication from complete duplication of the genome, complete duplication of a chromosome, partial chromosome duplication (sometimes referred to as tandem duplication), to complete and partial gene duplication.

Gene duplication and the related process gene conversion increase the size of the genome and the range of proteins expressed (for details of both processes see Ridley, 1993; Li & Graur, 1991). Two other processes, both of which increase the diversity of function of an organism without necessarily increasing the genome size, are alternative splicing and gene sharing. There are now many examples of alternative splicing found in a wide range of organisms (Breitbart, Andreadis, & Nadal-Ginard, 1987), but the number of examples of gene sharing is more limited (Wistow, 1993). In eukaryotes, DNA is initially transcribed as pre-mRNA, which, after having the 5′ end capped with 7-methylguanosine and the polyA tail added to the 3′end, is spliced at the intron/exon junctions to remove the introns. This is known as constitutive splicing. Since the mid 1980s,

Table 9.1. *Avian protein isoforms generated by alternative splicing of mRNA*

Protein	Species	Splicing mechanism[a]	Isoforms and possible functions	References
Troponin T	Domestic fowl	Cassette	One form with inclusion site present in embryonic cardiac muscle: role in sarcomere assembly; other form present in adult cardiac muscle	Cooper & Ordahl (1985)
Troponin T	Quail	Mutually exclusive	2 exons differ in 38 nucleotides coding for C-terminal region	Hastings, Bucher & Emerson (1985)
Myosin light chain	Domestic fowl	Mutually exclusive	LC_1 predominant in early embryo, LC_3 predominant in late embryo and adult: developmental regulation	Nabeshima et al. (1984)
Collagen $\alpha 1$(IX), and $\alpha 2$(IX)	Domestic fowl	Alternative promoters, alternative polyadenylation sites	2 forms differ by the presence or absence of globular N-terminal domain in cartilage and cornea; different macromolecular organisation in two tissues, several transcripts in chondrocytes	Nishimura, Muragaki & Olsen (1989); Swiderski & Solursh (1992a,b)
Neural cell adhesion molecule	Domestic fowl	Cassette	Isoforms expressed during muscle differentiation differ near C-terminus: glycoprotein-mediated nerve cell interaction	Yoshima et al. (1993)
Proteoglycan PG-M	Domestic fowl	Cassette	2 forms differing by 100 kDa: chondroitin sulphate proteoglycan involved in cartilage development	Shinomura et al. (1993)
Ovomucoid	Domestic fowl	Internal donor site	Proteinase inhibitor in egg-white	Stein et al. (1980)
HMG-14a	Domestic fowl	Cassette, alternative polyA sites	Chromatin structure	Browne & Dodgson (1993)

[a] see Fig.9.1.

it has become apparent that in addition to constitutive splicing, *alternative splicing* can occur. The latter process is important in the regulation of gene expression and generating isoform diversity. Through alternative splicing, several different isoforms of a protein can be expressed from a single gene. Isoforms of a protein generally have equivalent functions, have a large proportion of shared identity but vary at specific domains. Many of the isoforms are expressed at different stages of development and have different tissue distribution. The mechanism and regulation of alternative splicing is not well understood (Smith, Patton & Nadal-Ginard, 1989). The alternative splicing of pre-mRNA can take one of a number of forms (Fig. 9.1); examples are given in Table 9.1 and also discussed under contractile proteins (Section 9.7).

Gene sharing is where a gene acquires a second function without gene duplication. The best examples are the lens crystallins (Section 9.8). This group of proteins is present in high concentrations in the lens and is responsible for its high refractive index. A number of different crystallins have second functions, either as enzymes or as heat shock proteins (Wistow, 1993).

Gene duplication can give rise to families with as few as two members, or as many as hundreds. The result can be many copies of the same gene, e.g. rRNA families, or the copies can change by mutation and new functions arise, e.g. the globin family. In addition to multigene families (Sections 9.3–9.7), gene duplication can lead to the formation of multidomain proteins. A particularly good example of this is the proteinase inhibitor family (ovomucoid and ovoinhibitor) present in avian egg-whites, which is discussed in the next section.

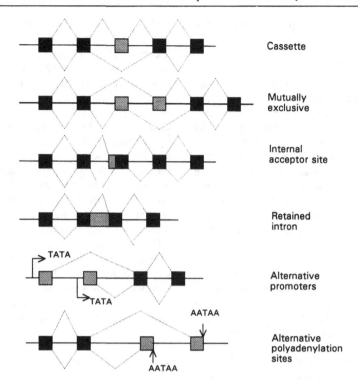

Fig. 9.1. Types of alternative splicing mechanisms (Breitbart *et al.*, 1987).

9.2 Ovomucoids, ovoinhibitors and positive evolutionary selection

Ovomucoid is an abundant protein in egg-white, making up about 10% of egg-white proteins. It has a structure comprising three domains, as is evident from the positions of the disulphide bridges (Fig. 9.2). Its physiological function is not clearly established, although its most striking biochemical property, namely that of inhibiting certain serine proteinases, has led to the suggestion that its role is that of protecting the developing oocyte from bacteria by inhibiting bacterial proteinases. Ovoinhibitor is a related egg-white protein but, by comparison with ovomucoid, has so far has been studied in relatively few avian species (Liu, Means & Feeney, 1971; Takahashi *et al.*, 1992; Wilson & Stevens, 1995). It is much less abundant, making up about 1% of egg-white proteins in those species studied so far. It also is a powerful inhibitor of serine proteinases. Ovomucoid has been studied in depth by Laskowski and co-workers over a number of years (see, for example, Laskowski *et al.*, 1987). Ovomucoid inhibits proteinases by combining specifically with the active site, forming a

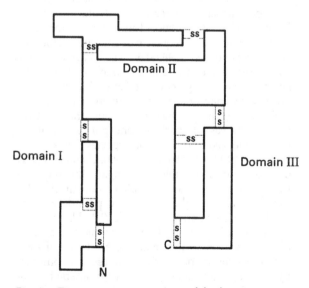

Fig. 9.2. Diagrammatic representation of the domain structure of ovomucoid.

complex in which a specific peptide bond in the ovomucoid domain is cleaved (Bode & Huber, 1992).

$$E + I \rightleftharpoons EI \rightleftharpoons E + I^*$$

where I is the native ovomucoid and I^* ovomucoid after cleavage. The rate of formation of EI from the enzyme and ovomucoid is very rapid compared with the rate of dissociation, either to form $E + I$, or to form $E + I^*$, and this accounts for its inhibitory properties (Empie & Laskowski, 1982). X-ray crystallographic studies have been made of both the native and the cleaved enzyme–inhibitor complex, and the only differences in conformations are that two amino acid residues (at P5 to P3') become more flexible, the rest of the molecule is unchanged (Bode & Huber, 1992).

The features of ovomucoid and ovoinhibitor that are of particular relevance in this chapter are their evolution as multidomain proteins by gene duplication and the hypervariability of the orthologous ovomucoids around the binding site. When the sequence of the three ovomucoid domains from domestic fowl are compared, the first and second domains are more closely related to each other than either is to the third domain (Apostol et al., 1993). In reptiles, ovomucoids are also present in egg-white, and in the alligator the ovomucoid has four domains (see Apostol et al., 1993) of which the first three domains are more closely related to each other than to the fourth. In ovoinhibitor, which has seven domains, the six N-terminal domains are closely related to each other and the C-terminal domain is distinct. Further evidence for an evolutionary relationship between ovomucoid and ovoinhibitor comes from the nucleotide sequence of their genes (Stein et al., 1980; Scott et al., 1987a), in which each domain is coded by two exons and in which introns occur between the coding sequence for each domain. The proposed evolutionary tree is given in Fig. 9.3.

Of the three ovomucoid domains, the third has proved the easiest to cleave and purify, and it has been sequenced from 153 avian species (Laskowski et al., 1987; 1990; Apostol et al., 1993). Using this large data base, it is possible to compare the sequence variability within the 56 residues constituting the third domain and the connecting peptide and the results are summarised in Table 9.2. Unlike most other orthologous proteins in which the most conserved parts of the sequence are those of most functional importance, e.g. active sites, in ovomucoid the binding site is a region of hypervariability. The amino acids comprising the consensus set of enzyme–inhibitor contacts are residues 13 to 21 and 32, 33 and 36, and these show most variability (Table 9.2). Residue 18 in the third domain is crucial in determining the specificity of proteinase inhibition, e.g. Lys-18 inhibits trypsin, Met-18 and Leu-18 generally inhibit chymotrypsin, elastase and subtilisin, Val-18 inhibits elastase, whereas Ala-18 or Gln-18 do not inhibit any of the proteinases so far tested (Kato, Kohr & Laskowski, 1978). This has been demonstrated both by comparing third domains from different species and by semi-synthetic replacement of residue 18 in turkey third domain (Komiyama et al., 1991). Creighton & Darby (1989) suggest that this hypervariability correlates with the hypervariability of certain proteinase active sites and is an example of coevolution under positive selective pressure. The proteinase inhibitor and the proteinase evolve in parallel. However, it is difficult to rationalise this until the physiological role of the proteinase inhibitors is clarified. It does not appear to support the neutral theory of evolution (Kimura, 1983), in which random mutations would be expected to occur throughout the gene.

9.3 Collagen

Collagen is the most abundant protein in the body, comprising about one third of total body protein. It is the main structural protein present in bone, tendons, skin, ligaments and blood vessels. It occurs in almost every tissue, but its properties vary according to its role. For some functions, flexibility is important, whereas others require it to be rigid. Collagen exists as a family of related proteins, which reflect these differences in functional requirements. It is classified into 15 different types, designated type I to type XV in order of their discovery, and these are encoded in a multigene family comprising about 25 genes (Vuorio & de Crombrugghe, 1990). Each type is coded for at a different genetic locus and generally has a characteristic tissue distribution. All types of collagen include the triple helical element as an important part of their structure, but they also have globular domains that vary considerably between different types. They have a high proportion of the amino acids glycine, proline, alanine, and lysine, a repeat sequence Gly–Xaa–Yaa, in which glycine recurs in every third position. Post-translational modification is important in the maturation of collagen fibres. It involves hydroxylation of some of the proline and lysine residues, oxidation of some of the lysine residues, which then form interchain cross-links, formation of disulphide bridges and glycosylation. The cross-links are important in maintaining

Table 9.2. *Amino acid sequences of ovomucoid third domains from 153 species of birds*

Residue number	Amino acids	Contact residue
6	V(151),I(2)	
7	D(150),N(3)	
8	C(153)	
9	S(153)	
10	E(67.5), D(58.5), G(27)	
11	Y(145), H(8)	
12	P(153)	
13	K(142),Q(6),R(2),M(1),T(2)	α β γ
14	P(150), H(1),S(1)	α β γ
15	A(107),V(30),D(8),G(3),T(2),S(1)	α β γ
16	C(153)	α β γ
17	T(111),S(28),L(5),P(4),M(3),A(1),R(1)	α β γ
18	L(65.5),M(60),A(6.5),Q(5),V(4.5),T(3),P(3),K(2),S(1.5),G(1),I(1)	α β γ
19	E(139),D(13),L(1)	α β γ
20	Y(124),D(8),Q(5),H(4),E(4),R(2),S(2),L(2),F(1),N(1)	α β γ
21	R(63),M(72),L(7),F(5),V(4),L(1),T(1)	α β γ
22	P(153)	
23	L(130),V(9),F(8),I(6)	
24	C(153)	
25	G(153)	
26	S(153)	
27	D(151),N(2)	
28	N(119),S(34)	
29	K(132),Q(13),I(4),T(3),E(1)	α
30	T(150),S(2),I(1)	α
31	Y(153)	
32	G(79.5),S(46.5),D(13),N(5),V(1),H(1),R(1),P(0.5)	α γ
33	N(151.5),D(1),S(0.5)	α β
34	K(151),R(1),E(1)	
35	C(153)	
36	N(118),D(18),S(6),A(4),G(2),Y(1),T(1),I(1)	α β γ
37	F(153)	
38	C(153)	
39	N(151),S(2)	β
40	A(153)	
41	V(149),A(3),F(1)	
42	V(141),A(10),M(1),L(1)	
43	E(71),D(74),K(5),Q(2),H(1)	
44	S(148),K(4),R(1)	
45	N(152),S(1)	
46	G(148),V(5)	
47	T(148), deletion(5)	
48	L(144),I(4),deletion(5)	
49	T(148),N(2),S(2),I(1)	
50	L(150),V(2),F(1)	
51	S(133),N(11),R(5),G(4)	
52	H(146),R(5),Y(1),N(1)	
53	F(148),L(3),I(1),P(1)	
54	G(150),E(3)	
55	K(128),E(20),Q(2),R(2),V(1),T(1)	
56	C(153)	β

Numbers in parentheses in the second column indicate the number of species having that amino acid residue. '0.5' is indicative of polymorphic forms. Contact residues between ovomucoid and elastase (α), chymotrypsin (β) and *Streptomyces* proteinase (γ) in the EI complexes.
Source: Apostol *et al.* (1993).

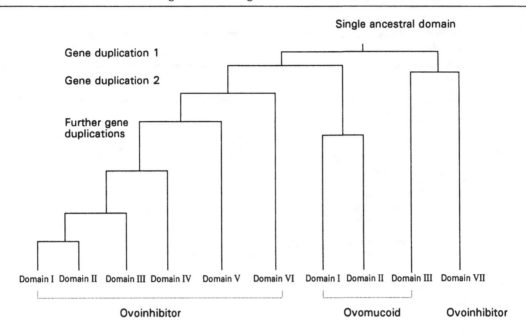

Single ancestral domain

Gene duplication 1

Gene duplication 2

Further gene
duplications

Domain I Domain II Domain III Domain IV Domain V Domain VI Domain I Domain II Domain III Domain VII

Ovoinhibitor

Ovomucoid   Ovoinhibitor

Fig. 9.3. Evolutionary tree for the ovoinhibitor and ovomucoid domains.

supramolecular organisation. The extent of cross-linking varies with the rigidity of the structure for which collagen is a component. Collagens are synthesised on the rough endoplasmic reticulum as the precursor procollagen and are subsequently secreted.

The triple helix may exist as a homotrimer with three identical chains or as a heterotrimer. The nature of the three chains is shown by giving the individual chains (α-chains) an Arabic numeral. Thus, $\alpha1(II)$ is the designation of a single $\alpha1$ chain present in type II collagen. The chain composition for a given collagen type shown as $[\alpha1(I)]_3$, or as $[\alpha1(I)]_2\alpha2(I)$ indicates the type I homotrimer comprising three $\alpha1$s, and the type I heterotrimer comprising two $\alpha1$s and one $\alpha2$, respectively. A further point to note is that the assignment of numbers to chains within a type does not indicate the degree to which these chains are homologous to chains in another collagen type. Thus $\alpha1(II)$ is not necessarily similar to $\alpha1(I)$ or $\alpha1(IV)$.

Collagens can be divided into two major classes: the fibril-forming collagens (types I, II, III, V and XI) and the non-fibril-forming collagens (types IV, VI, VII, VIII, IX, X and XII). Type XIII has been determined from cDNA clones, but the corresponding protein has not yet been isolated. The classification of type XV is not clear. The fibril-forming collagens have long continuous triple helices, aligned in parallel but staggered at one quarter phase relative to one

another. This gives rise to the microscopic banding pattern (Fig. 9.4). They provide structural support in the skeleton, skin, blood vessels, nerves, intestines and in fibrous capsules of organs. The non-fibril-forming collagens are more heterogenous, either forming extracellular matrix networks or interacting directly with fibril-forming collagens (Fig. 9.4). Fibre formation imposes greater constraints on the conformation of individual chains than matrix formation. Although in some tissues, one type of collagen predominates, most have more than one type. Type I collagen is the major type in most tissues, e.g. in bone 90% of the organic matrix is type I collagen. In hyaline cartilage, type II is the main type. Types III and V are found in association with type I in soft connective tissues. Type V is also found in association with type I in bone. A number of the minor types of collagen are thought to regulate the diameters of fibrils for certain specialised functions (see Section 9.3.1). During the differentiation of chondrocytes (cartilage-forming cells), there is a switch from synthesising type I to type II, IX and X procollagen. This is evident from the levels of the procollagen mRNAs during this period (Dozin *et al.*, 1990). Monoclonal antibodies have been used to study the location of different types of collagen within tissues. Swasdison *et al.* (1992) showed that type I and III collagens are colocalised in the connective tissue of domestic fowl heart, aorta, kidney, thymus, and skin,

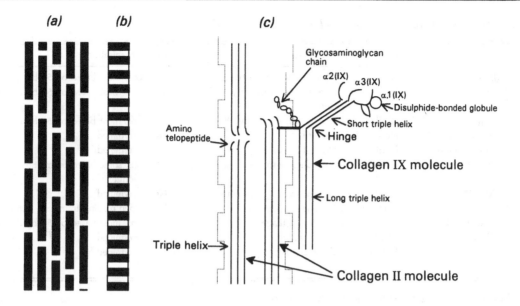

Fig. 9.4. Diagrammatic representation of collagen structure. (*a*) The arrangement of the triple helices in fibril-forming collagens with the quarter stagger of the triple helices giving rise to the microscopic banding pattern (*b, c*) The attachment of non-fibril-forming collagen (type IX) to a fibril-forming collagen (type II) (Nishimura, Muragaki & Olsen, 1989).

but in the gizzard and in tendons a distinct staining pattern for the two types was found. The properties of the main types of collagen isolated from the domestic fowl are summarised in Table 9.3. The fibril-forming collagens have been more extensively studied and will be considered next.

9.3.1 Fibril-forming collagens

The gene organisation for all of the fibril-forming collagens is very similar. They each have a total of 44 exons of similar sizes, suggesting a common evolutionary origin. The size of most of the exons are multiples of 54 bp, and the exceptions are multiples of 9 bp (23 have 54 bp, eight have 108 bp, one has 162 bp, five have 45 bp and five have 99 bp). The exons start with the codon for glycine and end precisely the third base for Yaa in the repeat structure, $(Gly–Xaa–Yaa)_n$. It has been proposed that the ancestral gene was 54 bp long and that the 108 and 162 bp exons have arisen by loss of introns (Vuorio & de Crombrugghe, 1990). The 45 and 99 bp exons may have arisen by recombination. The 54 bp unit probably arose from a 9 bp primordial ancestor, GGTCCTXCT, which underwent duplication and divergence to form a 27 bp domain that then condensed to form the 54 bp domain (Benveniste-

Schrode *et al.*, 1985). The α-chains are synthesised as pro- α-collagen, which includes both N- and C-terminal extensions (Fig. 9.5). The C-peptide is 243–247 amino acid residues and has a globular structure. It is cleaved leaving a short telopeptide of 11 to 27 amino acid residues. The telopeptide is important in forming the interchain disulphide bridges that link the triple helices to one another in staggered formation (Fig. 9.4). It is coded for by exons 49–52. The C-peptide is highly conserved among fibril-forming collagens. In contrast, the N-peptide is more divergent both in length and structure and is made up of a number of units including the signal peptide, a Cys-rich globular domain, three short triple helical domains and an N-telopeptide.

The four α-chains from the three most abundant fibril-forming procollagens in domestic fowl, namely, α1(I), α2(I), α1(II) and α1(III), have been sequenced and found to be closely related. The pair having least sequence differences are α1(I) and α1(II), suggesting they diverged most recently: around 6–9 million years ago (Deák *et al.*, 1985). The gene structures for α1(I), α2(I) and α1(III) from the domestic fowl are highly conserved. The size, sequence and distribution of the exons are very similar (Yamada *et al.*, 1984; Upholt & Sandell, 1986). There is evidence of polymorphism in the pro-α2(I) in White Carneau and

Table 9.3. *The types of collagen characterised in the domestic fowl*

Type	Composition	Class	Distribution	Gene organisation
I	[α1(I)]₂α2(I)	Fibril	Major type in most tissues	44 exons, most based on 54 bp domain
II	[α2(II)]₃	Fibril	Several, but particularly hyaline cartilage	44 exons, most based on 54 bp domain
III	[α1(III)]₃	Fibril	Mainly soft connective tissue	44 exons, most based on 54 bp domain
IV	[α1(IV)]₂α2(IV)	Non-fibril	Gizzard	Not determined
V	[α1(V)]₂α2(V)	Fibril	Highest in cornea	Not determined
VI	α1(VI)α2(VI)α3(VI)	Non-fibril	Cell adhesive to basement membrane	34 exons, some based on multiples of 9 bp, others longer
IX	α1(IX)α2(IX)α3(IX)	FACIT	Hyaline cartilage, vitreous humor	Not determined
X	[α1(X)]₃	Non-fibril	Hyaline cartilage	Long open reading frame, no introns
XI	α1(XI)α2(XI)α3(XI)	Fibril	Hyaline cartilage	44 exons, most based on 54 bp domains
XII	[α1(XII)]₃	FACIT	Chick embryo	Not determined
XIV	[α1(XIV)]₃ ?	FACIT	Chick embryo	Not determined

FACIT, fibril-associated collagens with interupted triple helices.

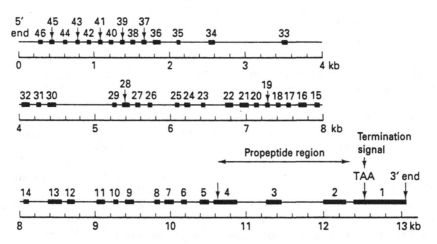

Fig. 9.5. Organisation of the gene for type II procollagen from the domestic fowl (Upholt & Sandell, 1986). The gene comprises 46 exons. Exons 5 to 46 and part of exon 4 encode a strand of the triple helix present in mature collagen. Exons 3 and 2 and parts of exons 1 and 4 encode the region of procollagen that is excised before its secretion to form the collagen matrix.

Show Racer pigeons (Boyd *et al.*, 1991). When *Eco*R1 restriction fragments of the pigeon DNA were separated by electrophoresis and probed with a domestic fowl pro-α2(I) cDNA, there was evidence for two alleles in both varieties. However, in the White Carneau pigeons, both alleles were present in similar amounts, whereas in the Show Racer one of the alleles was present in very much smaller proportions. White Carneau pigeons have a tendency to develop spontaneous aortic atherosclerosis, but this tendency did not correlate with the proportion of the second allele.

The collagen present in the avian corneal stroma is a striated collagen, but with a smaller and more uniform diameter (25 μm) than most fibril-forming collagens, and this is thought to be necessary for corneal transparency. Linsenmayer *et al.* (1993) proposed that this is achieved by a coassembly of collagen types I and V. The latter is present in a much higher proportion in the cornea (15–20% of total collagen) than in other tissues (<5%). Type V collagen is a heterotrimer [pro-α1(V)]$_2$pro-α2(V). Its cDNA has been sequenced.

Collagen type XI is a minor extracellular component of avian cartilage and exists as a heterotrimer, α1(XI)α2(XI)α3(XI). The α1(XI) and α2(XI) are products of distinct genes and have similar amino acid compositions to αI(V) and α2(V), in contrast to α3(XI), which is thought to be a post-translationally modified form of α1(II) (Nah, Barembaum & Upholt, 1992). The exon arrangement is homologous to that of the other fibril-forming collagens, i.e. types I, II and III. Collagen type XI occurs together with type II and may have a similar role to that proposed for type V, namely in fibre assembly and in the determination of their diameter. Collagen α1(XI) mRNA is detectable in a wide range of domestic fowl embryo tissues (Nah et al., 1992).

9.3.2 Non-fibril-forming collagens

These are a much more heterogenous group of collagens. Types IV, VI, VII, VIII and possibly X form extracellular matrix networks, and types IX, XII and XIV interact directly with fibril-forming collagens. There is considerable divergence in their gene structures, although a 9 bp pattern is evident. The proteins themselves generally have short triple helices compared with the fibril-forming collagens, interrupted by globular domains. The globular domains contain modules that have been found in other proteins. Bork (1992) has classified these modules into eight different classes. These modules are important since they account for both their tissue specificity and their binding specificity.

Collagen type IV has been isolated from domestic fowl gizzard, although its gene structure has not been determined. It is a heterotrimer [α1(IV)]$_2$$\alpha$2(IV) which has two globular domains (Mayne et al., 1983). Collagen type VI is one of the minor constituents of the extracellular matrix, comprising a heterotrimer of three different polypeptides α1(VI)α2(VI)α3(VI). The complete structure of α2(VI) from domestic fowl is known (Hayman, Köppel & Trueb, 1991). The overall structure is that of a short triple helix flanked by large globular domains, the latter contributing more than 80% of its molecular mass. Type VI collagen is unusual in that its triple helix has a number of Arg–Gly–Asp tripeptide sequences, which may be used in cell-binding sites. The globular domains are composed of several homologous repeats

of about 200 amino acid residues each. These features together have led to the suggestion that type VI acts as an adaptor molecule for cell adhesion. This may be achieved by cells binding to the central helix of type VI collagen, and this in turn might anchor these cells to the interstitial collagen fibres via its collagen-binding repeats in the globular domain. The genes encoding α1(VI) and α2(VI) comprise 34 exons and 33 introns (Walchli et al., 1992). Both have 19 exons coding for the triple helical component, each of which are multiples of 9 bp (27 bp, 36 bp, 45 bp, 54 bp, 63 bp and 90 bp). There is no evidence for a 54 bp domain, as is found in fibril-forming collagens. Larger exons encode the globular domains.

The gene organisation of type VIII has not been studied in any avian species, but the protein has been detected by immunocytochemistry in developing chick embryo. It is associated with cardiac myoblasts, being first detected at E19 and remaining detectable until hatching (Iruela-Arispe & Sage, 1991). Type IX is a heterotrimer α1(IX)α2(IX)α3(IX) and occurs on the surface of hyaline cartilage and in vitreous humour. The cDNA for each chain has been sequenced (Brewton et al., 1992; Harel et al., 1992). They have three triple helical domains interspersed with globular domains. It occurs cross-linked to type II collagen in cartilage (Fig. 9.4), and this may be important in regulating the fibril diameter. Two forms of type IX are coexpressed during avian limb chondrogenesis that are thought to arise from alternative splicing (see Table 9.1), including alternative promoter use and alternative polyadenylation (Swiderski & Solursh, 1992a,b).

The chrondrocytes of the epiphysial growth plate at the ends of long bones are sites of active synthesis and turnover of collagen. Five different collagen types (II, VI, IX, X and XI) are found at the epiphyseal growth plate. Type X is transiently present at the late stages of differentiation, specifically where cartilage is to be replaced by bone. Genomic and cDNA clones of type X have been sequenced. The collagen comprises a 460 amino acid residue helical domain, with a 170 amino acid residue globular domain at the C-terminus, and this is coded for by a long open reading frame with no introns (Ninomiya et al., 1986). Type XII is a homotrimer of α1(XII) and has a similar sequence to that of type IX. It has been isolated from chick embryo tendons (Dublet et al., 1989). Types IX, XII and XIV have interrupted triple helices and are thought to be involved in controlling the spatial arrangement of collagens within tissues. These are now being classified separately as FACIT (fibril-associated collagens with interrupted triple helices). Type XIV is similar to type XII except for having 16 individual domains rather than 28 (Walchli et al., 1993).

9.4 Keratin: the principal feather protein

Keratins are an important group of structural proteins and are the major gene products of the epidermal tissues that form feathers, scale and claws. They are also synthesised by other tissues but generally in much smaller amounts. Keratins are classified into two main groups, α and β. This terminology arose from the interpretation of X-ray photographs of normal hair (α-keratin) and stretched hair (β-keratin) (see Zubay, 1993). The latter has a combination of β-sheet and β-turn structure (for a general account of protein secondary structure, see Branden & Tooze, 1991), whereas the former comprise α-helices, pairs of which are coiled together to form protofilaments. The principal form of keratin present in feathers, scales and claws is β-keratin, although α-keratins have been detected in much smaller quantities in E17 scale tissue and these are believed to occur in the hinge regions between individual scales (Wilton, Crocker & Rogers, 1985). For a detailed account of the tertiary and quaternary structure of avian keratins, see Gregg & Rogers (1986). Keratins, present in feathers and scales, are insoluble in water, largely because of their covalent attachment to one another and to other matrix proteins, mainly through disulphide bonds. They can generally be made soluble by either reductive or oxidative cleavage of the disulphide bonds, followed by denaturation using either urea, guanidine hydrochloride or SDS. Solubilised keratins can be induced to self-assemble in vitro and form filaments ressembling native filaments (Brush, 1983).

Feather keratins are uniform in size (10.4 kDa), scale keratins are larger ($\approx$ 14.5 kDa) and claw keratins have an intermediate size of 12.2 kDa (Whitbread, Gregg & Rogers, 1991). The most abundant amino acid residues are glycine and serine. Keratins from scale, beak and claw have $\approx$ 30% glycine, whereas feather keratins have $\approx$ 10%. Feather keratins have been sequenced from a number of species (Gregg & Rogers, 1986). The total length of the keratins is highly conserved at 96–7 residues, as is the central part of the sequence from residues 27–93 (Fig. 9.6). Most variations occur at the N- and C-termini. Most of the half-cystine residues, involved in the interchain links, lie in the N-terminal region and none are present in the central region. The central region is rich in hydrophobic residues. The larger size of the scale keratins is accounted for by an additional 52 residues (Fig. 9.6), made up of four repeats of a 13 amino acid residue sequence that contains the Gly–Gly–Xaa motif (Xaa is most commonly Tyr or Phe). This motif accounts for the higher glycine content of claw, scale and beak β-keratins. The similarities in sequence of feather and scale β-keratins suggest they have common ancestry, and since scales are believed to have evolved before feathers, this might have been achieved by deletion of the 52 residue portion.

The most extensive gene sequencing of avian keratins is that of the domestic fowl feather β-keratins, comprising about 20 proteins (Presland et al., 1989a,b). A group of 18 different β-keratin feather genes, together with four claw keratin and three keratin-like genes, have been mapped on a continuous segment of DNA of 100 kb (Fig. 9.6). Each of the feather keratin genes together with the claw keratin and feather-like genes contains a single intron that interrupts the gene 21 or 22 bp prior to the initiation codon. There are almost certainly more scale keratin genes to be found, since a minimum of nine keratin genes are expressed in scale tissue (Wilton et al., 1985). The close linkage of this multigene family, together with the position of the intron, is probably indicative of common ancestry arising from gene duplication. The different keratin proteins and mRNAs expressed in embryonic chick feather suggest that almost all of these genes are expressed. The feather-like gene encodes 115 amino acid residues, 18 longer than those encoding the feather genes. The complete sequence of four of the feather keratin genes has been determined (Presland et al., 1989a) and they show > 90% homology.

High concentrations of these four feather keratin mRNAs are detectable at E15 (3×10^5 molecules per cell) and the same is true of claw keratins. There are several claw keratin genes, one of which has been sequenced (Whitbread et al., 1991). All types of avian β-keratin gene show close similarities, including the presence of the intron, suggesting their origin in a single ancestral gene, but claw keratin is more closely related to scale keratin than to feather keratin. The expression of claw and scale keratin mRNA in embryonic feathers may be the result of 'promoter leakage'. Feather keratin mRNAs are not detectable in other chick keratinising epidermises. Claw keratin genes show greater expression in claw tissue, lower levels in beak and lowest levels in feather (Whitbread et al., 1991). There is evidence of biphasic regulation of the expression of β-keratins in different regions of the embryonic epidermis of domestic fowl. Using two-dimensional gel electrophoresis, a different range

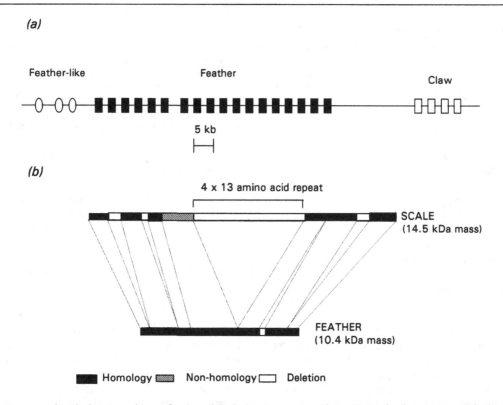

Fig. 9.6. Structure of (*a*) the keratin multigene family and (*b*) the keratin proteins. The multigene family comprises 18 feather keratins, three feather-like keratins and four claw keratins, contained within a 100 kb stretch of the genome. Scale keratin is larger than feather (*b*) largely because of the 4 × 13 amino acid residue repeat.

of β-keratins is detectable in different parts of the epidermis at different times. Accumulation of a specific group of β-keratins in the developing anterior metatarsal region occurs in two distinct phases: first an early region-specific expression in cells of the embryonic layers followed by a second phase in which expression occurs in conjunction with appendage morphogenesis (Knapp *et al.*, 1993).

9.5 The histone multigene family

Eukaryote chromosomes contain roughly equal amounts of DNA and protein. The proteins present are divided into two categories, histones and non-histone proteins. The former are small basic proteins, and in most cells there are five classes, known as H1, H2a, H2b, H3 and H4 (Table 9.4), although a sixth class, H5, is found in nucleated erythrocytes of certain vertebrates (see Section 9.5.2.). Each class from one species generally comprises less than ten different proteins, and in some cases only a single protein. By contrast, the non-histone proteins are much more diverse, both in number and properties. There may be > 300 different non-histone proteins, having a range of sizes and isoelectric points. The amount of non-histone proteins varies from one cell type to another, varying between 0.05–1 g/g DNA. The histones are principally concerned with maintaining the compact structure of chromosomes, while the non-histone proteins are more concerned with control of gene expression. Two molecules of four of the different histones, H2a, H2b, H3 and H4, form the octameric core around which the DNA is supercoiled to form the nucleosome. H1 histone forms a linker on the outside of the nucleosome. Nucleosomes are arranged higher orders of structure (for details, see Adams, Knowler & Leader, 1992). The nucleosomes are distributed throughout the chromosome with DNA running from one nucleosome to the next. When chromatin is incubated with DNAase, different regions of the DNA show varying sensitivities, and certain regions are said to be hypersensitive (see Section 8.5 and Gross & Garrard, 1988). It is the exposed regions

Table 9.4. *Properties of histones*

Histone	Function	Rate of evolution[a]	No. of genes for each type in domestic fowl	Expression
H1	Holds ends of nucleosomal DNA together	12	6	S-phase of cell cycle
H2a	Nucleosome core	1.7	9 in clusters *a* and *b*, 1 outside	S-phase of cell cycle
H2b	Nucleosome core	1.7	8	S-phase of cell cycle
H3	Nucleosome core	0.30	9 in clusters *a* and *b*, 2 outside	S-phase of cell cycle
H4	Nucleosome core	0.25	8	S-phase of cell cycle
H5	Maintains chromatin of avian erythrocyte in a condensed state, repressing replication and transcription	–		Expressed constitutively in avian erythrocytes throughout the cell cycle

[a] Accepted point mutations: No./100 residues per 10^8 years.

that are the control regions at which transcription is initiated. Since the total number of genes expressed in a eukaryote cell occupies a small fraction of the total DNA (see Section 8.2), the positioning of the nucleosomes in relation to the genes is important.

Both *in vitro* and *in vivo* studies have been carried out to examine the positioning of the nucleosomes (see Kornberg & Lorch, 1992). Two main factors seem to determine the position. The first of these is that certain non-histone proteins bind with very high specificity to particular DNA sequences, thereby excluding histone binding adjacent to that region. The second is that the histone octamer also shows some sequence specificity, although this is orders of magnitude lower than non-histone specificity. The developmental regulation of β-globin genes in avian erythrocytes provides good evidence for specific nucleosome positioning (see Section 9.6). In addition, histones may become acetylated, methylated or phosphorylated, and these covalent modifications reduce their affinity for DNA and act as a control mechanisms.

In many eukaryotes, there are multiple copies of histone genes and this may allow a rapid rate of histone synthesis particularly during embryonic development. In the axolotl, there are as many as 1600 copies, whereas in yeast there are only two. Avian species do have multiple copies, although the number is modest (see below) compared with many other eukaryotes. The rapid rate of histone synthesis required during embryonic chick development appears to be achieved by having a high steady-state level of

histone mRNA during early development (Sugarman, Dodgson & Engel, 1983). The synthesis of about 90% of histones occurs during the S-phase of the cell cycle, when the steady-state levels of histone mRNAs increase 10- to 50-fold (Table 9.4), and there are complex transcriptional control mechanisms regulating their synthesis (Osley, 1991). The remaining 10% persists throughout the cell cycle but is selective in involving certain histone variants (Wu & Bonner, 1981), which, in domestic fowl, include H2A$_F$ and H3.3. Histone genes are unusual for eukaryote genes in two respects. First, the majority lack introns and, second the mature mRNA from histone genes lacks the 3'-polyA typical of most mRNAs transcribed by RNA polymerase II. H2A$_F$ and H3.3 genes are in the minority in having introns and polyadenylated mRNAs (Dalton *et al.*, 1989).

9.5.1 Organisation of histone genes

The organisation of the histone genes from the domestic fowl has been extensively studied. A total of 43 histone genes occur within a 180 kb span of the genome and these comprise six H1, ten H2A, eight H2B, eleven H3 and eight H4 (D'Andrea *et al.*, 1985; Nakayama, Takechi & Takami, 1993). Forty of these genes occur within two clusters (*a* and *b*) located on chromosome 1 (van Hest *et al.*, 1994), as shown in Fig. 9.7, and three occur outside (one H2A and two H3). There is no evidence for pseudogenes. The organisation into the two clusters is clearly not a case of simple gene duplication, which would have

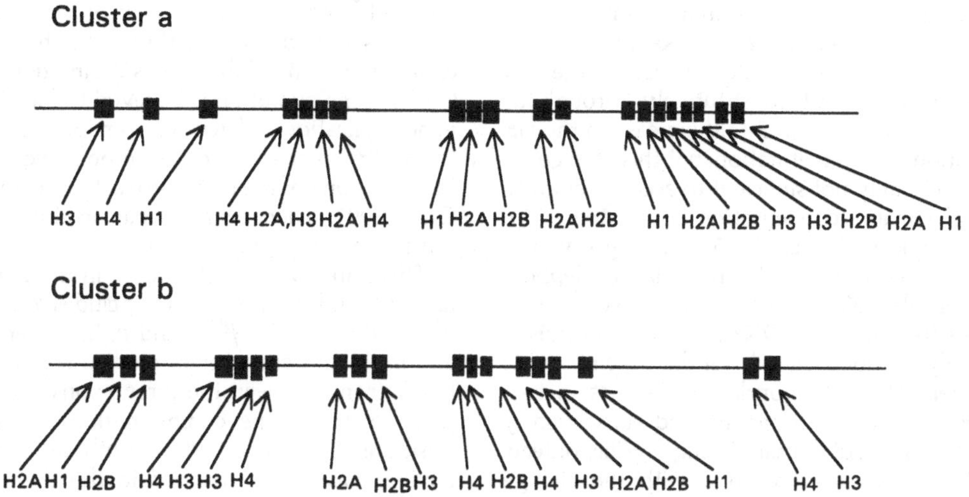

Cluster a

Cluster b

Fig. 9.7. Organisation of the histone gene family in domestic fowl. Clusters a and b are located on chromosome I (Nakayama *et al.*, 1993).

given rise to tandem repeats as is found in sea-urchin and *Drosophila*. The direction of transcription varies between genes and this means that transcription occurs on different strands of the DNA. With each type of histone there is a number of variants, but the differences between variants within each type in most cases amounts to a single amino acid replacement; many of the replacements are silent mutations (there is a change in the nucleotide sequence but no change in the amino acid sequence). Domestic fowl is the only avian species in which the organisation of the histone genes has been studied, but histone proteins have been isolated from a number of species. Histone H1 has been separated into six subtypes with comparable electrophoretic mobilities from 41 avian species including turkey, duck, goose and quail (Berdnikov, Gorel & Svinarchuk, 1978; Palyga, 1990; 1991). There is one example of four amino acid replacements in one of the H3 variants (Nakayama *et al.*, 1993). In contrast, the 5' non-coding regions show marked differences, and Nakayama *et al.* (1993) suggest that these differences enable each member of the gene families to be differentially controlled by its own *cis*-acting element.

Histones are among the most highly conserved proteins, and within this class, H4 are the highest and H1 the least conserved (Wilson, Carlson & White, 1977 and Table 9.4). The DNA sequence of four of the eight members of the H4 family in the domestic fowl has been determined and each codes for an identical amino acid sequence. At one time, it was assumed that this was because the nucleosome

structure was so highly specified that very few mutations could be tolerated, but recent evidence has challenged this view (Behe, 1990).

9.5.2 The role of H1 and H5

A histone that is only expressed in nucleated erythrocytes of birds, fish, reptiles and amphibia is histone H5. It is a linker histone like H1 and was originally isolated as a specific fraction of lysine-rich histones from avian erythrocytes (Neelin *et al.*, 1964). It is similar to H1 in having a central hydrophobic core and basic N- and C-terminal regions, which enable it to act as a linker, but there is little sequence homology with H1 (Sugarman *et al.*, 1983). H5 histone is not, however, merely a linker histone, but in common with H1⁰, found in *Xenopus*, it is developmentally regulated and is involved in terminal differentiation of erythrocytes (Khochbin & Wolffe, 1994). The levels of H5 protein increase during the differentiation and maturation of avian erythrocytes, where it displaces histone H1, and this increase is correlated with chromatin condensation and the shut down of replication and repression of most genes, but not those encoding β-globin and H5 itself until maturation is complete (see Penner & Davie, 1992). H5 has a large number of arginine residues, thereby strengthening its binding to DNA, and this may be necessary to enable effective repression of transcription. Transcription factor IIA is also involved in the general repression of transcription for most genes, except those having alternative initiation mechanisms, such as H5 (Bungert *et al.*, 1992). Unlike histone H1, which is expressed during the S-phase of the cell cycle, H5 is expressed constitutively throughout the cell cycle. There is a 3–5-fold increase in the H5 concentration during maturation, but this change on its own is probably not sufficient to account entirely for the chromatin condensation, since there are significant concentrations of H5 in young dividing erythroblasts. Histone H5 phosphorylation/dephosphorylation also plays a role. Phosphorylation of H5 lowers its affinity for DNA. Newly synthesised H5 is highly phosphorylated, but during embryonic development H5 phosphorylation declines in two stages, first between E5 and E6 and subsequently between E17 and E18 (Pikaart, Irving & Villeponteau, 1991). These stages correspond with the change from primitive to definitive erythroid cells (see Section 9.6), and with the maturation of definitive erythroid cells. The expression of H1 is controlled by a core sequence 5'-AAACACA-3' (H1-box) which is present ~ 100 bp upstream of the H1 coding sequence (Dalton & Wells, 1988). H5 mRNA transcription is controlled by erythroid-specific and ubiquitous elements, including two 5' enhancers, a 3' enhancer and a silencer (Sun, Penner & Davie, 1993; Khochbin & Wolffe, 1994). The 3' enhancer region of histone H5 binds tightly to nuclear matrices from avian liver or erythrocytes (Sun, Hendzel & Davie, 1992) and in the presence of various transcriptional factors also binds to the 5'promoter region (Sun, Penner & Davie, 1993). These may be important in organising the chromatin into higher orders of structure in which the chromatin is looped onto the nuclear matrix, and also in maintaining the chromatin in the condensed state.

The organisation of the core histones together with the H1 in clusters may have evolved to enable a coordinate expression through the cell cycle. The expression of histone H5 and the intron-carrying variants H2A$_F$ and H3.3 are separately controlled and have not been shown to be linked to the other histone genes (Sugarman *et al.*, 1983; Dalton *et al.*, 1989).

9.6 The globin multigene family and its developmental regulation

Haemoglobins are amongst the most extensively studied proteins. Their importance as oxygen carriers in birds living in different environmental conditions has already been discussed (Section 6.3); in this section the genetic organisation of the avian globin family is considered together with the control of expression of the globin genes during development. Globin gene evolution is believed to have involved gene duplication and gene conversion (Reitman *et al.*, 1993). By comparison of the globin sequences and the use of one of the tree construction methods (Li & Graur, 1991), a dendrogram for avian globin evolution can be constructed (Fig. 9.8).

There are two main branches in the globin gene family, α and β. The α family includes α^A, α^D, π and π', the β-family β^A, β^H, ε and ρ. In all vertebrate α- and β-globin gene families studied so far, there are two introns between the three exons encoding the genes, splitting the coding regions in equivalent positions (Hardison, 1991). Both α- and β-gene families have been mapped in the domestic fowl (Dolan *et al.*, 1981). After a gene duplication event, as a result of unequal crossing over, the duplicated genes will initially occur in clusters. In amphibians, both α- and β-genes are present in a single cluster; in birds and mammals the two families have become separated in different chromosomes, present as an α- and a β-cluster (Fig. 9.9). In domestic fowl the β family is either on chromosome 1 or 2 and the α family is either on a small macrochromosome or a large microchromosome (Hughes *et al.*, 1979). In many other vertebrates the order of the genes in the cluster reflects the temporal sequence of their expression, e.g. embryo→foetus→adult. This also occurs with the avian α family, but the β family is atypical with the adult β-genes flanked by embryonic ε and ρ genes.

In adult domestic fowl, there are four types of haemoglobin present: HbA, HbD, HbH and HbL (referred to as 'definitive'), although HbH and HbL are only minor constituents. The polypeptide composition and the proportions of each are summarised in Table 9.5. There are three forms of embryonic (primitive) haemoglobin (HbP, HbE and HbM). Erythroid cells of the primitive lineage arise from blood islands of the area vasculosa and enter the circulation by E2. They mature and are recognisable as erythroid cells by E5, which is the stage at which the definitive cells enter the circulation. The primitive cells and definitive cells arise from distinct cell lineages. The number of definitive cells increases and rapidly outnumbers the primitive cells, so that by E16 the primitive cells are scarcely detectable. Between E5 and E16, a corresponding replacement of primitive haemoglobins by definitive haemoglobins occurs. The control of globin synthesis occurs at the transcriptional level. By about E9, those primitive

Table 9.5. *Adult and embryonic haemoglobins from the domestic fowl*

Haemoglobin	Composition	Proportion (%)
Primitive haemoglobins		
E	$\alpha_2{}^A\varepsilon_2$	20
P	$\pi_2\rho_2$, $\pi_2{}'\rho_2$	70
M	$\alpha_2{}^D\varepsilon_2$	10
Definitive haemoglobins		
H	$\alpha_2{}^A\beta_2{}^H$	Trace[a]
A	$\alpha_2{}^A\beta_2$	70
D	$\alpha_2{}^D\beta_2$	30
L	$\alpha_2{}^D\beta_2{}^H$	Trace

[a] A trace at E7 but not adults.

Source: Stevens (1991a)

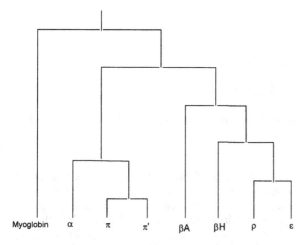

Myoglobin α π π' βA βH ρ ε

Fig. 9.8. Evolutionary tree for the globin genes in domestic fowl (Chapman *et al.*, 1980; 1981; 1982a; 1982b; Reitman *et al.*, 1993).

cells remaining become senescent and no longer synthesise primitive globins (Lois & Martinson, 1989). Developmental regulation of globin gene transcription occurs. The α^A- and α^D-globins are present in both primitive and definitive haemoglobins, but ε-, ρ-, and π-globins are only present in primitive globins. The β^A-globin is only expressed in definitive globins, and β^H-globin (H, hatching) is detectable during the transition but is completely repressed in adult domestic fowl (Rupp *et al.*, 1988).

9.6.1   The β-globin cluster

The developmental control of β-globin expression has often been regarded as a model for studying developmental expression. Promoters and enhancers have been identified adjacent to the β-globin genes. A typical β-globin promoter contains the elements TATA (-30), CCAAT (-70 to -90) and CACC (-95 to -120). (NB Negative numbers indicate the number of nucleotides upstream of the coding sequence, positive numbers downstream of the coding sequence.) In addition, there is a region at 456 bp upstream of the ρ-gene that contains sites for the general transcriptional factor Sp1 (see Sections 8.5 and 10.1), and there are three enhancer sites for a specific haemopoietic factor GATA-1 (Fig. 9.9), which are required for the globin gene expression in primitive erythroid cells. Two of these are ≈ 6.4 and ≈ 4.2 kb upstream of the ρ-gene and the third is between the β^A- and ε-genes. (GATA-1 is so called because of the nucleotide sequence 'GATA' it recognises. It was previously known as Eryf1, NF-E1 and GF-1; the new nomenclature was agreed in 1990, see Orkin, 1990.) GATA-1 factor is a 304 amino acid residue polypeptide having two zinc finger motifs (see Branden & Tooze, 1991) enabling it to bind DNA (Minie *et al.*, 1992a). This GATA-1 polypeptide is present in the primitive cell nucleus at about 10-fold higher concentration than in the definitive cell nucleus, and this may account for the expression of ρ-globin in the primitive cells (Minie, Kimura & Felsenfeld, 1992b). The gene encoding domestic fowl GATA-1 polypeptide has been sequenced and its promoter region has three strong GATA-1 binding sites, which may autoregulate its own expression. However, the activation of expression of the GATA-1 polypeptide may involve interaction with other factors at sites near the GATA-1 promoter. There is a protooncogene c-*myb* binding sequence (see Section 11.4) and a steroid hormone receptor-binding sequence (see Section 10.2) in the first intron of the GATA-1 gene, which may be important in activation (Minie *et al.*, 1992a). The enhancer sequence, about 400 bp downstream from the polyA site for β^A and about 1500 bp upstream of the ε-globin gene (Fig. 9.9), is able to stimulate a 5-fold increase in transcription. It has been proposed that competition between β^A and ε promoters for interaction with this enhancer determines which of the two genes is transcribed at each developmental stage (Reitman *et al.*, 1993). The enhancer may increase the binding of transcriptional factors to the promoter, causing a looping out of the region of DNA between these

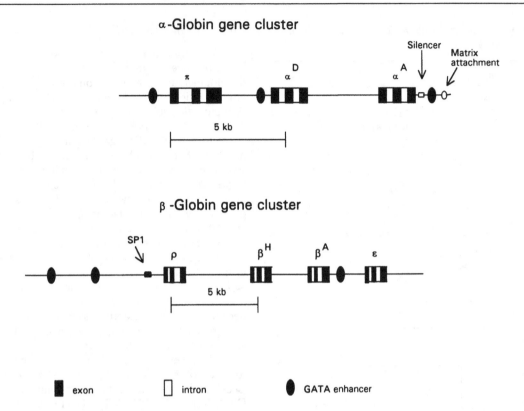

Fig. 9.9. The α- and β-globin multigene families and their control elements in domestic fowl. The α- and β-clusters are located on separate chromosomes.

two elements (Cvekl & Paces, 1992). The enhancer sequence has been identified in both domestic fowl and duck and is highly conserved (Kretsovali et al., 1987).

The β^A-globin gene is transcribed in definitive erythrocytes, although it is inactive in early development. However, both the flanking genes for primitive ε- and ρ-globins are active, and the rate of transcription of ρ-globin is two to five times that of ε-globin in the early embryo (Brotherton, Reneker & Ginder, 1990). It has been demonstrated, using a number of different methods including cross-linking agents (Postnikov et al., 1991) and DNAase hypersensitivity, that the expression of these globin genes is related to the regions of chromatin that are nucleosome free (Buckle et al., 1991).

9.6.2 The α-globin cluster

The α-globin gene cluster also has a number of promoter and enhancer sites flanking the genes.

Binding sites for silencer factors and matrix attachment regions have been identified (Fig. 9.9). The polypeptide GATA-1 binds upstream of the π- and α^D-globin genes and downstream of the α^A-gene. Only the last of these sites has been identified as an enhancer (Knezetic & Felsenfeld, 1989). Three other protein factors, an NF1 family member, a Y-box factor and an Sp1-like factor, interact to stimulate or inhibit the transcription of the π-gene. The concentrations of these three protein factors, together with GATA-1 polypeptide, control the stage-specific expression of the π-gene in primitive erythrocytes, probably by altering the conformation of the control region upstream of the gene (Knezetic & Felsenfeld, 1993). The α-globin genes are expressed in both primitive and definite haemoglobins in the domestic fowl and duck, but the ratio α^A:α^D is approximately 1:1 in primitive haemoglobins and about 3:1 in definitive haemoglobin. The enhancer may, therefore, act to increase the proportion of α^A during the transition. In some avian species, e.g. pigeon, only one α-globin gene appears to be expressed (Eguchi et al., 1990).

However, it is likely that more than one protein is able to bind at or near this enhancer region. A detailed analysis of both duck and domestic fowl enhancer regions revealed three closely linked GATA-1 motifs, an AP-1/NF-E2-binding site and a fourth GATA-1 motif interspersed by several GGTGG motifs (Targa, Gallo & Scherer, 1993a). (AP-1 and NF-E2 are also protein factors binding to specific DNA sequences, the former recognising TGATCA.) These motifs probably enable differential binding of different protein factors during the stages of erythroblast differentiation. Between the domestic fowl α^A-globin gene and the enhancer lies a silencer region (Fig. 9.9) that binds a nuclear factor of 50 kDa (Targa, Huesca & Scherrer, 1992). A similar silencer element has not been found in the corresponding position in the duck (Targa *et al.*, 1993b). Downstream of the enhancer lies one of the matrix attachment sites, which enables the cluster to bind the nuclear matrix. Upstream of the α-globin cluster lies the replication origin, a non-tissue-specific enhancer and a further matrix attachment site (Gallo *et al.*, 1991). Chromatin is arranged in loops composed of 10 to 100k bp of DNA. Evidence suggests that there is a functional link between the matrix–DNA contact points and the functional activity of the chromatin (Brotherton *et al.*, 1991).

The picture emerging is that of a sophisticated set of control mechanisms that regulate tissue-specific expression and developmental expression through a number of *cis*-acting control elements adjacent to the structural genes; a number of proteins then bind specifically to these control elements.

9.7 Contractile proteins: tissue specific and developmentally regulated gene expression

Actin and myosin are the principal contractile proteins of the three different types of vertebrate muscle: skeletal (or striated) muscle, cardiac muscle and smooth (non-striated) muscle. Muscle protein makes up approximately 40% of body protein. Actin comprises about 25% and myosin about 50% of myofibrillar protein. In striated muscles, myosin is the principal protein of the thick filaments, and actin the thin filaments. Shortening of the myofibrils is achieved by sliding the thick and thin filaments to increase the overlap between them (see Rayment & Holden, 1994). Many of the other proteins assist in this process (see Harold, 1986). Actin and myosin also occur in non-muscle cells, but their spatial organisation is more variable and less regular and

there is a much lower myosin:actin ratio. The actomyosin filaments are able to change the shape of cells. There are at least six different *isoforms* of actin, three α-actins, one β-actin and two γ-actin. The α-actins occur mainly in striated muscle, cardiac muscle and smooth muscle. The β- and γ-actins are found in the cytoplasm of all muscle and non-muscle cells. Many of the muscle proteins exist as isoforms expressed from multigene families. The different isoforms impart different characteristics to the contractile fibres and account for some of the tissue differences. Embryonic muscles also have different physiological characteristics from adult muscles and this is reflected in their molecular biology.

The development and differentiation of muscle cells has been extensively studied in a number of species, particularly the domestic fowl, since it can be examined in cultured cells. Mononucleated myoblasts, the progenitors of myofibrils, can be removed from chick embryo muscle and allowed to undergo the terminal stages of myogenic differentiation in culture. *Myoblasts* are actively dividing cells that do not express differentiated muscle characteristics such as functioning contractile proteins, although they already possess a distinctive phenotype. After their terminal division prior to fusion into multinuclear myotubes, they accumulate specific mRNAs such as those for myosin, actin, tropomyosin and troponin. The isoform families that occur with many of the muscle proteins generally arise in one of two ways, either by gene duplication or by alternative splicing of the RNA transcript. Myosin heavy chain is the one extreme where much of the diversity is generated by gene duplication, whereas a single troponin T gene has the potential to generate as many as 64 different transcripts by alternative splicing (Bandman, 1992). The multigene families for the proteins listed in Table 9.6 are described below.

9.7.1 Myosin heavy chain

It is estimated that there are more than 30 genes encoding the myosin heavy chain (MyHC), but no pseudogenes have been found (Moore *et al.*, 1993). Because of their complexity, only a few have been completely sequenced. The genes encode protein sequences of 1940 amino acid residues and comprise a total of 40 exons. (Many authors abbreviate myosin heavy chain to MHC. MyHC is used here since MHC is also the standard abbreviation for major histocompatibility complex, described in Chapter 13.) Their expression is developmentally and tissue-specifically regulated. It is generally assumed that

Table 9.6. *Contractile proteins of the domestic fowl*

Protein	Location	Subunit size	Isoforms	Mechanisn of isoform generation	Function
Myosin	Thick filament	2MyHC > (200 kDa) + 2MyLC1/3 (20 kDa) + 2MyLC2 (20 kDa); total $\approx$ 500 kDa	5 fast, $\geq$ 3 slow and 2 smooth muscle isoforms	> 30 genes; mainly gene duplication, some alternative splicing	Contraction
Actin	Thin filament	42 kDa	$\geq$ 6 including 3α, 1β and 2γ	6–11 genes; gene duplication	Contraction
Tropomyosin	Binds actin in thin filament	α, 34 kDa; β, 36 kDa	4 classes: fast α, slow α, β and platelet-like; but many more transcripts	$\geq$ 4 genes; gene duplication plus mainly alternative splicing	Regulation of contraction
Troponin	Thin filament	T, 30 kDa; I, 30 kDa; C, 18 kDa	$\approx$ 64 variants of troponin T	Mainly alternative splicing	Regulation of contraction
α-Actinin	Z disc	2 × 100 kDa	4	2 genes, 2 transcripts from each	Structural: binds F-actin
Titin	Thick filament	2800–3000 kDa	> 3	Alternative splicing	Regulating length and assembly of myofibrils

Source: Chang *et al.* (1984); Forryschaudies & Hughes (1991); Hastings *et al.* (1991); Bandman, 1992; Fujisawasehara *et al.* (1992); Inoue *et al.* (1992); Tan *et al.* (1993).

particular isoforms are best adapted for particular physiological conditions, e.g. the MyHC isoforms in fast-twitch muscle have intrinsically higher ATPase activity than those from slow-twitch muscles. From a study of the nucleotide sequences of five isoforms of fast MyHCs from domestic fowl together with their clustering on the genome, Moore *et al.* (1993) conclude that the closely related isoforms have arisen from unequal crossing-over and gene conversion. Monclonal antibodies have been raised against a number of these domestic fowl isoforms. These antibodies have identical specificity with the quail and turkey MyHC but not with ducks. Moore *et al.* (1993) suggest that some of the isoforms studied arose before the radiation of domestic fowl, quail and turkey from their common ancestor (35–65 million years ago) and others arose before the duck and domestic fowl split (70–100 million years ago). At least three slow MyHC isoforms are expressed in skeletal muscle, and muscle fibres from different skeletal muscles have different proportions of the isoforms. The proportions also vary during development. Embryonic muscles destined to become either fast or slow muscles express a particular slow isoform (SMyHC1). This isoform is lost from the fast muscles but not the slow muscles in the adult. Another isoform (SMyHC3) is found only in those embryonic muscles destined to form adult slow fibres (Page *et al.*, 1992). Using a DNA probe corresponding to part of domestic fowl fast muscle MyHC, Dominguez-Steglich, Robbins & Schmid (1993) mapped an MyHC gene by *in situ* hybridisation to one of the microchromosomes. There are at least two isoforms of smooth muscle MyHC (vascular and intestinal) that have been detected in gizzard, aorta and jejunum. They are distinct from those of the fast- and slow-twitch skeletal muscle just mentioned. In contrast to the latter, they are encoded by a single gene and generated by alternative splicing (Hamada *et al.*, 1990). Two different mRNAs encoding two different non-muscle MyHC have been detected, the proportions varying between tissues. The ratios of MyHC-A mRNA to MyHC-B mRNA are 9:1 in spleen and intestine, 6:4 in kidney and 2:8 in brain (Kawamoto & Adelstein, 1991).

9.7.2 Myosin light chain

Myosin light chain genes have been cloned and sequenced in skeletal, smooth, cardiac and embryonic muscle (Fujisawasehara *et al.*, 1992). The principal

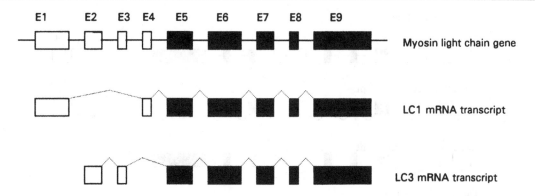

Fig. 9.10. Alternative splicing of myosin alkali light chain gene in domestic fowl (Nabeshima *et al.*, 1984). The filled boxes are exons common to both transcripts, and the open boxes are alternately spliced exons.

embryonic gene MyLC-L23 (Uetsuki *et al.*, 1990) is transcribed in embryonic skeletal muscle, gizzard and heart but is undetectable in adult except in the brain. A silencer element has been located between 3.7 and 2.7 kb upstream of the mRNA initiator site, which may be important in inhibiting transcription. The principal adult light chain gene MyLC-LC1/3 is expressed in the embryo at E11, and transcripts of LC1 appear in skeletal muscle. A few days later transcripts encoding LC3 appear. In adult, both LC1 and LC3 are found in skeletal muscle, whereas adult heart contains only LC1. Therefore, there is both developmental and tissue-specific regulation of expression. Nabeshima *et al.* (1984) have shown that MyLC1 and MyLC3 are generated by alternative splicing, as in Fig. 9.10. Two alternative transcripts are generated of 17 kb and 8 kb by use of alternative initiation sites and mutual exclusion mechanisms.

Isoform forms of MyLC2 have also been found. At least three separate genes exist for smooth muscle/non-muscle for myosin regulatory light chains (MyLC2), L_{20}-A, L_{20}-B1 and L_{20}-B2, in addition to that of skeletal muscle. Both L_{20}-B1 mRNA and L_{20}-B2 mRNA were found present in skeletal, cardiac and smooth muscle and non-muscle (Inoue, Yanagisawa & Masaki, 1992), although none of the corresponding protein was detectable in skeletal and cardiac muscle, suggesting tissue-specific regulation at the translational level. L_{20}-A is the predominant form in smooth muscle, whereas L_{20}-B is the predominant form in domestic fowl brain.

9.7.3 Actin

The actin genes also exist as a multigene family in which expression is both developmentally controlled and tissue specific. At least six different actin polypeptides have been detected in the domestic fowl: three α-actins, one β-actin and two γ-actins. The three isoforms of α-actin are the predominant forms in skeletal muscle, cardiac muscle and vascular smooth muscle, respectively. The cardiac and skeletal muscle forms only differ by four amino acid residues. The β-isoform and one of the γ-isoforms occur in non-muscle cells. The second γ-actin occurs in the smooth muscle of visceral tissues such as gizzard, in contrast to vascular tissues such as aorta, which have predominantly α-actin. There are estimated to be between 6 and 11 actin genes per haploid genome in domestic fowl, and six of these have been isolated from a cDNA library (Chang *et al.*, 1984). Each isoform appears to be encoded by a single gene. During the terminal differentiation, when the multi-nuclear myotubes form, there is an increase in the synthesis of muscle protein, and this is preceded by a 25-fold increase in the level of α-actin mRNA. This developmental control of α-actin gene expression is regulated by *cis*-acting DNA sequences within 200 bp upstream of the mRNA cap site (Bergsma *et al.*, 1986). The β-actin gene sequences have been found by *in situ* hybridisation using a β-actin cDNA probe on chromosome 2, and on one of the microchromosomes (Shaw, Guise & Shoffner, 1989).

9.7.4 Tropomyosin

Tropomyosin exists in a number of different isoforms that arise partly from having more than one genetic locus but mainly from alternative splicing. Tropomyosin is a good example of the large number of isoforms that can be generated by alternative splicing. There are four to five classes of tropomyosin

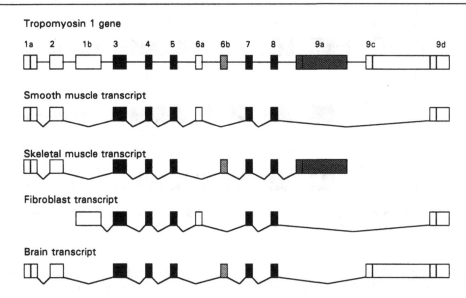

Fig. 9.11. Alternative splicing of tropomyosin 1 gene showing alternative transcripts produced in different tissues (Forryschaudies & Hughes, 1991). The solid boxes are exons transcribed in all tissues, others are alternatively spliced. The lines represent introns.

in vertebrates: (i) fast skeletal muscle α-tropomyosins, (ii) slow skeletal muscle α-tropomyosins, (iii) skeletal muscle β-tropomyosins, and (iv) platelet-like low M_r tropomyosins, and (v) a possible fifth group, cardiac tropomyosins. The diversity is generated from at least four separate tropomyosin genes, each of which generates multiple transcripts through alternative splicing. The gene encoding β-tropomyosin in the domestic fowl generates nine different mRNA products by use of alternative promoters, mutual exclusion and cassette mechanisms (Fig. 9.11). Two sizes of tropomyosin polypeptide are found in different tissues, the so-called low M_r (247 amino acid residues) and the high M_r (284 amino acid residues). The difference results from the presence of exon 1b, which encodes the additional N-terminal amino acid residues together with a 5'-untranslated region (Libri et al., 1990). Exons 6A and 6B are mutually exclusive, and regulation of the splicing process depends on the secondary structure of the primary RNA transcript. Smooth muscle and non-muscle isoforms include exon 6A, whereas skeletal muscle includes exon 6B. Myogenic cells express exon 6A before differentiation, but switch to exon 6B on maturation (Libri, Piseri & Fiszman, 1991). The proportions of α- and β-tropomyosin vary during embryonic development, so that at E10 there is 25% α- and 75% β-tropomyosin; by the time of hatching there are roughly equal amounts of each, and in the adult muscle only α-tropomyosin is

present (Montarras, Fiszman & Gros, 1982).

The following isoforms have been isolated in quail: skeletal α-tropomyosin, smooth muscle γ-tropomyosin, cardiac tropomyosin and cytoskeletal tropomyosin (Fleenor et al., 1992).

9.7.5 Troponin

The troponin complex consists of three proteins: troponin I, troponin C and troponin T. Three isoforms of troponin I have been identified in domestic fowl: cardiac, fast skeletal and slow skeletal. The fast skeletal isoform has been cloned and sequenced (Nikovits, Kuncio & Ordahl, 1986). It comprises eight exons, of which the first is not translated, and the coding sequence begins with the triplet ATG in the second exon. The cardiac isoform of troponin I is expressed in the domestic fowl and quail heart at E4, but not in embryonic skeletal muscle (Hastings et al., 1991). It is with troponin T that greatest diversity is shown. More than 40 troponin T variants have been identified in domestic fowl embryonic muscle (Bandman, 1992). (Alternative splicing allows for such a large number of variants. If n is the number of exons that may be individually included or excluded by alternative splicing, then the potential number of variants is 2^n.) The cDNAs corresponding with four isoforms have been studied in one-week-old domestic fowl (Smillie, Gologinska & Reinach, 1988). The troponin T gene has 17 exons, and the four isoforms studied were identical in exons 1–3 and 9–15, the variation occurring between exons 4–8 and 16/17. In the quail, two closely related isoforms of troponin T have arisen by alternative splicing, and the corresponding isoforms have also been found in the rat (Hastings, Bucher & Emerson, 1985). This suggests that this splicing event evolved before the divergence of mammals and birds.

9.7.6 Other contractile proteins

Titin is expressed in embryonic and adult cardiac and skeletal muscle. Isoforms with different M_r values are expressed differentially during embryonic development. Evidence suggests that there is a single gene in domestic fowl and that the isoforms are generated by alternate splicing (Tan et al., 1993).

Four forms of α-actinin occur; two mainly occur in skeletal and smooth muscle, in which the binding to F-actin is insensitive to Ca^{2+}, and two predominantly in non-muscle tissue, in which binding to F-actin is inhibited by Ca^{2+}. There are two genes encoding these actinins, designated 1 and 2, and two isoforms generated from each: 1a and 1b by mutually exclusive splicing of gene 1 transcript, and 2a and 2b by alternative splicing of gene 2 transcript (Parr et al., 1992). Types 1a and 2a are expressed in muscle, whereas 1b and 2b are expressed in non-muscle tissues such as brain, liver and spleen. Types 2a and 2b are identical except that part of the C-terminal EF-hand motif is absent from the Ca^{2+}-insensitive form, as the result of two exons being alternatively spliced. (The EF-hand motif forms the Ca^{2+}-binding site in a number of proteins; for details see Branden & Tooze, 1991.) Isoforms of other muscle proteins have been detected, e.g. β-actinin, C-protein, M-line protein, filamin and myomesin, but they have not yet been well characterised.

Contractile proteins and those associated with them show a greater abundance of isoforms than other types of protein. It is generally assumed that these isoforms enable different types of muscle and cytoskeletal protein to have different characteristics appropriate to their physiological functions. However, except in a few cases, e.g. MyHC, it is not yet clear what characteristics are endowed by particular isoforms. It is also clear that expression of these isoforms is regulated both on a temporal basis and spatial basis. Several control elements associated with these multigene families have been identified, but the details of the regulation has still to emerge.

9.8 Crystallins and gene sharing

Crystallins are a group of proteins that make up about 90% of the lens protein. The lens is an unusual tissue. It is formed from cuboidal epithelial cells that terminally differentiate to form fibre cells. They then lose their nuclei and other subcellular organelles to achieve the degree of transparency necessary to function as a lens. The protein components of these cells remain for the life of the organism, i.e. there is very little protein turnover. In order to fulfil their role of being highly refractive to light and also transparent, the crystallins must be present in high concentrations and be closely and regularly packed. They must also be very stable proteins, being the longest lived of vertebrate proteins. The protein concentration of lens fibre cells is between 20 and 60% of wet weight. Avian lenses are soft by comparison with other vertebrates, particularly mammals, and have protein concentrations at the lower end of the range. The softer lens is believed to account for the visual acuity of birds. Crystallins are classified on the basis of their properties into four main groups: α, β, γ and δ, but recently other minor groups have been discovered. In most vertebrates α, β and γ are the main groups. In avian lenses, δ-crystallins are the most abundant, replacing γ-crystallins, which are either present in very small amounts or are undetectable. The γ-crystallins bind very little water and are most abundant in hard lenses. The δ-crystallins have properties more suited to soft lenses and have predominantly α-helical secondary structure; whereas α-, β- and γ-crystallins are mainly β-pleated sheets. The proportion of α-helical structure in avian δ-crystallins appears to vary considerably between species and also between embryo and adult, ranging from 28% to 80% (Horwitz & Piatigorsky, 1980; Chiou et al., 1991). The properties of the main avian crystallins are summarised in Table 9.7.

An interesting feature of crystallins is that apart from having an important role in providing the refractive components of the lens, most of them also have second functions, usually enzymic. A process known as gene recruitment leading to *gene sharing* has occurred. Wistow (1993) has proposed that this process may have occurred in the following stages. First, gene recruitment occurs, probably by insertion of a new promoter and enhancer upstream of the crystallin coding sequences. This leads to increased crystallin mRNA expression, particularly tissue-specific expression in the lens. The crystallin serves both the enzyme function and also, being present in high concentrations in the lens, contributes to the refractive

Table 9.7. *Properties of avian crystallins*

Type	% of total lens protein	Molecular composition	Associated enzyme activity	Genetic characteristics	Avian species in which studied
Alpha (αA, αB)	Up to 50%	20 kDa subunits→ ≈ 800 kDa aggregates; mainly β-pleated sheet structure	Small heat shock proteins (hsp-27), also more distantly related to large (hsp-70), may act as molecular chaperones in non-lens tissue	3 exons; αA expressed as 14S mRNA, exon shuffling and alternative splicing involved in gene expression; αA expressed in lens only; αB in lens and non-lens	Domestic fowl, duck
Beta, at least 4 types	Up to 70%	20–30 kDa subunits →50–100 kDa aggregates; mainly β-pleated sheet structure, no cysteine and abundant leucine	Protein S of *Micrococcus xanthus*	βA1/A3 has 6 exons and 2 alternative start codons	Domestic fowl
Gamma	Very low levels, if detectable	Cysteine-rich, binds little H$_2$O			
Delta (δ_1, δ_2)	Up to 70%	Tetramer, 48 and 50 kDa subunits; mainly α-helical structure, no cysteine	δ_2 is argininosuccinate lyase	17 exons, δ_2 expressed more extensively in non-lens tissue	Domestic fowl, duck, goose, pigeon, ostrich, rhea, house finch, starling, house sparrow, swallow, Anna's humming bird Absent from chimney swifts
Epsilon	Up to 50%	38 kDa subunits→120 kDa aggregates; α-helical and β-sheet	Lactate dehydrogenase, also malate dehydrogenase	8 exons, 2 transcriptional starts, GC-rich promoter	Duck, goose, grey heron, curlew, Palaearctic oystercatcher, northern gannet, white stork, flamingo, Eurasian buzzard, chimney swift, gull, Anna's humming bird
Tau	Up to 10%	48 kDa	Enolase	Regulated by proto-oncogene, c-*myc*	Duck, emu, penguin, northern gannet, gull, cuckoo

Source: Yasuda & Okada (1986); Wistow *et al.* (1987); Wistow & Piatigorsky (1988); Wistow *et al.* (1990); Chiou *et al.* (1991; 1992); Piatigorsky & Wistow (1991); Wu *et al.* (1992); Wistow (1993); Yu & Chiou (1993).

properties of the lens. The high enzyme activity associated with the high concentration of this protein in the lens may cause adaptive conflict, i.e. high activity of this enzyme in lens may not be 'desirable'. In some cases, this may produce the selective pressure for gene duplication, as in the case of avian δ_1- and δ_2-crystallins, and also αA and αB-crystallins. Subsequent mutations in the δ_1-gene cause it to lose

its enzyme activity. The second functions of the different avian crystallins are listed in Table 9.7.

9.8.1 The δ-crystallins

Since δ-crystallins are the most abundant in avian lenses, this group has been most thoroughly investigated and will be discussed first. They have been detected in a wide range of birds, and the only species in which they have so far not been detected is the chimney swift, where they are replaced in the lens by a high concentration of ε-crystallin. In Anna's humming bird, which belongs to the same order (Apodiformes), δ-crystallin is present in very low concentrations and ε-crystallin is abundant. In the swallow, which is a fast-flying insectivore like the chimney swift, but from a different order (Passeriformes), the more typical crystallin distribution of mainly δ-crystallin and only low levels of ε-crystallin is found. On this basis, Wistow, Anderson & Piatigorsky (1990) suggest that phylogeny is more important than environmental selection in recruitment of crystallins.

The δ-crystallins have been most studied in the domestic fowl. It is the first to be detected in the developing embryonic lens by E4. It accumulates to become 70–80% of the lens protein by E19 and its synthesis then decreases. The domestic fowl δ-crystallins occur in two closely related (c. 90% homologous) but non-allelic forms, δ_1 and δ_2, with their genes closely linked, separated by 4 kb, both with 17 exons and 16 introns. Gene conversion could be responsible for their high sequence similarity. The duck also has two δ-crystallins genes with very similar sequences (94% homology) (Wistow & Piatigorsky, 1988; 1990). The cDNA sequences have also been determined for pigeon (Lin & Chiou, 1992) and goose (Yu & Chiou, 1993). Goose δ-crystallin has extensive homologies with pigeon (88%), duck δ_1 (97%), duck δ_2 (94%), domestic fowl δ_1 (88%), domestic fowl δ_2 (90%) and human argininosuccinate lyase (69%). The significance of the last of these is that gene recruitment of δ-crystallins from argininosuccinate lyase is believed to have occurred. The ratio of δ_1-crystallin mRNA to δ_2-mRNA varies between tissues and the stage of development. In embryonic chick tissues it is between 100 and 0.07 (lens, 50–100; cornea, 3–4; neural retina, 2–20; heart, 0.5; and brain, 0.07). The amount of δ_1 mRNA decreased in all tissues after hatching, favouring δ_2-mRNA in the adult (Li, Zelenka & Piatigorsky, 1993), although in the duck both forms are abundantly expressed (Wistow & Piatigorsky, 1990). Since δ_2-crystallin has high argininosuccinate lyase activity compared with δ_1-crystallin (Piatigorsky et al., 1988), this accounts for a 1500-fold higher argininosuccinate lyase activity in duck lenses compared with domestic fowl lens (Piatigorsky & Wistow, 1991). Ostrich and goose δ-crystallin also have high argininosuccinate lyase activity, in contrast with the pigeon, which has low activity (Chiou et al., 1991; Yu & Chiou, 1993).

Most of the crystallins are expressed in much higher concentrations in the lens than in other tissues of the body. This has led to a search for control elements that might account for lens-specific expression. Although a number of control elements have been identified, so far no features common to all crystallins have been found. Domestic fowl δ-crystallin promoter-like sequences have been found within the first intron (Wistow, 1993). A 120 bp enhancer sequence has been found within the third intron of δ_1. This δ_1-crystallin enhancer activates transcription 20- to 40-fold (Hayashi et al., 1987). Comparison of the lens- and non-lens-specific duck δ_1 and δ_2 mRNAs suggests that the same transcriptional starts are used in both, rather than alternative promoters being required (Hodin & Wistow, 1993).

9.8.2 The ε-crystallins

One of the taxon-specific crystallins that has been detected in a large number of birds, especially waterfowl and birds living at high altitudes, is ε-crystallin (Table 9.7); however, it is absent from a number of terrestrial birds (Wistow et al., 1987). It makes up ≈ 20% the crystallins in the northern gannet and grey heron and ≈ 10% in duck and flamingo. Duck ε-crystallin has 30% α-helix, 26% β-sheet and 15% β-turn (Stapel et al., 1985) and is very closely related to the lactate dehydrogenase-B4 (LDH-B4) isolated from duck heart (Wistow et al., 1987). A single copy of the Ldh-B gene encodes both ε-crystallin and lactate dehydrogenase (Hendriks et al., 1988). The small differences may be the result of post-translational modification. Chiou, Lee & Chang (1990) compared the kinetics of duck (Cairina-Anas hybrid) ε-crystallin with those of domestic fowl heart LDH-B4 and found small but significant differences in kinetic parameters. These differences could have arisen because of species differences between duck and domestic fowl. A phenylalanine residue in domestic fowl heart LDH-B4 is replaced by glycine in duck lens ε-crystallin/LDH-B. The smaller glycine residue may facilitate the close packing of crystallins important in the lens.

The control of ε-crystallin gene expression has been studied by analysing its promoter region (Kraft et al., 1993). It lacks the TATA box (see Section 8.5) but is very GC-rich, typical of the control region of housekeeping enzymes (Kim & Wistow, 1993), and has multiple binding sites for the general transcription factor Sp1. The gene has two discrete transcription start sites located 28 bp apart. Both sites are used equally in heart, but the downstream site predominates in lens and this is the basis for its differential expression.

9.8.3 The τ-crystallins

Duck, emu, northern gannet, gull and Eurasian cuckoo contain τ-crystallin, but it is not found in Eurasian buzzard, black coot, budgerigar or northern eagle owl (Stapel & de Jong, 1983). It has been most studied in the duck where gene sharing occurs with α-enolase. Its promoter is regulated by the proto-oncogene c-*myc* (see Section 11.4), which promotes its expression in lens epithelia. Unlike the ε-crystallin duck gene, its promoter has a TATA box (Wistow, 1993).

Since both ε- and τ-crystallins are abundant in duck lens, and both have glycolytic enzyme activity (enolase and lactate dehydrogenase), the question arises as to whether they have a significant effect on lens metabolism. Reddy *et al.* (1993) used NMR spectroscopy to compare organic phosphate metabolism in duck lens with that in calf lens, since the latter has no known recruited crystallins. They found significant changes in organic phosphate metabolism, particularly in the ratio of $NAD^n/NADH$ and concluded that the crystallins have significant impact on lens metabolism. Wistow (1993) suggests that NADH may be sequestered by ε-crystallin and this could protect against oxidation or possibly against ultraviolet light.

9.8.4 The α-crystallins

There are two closely related α-crystallins in the domestic fowl, αA and αB. Both are encoded in closely related genes, each having three exons and two introns, but αA-crystallin is only expressed in the lens, whereas αB is expressed in a number of tissues. The sequence of duck αB-crystallin is very similar to both the small heat shock protein (hsp-27) and large heat shock protein (hsp-70), suggesting that it may act as a molecular chaperone in common with other heat shock proteins (Lee, Kim & Wistow, 1993). The αA-crystallin gene has an enhancer element 84 bp long, at positions -162 to -79 upstream of the coding region. This region contains a number of motifs that confer lens-specific expression on the gene (Matsuo & Yasuda, 1992). Alpha-crystallin is the last crystallin to be synthesised during embryogenesis in the domestic fowl (Zwaan & Ikeda, 1968).

9.8.5 The β-crystallins

Avian β-crystallins form the least widely studied group. They make up approximately half of the soluble lens proteins in domestic fowl. Seven β-crystallins have been divided into two groups: βA (acidic) and βB (basic). They are expressed principally in the lens but also at low levels in other tissues such as brain and retina. Four cDNAs have been isolated for β-crystallins in the domestic fowl (Fielding Hejtmancik *et al.*, 1986). One of these, βB1, is specifically expressed in the elongated fibrous cells of the lens but not in the cuboidal epithelial cells. Beta-crystallin is synthesised in specific regions of the developing retina, and βA3/A1- and βB2-crystallin mRNAs have been detected within a subpopulation of developing retina cells (Head, Peter & Clayton, 1991). The βA1- and βA3-crystallins have identical sequences except for 17 N-terminal amino acid residues that are present in βA1. Both proteins are encoded in the same gene, which is able to use two different start codons (McDermott, Peterson & Piatigorsky, 1992). Beta-crystallins are related in structure to two proteins that play important roles in the formation of dehydrated spores, but the significance of this is not clear.

10

Avian steroid hormones and their control of gene expression

10.1 Introduction

Steroid hormones have a common mechanism of action: they enter the target tissue and become bound to nuclear protein receptors that interact with chromatin to change gene expression. This property is also shared with the thyroid hormones and morphogens such as retinoic acid. In this chapter, the hormones which will be discussed are the oestrogens, progesterone, androgens, glucocorticoids and dihydroxycalciferol (Fig. 10.1). Dihydroxycalciferol differs from the rest in that it is not synthesised in an endocrine gland but is derived from calciferol (vitamin D, see Section 2.4.2) by successive hydroxylations in the liver and kidney. However, its mechanism of action is similar to the others. Table 10.1 summarises the site of synthesis of the steroid hormones, their principal target tissues and the genes expressed as a result of their interaction. These hormones regulate gene expression by affecting transcription of specific genes, catalysed by RNA polymerase II. The increased amount of mRNA generally results in a corresponding increase in protein synthesis. The most extensively studied avian proteins that are hormonally controlled by gene expression are egg-white proteins and the egg-yolk proteins. The former are synthesised in the oviduct and the latter in the liver. The control of protein synthesis in hen oviduct is often regarded as a model for understanding hormonal gene expression in general. The general features of steroid hormone gene expression are first described, and this is followed by a more detailed consideration of each different system.

The essential steps in steroid hormone action are as follows. First, the circulating hormone is taken up by the target tissue; being a hydrophobic molecule it is able to pass through the cell membrane without specific membrane receptors. The hormone then binds to a specific intracellular receptor. Until recently, hormone receptors were difficult to isolate, since they are present in such small quantities in target tissues. It was originally believed that, in unstimulated cells, the hormone receptors were present in the cytoplasm, and that the hormone first bound to the cytoplasmic receptor which then underwent some activation process before entering the nucleus. However, it now seems more likely that for the majority of steroid hormones the receptors are located in the nucleus even in unstimulated cells, the exceptions being glucocorticoid and mineralocorticoid receptors which may also be present in the cytosol (Hardie, 1991). Using immunohistochemical techniques, progesterone receptors have been located exclusively in the nuclei of target cells of quail oviduct (Pageaux et al., 1989) and glucocorticoid receptors in the nuclei of a number of different regions of the quail brain cells (Kovács, Westphal & Péczely, 1989).

In the plasma, some of the hormones are bound to specific proteins having high affinities for the hormones and some are bound to albumin with much lower affinity. Proteins with a high affinity for binding corticosterone (K_d, 10^{-7} to 10^{-9} M) have been detected in the plasma of eight different orders of bird (Wingfield, Matt & Farner, 1984). These proteins also bind progesterone with similar affinity and testosterone and oestradiol with lower affinity. The flux of hormone from the plasma to the nucleus is assumed to depend on the higher affinity of the hormone for the receptor than for the plasma binding proteins. The specific hormone receptors are only present in the target tissue. The binding of the hormone to the receptor causes a conformational change in the latter, which enables it to enter the nucleus and bind to a specific DNA sequence known as the hormone response element. A region of the receptor recognises specific sequences on the 5'-flanking promoter region of a hormone-sensitive

Table 10.1. *The principal avian steroid hormones*

Hormone	Site of synthesis	Typical plasma concentration ranges[a] (nM)	Target tissue	Genes identified as being expressed in some target tissues
Progesterone	Granulosa cells of the ovary	Female 1–15	Oviduct, liver, hypothalamus, pituitary	Proteins of egg-white and yolk
Oestradiol-17β	Thecal cells of the ovary	Female 0.02–0.9	Oviduct, liver	Proteins of egg-white and yolk
Oestrone	Thecal cells of the ovary	Female 0.6–1.3	Oviduct, liver	Proteins of egg-white and yolk
Testosterone	Interstitial cells of the testis, also small amounts in the ovary	Laying female, up to 3; male 1–20	Brain, hypothalamus, pituitary, skin, deferens ducts	
Corticosterone	Adrenal cortical tissue	1–25	Liver, kidney, retina, adipose tissue	Tyrosine aminotransferase, tryptophan oxygenase, glutamine synthetase, metallothionein, lipoprotein lipase, egg proteins
Aldosterone	Adrenal cortical tissue	1–5	Kidney, intestine	
1,25-Dihydroxyvitamin D$_3$	Vitamin D$_3$ hydroxylated in liver and kidney	Female, laying 0.3–0.7, non-laying 0.07–0.1; 1–3 week old male 0.9	Principally intestine, kidney, bone and shell gland	Calbindin D$_{28K}$, tubulin, osteocalcin

[a] Data from a range of species, including domestic fowl, turkey, penguins, ostrich, pigeon, Harris hawk, zebra finch and junco.

gene, either activating or inhibiting its transcription. To initiate transcription of a given gene requires RNA polymerase II, a set of general initiation factors and, in some cases, sequence-specific transcriptional factors. The general initiation factors and the sequence-specific transcriptional factors are DNA-binding proteins that bind to the promoter region of the gene (Fig. 10.2). The general initiation factors are TFIIA, TFIIB, TFIID and TFIIE. These, together with RNA polymerase II, form a complex with the TATA box of the promoter and are required for all genes constitutively expressed in each cell. Sp-1 is another general initiation factor that stimulates the expression of defined sets of genes and binds the GC box (see Section 8.5). The sequence-specific transcription factors bind to specific regions upstream from the general initiation factors. There are also enhancer regions (see Section 8.5), which may lie up to 1000 bp either upstream or downstream from the coding region, but which influence the specific promoter.

The steroid hormone DNA-binding receptors are sequence-specific DNA-binding proteins. They generally have three domains (Fig. 10.3) a modulator domain at the N-terminus, a DNA-binding domain in the centre and a hormone-binding domain, which

also contains a dimerisation site, at the C-terminus (Orti, Bodwell & Munck, 1992). The DNA-binding domain is the region that enables the receptor to bind DNA adjacent to the initiation complex (Fig. 10.2) and allows transcription to proceed. A structural feature of DNA-binding domains is the presence of zinc fingers, which enable them to bind in the major groove of DNA (see Branden & Tooze, 1991). The function of the modulator domain is not clear, although it may increase the discrimination of the DNA-binding region for the hormone response element. The steroid hormone DNA-binding receptors are regarded, on the basis of their structures, as belonging to a superfamily. The three-dimensional structures of the DNA-binding domains of two members of the family, namely the glucocorticoid and oestrogen receptor, have been determined by two-dimensional NMR spectroscopy (see Schwabe & Rhodes, 1991). Although these receptors were from mammalian sources, sequence data suggest that the receptor structures are highly conserved (see Section 10.2), and so it seems unlikely that avian receptors will have a significantly different structure.

In addition to the members of the superfamily already mentioned, there are also 'orphan members'.

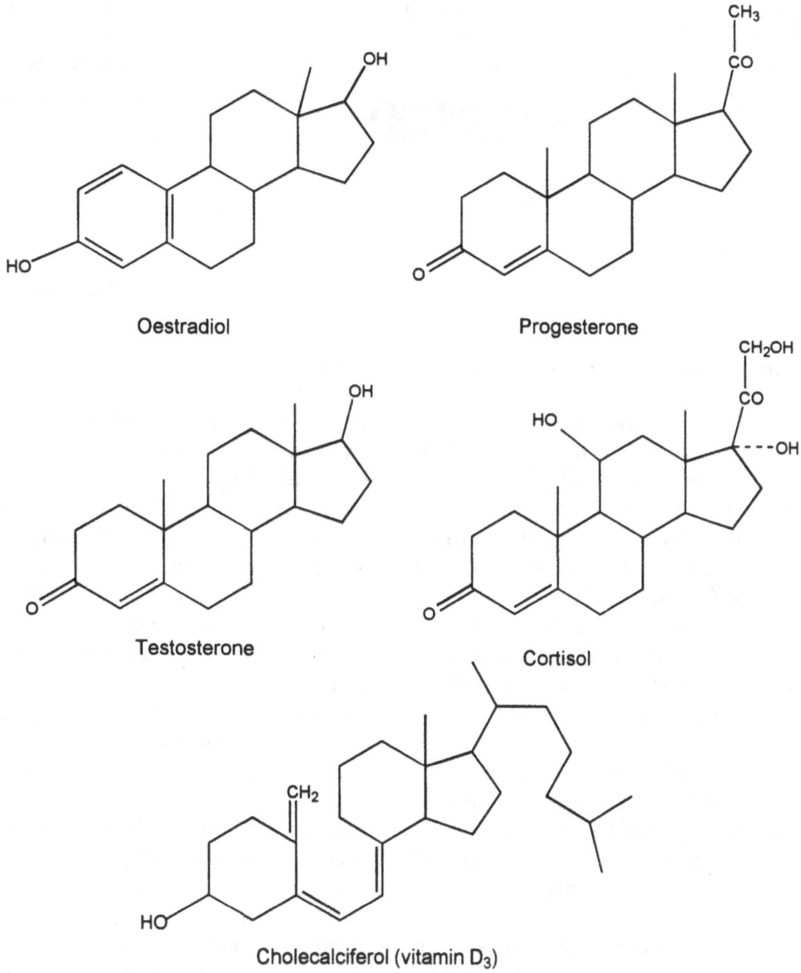

Oestradiol

Progesterone

Testosterone

Cortisol

Cholecalciferol (vitamin D₃)

Fig. 10.1. Steroid hormone structures.

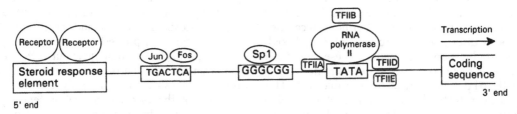

Fig. 10.2. The promoter region of a hormone-sensitive eukaryote gene, showing the transcriptional factors that interact with different elements of the promoter. TF, transcription factor; Sp1, general initiation factor.

These are DNA-binding proteins that show homologies with steroid hormone receptors and possess the double zinc finger motif, but for whom the true physiological ligands are unknown. One of these, COUP-TF, and a structurally very similar COUP-TF II (also called ARP-1) were originally characterised by their ability to bind to an element at −70 to −90 bp upstream of the ovalbumin-coding region (chick ovalbumin upstream promoter-transcription factor) and activate transcription of ovalbumin mRNA

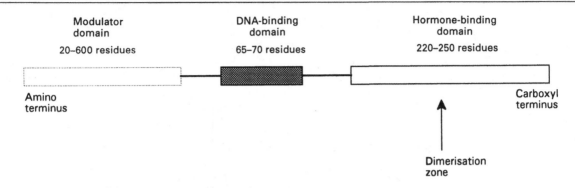

Fig. 10.3. General structural features of steroid hormone-receptor proteins.

(Tsai, Tsai & O'Malley, 1991). These elements have a core sequence of an imperfect repeat A/GGGTCA, separated by a single nucleotide (GTGTCAAAG-GTCAAA). COUP-TF is a 45 kDa protein able to bind to many different A/GGGTCA repeats, which are also bound by dihydroxycalciferol, thyroid and retinoic acid receptors (Cooney et al., 1993) and can activate or repress transcription depending on the promoter context. It has been found in a number of different tissues in different species. Hepatic nuclear factor (HNF-4) is another orphan member of this superfamily.

When a steroid hormone binds the DNA-binding receptor protein, this induces a conformational change that enables the protein to bind to the hormone-response element adjacent to the target gene. Steroid hormones acting in this way can only affect transcriptionally active genes. Although, in the absence of the hormone the rate of transcription may be very low or undetectable, the region of the chromatin to which the receptor binds is a DNAase-hypersensitive region (see Section 8.5). On binding to the DNA, further DNAase-hypersensitive sites are exposed. So far, three classes of hormone-response element, all belonging to a superfamily of hormone receptors, have been discovered: oestrogen-, glucocorticoid- and thyroid-response elements. These are short palindromic sequences of 6 bp per half site, with 4 out of the 6 bp common to all three (Table 10.2). This enables them to bind dimeric forms of the DNA-binding receptor proteins. The spacing between the two half sites varies between members of the family.

10.2 Avian DNA-binding receptor proteins

10.2.1 The progesterone receptor

The avian steroid hormone receptor which has been most extensively studied is that for *progesterone*. It

exists in two forms and has been purified from the oviduct of domestic fowl. It is encoded by a gene comprising eight exons ($\approx$ 38 kb long) and is located on the long arm of chromosome 1 (Jeltsch et al., 1986; Huckaby et al., 1987; Dominguez-Steglich et al., 1992). It has strong sequence homologues with the oestrogen receptor from domestic fowl, and also the human glucocorticoid receptor. In the absence of progesterone stimulus, the progesterone receptors can be isolated from domestic fowl oviduct as a complex of 250–300 kDa. The complex is made up of progesterone receptor A (79 kDa), progesterone receptor B (110 kDa) and heat shock protein 90 (hsp-90, 90 kDa). Other proteins have also been found associated with the complex; these include a second heat shock protein (hsp-70) together with three proteins of unknown function, p54, p50 and p23 (Smith & Toft, 1992). Under physiological conditions, administration of progesterone promotes ATP-dependent phosphorylation of progesterone receptors A and B and the release of hsp-90 from the complex (Nakao et al., 1992). Both heat shock proteins may have roles in protein folding and assembly. A similar type of dissociation occurs with the glucocorticoid receptor and this releases hsp-90. The two forms of progesterone receptor (A and B) are the product of a single gene but with alternative initiation sites for the primary transcript. Although both forms can bind to the progesterone-response element of the tyrosine aminotransferase gene *in vitro*, only the A form has functional activity towards the ovalbumin promoter (Tsai & O'Malley, 1991). The phosphorylation sites on the progesterone receptor have been identified as Ser-211, Ser-260, Ser-367 and Ser-530, and phosphorylation is thought to play a role in regulation of receptor function (Poletti et al., 1993). Protein kinase A phosphorylates both A and B forms of domestic fowl oviduct progesterone receptor (Orti et al., 1992).

Table 10.2. *Comparison of DNA sequence in response elements of hormone receptors*

Glucocorticoid-response element	AGAACANNNTGTTCT
Oestrogen-response element	AGGTCANNNTGACCT
Thyroid-response element	AGGTCATGACCT
Dihydroxy vitamin D_3-response element	GACTCANNNNNTGAACG
COUP element (orphan-response element)	GTGTCANAGGTCA

N, unspecified nucleotide.
Source: Tsai & O'Malley (1991).

The progesterone receptor, together with a receptor-binding factor (RBF-1), is localised in the nuclear matrix of cells of the oviduct. RBF-1 appears to enable the progesterone receptor to bind with high affinity to the nuclear matrix (Schuchard *et al.*, 1991). Using immunohistochemical methods, Zhuang *et al.* (1993) have shown that both progesterone receptor and RBF-1 are localised in nuclei of specific cells of the avian oviduct. The binding of the progesterone receptor to the progesterone-response element on chromatin initiates the primary hormone response. The early regulated genes that have been identified are the proto-oncogenes c-*myc*, c-*jun* and c-*fos* (see Section 11.4). These proto-oncogenes are members of a multigene family that are implicated in signal transduction cascades associated with growth and differentiation. c-*jun* encodes a DNA-binding protein with a leucine zipper (see Branden & Tooze, 1991), which enables it to form either Jun–Jun homodimers or Jun–Fos heterodimers (Landschulz, Johnson & McKnight, 1988). These dimers are sequence-specific transcription factors, binding consensus sequence TGACTCA present in the promoter region (Fig. 10.2). The proto-oncogenes may then bind the oestrogen-response elements, which in turn regulate late-regulated genes such as the ovalbumin gene (Schuchard *et al.*, 1991).

10.2.2 The oestrogen receptor

The oestrogen receptor has also been purified from the domestic fowl oviduct, where it exists in three states (Raymoure *et al.*, 1986). The three forms are interconvertible; R_x, which has a high affinity for oestrogens (K_d 0.06 nM), can be phosphorylated to R_y, a lower affinity form (K_d 0.8 nM), and R_y can be converted to a non-binding form (R_{nb}).

$$R_x \underset{ADP/Mg^{2+}}{\overset{ATP/Mg^{2+}}{\rightleftharpoons}} R_y \overset{dialysis}{\rightleftharpoons} R_{nb}$$

All three have $M_r \approx 66\,000$. Although the significance of the three forms is not yet clear, it seems likely that phosphorylation is involved in regulation of binding. The complete cDNA sequence has been determined (Krust *et al.*, 1986) and it shows a high degree of homology with those from other species. For example, the DNA-binding domain is identical with that from human oestrogen receptor, the hormone-binding domain shows 94% identity with human oestrogen receptor and the overall homology is 80% (Green & Chambon, 1991). Its genomic organisation is like that of the progesterone receptor and other steroid hormone receptors so far analysed. It is divided into eight exons with the intron positions completely conserved (Nestor *et al.*, 1994).

10.2.3 The androgen receptor

A good source of the androgen receptor is domestic fowl oviduct, from which it has been purified (Ellis & Danzo, 1989). It is induced in the oviduct by oestrogen administration. Both androgens and oestrogens activate the transcription of the ovalbumin gene (see Section 10.3). It has a M_r of 61 200, and binds ligands with decreasing affinities as follows: 5α-dihydrotestosterone (K_d 0.13 nM) > progesterone > oestradiol. Androgens, as well as oestrogens and progestins, play important roles in the regulation of ovarian function. They are believed to exert paracrine or autocrine effects on steroidogenesis in the granulosa cells of the ovary. Using anti-human androgen receptor polyclonal antibodies, androgen receptors have been detected in the granulosa cell nuclei of the small pre-ovulatory and post-ovulatory follicles of domestic fowl (Yoshimura *et al.*, 1993).

10.2.4 The 1,25-dihydroxycalciferol receptor

The 1,25-dihydroxyvitamin D_3 (1,25-$(HO)_2D_3$) receptor has been purified from the small intestine of domestic fowl. The major component has a M_r

60 300 and there is also a minor component of M_r 58 600 (Pike, Sleator & Haussler, 1987). The latter probably arises either from the use of an alternative initiation site or as a result of proteolysis during purification. The receptor binds 1,25(HO)$_2$D$_3$ very strongly (K_d 1–50 × 10^{11} M), with the three hydroxyl groups (C$_1$, C$_3$ and C$_{25}$) playing a crucial role (Norman, 1987). The receptor can be cleaved by trypsin into a 20 kDa N-terminal fragment that has DNA-binding activity and a 40–44 kDa C-terminal fragment that has 1,25-(HO)$_2$D$_3$-binding activity (Allegretto, Pike & Haussler, 1987). It has a cysteine-rich region containing the zinc finger motifs that bind DNA. Besides being present in small intestine, the receptor has also been detected in kidney but not in liver, and this is consistent with the presence of the Ca^{2+}-binding protein in the former but not the latter. The mRNA for 1,25(HO)$_2$D$_3$ receptor is most abundant in intestine, kidney and brain, present in low amounts in oviduct and heart, but absent from liver (McDonnell et al., 1987). The complete amino acid sequence has been deduced from its cDNA (McDonnell et al., 1987). The 1,25(HO)$_2$D$_3$ receptor is highly homologous with avian oestrogen, proges-terone and glucocorticoid receptors and also with the gag–erbA gene product (see Section 11.4.4.1).

Phosphorylation of the 1,25-(HO)$_2$D$_3$ receptor is induced by 1,25-(HO)$_2$D$_3$ in embryonic chick duodenal organ culture. The phosphorylation occurs prior to initiation of Ca^{2+} uptake and induction of synthesis of Ca^{2+}-binding protein, which suggests a causal relationship (Brown & DeLuca, 1990). Phosphorylation appears to be generally involved in the modulation of steroid receptor activity.

10.2.5 The mineralocorticoid receptor

The domestic fowl small intestine is a good source of the mineralocorticoid receptor, since large fluxes of ions and water occur in the epithelial cells of the small intestine. The mineralocorticoid receptor from the domestic fowl intestine exists as a complex of ≈ 300 kDa associated with hsp-90 and has a high affinity for mineralocorticoids, e.g. aldosterone K_d ≈ 0.3 nM (Rafestin-Oblin et al., 1989).

10.3 Steroid hormone regulation of the synthesis of egg-white protein

Egg-white serves the roles of a nutrient for the developing embryo and of a stabilising cushion, against both physical stress, such as sudden vibration, and biological stress caused by microbial infection.

The principal organic solutes present in egg-white are proteins; their characteristics and possible functions are summarised in Table 10.3. Egg-white proteins are synthesised in the oviduct; most of them in the tubular gland cells, but avidin is synthesised in the goblet cells. Their synthesis in under hormonal control and has been extensively studied as a model of hormonal control of protein synthesis. The gene organisation (Fig. 10.4) is known for ovalbumin (Royal et al., 1979), located on the long arm of chromosome 2 (Dominguez-Steglich et al., 1992), ovotransferrin (Cochet et al., 1979), lysozyme (Quasba & Safaya, 1984), ovomucoid (Stein et al., 1980) and ovoinhibitor (Scott et al., 1987a). There is a family of avidin-related genes (avr1–avr5) in addition to the avidin gene. The avidin gene is expressed in both the oviduct and in the intestine in response to progesterone and to inflammation. The avidin-related genes are also expressed at a low level (Kunnas, Wallén & Kulomaa, 1993).

The oviduct shows two types of response to steroid hormones. When the immature hen is exposed to oestrogen (either naturally or precociously by injection), the primary response is elicited. This involves cell proliferation and differentiation in preparation for egg-laying and synthesis of egg-white proteins by the tubular gland cells. Luminal epithelial cells differentiate into three cell types; ciliated cells, goblet cells and tubular gland cells. Before the different cell types are visible, two subpopulations of cells can be distinguished. One of these expresses progesterone receptor before the onset of sexual maturity (constitutive expression) and the other during sexual maturity or after oestrogen administra-tion (inducible expression). The former are believed to be progenitors of the tubular gland cells and the latter of the ciliated and goblet cells (Pageaux et al., 1989). If oestrogens are withdrawn after this primary response, egg-white protein synthesis ceases, but regression of the differentiated tissue does not occur. If hormone is readministered, there is more rapid and pronounced induction of egg-white protein synthesis, and this is known as the secondary response. The primary response can only be induced by oestrogen, but the secondary response can be induced by oestrogens, progestrins, glucocorticoids and andro-gens. In studying the hormonal induction of protein synthesis in the oviduct, it is the secondary response that has been studied.

The major female sex hormones are oestradiol-17β and progesterone. Their interactions are complex, at times acting synergistically and at times antagonisti-cally. Oestrogen is required for differentiation of the

Table 10.3. *Properties of the principal egg-white proteins*

Protein	Amount in egg white (%)	Size (kDa)	Possible function
Ovalbumin	54	45	Nutrient, ligand binding
Ovotransferrin (conalbumin)	12	76	Fe^{3+} binding
Ovomucoid	11	28	Proteinase inhibitor
Lysozyme	3.4	14.3	Enzyme, hydrolysing peptidoglycan links in bacterial cell wall
Ovoinhibitor	1.5	49	Proteinase inhibitor
Ovomucin	1.5	210 and 720	Structural, increases viscosity
Ovoglobulin G2	1.0	47	?
Ovoglobulin G3	1.0	50	?
Riboflavin-binding protein	1.0	29.2	Riboflavin transport
Avidin	0.5	68.3	Binding biotin
Ovostatin	0.5	780	Proteinase inhibitor
Cystatin	0.05	13.1	Proteinase inhibitor
Thiamin-binding protein	?	38	Thiamin transport

Source: Stevens (1991b).

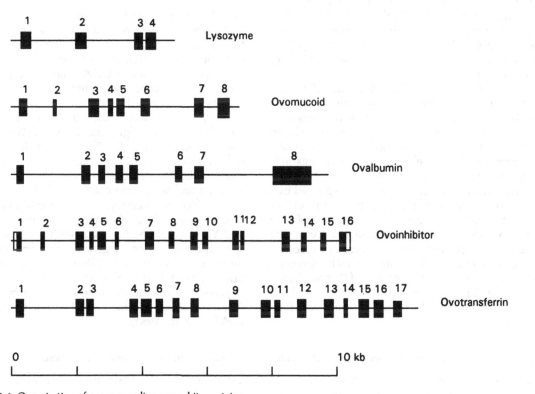

Fig. 10.4. Organisation of genes encoding egg-white proteins.

goblet cells, but further stimulation by progesterone is required to secrete avidin. The rate of synthesis of the four principal proteins (ovalbumin, ovotransferrin, ovomucoid and lysozyme) is not coordinated, although synthesis occurs in the same cell type. The proportion of ovomucoid and ovotransferrin relative to ovalbumin is increased when progesterone and testosterone are given in addition to oestrogens (Palmiter, 1972). Synthesis of ovotransferin mRNA occurs without a lag, to reach a maximum 4 h after stimulation, in contrast to ovalbumin, ovomucoid and lysozyme where increased mRNA is detectable after 2 h (Fig. 10.5). The secondary response involves a rapid increase in protein synthesis in the tubular gland cells of the oviduct. Ovalbumin accounts for about 60% of the total protein synthesis, ovotransferrin 10%, ovomucoid 8% and lysozyme 2–3% (de Pomerai, 1990). Cystatin, a thiol proteinase inhibitor, is a minor constituent of egg-white (0.05% egg-white protein). Its mRNA has been detected in a number of tissues, including oviduct (Colella et al., 1989). Although the levels of cystatin mRNA increase in liver upon maturation, there is no evidence for an increase in the oviduct during ovulation or in response to oestradiol, suggesting that unlike the other proteinase inhibitors in egg-white, it is not regulated by oestrogens (Colella, Johnson & Bird, 1991). The rates of synthesis of specific proteins in the oviduct relate closely to the final protein composition of egg-white. There are very large increases in the rates of transcription and translation. For example, in an unstimulated tubular gland cell, there are 10–30 molecules of ovalbumin mRNA; this rises to about 10 000 at 12 h after hormone stimulation, and 70 000 at the point of maximum stimulation. After binding their receptors, hormones interact with the hormone response elements on the chromatin to stimulate transcription. The different time course for ovotransferrin synthesis can only be explained by a detailed examination of the gene structures and their associated control regions. These have been studied most extensively with ovalbumin and lysozyme. One that has proved useful in identifying control regions is to examine chromatin from stimulated and unstimulated tissue for regions of DNAase hypersensitivity (Gross & Garrard, 1988). The results of these analyses are discussed in the next three subsections.

10.3.1 Control of ovalbumin biosynthesis in the oviduct

Ovalbumin synthesis in the oviduct is regulated by oestrogens, progestins, glucocorticoids and androgens.

Although oestrogen receptors are present in both liver and oviduct, ovalbumin is not expressed in the liver (Bloom & Anderson, 1982). In contrast, ovotransferrin is expressed in both oviduct and liver, although it is regulated in the liver primarily by serum iron levels (Dierich et al., 1987). The lack of expression of ovalbumin in the liver is not through a lack of cell-specific transcription factors. There are four DNAase-hypersensitive regions upstream of the ovalbumin gene and within 6.2 kb of the coding sequence (Fig. 10.6). Of these sites, II and III can be detected after oestrogen treatment but disappear on oestrogen withdrawal. Altogether five sequences appear to be involved in oestrogen-induced ovalbumin synthesis: a single GGTCA (-43 to -47) upstream of the TATATA (-26 to -32), which together form part of the promoter, and four TGACC sequences forming the enhancer element. The enhancer element acts synergistically with the promoter to enhance the binding of the oestrogen receptor. An unusual feature is that the four TGACC half-palindromic motifs are each separated by more than 100 bp. In addition there is a COUP-TF-binding site at -69 to -84, which overlaps with one of the oestrogen-response elements, negative control elements at -132 to -425, and also a possible NF-1 binding site further upstream (Fig. 10.6).

Another approach to understanding the nature of the control regions is to study the in vitro alignment of nucleosomes on DNA fragments. Using a 12 kb domestic fowl genomic DNA fragment containing the ovalbumin gene and adding the core histones together with histone H5 (see Section 9.5), Lauderdale & Stein (1992) were able to show that DNA contained discrete alignment signals on two of the large introns, whereas the 5′ and 3′ flanking regions did not undergo nucleosome alignment. This suggests that the flanking regions are most likely to remain exposed and so are more able to bind the various protein factors necessary for their transcription.

Upstream of the ovalbumin gene are two genes, X and Y, which are homologous with the ovalbumin gene and have arisen by gene duplication. All three are contained within 46 kb DNA. All three genes are under steroid hormonal control, and all three are expressed in the oviduct of the laying hen, but X and Y at only about 2% the level of ovalbumin mRNA (LeMeur et al., 1981).

10.3.2 Control of lysozyme biosynthesis

Lysozyme biosynthesis in the tubular gland cells of the oviduct is regulated by steroid hormones in

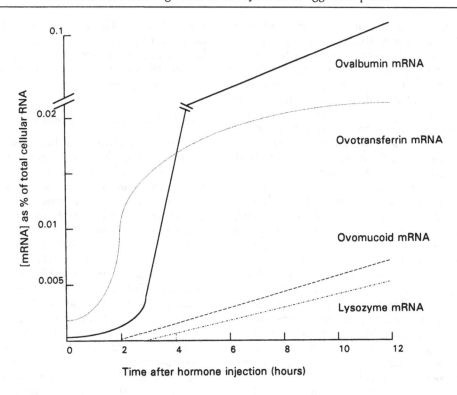

Fig. 10.5. Oestrogen induction of egg-white protein mRNAs (de Pomerai, 1990).

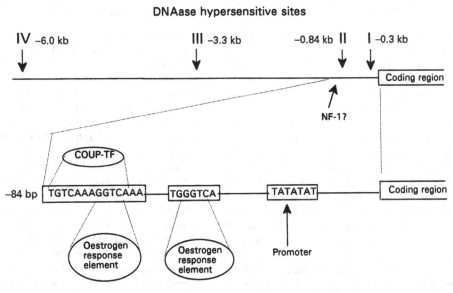

Fig. 10.6. Control elements and DNAase I-hypersensitive sites upstream of the ovalbumin-coding sequence (Deeley *et al.*, 1992).

much the same way as that of ovalbumin. In addition, it is synthesised constitutively in macrophage cells upon maturation. It is, therefore, possible to compare the promoter and enhancer regions of the lysozyme gene in tissues in which lysozyme synthesis is inducible, constitutive or does not occur (see Section 8.5). Lysozyme is not synthesised in liver or in erythrocytes, although liver does have steroid hormone receptors, in contrast to erythrocytes. When chromatin fragments containing the lysozyme gene (≈ 20 kb) are incubated with DNAase, nine hypersensitive sites are found (see Table 8.4). Seven of these are upstream of the coding region and include three enhancers (at -6.1, -2.7 and -0.7 kb), a silencer (-2.4 kb), a promoter (-0.1 kb) and a progestin-response element (-1.9 kb). No function has yet been attributed to the hypersensitive site at -7.9 kb. The promoter region itself, in addition to mediating the effects of the upstream regulatory elements (enhancers and silencers), contains sequences of regulatory significance. A hormone-response element controlling induction by progestins within the promoter region has been localised at -164 to -208 bp (Dölle & Strätling, 1990). The glucocorticoids bind to the same region, but footprinting also reveals a second region within the promoter (-49 to -80 bp) to which the glucocorticoids bind. By comparison of hormone-induced and uninduced oviduct, it is found that the -1.9 kb progestin-response element disappears from the former (Hecht *et al.*, 1988).

By contrast, only the -2.4 kb site is detectable in liver. Macrophages have the -6.1, -2.7, -0.7 and -0.1 kb sites; the -2.4 kb silencer element, present in immature macrophages, disappears on maturation and so is concerned with developmental regulation. The functioning of the promoter is complex, since it is influenced by both *cis*- and *trans*-acting elements. (*Cis*-acting elements are elements acting on an adjacent region of DNA, whereas *trans*-acting elements are not adjacent and act either via a diffusible intermediate or by looping of the back of the DNA.) Each enhancer element may also interact with a number of protein factors. The -6.1 kb enhancer element consists of ≈ 150 bp, and within this are at least six elements contributing to the enhancer function (Grewal *et al.*, 1992). A nuclear transcriptional regulatory protein, HNF-1α, first isolated from liver was believed to have a liver-specific function as an essential factor for activation of a number of liver-specific proteins, including vitellogenin. In mammals it is expressed in liver but not oviduct, but in birds it is expressed in both. It binds strongly within the -2.4 kb silencer element next to the hormone-response elements and may, therefore, have a role in oestrogen induction of lysozyme, although this still has to be proved (Grajer, Hörlein & Igo-Kemenes, 1993).

10.3.3 Ovomucoid and ovoinhibitor

Although ovomucoid has been sequenced from a large number of different avian species (see Section 9.2), the control of its biosynthesis has been less extensively studied than that of ovalbumin and lysozyme. Genes for both ovomucoid and ovoinhibitor are expressed in the domestic fowl oviduct, but only ovoinhibitor is expressed in the liver (Scott *et al.*, 1987a). The DNAase sensitivity of a 47 kb region of chromatin from domestic fowl oviduct that contains both ovomucoid and ovoinhibitor genes has been examined (Scott *et al.*, 1987b). The region spanning both genes is DNAase sensitive, with highest sensitivity through the region of the ovomucoid gene (Fig. 10.7). Within this 47 kb region, there are five CR1 sequences, three downstream of the ovomucoid gene, one between the two genes and the fifth within the ovoinhibitor gene. CR1 are SINES (see Section 8.4), of which there are about 1500–7000 copies per haploid domestic fowl genome.

10.4 Steroid hormone regulation of the biosynthesis of egg-yolk protein

In contrast to the egg-white proteins that are synthesised in the oviduct, the egg-yolk proteins are synthesised in the liver and transported to the developing oocyte via the plasma. Four types of protein make up $> 95\%$ of the total egg-yolk proteins: namely low-density lipoprotein, lipovitellin, phosvitin and livetin (Table 10.4). Egg-yolk protein synthesis can be as high as 1.0–1.5 g protein per day at peak laying. LDL accounts for 93% of yolk lipids, the bulk of the remainder being present in lipovitellin. For a review of lipovitellin structure see Banaszak, Sharrock & Timmins (1991). Because of the high lipid content of egg-yolk, it is more difficult to purify and characterise the proteins and so these have been less extensively studied than those of egg-white. Besides the proteins mentioned above, there are also small amounts of enzymes and binding proteins. The best characterised of the binding proteins are ovotransferrin, which binds Fe^{3+}, and riboflavin-binding protein. Both of these proteins are present in egg-yolk and egg-white. Those in the former are synthesised in the liver, and in the latter in the oviduct. They are each

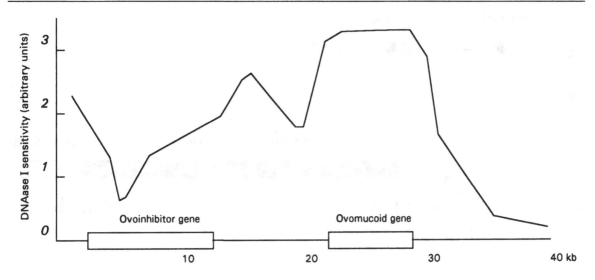

Fig. 10.7. DNAase I sensitivity of the ovomucoid and ovoinhibitor complex genes (Scott *et al.*, 1987b).

products of single genes but differ slightly in structure. Ovotransferrin has slight differences in the glycosylation pattern (Williams, 1968), and riboflavin-binding proteins have differences in the glycosylation pattern and proteolytic cleavage (White & Merrill, 1988).

Hormonal control of egg-yolk protein synthesis has been most extensively studied with vitellogenin and VLDL (Burley & Vadehra, 1989). Unlike egg-white protein synthesis, there is only a primary response, and this does not require cell proliferation or DNA synthesis (Deeley *et al.*, 1992). The livetins differ from the other major yolk proteins in that their synthesis is not under hormonal control (Burley, Evans & Pearson, 1993). Both α- and β-livetins are synthesised in the liver. They correspond in size and amino acid composition to serum albumin and αII globulin, respectively, which are probably their precursors. Similarly γ-livetin resembles γ-globulin and it is synthesised in the bone marrow.

10.4.1 Vitellogenin biosynthesis

Vitellogenins are synthesised in the parenchymal liver cells in response to oestrogens. They are secreted into the plasma and are taken up by the developing oocytes, where they are cleaved into lipovitellin, phosvitin and phosvettes. Vitellogenins are composed of two identical monomers, each of 240 kDa. There are three distinct vitellogenin poly-peptide chains VTGI, VTGII and VTGIII which, in domestic fowl, occur in the ratio 0.33:1.0:0.08. Three chains have also been detected in quail and two in

ducks (Burley & Vadehra, 1989). The three polypept-ides are encoded in three separate genes *vtgI*, *vtgII* and *vtgIII*. The complete sequence of the *vtgII* gene (20 342 bp) has been determined and comprises 35 exons (Fig. 10.8). The minor gene *vtgIII* has been cloned and it has a pseudogene Ψ*vtgII* 1426 bp upstream. Similarly *vtgII* has a pseudogene Ψ*vtgIII* 1345 bp downstream (Silva, Fischer & Burch, 1989). There are both oestrogen- and glucocorticoid-response elements upstream of the coding region for *vtgII*, and oestradiol exposes DNAase-hypersensitive sites in the promoter region. This is associated with de-methylation of specific methyl-CpG dinucleotides within the DNAase-hypersensitive site of the promoter and the oestrogen-response element. The oestrogen-response elements are at -613 and -335 and contain the 13 bp motif 5′-GGTCANNTGACC-3′. DNA footprinting of the oestrogen-response element revealed that it is protected by protein binding in both hen and rooster (McEwan, Saluz & Jost, 1991). This suggested that a protein, distinct from the oestrogen receptor (which is absent in unstimulated roosters), is also able to bind specifically to the oestrogen-response element associated with the vitellogenin genes. Such a non-histone protein with $M_r \approx 75\,000$–$80\,000$ has been found in nuclear extracts from both hen and rooster liver. McEwan *et al.* (1991) suggest that the protein might stabilise the dimer contacts of the oestrogen receptor on binding DNA. Slater, Redeuilh & Beato (1991) propose, on the basis of DNAase I footprinting and methylation protection experiments, that there are four hormone

169

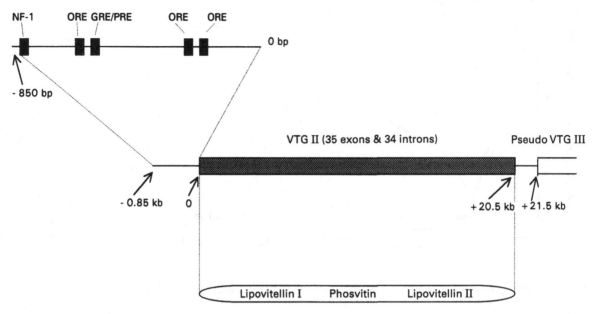

Fig. 10.8. The principal vitellogenin gene (*vtgII*). The gene comprises 35 exons and 34 introns spanning 20.5 kb. It is shown as a single block, with the control regions upstream and Ψ *vtgIII* downstream. Vitellogenin (VTG) is cleaved into three main proteins, lipovitellins I and II and phosvitin, in the oocytes. The regions encoding each of these proteins are shown. GRE, PRE, ORE, glucocorticoid-, progestin- and oestrogen-response elements, respectively (Schip *et al.*, 1987; Silva *et al.*, 1989; Jost *et al.*, 1990).

receptor-binding sites between -721 and -591. Two distal and the most proximal are recognised by the oestrogen, glucocorticoid and progesterone receptors, whereas the central binding site is only recognised by the oestrogen and glucocorticoid receptors. Progesterone and oestrogens act synergistically.

Vitellogenin translation occurs on the rough endoplasmic reticulum of the liver after induction by oestrogens, which regulate synthesis at the transcriptional level. Induction occurs at the onset of egg-laying, but can also be induced in roosters by a single injection of oestradiol. In immature chicks, an initial injection induces primary stimulation and there is a lag of about 12 h before the appearance of vitellogenin mRNA, which eventually returns to background levels. If at this stage another injection is given, the increase in mRNA occurs without any lag. The completed vitellogenin polypeptide is glycosylated prior to phosphorylation, which in turn occurs immediately before secretion from the liver. Much of the phosphorylation occurs on serine residues, which become the phosvitin moiety (Byrne, Gruber & Ab, 1989). The secreted vitellogenins circulate in the plasma as one of the HDLs and are taken up by the oocyte, which has a high-affinity Vtg-specific receptor

(M_r 96 000, K_d 2×10^{-7} M, Stifani, George & Schneider, 1988). Vitellogenin undergoes proteolytic cleavage catalysed by cathepsin D, yielding lipovitellin I and II, phosvitin and phosvettes (Retzek *et al.*, 1992).

Phosvitins are unusual proteins; 54% of their amino acid residues are serine, and 95% of these are phosphorylated. Two different phosvitins (28 and 34 kDa) have been detected in oocytes of domestic fowl, duck and turkey (Burley & Vadehra, 1989). Their principal role is thought to be that of a phosphate reservoir. The two lipovitellins have similar amino acid compositions, 15–20% lipid and slightly different phosphate contents (0.5% α and 0.27% β). Phosvitin appears to be encoded in the central part of the vitellogenin genes, with the lipovitellins on either side (Fig. 10.8). The lipovitellins contain the binding site recognised by the 95 kDa oocyte receptor (Steyrer, Barber & Schneider, 1990).

10.4.2 *ApoVLDL-II biosynthesis*

Most of the *de novo* lipid biosynthesis in birds occurs in the liver (see Section 4.3), and this is transported to peripheral tissues (including the developing oocytes) in the form of VLDL. VLDL comprises about 13% protein and the rest is triglyceride, phospholipid,

Table 10.4. *Properties of the principal egg-yolk proteins*

Protein	% of total protein	Proportions polypeptide: lipid: phosphorus	Mass of the polypeptide (kDa)	Precursor
VLDL	22	13:86:0.2	Apo-B, 500 ApoVLDL-II, disulphide-linked homodimer of 9.5 kDa subunits	Serum VLDL
Phosvitin	12	80:0:9	28–35	Vitellogenin
Lipovitellins α, β	36	α: 75:24:0.5 β: 73:27:0.3	120 and 30	Vitellogenin
Livetins α	36	100:0:0	70	Serum albumin
β			42	Serum αII globulin
γ			150	Serum γ globulin
δ			200	Serum γ globulin
Ovotransferrin	<2	100:0:0	71	Serum globulins
Riboflavin-binding protein	<2	99:0:0.2	36	Serum riboflavin-binding protein
Biotin-binding protein	<2		56	–
Thiamin-binding protein	<2		38	–
Retinol binding protein	<2		21	Serum retinol-binding protein

Source: Burley & Vadehra (1989); Feeney & Osuga (1991); Burley *et al.* (1993); Vieira & Schneider (1993).

cholesterol and cholesteryl esters. The principal protein component is apolipoprotein B (apoB or apoVLDL-I) and a second smaller component is apovitellenin I (apoVLDL-II) (Table 10.4). Although apoB is much more abundant than apoVLDL-II in VLDL from laying hen (Burley & Vadehra, 1989), because of the smaller size of apoVLDL-II (Table 10.4) the hormonal control of its synthesis has been more extensively studied. The gene organisation including the 5′-flanking region is shown on Fig. 10.9. The coding region comprises four exons and three introns (Schip *et al.*, 1983). Its expression is liver specific and oestrogen induced. The induction is very dramatic: it is undetectable in the livers of untreated roosters and is expressed at $\approx$ 400 000 copies of mRNA per cell in the laying hen. This can result in a 400-fold rise in the plasma concentration of VLDL (Kudzma, Swaney & Ellis, 1979).

There are a number of response elements in the 5′-flanking region (Fig. 10.9). The two-oestrogen response elements (-163 to -177 and -208 to 220) act synergistically, producing a potent response element. In addition, another important *trans*-acting protein is the liver-enriched transcription factor (C/EBP) that binds elements B_1, B_2 and D (Beekman

et al., 1991). These two together are sufficient to account for both hormone dependence and tissue specificity. There are, however, at least two more factors involved. COUP-TF (see Section 10.1) binds at -54 kb, and liver factor A1 (LF-A1, also known as HNF-4), also binds at -54 kb but has, in addition, a weak recognition site at C (Fig. 10.9). There are also several potential glucocorticoid response elements between -1.2 kb and -4.5 kb (Deeley *et al.*, 1992). These binding sites are only observed *in vivo* in chromatin from the laying hen or oestrogen-stimulated rooster. Beekman *et al.* (1991) suggest that in livers not expressing apoVLDL-II, DNA is folded in a nucleosome structure and that activation of the oestrogen receptor causes unfolding which exposes the various response elements. Initiation of transcription can occur at three alternative start points, the main site and two minor sites 1105 and 1530 bp upstream from the apoVLDL-II gene. The function if any, of the minor sites is not clear, especially as their RNAs do not appear in the polysomal RNA fractions and so do not appear to be translated (Strijker *et al.* 1986). There is also evidence of a control region much further upstream at -2600 bp, which resembles binding sites for factors such as C/EBPα. The

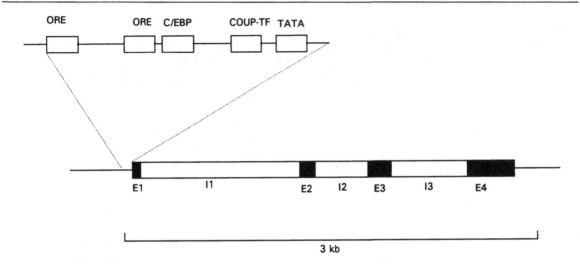

Fig. 10.9. ApoVLDLII gene organisation showing four exons, three introns and the upstream control elements (Deeley *et al.*, 1992). ORE, oestrogen-response element; C/EBP, liver enriched transcription factor.

domestic fowl C/EBPα is a transcription factor of 324 amino acid residues, containing a DNA-binding site and a leucine zipper (see Branden & Tooze, 1991), enabling it to dimerise like many other DNA-binding proteins involved in transcription (Calkhoven, AB & Wijnholds, 1992).

10.4.3 Developmental expression of apoVLDL-II and vitellogenin

Demethylation of nucleotides at the control region −2600 bp upstream of the apoVLDL-II gene occurs between E7 and E9, and this stage corresponds with the stage at which the apoVLDL-II gene can be first induced prematurely by oestrogen administration (Hoodless *et al.*, 1992). This site is most likely to be concerned with developmental regulation, which is necessary before the liver becomes oestrogen responsive. At least five yolk protein genes (*vtgI, vtgII, vtgIII, apoVLDL-II* and *apoB*), which are responsive to oestrogens, are developmentally regulated (Evans *et al.*, 1987). Their developmental regulation correlates with the number of oestrogen receptors per cell, which continues to increase up to ≈ 3500 per cell at 6 weeks after hatching. The times of embryonic development, at which the genes for yolk proteins can become activated by oestrogens, varies as follows: *apoB* at E7, *apoVLDL-II* at E8, *vtgII* at E11, and *vtgI* and *vtgIII* at E13 (Deeley *et al.*, 1992). ApoB and apoVLDL are synthesised on the rough endoplasmic reticulum. VLDL is then assembled in the

Golgi apparatus and secreted into the plasma. After VLDL has been taken up by the developing oocyte, apoB is cleaved into smaller polypeptides, the principal ones known as apovitellenins III, IV, V and VI.

10.5 Glucocorticoid regulation of protein synthesis

Glucocorticoids modulate the expression of a variety of genes. In the most well-characterised examples, positive modulation of gene expression is found, but negative regulation also occurs. The induction of tyrosine aminotransferase and tryptophan oxygenase in mammalian liver were two of the earliest reported examples of hormone-induced protein synthesis (see Lucas & Granner, 1992). Other proteins induced include phosphoenolpyruvate carboxykinase, glutamine synthetase, α_2-globulin and metallothionein. Negative regulation by glucocorticoids occurs with the α-subunits of glycoprotein hormones such as pro-opiomelanocortin and α-fetoprotein. The glycoprotein hormones are heterodimers comprising α- and β-subunits. The α-subunit is smaller and is almost identical for all such hormones from one species, whereas the larger β-subunit determines the specificity of the hormone. These hormones include thyroid-stimulating hormone, luteinising hormone, follicle-stimulating hormone and chorionic gonadotrophin. Thus glucocorticoids regulate a number of other hormones. Most of the studies on the effects of glucocorticoid hormones have been carried out in mammals, although many of the effects are probably similar in birds. Apart from the secondary response in the hen oviduct (see Section 10.3) causing induction of egg-white protein synthesis, glucocorticoids have been shown to induce glutamine synthetase in avian retina and metallothionein in the liver of domestic fowl (Wei & Andrews, 1988). Metallothioneins are small cysteine-rich proteins that bind heavy metals with high affinity and provide protection from metal toxicity. A single isoform has been isolated from domestic fowl and turkey, but two have been described in quail (Wei & Andrews, 1988). cDNA encoding metallothionein from domestic fowl has been sequenced, and by Northern blot hybridisation it has been shown that glucocorticoid injection induces ≈ 50-fold increase in metallothionein mRNA in domestic fowl liver. Its transcription is also induced by heavy metal ions (Cd^{2+}, Zn^{2+}, Cu^{2+}). Glutamine synthetase catalyses the reaction

$$glutamate + NH_3 + ATP \rightarrow glutamine + ADP + P_i$$

which occupies a pivotal position in nitrogen metabolism, being involved in the uptake of ammonia and in the formation of precursors for neurotransmitters, glutamate and γ-aminobutyric acid (GABA). Its catalytic activity is controlled by effectors such as 2-oxoglutarate. It can be induced by glucocorticoids in tissues such as embryonic chick retina and skeletal muscle. In the chick retina, it is inducible by E8, and a glucocorticoid response element has been identified at − 1849 to − 2120 upstream of the coding region (Zhang & Young, 1991). An increase in specific activity of glutamine synthetase of > 100-fold occurs between E7 and E14, accompanied by a parallel increase in glutamine synthetase mRNA. There is an increase in glucocorticoids produced by the adrenals at this stage of development. Glutamine synthetase is produced constitutively, but glucocorticoids potentiate its activity during development (Patejunas & Young, 1990). Increased transcriptional activity of the glucocorticoid receptor in the embryonic retina is developmentally controlled and is linked to the competence for induction of glutamine synthetase, which increases rapidly from a minimum at E6 to a high level at E10 (Ben-Dror, Havazelet & Vardimon, 1993).

Glucocorticoids also promote the lipolysis of triglycerides from adipose tissue. The lipoprotein lipase gene from domestic fowl has four glucocorticoid-response elements upstream of the coding sequence, accounting for its sensitivity to glucocorticoids (Cooper *et al.*, 1992).

10.6 1,25-Dihydroxyvitamin D₃ regulation of avian calcium metabolism

The steroid modulator $1,25(HO)_2D_3$ is not secreted by an endocrine gland but is synthesised from vitamin D_3 by two successive hydroxylations, occurring in the liver and kidney (Fig. 2.5, p. 21). The dietary role and metabolism of vitamin D is discussed in Section 2.4.2. In common with the steroid hormones, $1,25(OH)_2D_3$ acts principally by activating gene transcription. The main role of $1,25(OH)_2D_3$ is in the regulation of calcium metabolism, along with two other hormones, the parathyroid hormone and calcitonin. The latter is secreted by the ultimobranchial glands adjacent to the parathyroid gland. Generally, $1,25(HO)_2D_3$ and the parathyroid hormone act synergistically in increasing serum Ca^{2+}, whereas calcitonin decreases serum Ca^{2+} by inhibiting the resorption of Ca^{2+} from bone and kidney. $1,25(HO)_2D_3$ also participates in the regulation of cellular proliferation and differentiation, secretion of polypeptide hormones, induction of enzymes catalysing the degradation of vitamin D and modulation of certain aspects of the immune system. However, the molecular basis of these effects are not well understood and will not be discussed further. The three tissues that play a major role in the movement of Ca^{2+} are the small intestine, which regulates the dietary uptake of Ca^{2+}, bone, which is the largest body reserve of Ca^{2+}, and the uterus of the laying hen. The last of these three is particularly important in egg-laying vertebrates. Typically, about 10% of the total body Ca^{2+} is used in a single egg.

The mode of action of $1,25(HO)_2D_3$ is similar to that of other steroids; $1,25(HO)_2D_3$ is taken up by the target tissue, where it binds to the receptor. The receptor combines with the response element lying upstream of the coding region of the particular genes that are $1,25(HO)_2D_3$ sensitive. $1,25(HO)_2D_3$ receptors have been detected in the following avian tissues: intestine, kidney, bone, parathyroid gland, pancreas, pituitary, chorioallantoic membrane and egg-shell gland (Norman, 1987). The receptor (see Section 10.2.4) is present in very small amounts even in the small intestine, the tissue in which it is most abundant; the levels are generally an order of magnitude lower than for other steroid receptors. $1,25(HO)_2D_3$ is able to modulate the production of its own receptor by causing a direct increase in mRNA for $1,25(HO)_2D_3$ receptor.

A number of proteins are expressed as the result of $1,25(HO)_2D_3$ stimulation, but the one most studied is the Ca^{2+}-binding protein *calbindin D_{28K}*. It is referred to as calbindin D_{28K}, because, unlike mammalian calbindins which have a mass of $\approx 10\,kDa$, avian calbindins are $28\,kDa$. A $9\,kDa$ calcium-binding protein has been detected in chick embryo myoblasts, using antibody raised against rat calbindin D_{9K}, but its importance has still to be assessed (Lila, Susana & Ricardo, 1994). Other proteins expressed include α-tubulin, osteocalcin, alkaline phosphatase and Ca^{2+}-ATPase. Calbindin D_{28K} has been detected in avian intestine, kidney, bone, parathyroid gland, pancreas, chorioallantoic membrane, egg-shell gland and pineal gland (Norman, 1987; Pochet et al., 1994). Its gene spans $19\,kb$ and comprises 11 exons and 10 introns. A putative $1,25(HO)_2D_3$-response element has been located by footprint analysis (Boland et al., 1991).

The avian small intestine is able to respond rapidly to changes in both dietary Ca^{2+} and serum levels of $1,25(HO)_2D_3$. When dietary Ca^{2+} is lowered, the concentration of calbindin D_{28K} in intestinal epithelium increases and vice versa. An increase in the serum $1,25(HO)_2D_3$ results in increased calbindin D_{28K} and increased transport of Ca^{2+} from the lumen of the intestine across the epithelial cells within 30 min (Nemere et al., 1991). Calbindin D_{28K} and tubulin, a component of microtubules, are actively involved in this movement. The intestinal epithelial cells may be in either an active or an inactive state with regard to transcellular Ca^{2+} transport, depending on the Ca^{2+} status of the bird. When actively transporting Ca^{2+}, calbindin D_{28K} and tubulin are located in the villus core, but, when transport is inactive, these molecules are found in the brush border and basal-lateral membrane (Nemere et al., 1991). The sequence of events following the stimulation of transcellular Ca^{2+} transport by increased serum $1,25(HO)_2D_3$ are thought to be as follows: (i) uptake of $1,25(HO)_2D_3$ and binding to the receptor, resulting in increased expression of calbindin D_{28K}, (ii) energy-dependent uptake of Ca^{2+} across the brush border, (iii) Ca^{2+} bound to calbindin D_{28K} within endocytic vesicles fuses with lysosomes, (iv) transfer of organellar Ca^{2+} across the cell along microtubules, and (v) exocytosis of Ca^{2+} at the basal–lateral membrane, probably involving Ca^{2+}/ATPase (Norman, 1987; Nemere et al., 1991). There is also evidence for a second mode of action of $1,25(HO)_2D_3$, in which Ca^{2+} channels are opened in the plasmalemma allowing an influx of Ca^{2+}, which may function as a second messenger. A $1,25(HO)_2D_3$ receptor has been identified in the plasmalemma of the cells of the small intestine that stimulates Ca^{2+} transport by a mechanism which does not involve gene expression (Nemere et al., 1994).

Calbindin D_{28K} in the shell gland of the uterus is also responsive to serum $1,25(HO)_2D_3$. Egg-laying causes a high loss to the body Ca^{2+} reserves, which must be met either by increased dietary uptake or by mobilisation of bone Ca^{2+}. A diet deficient in vitamin D causes a decline in the calbindin D_{28K} concentration of the uterus. However, other steroids are also involved in controlling its level. During the pauses between egg-laying, there is a decrease in the concentration of calbindin D_{28K} and its mRNA in the uterus, although the serum concentration of $1,25(HO)_2D_3$ remains relatively constant. Both oestradiol and progesterone influence calbindin D_{28K} mRNA transcription in the uterus (Corradino, 1993).

$1,25(HO)_2D_3$ also regulates the expression of osteocalcin. Osteocalcin is a small Ca^{2+}-binding protein (M_r 5500) of 49 amino acid residues, of which three are γ-carboxyglutamic acid (see Section 2.4.3). It is the most abundant non-collagenous protein associated with bone matrix. Its function is unclear, but it is thought to be involved the mineral turnover of bone (Nys, 1993). The $1,25(HO)_2D_3$ receptor binds to response elements associated with the osteocalcin gene, promoting increased transcription. This differs from its action on calbindin D_{28K}, in that it modulates rather than initiates osteocalcin synthesis, since there is significant osteocalcin synthesis in vitamin D-deficient chicks, where serum $1,25(HO)_2D_3$ levels are low.

$1,25(HO)_2D_3$ also has a role in bone resorption by osteoclasts. For bone resorption to occur, the osteoclasts must attach to the bone matrix. The attachment occurs through a transmembrane adhesion molecule, integrin. Integrin is a heterodimer, which in domestic fowl osteoclasts has the subunit structure $\alpha_V\beta_3$. The subunit β_3 is generally present in lower concentrations than α_V. At concentrations of $10^{-11}\,M$, $1,25(HO)_2D_3$ activates the β_3-subunit gene, and this results in a higher concentration of integrin ($\alpha_V\beta_3$) expressed on the cell surface of the osteoclasts and promotes the attachment of the osteoclasts to the bone matrix (Mimura et al., 1994).

10.7 The formation of oestradiol from testosterone and its physiological significance

Testosterone is an intermediate in the biosynthesis of oestradiol. The conversion, which involves three successive hydroxylations and the loss of a methyl group, is catalysed by the enzyme aromatase. Aromatase has three catalytic activities: 19-hydroxylase, 19-hydroxysteroid dehydrogenase and

10,19-lyase. These reactions require six reducing equivalents from three NADPHs and three molecules of molecular oxygen. The mechanism is not known for certain, but Fig. 10.10 is a proposed scheme. Aromatase is a microsomal enzyme and one of the cytochrome P-450 family. The amino acid sequence (503 residues) of aromatase from domestic fowl has been deduced from the cDNA sequence (McPhaul *et al.*, 1988); it is similar to that from mammals and fishes (Chen & Zhou, 1992). From comparison of the sequences, a model of its structure has been proposed in which the bulk of the enzyme is on the cytosolic side of the endoplasmic reticulum, but with the N-terminus forming a transmembrane anchor. Aromatase activity is present in traces in male avian tissues and is present in high concentrations in the ovaries of female birds. The aromatase gene is located on the long arm of chromosome 1 in the domestic fowl (Tereba, McPhaul & Wilson, 1991). The enzyme is important in the sex-determining mechanism. In embryos less than 7 days old, the gonads are bipotential, but after this period they begin to develop into either ovaries or testes depending on the genotype of the embryo. If, at E5, embryos are treated with an inhibitor of aromatase, preventing the formation of oestradiol, then at hatching 100% are vent-sexed as males. Whether they are male or female genotypes, they will appear as male phenotypes in the absence of oestradiol (Elbrecht & Smith, 1992). These female genotypes were capable of spermatogenesis. They also had the physical appearance of male birds, i.e. the male hackles, comb and wattle. Therefore aromatase activity plays a key role in the developmental switch.

A further manifestation of the role of aromatase is seen in the henny feathering traits in the males of certain breeds of domestic fowl, e.g Sebrights and Campines. Most breeds of domestic fowl show sexual dimorphism in feathering. The main differences between cock and hen feathering are that in the male the tail and neck feathers are long and curved, whereas in the female the corresponding feathers are short and stand erect. The henny feathering trait in the males results from a mutation that causes an increase in aromatase activity in the skin (George & Wilson, 1982). The mutation limits the rate of oestrogen production in the skin but not in the ovaries. The difference has been clearly demonstrated by culturing fibroblasts from the skin of normal domestic fowl and from those with the henny feathering gene (Leshin *et al.*, 1981). The rate of oestrogen formation was several 100-fold higher in the latter. The aromatase gene comprises nine exons,

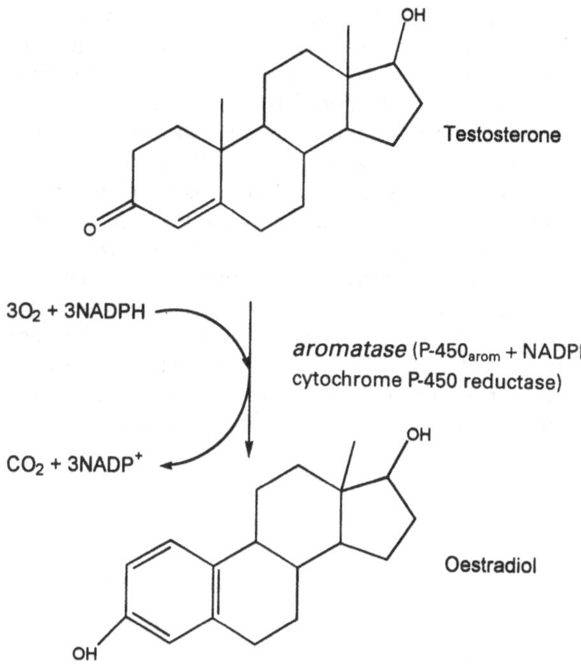

Fig. 10.10. Probable mechanism of action of aromatase.

and the site of transcription initiation is -147 upstream from the initiator methionine. Matsumine *et al.* (1991) have shown that the gene is organised in an identical fashion in breeds of domestic fowl both with and without the henny feathering trait. However, there is a difference in transcription. In strains having the henny feathering trait, there are two mRNAs detectable in the ovaries compared with one in the normal strains. The second of these mRNAs is also detectable in skin fibroblasts of Sebrights and Campines, but not in the Leghorn, which lacks the henny feathering trait. The transcription of the second type of mRNA is controlled by a different promoter, which is the result of a splicing event -25 upstream of the initiator methionine. This accounts for the expression of the aromatase gene in the skin of henny feathering breeds.

Another area in which aromatase activity is important is reproductive behaviour (Schlinger & Arnold, 1992). The copulatory and singing behaviour has been intensively studied in the quail and the zebra finch. In both species, aromatase activity is high in the ovaries and in certain regions of the brain, but very low in the testis and adrenals. In the zebra finch, no significant sex differences are found in the levels of circulating oestrogens at any age (Schlinger

& Arnold, 1992). The circulating oestrogens are generated by the ovary and brain, and, therefore, in the male the brain is likely to be an important contributor. Oestrogens have been shown to masculinise singing behaviour in the zebra finch. In the male quail, aromatase activity in the brain provides sufficient oestrogen to bind local oestrogen receptors and activate reproductive behaviour. It appears that brain-derived oestrogen, rather than plasma levels, are important in activation of reproductive behaviour. However, it is puzzling that no significant differences are found in the aromatase activities in different regions of the male and female brain. In the ring-necked dove, neural aromatase is related to behaviour, which in turn is synchronised with the reproductive cycle in both males and females (Hutchison, Wozniak & Hutchison, 1992). Aromatase from the preoptic area of the brain of the ring-necked dove has a higher substrate affinity than that from the ovaries (K_m 7×10^{-9} M and 29×10^{-9} M, respectively). These may be distinct enzymes, although the differences in K_m could arise through modifiers etc. In the male, the preoptic area aromatase appears to regulate the transition from aggressive behaviour to nest-oriented behaviour. In the brains of both male and female ring-necked doves, steroids are able to induce higher levels of aromatase activity. The brain aromatase activity appears to be important in the local generation of oestrogens, which are important sexual behavioural determinants.

11

Avian oncogenes

11.1 Introduction

This chapter focuses on the oncogenes that have been discovered in avian species and what is known about their mode of action. Since normal cells contain homologous genes known as proto-oncogenes, these are also discussed. Normal somatic cells grow and divide in a controlled manner, such that in an adult animal, the rate of cell division in a given tissue is controlled to enable cell replacement to occur without any large increase in the total number of cells. A number of features distinguish normal cells from transformed or cancerous cells; these are most easily seen in cells growing in culture, but they are also manifest in tumours present in whole tissues. The main differences evident in transformed cells grown in culture are (i) cells are more rounded and often appear in disorganised clusters, (ii) they proliferate in the absence of exogenous mitogenic stimuli such as growth factors, (iii) proliferation is not inhibited by contact inhibition, which occurs with normal cells, (iv) they will grow in liquid medium without adhering to a solid surface, and (v) when injected into animal hosts they give rise to tumours. To understand the basis of oncogenesis, it is first necessary to understand the process of cell division and the normal controls which regulate the cell cycle.

The first evidence for a tumour-inducing virus was from sarcomas in domestic fowl. Rous discovered in 1907 that a cell-free extract prepared from the minced extract of a sarcoma found in a Plymouth Rock caused sarcomas when injected into other domestic fowl. He postulated that the tumour was transmitted by a virus. The discovery received little attention then since it did not fit with the generally held theories of cancer at that time. Many years later, when a number of oncogenic viruses had been identified, the importance of Rous' early discovery became apparent. Rous was awarded the Nobel prize

for Medicine in 1966 at the age of 87 and died of bladder cancer in 1970. The arrangement of the four genes making up the Rous sarcoma virus (RSV) was established in 1976, and the complete 9312 nucleotide sequence of the genome solved by Schwartz, Tizard & Gilbert in 1983. The importance of RSV as a model system for understanding oncogenesis was by then unquestioned. Many other oncogenic viruses have been discovered, and an overall pattern is emerging. Many of the oncogenes so far characterised are from the domestic fowl. In a review of oncogenes, Bishop (1983) listed 25 oncogenes of which nine were from the domestic fowl. At least 50 oncogenes have now been discovered, and it is thought the number may rise to the region of 100 (Watson *et al.*, 1987) although the number of classes appears to be less than 30.

This chapter begins with an outline of the cell cycle and the importance of normal transcriptional factors and proto-oncogenes, emphasising those characterised in avian species (almost entirely from the domestic fowl). The different classes of avian oncogene are then described, together with their relationship to proto-oncogenes.

11.2 The cell cycle

The cycle of events that leads to the formation of two daughter cells from a single parent cell is known as the cell cycle. During the early 1950s, with the development of microspectrophotometry and autoradiography, it became apparent that DNA synthesis occurred during a restricted period of interphase. This led Howard & Pelc (1953) to introduce the nomenclature G_1-, S-, G_2- and M-phase to describe the phases of the cell cycle. For most of the life of the average eukaryote cell, it is in interphase. Interphase is subdivided into the G_1-phase, or the pre-synthetic gap phase, which preceeds the

177

S-phase, the phase during which DNA synthesis occurs, and this is followed by the post-synthetic gap phase (or G_2-phase), the gap before mitosis occurs. In cells that have ceased to divide and which require an external stimulus such as a hormone or mitogen, the term G_0-phase is used. An external stimulus is also required for a cell that has entered G_1-phase after completion of mitosis, except in the case of a tumour cell where the requirement is bypassed by the constitutive expression of oncogene proteins. Since this early period, many of the molecular events occurring during the cell cycle have become known, although understanding of its control is far from complete. Only a brief outline of the complex events is given here (for review, see Murray & Hunt, 1993).

Many studies of the cell cycle have used cells grown in culture that have been stimulated to divide by an external stimulus or mitogen. The commitment to grow and differentiate is made in G_1, and the point of the cycle at which they become committed to cell division is often referred to as the restriction point. It is one of three important control points in the cell cycle, the other two being the entry to mitosis and the exit from mitosis. Progression through the cell cycle after the restriction point does not depend on extracellular signals. The timing and initiation of mitosis occurs in G_2, after the complete duplication of the cells' genetic information. The transition from G_2- to M-phase is the best understood.

11.2.1 Transition from G_2- to M-phase

Two types of polypeptide that have key roles as regulators of the cell cycle are the *cdc* gene product, p34, and proteins called cyclins. Many recent developments in understanding the cell cycle have come from work on either budding or fission yeasts. A number of conditional mutants affecting the cell cycle have been isolated in yeasts; these are referred to as cell division cycle (*cdc*) mutants. The gene *cdc2*$^+$ encodes the p34 polypeptide, M_r 34 000, that associates with a second polypeptide, cyclin, to form a heterodimer known as cyclin-dependent kinase or, alternatively, maturation-promoting factor (MPF). The protein p34 is sometimes referred to as Cdc2 or p34^{cdc2}, but in this chapter we use p34. *Cyclins* are a group of proteins that are synthesised periodically during the cell cycle, appearing at highest concentration during M-phase. No enzyme activity has yet been attributed to them. It has been suggested that their role is either to determine the subcellular location or the specificity of p34. Since the initial discoveries in yeast, these polypeptides have been found in both

unicellular and multicellular organisms. There are a number of different cyclins (A, B1, B2, C, D1, D2, D3 and E) and also different catalytic subunits, related to p34, for example Cdk2, Cdk3, Cdk4 and Cdk5. They are able to interact with multiple partners to give rise to a wide range of functional cyclin-dependent kinases (Nigg, 1993), although it has not been established *in vivo* for many of these which interactions are important at particular stages of the cell cycle. p34–cyclin B is required for entry to mitosis, and cyclin A appears to function during S phase and the G_2–M transition (Nigg, 1993).

Two important questions about p34 for the overall understanding of cell cycle control are: (i) how is its activity regulated, and (ii) what are its physiological substrates? The activity of p34 itself appears to be controlled by phosphorylation. p34 from vertebrates has phosphorylation sites at Thr-14, Tyr-15 and Thr-167. During late G_2-phase the heterodimer (p34–cyclin B) accumulates in an inactive form. On entering M-phase there is a rapid dephosphorylation of Thr-14 and Tyr-15 which activates the cyclin-dependent kinase (Fig. 11.1). Dephosphorylation is catalysed by a phosphatase, and at least three different phosphatases are known that act on cyclin-dependent kinase in higher vertebrates. The cDNA encoding p34 from domestic fowl has been cloned and its expression studied in developing chick embryos. When whole embryos were examined, Krek & Nigg (1989) found a correlation between the rate of cell proliferation and the levels of both p34 and its mRNA. For example, between E3 and E11 there was a $\approx$ 3-fold reduction in p34 and a $\approx$ 4-fold reduction of its mRNA. However, when individual tissues such as liver and brain were compared at the same stage of development, whilst the amounts of p34 were very similar, the amounts of the corresponding mRNA were widely different. Krek & Nigg (1989) were unable to detect p34 mRNA in adult tissues and assumed that only very low levels would be present in tissues that were not proliferating. They suggest that multiple controls regulate the cyclin-dependent kinase activity, the long-term changes being effected by transcriptional regulation, and the short-term changes by post-translational modification, the latter accounting for the differences between embryonic liver and brain.

Several substrates of cyclin-dependent kinase have been identified from *in vitro* studies, including nuclear lamins, vimentin, caldesmon, histone H1 and RNA polymerase II (Table 11.1) and these are the most likely physiological substrates. The onset of mitosis is marked by at least four morphological changes:

Table 11.1. *Some of the possible substrates phosphorylated by cyclin-dependent kinases*

Substrate	Possible role	Phase of the cell cycle in which phosphorylated
Lamins	Nuclear lamina disassembly	G2-M
Vimentin	Intermediate filament disassembly	G2-M
Caldesmon	Microfilament contraction	G2-M
Nucleolin and NO38	Nucleolar function and disassembly	G2-M
Histone H1	Chromosome condensation	G2-M
HMG	Chromosome condensation	G2-M
Myosin II light chain	?	G2-M
Tyrosine kinases (p60, p150)	Cytoskeletal reorganisation	G2-M
SV40 T antigen	Initiation of DNA replication	G1-S
RPA	Initiation of DNA replication	G1-S
pRb	Release of transcription factors	G1-S
p53	Regulation of nuclear localisation	G1-S
RNA polymerase II	Transcription inhibition	

Source: Norbury & Nurse (1992); Nigg (1993).

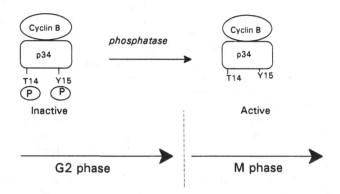

Fig. 11.1. Activation of cyclin-dependent kinase at the end of G2 phase (Norbury & Nurse, 1992).

breakdown of the nuclear envelope, chromatin condensation, reorganisation of the cytoskeleton and reorganisation of the nucleoli. At the onset of mitosis, the nucleoli are not visible, but the nucleolar regions are visible on certain chromosomes. After completion of mitosis, organised nucleoli reappear. These morphological changes can be understood in relation to the substrates of cyclin-dependent kinase.

Nuclear *lamins* are intermediate filament proteins that polymerise to form the nuclear lamina, the structure underlying the nuclear membrane. They also form attachments to chromatin. In avian species, lamins have been classified on the basis of the amino acid sequence and their chemical properties into three groups: A, B_1 and B_2. The B lamins are expressed in all somatic cells, and in birds B_2 lamins are far more abundant than B_1 (Heitlinger *et al.*, 1991). The amount of lamin B_2 per cell is roughly proportional to the surface area of the nuclei. The A-type lamins are developmentally expressed. In early chick embryos, most of the lamin is B_1 and B_2 with very little lamin A. During embryonic development, lamin A becomes more evident whereas the amount of lamin B_1 declines; the transition occurs at different stages in different tissues (Lehner *et al.*, 1987). Lamins have M_r values between 60 000 and 75 000 and comprise a central α-helical domain made up of four α-helices connected by linker regions and flanked by non-helical domains (Fig. 11.2). Other important structural features are a

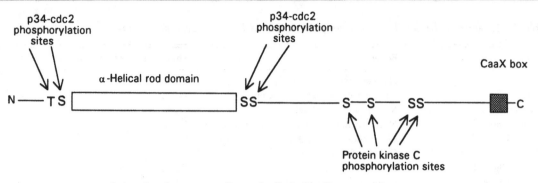

Fig. 11.2. Lamin structure and phosphorylation sites; cdc2, cyclin-dependent kinase.

nuclear localisation signal and a C-terminal motif known as the CaaX box (C, cysteine; a, aliphatic amino acid; and X, any amino acid). The CaaX box is subject to post-translational modification (farnesylation, proteolytic trimming and carboxymethylation). Phosphorylation of lamins at the onset of mitosis causes their depolymerisation, which results in the A-type lamins being solubilised, whereas the B-type remain associated with the membranes. The farnesylation of the B_2 lamins gives them the necessary hydrophobicity to remain associated with the membranes. Phosphorylation has been demonstrated *in vitro*, where cyclin-dependent kinase phosphorylates B-type lamins from domestic fowl hepatoma cells with concomitant depolymerisation (Peter *et al.*, 1990). This phosphorylation of the central α-helical domain occurs during mitosis (Fig. 11.2). Phosphorylation of lamins also occurs during interphase, but this is catalysed by protein kinase C and it is serines 400, 404, 410 and 411 on the C-flanking region that are phosphorylated (Hennekes *et al.*, 1993).

Vimentin is a constituent of the intermediate filaments. Its gene exists as a single copy in the domestic fowl haploid genome, from which it has been cloned and sequenced. It comprises nine exons and has an unusual initiation sequence lacking the TATA box but possessing five GC boxes and a CAAT box (see Fig. 8.1, p. 119) (Zehner *et al.*, 1987). It produces at least three different mRNA transcripts *in vivo* (Zehner & Paterson, 1985), through differential utilisation of multiple polyadenylylation sites (see Fig. 9.1, p. 137). *Caldesmon* is a non-structural component of the intermediate filament and is thought to inhibit actomyosin ATPase. Caldesmon is phosphorylated by cyclin-dependent kinase during mitosis. Cyclin-dependent kinase is known to phosphorylate (Ser/Thr)–Pro sequences, of which there

are four in the C-terminal domain of caldesmon. However, only one of these, which has a basic amino acid in the third position, is phosphorylated. This phosphorylation causes caldesmon to dissociate from actin, and this is thought to promote the disassembly of the actin cytoskeleton that occurs during mitosis (Marston & Redwood, 1991).

Lamins, vimentin and caldesmon all become phosphorylated when cells enter M-phase, and this coincides with the nuclear envelope breakdown. Phosphorylation of RNA polymerase II inhibits transcription. Two abundant nucleolar proteins, *nucleolin* (92 kDa) and *NO38* (38 kDa) become highly phosphorylated during mitosis, and this facilitates the dispersal of the nucleoli. Nucleolin and NO38 have been sequenced using a cDNA library prepared from E10 RNA (Maridor & Nigg, 1990). After the completion of M-phase, cyclins are degraded by a ubiquitin proteolytic pathway, and phosphatases remove the phosphate residue from Thr-167 of p34.

11.2.2 G_1- to S-phase transition

The events triggering the transition from G_1-phase to S-phase are much less clear, although evidence suggests that the *cdc2* gene is involved. A number of genes are transcribed in response to mitogens, and these are divided into immediate early mRNAs and late mRNAs. The induction of the synthesis of the early mRNAs does not require prior protein synthesis, only the modification of pre-existing transcriptional modulators. This is in contrast to the late mRNAs. The early proteins include a number of transcription factors, e.g. *c-myc, c-fos, fra-1, fos-B, egr-1, KROX 20, c-jun* (also known as *junA*), *junB*, and *junD* and also some DNA-binding receptor proteins (for a more complete list, see Herschman, 1991). Most of these immediate early genes encode nuclear proteins of oncogenic potential. The three most studied are *myc*, *jun* and *fos*. All three are rapidly induced by mitogens, which promote the transition from G_0- to G_1-phase. Their protein products all have short half-lives, and their concentrations in normal cells increase only transiently during mitosis. The levels of their mRNAs and proteins are controlled by their rates of both synthesis and degradation.

The c-myc gene

myc was first identified as the transforming gene in avian leukosis virus and avian myelocytomatosis virus (myelocytomatosis) in 1979 (Marcu, Bossone & Patel, 1992), and subsequently mammalian homologues were discovered. It is involved in gene regulation, DNA replication, cell proliferation and programmed cell death. These last two roles appear as opposites, and the underlying molecular mechanism is not clear. It is thought that cells in growth arrest, through a deficiency of growth factors, will undergo cell death when the expression of c-*myc* is activated. Two homologues of c-*myc*, found in human neuroblastomas and lung carcinomas were termed N-*myc* and L-*myc*, respectively. There is a family of *myc* genes, all of which encode nuclear phosphoproteins of 65–68 kDa and having short half-lives, e.g. about 45 min in avian lymphoid cells (Moore & Bose, 1989). N-*myc* and c-*myc* from domestic fowl have a three exon structure with the coding sequence located in the second and third codon and many other features in common with mammalian *myc* genes (Sawai *et al.*, 1990). (NB The convention is to use a three-letter italicised abbreviation for the oncogene, e.g. *myc*, and Roman letters starting with upper case for the oncogene protein product, e.g. Myc.) The C-termini (about 85 residues) have structural features similar to those of a number of transcriptional activation factors. These include two structural motifs: the leucine zipper and the helix–loop–helix. The leucine zipper has a α-helical region of 20–30 residues with leucine every seventh residue and a basic region that binds the DNA (see Branden & Tooze, 1991). These features facilitate the formation of a homodimer or heterodimer with a specific DNA sequence (CACGTG). Although the functions of the Myc proteins are not known for certain, there is substantial evidence to suggest a role in relation to cell proliferation. In the lens epithelium of developing chick embryos, there is a correlation between the c-*myc* mRNA levels and the proportion of proliferating cells. By contrast, the level of N-*myc* mRNA increases at terminal cell differentiation (Harris, Talian & Zelenka, 1992). In domestic fowl liver and skeletal muscle, both c-*myc* and c-*fos* expression correlate with the rate of growth (Kim *et al.*, 1992).

The c-jun gene

c-*jun* was first characterised as a proto-oncogene, the cellular counterpart of the v-*jun* oncogene of avian sarcoma virus ASV17 (ju-nana is the Japanese for seventeen). Both avian v-*jun* and c-*jun* DNAs have been sequenced and the differences are discussed in Section 11.4.4 (Nishimura & Vogt, 1988). Jun is one of the components of the sequence-specific transcriptional factors that bind to the promoter region (Fig. 8.1) of a number of genes, mediating a pleiotropic cellular response to growth factors and tumour promoters. Since its discovery, two other related proteins have been discovered as part of the same family, and these are encoded by *junB* and *junD*. All three proteins have almost identical C-termini, encompassing the DNA-binding domain that recognises the consensus sequence TGATCA and the dimerisation zone having the leucine zipper motif. The transcriptional activation sites are at the N-terminal end. All three *jun* proteins can form dimers either with each other or with the *fos* protein. In contrast the *fos* family ('*fos*' was first used to describe the oncogene carried by the Finkel–Biskis–Jinkins murine osteogenic sarcoma virus) is able to act as dimerisation partners but is unable to self-dimerise. The reason for this is apparent from Fig. 11.3. The role of *jun* and *fos* in transformation is discussed in Section 11.4.4. When quiescent fibroblasts (G_0-phase) from domestic fowl are stimulated by serum factors to enter the cell cycle, c-*jun* and *junB* are rapidly activated, but *junD* is only weakly activated (Hughes *et al.*, 1992).

Cell proliferation and differentiation tend to be mutually exclusive processes. Whilst cells are actively proliferating they tend not to differentiate. It is only when the rate of proliferation begins to decline that differentiation begins to accelerate. The c-*jun* proteins and the glucocorticoid receptor protein have antagonistic effects on these processes. Heterodimers of Jun and Fos constitute active transcriptional factors, but the complex of glucocorticoid receptor and Jun inhibits transcription (Berko-Flint, Levkowitz & Vardimon, 1994). c-*jun* is actively expressed in the early chick embryo, and the rate of expression declines by E11 when the amount of glucocorticoid receptor increases. A different situation prevails in the development of the lens epithelium of domestic fowl. As the rate of proliferation decreases and the proportion of proliferating cells decreases, so c-*fos*

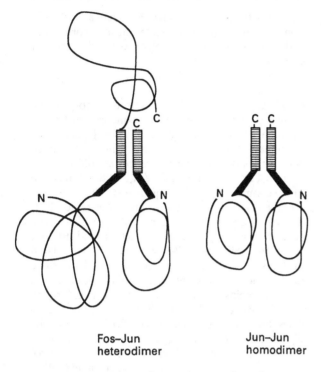

Fos–Jun
heterodimer

Jun–Jun
homodimer

Fig. 11.3. Fos–Jun heterodimer and Jun–Jun homodimer.

and c-*jun* mRNAs accumulate. Therefore, the cell cycle arrest appears to be at G_1, after mRNA accumulation of the early gene mRNAs (Rinaudo & Zelenka, 1992).

Other genes

Other genes that are expressed at the G_1/S-phase transition are the secondary response genes, and these include the heat shock proteins hsp-90α and hsp-70. Increased transcription of the hsp-90α and hsp-70 genes occur at mid G_1. There is however a difference in the subcellular localisation of the two during the cell cycle. Hsp-70 enters the nucleus at the G_1/S boundary, but hsp-90 does not show cell-cycle-dependent nuclear translocation (Jérôme *et al.*, 1993).

The rate of histone synthesis varies with the phase of the cell cycle. Most histone mRNAs accumulate during S-phase (Table 9.4, p. 146). mRNA for histone H2b in the chick embryo epithelium increases its rate of synthesis 5–10 times above the baseline value during S-phase (Brewitt, Talian & Zelenka, 1992). Dalton & Wells (1988) have shown that histone H1 gene expression in domestic fowl is regulated by a control element 5'-AACACA-3' (H1 motif), approximately 100 bp upstream of the coding region. Between this control element and the coding region lie the

G/C, CCAAT and TATA boxes. There are a number of copies of each of the histone genes, organised as a multigene family (Section 9.5).

11.3 Cell transformation

Much of our understanding of the molecular processes underlying the transformation from normal to cancerous tissue comes from studies made on cells grown in culture. The principal differences between normal and transformed cells are listed in Section 11.1. Transformation can be induced by a variety of carcinogens, but the transformations best understood at the present time are those induced by retroviruses. Although the majority of cancers may not be of viral origin and may be induced by a variety of carcinogens, transformations induced by retroviruses provide the simplest model system available (for reviews, see Weiss *et al.*, 1982; Varmus & Weinberg, 1993). Retroviruses are RNA viruses that on entering the host cell become transcribed into cDNA, catalysed by the enzyme reverse transcriptase. The cDNA subsequently becomes integrated into the host chromosome. These oncogenic viruses usually contain only three or four genes.

Evidence suggests that at least two mutations, one involving a nuclear protein and one involving a cytoplasmic protein, e.g. Myc and Ras, are necessary to transform a normal cell into a cancer cell, although for a number of cell lines that can be grown in culture, a single mutation can induce the transformation. The reason for this is that so called 'immortalised cells' have already undergone at least one mutation. Cell cultures are initially established from excised tissue. These primary cells are dissociated by various techniques. Cells arising from the transfer and multiplication of these cells are referred to as secondary cells. The vast majority of these cells will multiply only a limited number of times before they die. This is usually between 20 and 40 times for cells from the domestic fowl. The number of multiplications is close to the number that would have occurred in the intact animal from which they were excised. With certain tissue explants, a few cells continue to divide, passing what is known as the 'crisis period', and it is these cells that are said to be immortalised and are used in established cell lines. Most of the established cell lines in current use are of mammalian origin (Watson *et al.*, 1987). It is in these established cell lines that a single mutation can result in cell transformation. Some transformations arise from a mutation in a tumour supressor gene (Varmus & Weinberg, 1993).

11.4 Avian retroviruses, oncogenes and proto-oncogenes

The Rous sarcoma virus (RSV) is one of the most extensively studied retroviruses. It can rapidly transform (within 24 hours) cultured domestic fowl fibroblasts. This is readily seen under the light microscopy as the cells change from their elongated shape to an almost spherical shape. By the use of temperature-sensitive mutants of RSV, it was possible to identify the transforming gene. The RSV genome comprises four genes, known as *gag* (encoding an internal structural protein of the virion), *pol* (encoding reverse transcriptase), *env* (encoding a surface protein of the virion envelope) and the oncogene *src*. *gag*, *pol* and *env* are necessary for virus replication, and *src* is necessary for transforming the host cell. *gag* encodes the precursor polyprotein that is cleaved to yield the capsid proteins, and *pol* is also cleaved to yield reverse transcriptase. Gag has a protease domain at its C-terminus, and this activity is necessary for the activation of reverse transcriptase (Craven, Bennett & Wills, 1991). The *env* gene encodes surface proteins that are required for recognition of cell-surface receptors and are important in determining the host range. RSV Prague strain (PS-RSV) infects and transforms duck embryo fibroblasts *in vitro*, but less efficiently than chick embryo fibroblasts. After several passages in duck embryo fibroblasts, PR-RSV replicates as well as in chick embryo fibroblasts. Kashuba *et al.* (1993) have shown that during this adaptation changes occur in the coding sequence, mainly in the gp85-coding domain of the *env* gene. The C-terminus of the glycoprotein encoded by *env* contains the receptor-recognition site, and the N-terminus a hydrophobic membrane anchor.

At both ends of the retrovirus are the long terminal repeats (LTR). These contain sequences regulating initiation and termination of transcription. In addition to promoting transcription of viral genes, the LTRs of a number of retroviruses are capable of activating contiguous cellular genes. Many other transforming retroviruses have since been discovered, and they have features in common with RSV. They can be classified according to whether they are acutely (highly oncogenic) or non-acutely transforming (weakly oncogenic), and they may be replication defective or non-defective (Table 11.2). Many of the weakly oncogenic are avian leukosis viruses, and studies of these retroviruses have been important in understanding the basic molecular mechanisms of retrovirus infection in general. In many of the retroviruses, the oncogene replaces or partially replaces one of the genes necessary for replication (Fig. 11.4), and the virus can only multiply in the presence of a helper virus by using some of the latter's proteins. It is believed that the acute transforming viruses have arisen by recombination, in which part of a proto-oncogene replaces one, or part of one, of the *gag*, *pol* or *env* genes necessary for replication.

After the discovery of the *src* gene in RSV (Martin, 1970) required for transformation but not replication, Stehelin *et al.* (1976) found that cDNA *src* hybridised to DNA extracted from uninfected domestic fowl and other avian species. The melting temperature of the DNA–DNA hybrids with quail, turkey, duck and emu decreased in relation to phylogenetic distances separating them. This homologous sequence of DNA in uninfected cells was referred to as a proto-oncogene, and since the initial discovery many others have been identified. The differences between acutely transforming retroviruses and those that transform only after a long latent period, typically 4–12 months, lies in the presence or absence of an oncogene in the viral genome. Viruses possessing an oncogene are acutely transforming, since they transform with high efficiency once the retroviral DNA has been integrated into the host genome. Those lacking an oncogene only transform when they integrate into the host genome upstream of the proto-oncogene. The provirus replaces the native regulatory elements by inserting a viral promoter into a c-*onc*, which would then become equivalent to a v-*onc* (Zhou & Duesberg, 1989). Such changes generally induce leukaemias. The position of integration of the viral cDNA appears to be somewhat random, and this accounts for the low efficiency of transformation. When a number of avian leukosis viruses were injected into the blastoderm of fertile eggs, the proviral inserts appeared scattered throughout the genome (Crittenden, Salter & Federspiel, 1989). It is estimated that only 1 in $\approx 10^{11}$ infected lymphocytes initiates leukaemia in susceptible animals (Zhou & Duesberg, 1989).

RSV is rather an unusual transforming retrovirus in possessing the complete set of genes for replication, and also an oncogene. Most of the avian transforming viruses lack at least one or part of one of the other genes necessary for replication (see Fig. 11.4).

The abbreviation used to distinguish proto-oncogenes from viral oncogenes are the prefixes c- and v-, e.g. c-*src* and v-*src*. Proto-oncogenes have usually been identified from their sequence relationship to the viral oncogene, and so they have generally been named after the virus to which they relate. They differ from one another in two ways; only

Table 11.2. *Transforming retroviruses in domestic fowl and other avian hosts*

Oncogene	v-onc	Virus strain	Hosts[a]	Defective	Acute	Tumours
erb-A	AEV-erb-A	Avian erythroblastosis virus-ES4 and R strain		+	+	Erythroleukaemia, sarcoma
erb-B	AEV-erb-B	Avian erythroblastosis virus-ES4 and R strain		+	+	Erythroleukaemia, sarcoma
ets	E26-ets	Avian leukosis virus E26	Quail, turkey, guinea-fowl, ducks	+	+	Erythroblastoma, myeloblastoma
fos	NK24-fos	Avian retrovirus NK24				Nephroblasoma, osteogenic sarcoma
fps	FuSV-fps	Fujinami sarcoma virus		+	+	Fibrosarcoma, myxosarcoma
fps	PRCII-fps	PRCII sarcoma virus		+	+	Fibrosarcoma, myxosarcoma
fps	PRCIV-fps	PRCIV sarcoma virus		+	+	Fibrosarcoma, myxosarcoma
fps	URI-fps	Rochester sarcoma virus		+	+	Fibrosarcoma, myxosarcoma
fps	16L-fps	16L recovered sarcoma virus		+	+	Fibrosarcoma, myxosarcoma
jun	ASV17-jun	Avian sarcoma virus 17				Fibrosarcoma
mil	MH2-mil	Avian carcinomatosis virus strains MH2, IC10 and IC11	Quail	+	+	Carcinoma
myb	AMV-myb	Avian myeloblastosis virus BAI-1		+	+	Myeloblastoma
myb	E26-myb	Avian leukosis virus E26				Myeloblastoma
myc	M29-myc	Avian myelocytomatosis virus MC29	Quail	+	+	Myelocytomatosis, endothelioma, rhabdomyosarcoma
myc	CMII-myc	Avian myelocytomatosis virus CMII	Quail	+	+	Myelocytomatosis, endothelioma, rhabdomyosarcoma
myc	MH2-myc	Avian myelocytomatosis virus MH2	Quail	+	+	Myelocytomatosis, endothelioma, rhabdomyosarcoma
myc	OK10-myc	Avian myelocytomatosis virus OK10	Quail	+	+	Myelocytomatosis, endothelioma, rhabdomyosarcoma
rel	REV-T-rel	Reticuloendotheliosis virus T	Quail, turkey, pheasant, duck, geese	+	+	Leukaemia
sea	AEV-sea	Avian sarcoma virus S13				Myxosarcoma
ski	SKV-ski	Sloan Kettering viruses 770, 780 and 790	Quail			Sarcoma, erythroblastoma
src	RSV-src	Rous sarcoma virus	Turkey, pigeon		+	Fibrosarcoma
src	B77-src	B77 avian sarcoma virus	Turkey, pigeon		+	Fibrosarcoma
src	rASV-src	Recovered avian sarcoma virus	Turkey, pigeon		+	Fibrosarcoma
src	PR-RSV-src	Prague strain RSV	Turkey, pigeon		+	Fibrosarcoma
yes	Y73-yes	Y73 avian sarcoma virus				Fibrosarcoma, myxosarcoma
yes	ESV-yes	Esh sarcoma virus				

[a] Known hosts other than domestic fowl.

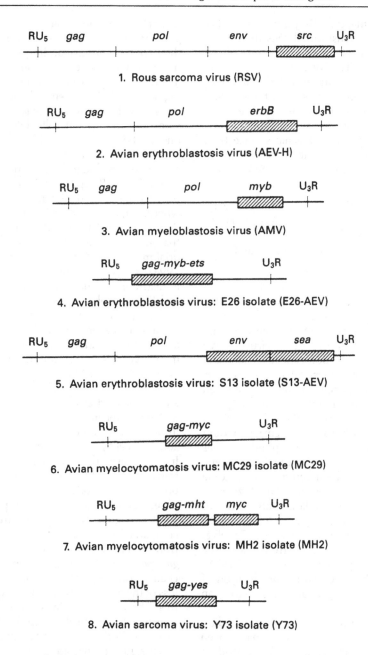

Fig. 11.4. Genomic structure of transforming viruses in the domestic fowl.

proto-oncogenes have introns and viral oncogenes are often truncated versions of proto-oncogenes. In order to understand how viral oncogenes may transform cells and what are the functions of proto-oncogenes, it is necessary to consider the product of oncogene expression, namely the onco-proteins.

11.4.1 Oncogene and proto-oncogene products

The oncogene products so far identified are all regulatory proteins, which are usually each able to regulate the actions of a number of other proteins. Many are known to have important roles in cell growth and division, e.g. Myc, Jun and Fos (see

Section 11.2.2). In normal cells, they are expressed at low levels, whereas in transformed cells the rate of expression is increased and so also is the rate of cell growth and division. The oncoproteins can be classified according to their intracellular location: (i) nuclear proteins, (ii) proteins present in the cytosol, (iii) transmembrane proteins spanning the plasmalemma, and (iv) proteins loosely associated with the plasmalemma on the cytoplasmic surface (Fig. 11.5). They can also be classified according to probable mode of action (Reddy et al., 1988) into the following:

1. Protein kinases, most of which phosphorylate tyrosine residues but a few phosphorylate serine/threonine residues
2. Growth factors and growth factor receptors
3. GTP-binding proteins
4. Nuclear proteins that bind to DNA.

The classification of oncogenes that have been identified in avian species is given in Table 11.3. Table 11.2 lists the strains of retroviruses that transform avian cells. Most of these viruses have been initially detected in the domestic fowl host, and in some cases in other domesticated species. It is probable that many also occur in non-domesticated species, but this has so far not been investigated. Some proto-oncogenes have been detected in normal uninfected avian species, but as yet no corresponding avian retroviruses carrying the corresponding v-onc have been discovered. For example, the ras oncogene is present in Harvey- and Kirsten-murine sarcoma virus, and it has been detected in about 30% of all human tumours. There are a large family of ras proteins, all of which bind GTP and are believed to function as transducers of mitogenic signals. Trueb & Trueb (1992) isolated a cDNA clone from a domestic fowl expression library that encodes a Ras-like protein, which may be involved in coordination of the cell cycle, but it has not yet been isolated from an avian virus.

Two other domestic fowl cDNA clones, encoding the proto-oncogenes snoN and bcl-2, have been detected by using human cDNA probes encoding the sequences (Cazals-Hatem et al., 1992; Boyer et al., 1993). For neither of these has an avian v-onc yet been detected. sno is a member of the ski family of nuclear proteins (Table 11.3). bcl-2 is unusual in that it encodes a 26 kDa integral membrane, p26^{bcl-2} which acts as an oncoprotein not by increasing cell proliferation but by blocking c-myc-induced apoptosis (Givol et al., 1994; Nunez & Clarke, 1994). Apoptosis is the term used to describe cell death showing a distinct set of morphological features. A range of gene products are expressed during this process and they control the dismantling of the cell. Apoptosis occurs in a number of situations, e.g. during development, when a particular group of cells are no longer required, and also when cells become infected with virus. Self-destruction in the latter prevents the virus spreading to adjacent cells. Like a number of other oncoproteins p26^{bcl-2} is able to form heterodimers with a second protein Bax (cf. Jun–Fos, Section 11.2.2). Bax is able to form homodimers Bax:Bax and heterodimers Bax:p26^{bcl-2}. It seems that the relative proportions of these dimers is a factor in determining whether the self-detruct programme is activated (Martin, Green & Cotter, 1994). Earlier work suggested that p26^{bcl-2} was associated with the inner mitochondrial membrane, but more recent work (Givol et al., 1994) using chick embryo fibroblasts indicates that it is distributed equally between nuclear membranes and the endoplasmic reticulum.

Most retroviral genomes encode a single oncogene, but avian leukaemia virus E26 has both myb and ets and is able to induce two types of tumour, namely myeloblastosis and erythroblastosis (Watson, McWilliams & Papas, 1988).

Studying the control of expression of oncogenes and the properties of their protein products helps to understand the mechanisms underlying growth control in normal and cancerous tissues. Mitogenic agents added to cells grown in culture stimulate cell division. In many cases, the oncogene product acts as a signal transducer. Embryonic tissues undergoing rapid proliferation are often found to contain higher levels of oncogene products than non-proliferating tissues. Viral oncogene products are often truncated when compared with the c-onc counterpart. Often the truncated gene product is fused to part of one of the gag, pol or env proteins. The overall effect of these changes on the functioning of the protein is often the loss of the regulatory element. Two examples illustrate this. The v-erbB protein is a truncated version of the transmembrane protein epidermal growth factor receptor. Both proteins have phosphorylation sites on the membrane side (Fig. 11.6) but the v-erbB protein lacks the extracellular ligand-binding domain and so can phosphorylate independently of epidermal growth factor control. The protein p59^{v-rel}, expressed by the virulant reticuloendotheliosis virus affecting several domesticated birds (Table 11.2), shows high levels of phosphorylation. The proto-oncogene product p68^{c-rel} (Fig. 11.8, below), however, is tightly regulated, and only low levels of phosphorylation are detectable in uninfected cells. The different types

Table 11.3. *Comparison of avian viral oncogene and proto-oncogene products*

Oncogene	v-*onc* product	c-*onc* product	Location	Functional properties	Comments	References
bcl-2	–	p26	Nuclei, ER, mitochondria	?	cDNA hybridised to human c-*bcl-2*	Cazals-Hatem *et al.* (1992)
erbA	$p75^{gag\text{-}erb}$	p46	Nucleus	DNA-binding protein	Resembles thyroid hormone receptor	Vennström & Damm (1988)
erbB		p75	Plasmalemma	Transmembrane glycoprotein	Homologous with EGF receptor	
ets	$p135^{gag\text{-}myb\text{-}ets}$	p68 and p54	Nucleus	DNA-binding protein recognising AGCAGGAAG	Alternative splicing accounts for 2 c-*onc*	MacLeod *et al.* (1992)
fos	p55	p55	Nucleus	Acidic phosphoprotein	Nuclear protein developmentally expressed	Mölders *et al.* (1987)
fps	$p130^{gag\text{-}fps}$	p98	Cytoplasm	Tyrosine kinase	Peripheral membrane protein	Hanafusa (1988)
jun		p34	Nucleus	DNA-binding protein	Acts as transcriptional activator	Nishimura & Vogt (1988)
mil	$p100^{\Delta gag\text{-}mil}$	p93.5 and p95	Cytoplasm	Serine/threonine protein kinase	Role in mitogen signal transduction to nucleus	Eychène *et al.* (1992)
myb	p48 and $p135^{gag\text{-}myb\text{-}ets}$	p75	Nucleus	DNA-binding protein	Transcriptional activator	Schuur & Baluda (1991)
myc	$p90^{\Delta gag\text{-}myc}$, $p110^{\Delta gag\text{-}myc}$ and $p200^{gag\text{-}\Delta pol\text{-}myc}$, $p57^{\delta gag\text{-}myc}$	p64	Nucleus	DNA-binding protein recognising CACGTG	Part of multigene family including N-*myc* and L-*myc*	Erisman & Astrin (1988)
nov		p32		Cysteine-rich polypeptide	*nov*, nephroblastoma-overexpressed gene	Martinerie *et al.* (1992)
ras		p20	Plasmalemma	GTP-binding protein	Transducer of mitogenic signals	Trueb & Trueb (1992)
rel	p59	p68	Nucleus	Serine/threonine protein kinase	p59 highly phosphorylated, cf. p68	Simek & Rice (1988)
ros	$p68^{gag\text{-}ros}$	?	?	Transmembrane protein kinase, mainly serine	c-*onc* probably larger having extended extracellular region	Wang (1988)
sea	$p137^{env\text{-}sea}$	p42	Plasmalemma	Transmembrane tyrosine kinase		Smith *et al.* (1989)
ski	$p110^{gag\text{-}ski}$, $p125^{gag\text{-}ski}$		Nucleus	Nuclear protein		Stavnezer (1988)
src	p60	p60	Plasmalemma	Protein tyrosine kinase	On cytoplasmic surface of plasmalemma	Golden & Brugge (1988)
yes	$p80^{gag\text{-}yes}$, $p90^{gag\text{-}yes}$		Plasmalemma	Protein tyrosine kinase	On cytoplasmic surface of plasmalemma	Shore (1988)

ER, endoplasmic reticulum.

187

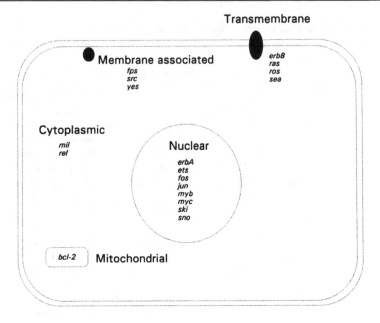

Fig. 11.5. Intracellular location of proto-oncogene products.

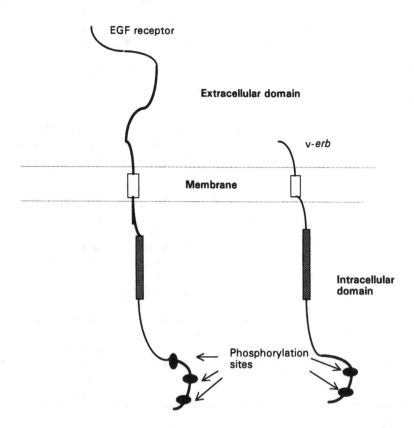

Fig. 11.6. Epidermal growth factor (EGF) receptor and erbB transmembrane proteins.

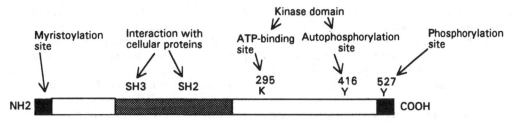

Fig. 11.7. Structure of p60 c-Src protein.

of oncogene product occurring in avian species are now considered in turn.

11.4.2 Protein kinases

Most of oncogene products that are protein kinases phosphorylate specific tyrosine residues. Examples of these are *src*, *fps*, *yes* and *sea*. RSV*src*, which occurs in Rous sarcoma virus, has been the most studied. The gene products of *src*, *fps* and *yes* are peripheral membrane proteins associated with the cytoplasmic surface of the plasmalemma (Fig. 11.5). Both p60^{v-src} and p60^{c-src} possess specific tyrosine kinase activity but with an important functional difference, namely, that p60^{c-src} kinase activity is regulated through normal cellular mechanisms and does not transform cells, whereas p60^{v-src} kinase is readily activated and is capable of transformation (Tanaka *et al.*, 1990). p60^{c-src} comprises 533 amino acid residues, whereas p60^{v-src} has 526. The 12 C-terminal residues on p60^{v-src} are distinct from the 19 C-terminal amino acids in p60^{c-src}; in addition, there are 12 amino acid differences throughout the rest of the chain (Golden & Brugge, 1988). The different regions of the p60^{c-src} are shown in Fig. 11.7. The N-terminal glycine is myristoylated, and this acts as an anchor to the plasmalemma. From comparisons of several mutants of *src*, the two regions of the oncoprotein SH2 and SH3 seem to be involved in the interaction with other cellular proteins (Hirai & Varmus, 1990). These interactions are important for transformation. Activity of p60^{c-src} is regulated by phosphorylation of Tyr-527. One of the proteins with which p60^{v-src} associates and activates is phosphatidylinositol-3 kinase, and intracellular levels of phosphatidylinositol phosphates increase upon stimulation with various mitogenic agents (Fukui, Saltiel & Hanafusa, 1991). p60^{v-src} also associates with the heat shock protein hsp-90. The level of expression of *src* is 15–30-fold higher in transformed cells, and many proteins show higher levels of tyrosine phosphorylation, e.g. fibronectin, vinculin, ezrin, talin, calpactin 1 heavy chain, paxillin,

p21^{ras}, p70 ribosomal protein S6 kinase, histone 1 kinase and several serine/threonine kinases (Sternberg *et al.*, 1993). These are probably physiological substrates of p60^{v-src}. The gene most strongly overexpressed upon RSV infection of domestic fowl neuroretina cells is gene T64. The product of this gene is capable of binding the protein clusterin, which is associated with programmed cell death. The T64 gene is not expressed in normal quail fibroblasts, but is actively expressed after their transformation by RSV (Herault *et al.*, 1992).

The protein products of *fps* and *yes* are similar to p60^{src}, although those from the viral oncogenes are gene fusion products (p130$^{gag-fps}$, p80$^{gag-yes}$ and p90$^{gag-yes}$). The tyrosine kinase activities of all three retroviruses require divalent metal ions, but *fps* tyrosine kinase differs in having a preference for Mn^{2+} over Mg^{2+}.

The most recently discovered oncogene in this group is *sea*. This occurs in S13 avian erythroblastosis virus, which causes sarcomas, erythroblastomas and anaemias when injected into chicks (Smith, Vogt & Hayman, 1989). Its oncogene product is a tyrosine kinase, but it differs from those of *src*, *fps* and *yes* in that it is a transmembrane protein. One of the products of expression of the S13 genome is a 155 kDa glycoprotein (gp115). This becomes cleaved to gp85 and gp70. The former is the product of the bulk of the *env* gene, and the latter contains the product of *sea* and part of *env*. This part contains the transmembrane region and accounts for the oncogene product spanning the membrane. The sequence of *sea* suggests that it most closely ressembles the insulin receptor family (see Fig. 7.4, p. 107).

Two avian oncogenes that encode protein kinases are cytoplasmic proteins not associated with membranes: *rel* and *mil*. *rel* occurs in avian reticuloendotheliosis virus (REV-T), which has been described as the most virulent of all retroviruses, causing a rapidly fatal lymphoma with a latent period of one to two weeks and a mortality approaching 100% (Moore & Bose, 1989). A modified form of REV has been used as a retroviral vector in avian transgenesis, by injection into the blastodermal embryo (Perry & Sang, 1993). REV has been detected in domestic fowl, turkey, geese, quail, ducks and pheasant. It transforms very immature haemopoietic cells, some of which are committed to B cell differentiation. REV-T is a replication defective transforming virus and is often associated with a non-transforming helper virus REV-A. The v-rel oncogene is presumed to be derived from the turkey proto-oncogene, since closely related sequences have been found in normal turkey DNA.

The Rel proteins are part of a family of homologous proteins, which include NF-κB and the product of the dorsal gene, all of which bind similar DNA sequences (Abbadie et al., 1993). NF-κB was originally identified as a nuclear factor that bound kappa (κ) light chain enhancer in B lymphoid cells and activated the κ-gene transcription. It has been the most extensively studied member of this family. It occurs as two subunits, p50 and p65. dorsal is a drosophila gene that encodes a transcription factor. Their structures are compared in Fig. 11.8. The v-onc is expressed as a nuclear phosphoprotein p59^{v-rel}, whereas the c-rel protein is about 68 kDa. Approximately 56 kDa (474 amino acid residues) of the c-Rel sequence is present in the central portion of p59^{v-rel}, with 12 N-terminal and 19 C-terminal amino acid residues added from the v-env protein (Simek & Rice, 1988). Approximately 300 of the 474 amino acid residues are conserved among members of the rel family. The C-terminal deletion in v-Rel removes the c-Rel sequences that are important for cytoplasmic retention and transcriptional activation. p59^{v-rel} is a nuclear protein in chick embryo fibroblasts, whereas p68^{c-rel} is a cytoplasmic protein (Sif, Capobianco & Gilmore, 1993). Rel proteins contain a nuclear-targeting sequence and also a potential phosphorylation site (Arg–Arg–X–Ser), recognised by cyclic AMP-dependent kinase, about 25 residues away (Rushlow & Warrior, 1992). Both p59 and p68 are associated with protein kinase activities that are Mn^{2+} dependent and serine/threonine specific and show little sequence identity with known kinases (Rice & Gilden, 1988). When p59^{v-rel} is isolated, it usually shows a high degree of phosphorylation, in contrast to p68^{c-rel}, which only shows low levels of phosphorylation. This suggests that there is tight regulation of phosphorylation of p68^{c-rel}, but not p59^{v-rel}. It has been suggested that the nuclear targeting site targets Rel to the nucleus, and the phosphorylation controls the rate of transport to the nucleus (Rushlow & Warrior, 1992). The activity of Rel appears to be controlled by its subcellular location rather than the rate of transcription. Once in the nucleus, Rel regulates the transcription of other genes whose promoters contain the consensus sequence. In the case of NF-κB, over 20 target genes have been identified, but much less is known about the targets of v-Rel and c-Rel, except that v-Rel increases expression at the Sp1 site of promoters (Sif et al., 1993). c-rel is expressed at low levels in most chick embryo tissues, but at high levels in cells undergoing programmed cell death by apoptosis. It may have a role in the induction of cell death. The low levels in embryonic tissues may represent cells that are no longer required during the process of differentiation.

The avian oncogenic retrovirus MH2 (Mill Hill 2) contains the v-onc known as mil. It is the avian equivalent of v-raf, originally identified in murine sarcoma virus. MH2 was first isolated from a spontaneous ovarian tumour in domestic fowl. Other more recent isolates, designated IC, are from the Institut Curie. The genome of MH2, 5'-Δgag–v-mil–v-myc–non-coding sequence-345', is unusual in containing two oncogenes and only part of one of the replication genes. No extensive studies have been made of the pathogenicity of strains MH2 with either defective mil or myc. The c-mil gene encodes cytoplasmic serine/threonine protein kinase, believed to play a key role in mitogen signal transduction from the cell surface to the nucleus, probably by interacting with some plasmalemma receptor. The c-mil protein is encoded in a gene having 19 exons and spanning about 100 kb DNA (Calogeraki et al., 1993). Two c-mil protein kinases, p93.5 and p95 (767 and 807 amino acid residues each) are produced as the result of alternative splicing (Eychène et al., 1992). These appear to be the longest protein kinases produced by the mil/raf proto-oncogene family. The v-onc protein is a gene fusion product p100$^{\Delta gag-mil}$, in which gag sequences make up over half of the protein (Rapp et al., 1988). The homologous parts of c-mil and v-mil are 100% identical (Calogeraki et al., 1993).

11.4.3 Growth factors and growth factor receptors

The discovery in 1984 that the erbB oncogene could be derived from the epidermal growth factor receptor (EGFR) gene was important in understanding the deregulation that occurs in cell transformation. It came shortly after the discovery that the sis oncogene, which occurred in monkeys, was related to the platelet derived growth factor (Hayman, 1987). The presence of an oncogenic factor in avian erythroblastosis virus (AEV) had been known since 1934. In

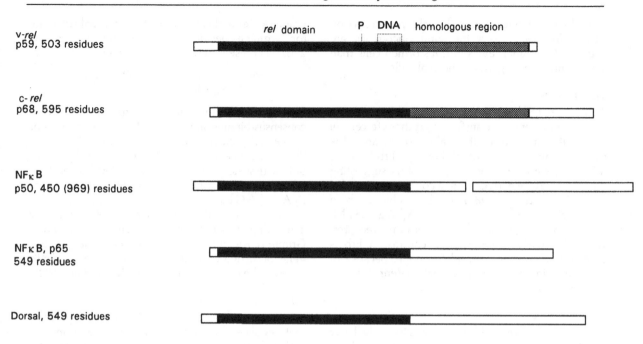

Fig. 11.8. Structures of Rel protein family.

domestic fowl it causes mainly erythroid leukaemias. *erbB* occurs together with *erbA* in AEV-ES4, but it also occurs in transducing viruses which lack *erbA*, e.g. AEV-H (Vennström & Damm, 1988). The v-*erbB* gene product is a transmembrane protein and closely resembles the EGFR but with the extracellular domain considerably shortened (Fig. 11.6). The v-*erbB* protein lacks all but 61 of the amino acid residues from the extracellular domain, and the cytoplasmic domain is shortened by 34 amino acid residues. The extracellular domain of EGFR contains the ligand-binding site for EGF, and a cytoplasmic domain having three phosphorylatable tyrosine residues. Binding EGF to the EGFR stimulates autophosphorylation on the three tyrosine residues. EGFR is also capable of phosphorylating other protein substrates, the nature of the physiological substrates is not clear. The v-*erbB* protein has two phosphorylatable tyrosine residues (Fig. 11.6), and the kinase activity appears to be deregulated, since phosphorylation does not require the presence of EGF.

The *ros* genes appear to resemble those of *erbB* and also the insulin receptor. They encode transmembrane proteins, and their nucleotide sequences suggest they have a number of structural features in common (Wang, 1988). c-Ros is a receptor-like protein kinase with an extracellular domain of about 2000 amino

acid residues and has extensive homology with the extracellular domain of the insulin receptor. Whether there is a physiological ligand that binds this extracellular domain is not clear. v-Ros is a truncated version of c-Ros, in which the extracellular domain has been considerable shortened. *ros* occurs in ASV-UR2, which causes myxosarcomas similar to those induced by RSV. ASV-UR2 encodes a fusion protein, p68$^{gag-ros}$, of which 150 amino acid residues are encoded in *gag* and 402 in *ros*. This transmembrane protein is capable of autophosphorylation or phosphorylation of other substrates. c-*ros* is expressed in domestic fowl most strongly in kidney and intestine, followed by lung, bursa, thymus and testis. In each of these, its expression is mainly restricted to specialised epithelial cells. Chen, Zong & Wang (1994) suggest that it has a role in embryonic organogenesis in certain organs, but it also has a function in mature organs, since in certain organs expression begins in late embryo but continues in the adult.

11.4.4 Oncogene products binding DNA

A number of oncogene products are DNA-binding proteins that act as transcriptional activators by binding to the promoter region upstream of the coding sequences. Some of these (*jun*, *fos* and *myc*) were discussed in relation to the cell cycle in Section 11.2.2. In many cases the viral oncogenes are expressed at higher levels in transformed cells than are the proto-oncogenes in normal cells.

The erb gene

AEV-strain ES4 contains *erbA* and *erbB*. v-*erbA* is weakly oncogenic; it transforms erythocyte cells *in vitro* but not *in vivo*. It is able to enhance the transforming efficiency of *erbB in vivo*. ErbA shows homology with steroid hormone receptors, suggesting that they may be derived from a common primordial gene (Krust *et al.*, 1986). c-Erb is able to bind tri-iodothyronine with an affinity (K_d 0.2–0.3 nM) similar to that of the thyroid hormone receptor, which it closely resembles. By contrast, v-Erb is defective in binding thyroid hormone, because of mutations in the binding domain (Vennström & Damm, 1988).

The myb gene

myb occurs in both AMV and ALV-E26, in the latter together with the *ets* oncogene. c-*myb* encodes a nuclear protein p75, which is a transcriptional activator. The v-*myb* is expressed as a truncated version, p45, in AMV-infected cells, and as a fusion protein, p135$^{gag-myb-ets}$, in E26-infected cells (Reddy, 1988). These proteins recognise the consensus sequence PyAAC^G/$_T$G. The DNA-binding region on each of these proteins is at the N-terminal end. p75 has a tripartite imperfectly conserved 52 residue repeat, and from computer-modelling studies of this repeat sequence, a helix–turn–helix motif is suggested. This is a common motif in DNA-binding proteins (see Branden & Tooze, 1991). Both p45 and p135$^{gag-myb-ets}$ have this N-terminal region truncated, p45 by 71 residues and p135$^{gag-myb-ets}$ by 80 residues (Gerondakis & Bishop, 1986), thus losing almost one set of the 52 residue repeats. The amino acid consensus sequence of the DNA-binding domain in all three proteins is YAACKGHH (Weston, 1992). v-*myb* is unusual in its high tissue specificity, restricted to transforming haemopoietic tissues. It is one of the few viral oncogenes unable to transform fibroblasts. c-*myb* is expressed at highest levels in immature haemopoietic cells, and it becomes down-regulated during terminal differentiation. v-*myb* has greater transcription-stimulating activity than c-*myb*.

A gene specifically activated by v-Myb is the *min-1* gene. It is a granulocyte-specific gene expressed during granulopoiesis, and encodes a 35 kDa secretable protein present in intracellular granules (Quéva *et al.*, 1992). It is one of the most abundant proteins in granulocytes and is probably the principal substrate for arginine-specific ADP-ribosyltransferase (EC 2.4.2.31) present in these cells (Yamada *et al.*, 1992).

$$\text{L-arginine} + NAD^+ \rightarrow N^2\text{-(ADP-D-ribosyl)-L-arginine} + \text{nicotinamide}$$

The *min-1* promoter region contains three Myb consensus-binding sites. The c-Myb is a poor activator of *min-1*. A gene related to c-*myb* is B-*myb*, but unlike the former its expression is not restricted to haemo-poietic tissues. It encodes a 110 kDa nuclear protein, which also recognises the consensus sequence PyAAC^G/$_T$G but fails to activate the *min-1* gene, and it may be an inhibitory member of the family competing for the same consensus sites (Foos, Grimm & Klempnauer, 1992). By contrast, the *ets-2* gene, occurring together with *myb* in E26, is able to increase the expression of *min-1* (Dudek *et al.*, 1992).

The ets gene

The *ets* gene family encode DNA-binding proteins that regulate transcription (Macleod, Le Prince & Stehelin, 1992). Some members cooperate with AP1 transcription factor or with *fos* or *jun* (see Section 11.2.2). The only v-*onc* belonging to this family so far detected in birds is that occurring in ALV-E26, which is unusual in being able to induce both erythroblastosis and myeloblastosis. The v-*onc* is expressed as a fusion protein p135$^{gag-myb-ets}$. Both v-*myb* and v-*ets* must cooperate to fully transform both erythrocyte and myelomonocyte precursor cells. Not only are the two gene products necessary for transformation, but they must occur as a fusion protein (Domenget *et al.*, 1992). The fusion protein p135$^{gag-myb-ets}$ has a DNA-binding domain at each end with the transcriptional regulator region towards the centre. Domenget *et al.* (1992) have constructed several mutant viruses expressing a fusion protein with deletions in either v-Myb or v-Ets. The integrity of the v-*ets* oncogene is necessary for transformation of erythrocytic cells, but neither the DNA-binding domain nor the transcriptional regulator region from v-*myb* is required for this transformation. The DNA-binding domain of v-*ets* is necessary to transform myelomonocytic cells.

Two *ets* genes are known in domestic fowl, c-*ets-1* and c-*ets-2*, each giving rise to two gene products as the result of alternative splicing. c-*ets-1* is expressed in the thymus and endothelial cells, and expression of c-*ets-2* can be induced following macrophage differentiation and T cell activation. Both c-Ets-1 and c-Ets-2 can become phosphorylated with the accompanying loss of DNA-binding properties. This is a feature they share with steroid hormone receptors

(Section 10.2). c-*ets-1* is expressed as p54 and p68, the result of alternative splicing. Both these proteins recognise the same consensus DNA-binding domain AGCAGGAAGT but differ in their *trans*-activating activities (Quéva *et al.*, 1993). Both are expressed in a wide variety of cells of mesodermal origin, but p54 and not p68 is expressed in T and B lymphocytes.

The genes jun, fos and myc

The proto-oncogenes *jun*, *fos* and *myc* have been discussed in relation to their role in the cell cycle in Section 11.2.2. Both c-Jun and c-Fos are sequence-specific transcription factors, binding to the promoter region of a number of genes and recognising the sequence TGATCA. c-Myc is also a sequence-specific DNA-binding protein recognising CACGTG. v-*jun* occurs in avian sarcoma virus 17 (ASV17), which induces fibrosarcomas in domestic fowl. ASV17 is replication defective, with a genome *gag–jun–env* (Maki *et al.*, 1987). The v-*jun* is truncated at both 5′and 3′ends when compared with c-*jun* (Nishimura & Vogt, 1988). Expression of v-*jun* from ASV17 is in the form of a fusion protein p55$^{gag–jun}$. When compared with c-Jun, there is a 27 amino acid residue deletion in the N-terminal portion of the protein and also three amino acid substitutions in the DNA-binding domain. At the C-terminal end of p$^{gag–jun}$, the leucine zipper (see Branden & Tooze, 1991) is conserved. The DNA-binding properties are not detectably different in c-Jun and v-Jun. It is not clear, at present, whether transformation is brought about by a higher level of expression of *jun*, or because of different properties of c-Jun and v-Jun.

The c-*fos* gene has been sequenced from the domestic fowl (Mölders *et al.*, 1987). The deduced amino acid sequence corresponds with a M_r of $\approx 38\,000$, but the protein, detected by SDS–PAGE from *in vitro* translation of the mRNA, has an apparent $M_r \approx 60\,000$. Mölders *et al.* (1987) attribute the anomalous result on SDS–PAGE to the high content of proline and of charged amino acids in c-Fos. The only avian retrovirus so far identified as having v-*fos* is Nishizawa-Kawai 24 avian sarcoma virus (NK24), isolated from a domestic fowl nephroblastoma (Nishizawa, Goto & Kawai, 1987). This is histologically similar to Wilms' tumour in humans. NK24 is also able to induce fibrosarcomas *in vivo* and to transform fibroblasts *in vitro*. The nucleotide sequence shows v-*fos* between *gag* and Δ*pol*, and the oncogene fusion product containing part of the *gag*-encoded region has a M_r of 95 000.

The oncogene c-*myc* is part of a multigene family that also includes N-*myc* and L-*myc*, all of which

function in cell growth and proliferation, and c-*myc* is expressed in a wide range of uninfected tissues throughout the cell cycle. It increases in concentration in quiescent cells stimulated to divide after the addition of serum factors. c-Myc is similar in structure to some transcriptional activators. It binds DNA by a basic region in its sequence and has both a leucine zipper and helix–turn–helix. Less is known about how *myc* acquires its transforming potential. Recently, a nuclear phosphoprotein Max has been isolated that is able to form heterodimers with Myc (Sollenberger, Kao & Taparowsky, 1994). DNA sequence analysis of *max* has shown that it lacks TATA and CAAT motifs, but has numerous GC-rich sequences. The protein has the basic region, helix–turn–helix and leucine zipper region common to DNA-binding proteins. Max is able to form homodimers (Max–Max) and heterodimers (Max–Myc). In normal cells, the Max–Max dimer is favoured and transcriptional activation by Myc is suppressed. Max–Myc heterodimers favour transcriptional activation by Myc, and these are important in transformed cells. v-*myc* occurs in at least four strains of avian myelocytomatosis virus (Table 11.2). In each of these, the v-*onc* product is one or more fusion protein: M29, p110$^{\Delta gag–myc}$; CMII, p90$^{\Delta gag–myc}$; OK10, p200$^{gag–\Delta pol–myc}$ and p57$^{\delta gag–myc}$; MH2, p100$^{\Delta gag–mht}$ and p57$^{\delta gag–myc}$. The fusion proteins are generally overexpressed in infected cells, and this may be important factor in transformation. Tikhonenko, Hartman & Linial (1993) introduced mutations into avian retrovirus MC29 in order to delineate the important regions of the fusion protein. Their results suggest that the binding of v-Gag–Myc to DNA down-regulates the expression of c-*myc*.

The ski gene

Retroviruses containing *ski* were first detected when chick embryo fibroblasts became transformed after being infected with what had previously been regarded as a non-transforming avian leukosis virus (ALV) tdB77. The isolates causing the transformation were referred to as Sloan Kettering viruses (strains SKV770, SKV780 and SK790), since that was the name of the institute in which they were isolated. The v-*ski* sequence is inserted into the viral *gag* gene and it is expressed as a Gag–Ski fusion protein, located principally in the nucleus (Sutrave & Hughes, 1989). The predicted protein sequence (Stavnezer, Brodeur & Brennan, 1989) includes a basic region similar to that found in the DNA-binding region of Myc, a proline-rich region and cysteine and histidine residues that might constitute zinc fingers. The c-*ski* gene has eight exons spanning 4.2 kb, whereas v-*ski* occupies

1.3 kb because it is truncated at both the 5' and 3'ends compared with c-*ski*. c-*ski* is expressed as a number of RNAs, as the result of alternative splicing (Grimes, Szente & Goodenow, 1992). Both c-*ski* and v-*ski* induce myoblast differentiation in quail embryos. *ski* is unusual for an oncogene in stimulating both cell differentiation and proliferation. The different isoforms of Ski are expressed at different stages of development and probably have differing stabilities and nuclear localisation, which enable them to function under different conditions. A gene related to *ski*, known as *snoN* (*ski*-related *novel* gene) has been detected in cDNA clones from 10-day chick embryos. When this gene was inserted into avian retroviral vectors, it was expressed as a 81 kDa protein that was able to induce muscle differentiation in quail embryo cells grown in culture (Boyer *et al.*, 1993).

11.5 Conclusions

The understanding of the mechanism of action of oncogenes in general has advanced considerably in recent years. Much of this understanding has been acquired from the study of avian retroviruses and their oncogenes. The first oncogenic virus to be studied intensively was the Rous sarcoma virus, and it was in the domestic fowl where the first evidence of a proto-oncogene arose. Unlike some other topics discussed in this book, where the basic discoveries have been made in mammalian species and then confirmed in birds, the converse is true here. The properties of many of the oncoproteins are now well understood, although the physiological substrates for many of the protein kinases are not yet clear. The viral oncogenes are generally truncated versions of the proto-oncogenes, often occurring as fusion proteins together with parts of other proteins, particularly Gag or Pol. Two principal mechanisms seem to account for the role played by v-*onc* in stimulating cell proliferation. The first is that the rate of transcription and translation of a v-*onc* is higher than that of the corresponding c-*onc*. The second is that the v-*onc* often has lost its control region and so is not subject to the modulation that occurs with c-*onc*.

12

Molecular genetics of avian development

12.1 Introduction

Two basic problems which have to be addressed in order to understand the molecular basis of development are the temporal and spatial control of gene expression. An adult bird develops from a single haploid ovum and sperm. All the genetic information is, therefore, contained within the two genomes. The development of the embryo involves cell division and cell differentiation. From an initial undifferentiated cell, specialised cells arise at different times and different positions within the developing embryo. Since all the cells have the same DNA, the main problem is to understand how different genes are expressed in different cells at different times. During the earlier part of this century, the avian embryo was one of the most studied species by embryologists, and the morphological stages of development were described in detail (Lillie, 1919). The avian embryo has many advantages over other vertebrates. It is a convenient size and can be developed in an incubator, so all stages are readily accessible. However, when it comes to understanding the molecular basis underlying these morphological changes, *Drosophila* is the species of choice. Its genome had been studied in more detail, and a wide range of developmental mutants are available, which made it the species of choice in which to study temporal and spatial gene expression. Work on a number of vertebrates has shown that many of the basic control mechanisms found in *Drosophila* are common to vertebrates. Therefore, knowledge gained from the study of *Drosophila* provides useful pointers to the study of avian development. This chapter will focus on the control of gene expression during avian development and will mention morphological details only where necessary to put the gene expression in context, since embryology of the domestic fowl is well described elsewhere (Patten, 1951; Bodemer, 1968;

Balinsky, 1981). It is often useful to refer to particular stages in development. Hamburger & Hamilton (1951) have divided the development of the chick embryo into 46 stages, which are well defined, and these are used as reference points in the study of development (see p. xiv).

For a fully differentiated adult to develop from a single cell, successive cell divisions must result in asymmetric daughter cells in which different genes are expressed and different proteins produced. Evidence shows that the control occurs largely at the transcriptional level. The assymetry is assumed to be generated by concentration gradients, probably of proteins and RNA within the cytoplasm. These intracellular concentration gradients are initially continuous, but after cell divisions will become discontinous between cells. Development occurs along a pre-determined pathway and involves genes acting in a hierarchical manner. As development procedes, cells in different parts of the embryo are clearly destined to become particular structures in the adult. When the avian embryo has reached the blastula stage, it is possible to draw a *fate map* (Fig. 12.1), since the destination of cells according to their position has been determined.

Avian chimeras between the domestic fowl and the quail have been useful in drawing such maps (Dieterlen-Lièvre & le Douarin, 1993). A prominent heterochromatic mass in the nuclei of quail cells enables them to be distinguished from those of the domestic fowl. By transplantation of quail cells into domestic fowl embryos, the location of the quail cells can be traced throughout development using DNA staining (le Douarin, 1993). This marker is not easy to identify at low magnification under a light microscope, especially if only a few quail cells are present among a large number of chick cells. Also, it is not possible to detect cytoplasmic processes extending far from a nucleus, such as axons and

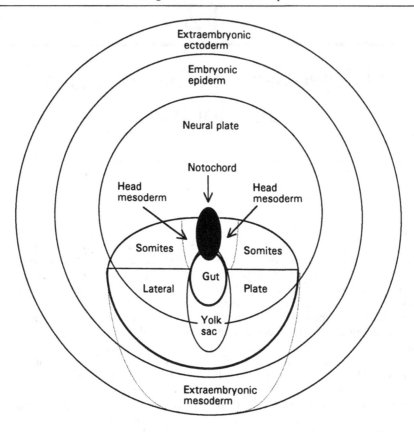

Fig. 12.1. Fate map of an avian embryo before the beginning of gastrulation.

dendrites of neurons. This problem has been overcome by using monoclonal antibodies raised against quail tissues (Aoyama *et al.*, 1992). Transplants made between chick and quail, providing the parts exchanged are matched, are highly compatible and chimaeras develop successfully. Chick–quail neurochimaeras are able to hatch and develop to a stage at which they can walk, fly and compete for food. A limitation to their use is that when the host immune system reaches maturity, it reacts to the presence of foreign nervous tissue.

The importance of particular genes at various stages of the development programme is evident from the phenotypic expression of defective mutants. This approach has been particularly valuable in the case of *Drosophila*, and comparative studies indicate that many features of the developmental plan are common to a wide range of organisms. The genes regulating development can be divided into three broad categories on the basis of studies with *Drosophila*. These groups act at successive stages of development. The first group is *maternal genes*, which are expressed during oogenesis before fertilisation. The second group is *segmentation genes*, expressed after fertilisation and concerned with the polarity of the segments of the developing embryo. The third group is *homeotic genes*, which control the identity of a segment, but not its number, size or polarity. The avian oocyte is quite different in morphology from either that of *Drosophila* or mammals, and also the segmentation pattern is much less obvious, especially in the adult. A brief outline of the stages up to the end of gastrulation in the avian embryo is described in the next section.

12.2   Chick embryo development up to the stage of gastrulation

The ovum develops in the ovary into a very large cell, which at the time of ovulation may be as large as 35 mm (Fig. 12.2*a*). The bulk of this volume is the large nutrient reserve in the form of yolk, with the nucleus and the small volume of cytoplasm sitting on top. At the time of ovulation, the single nucleus is

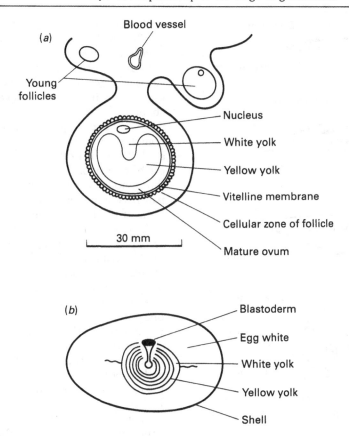

Fig. 12.2. (a) Structure of the ovum present in the ovary. (b) Hen's egg, showing the position of the blastoderm.

arrested at the diplotene stage of meiosis. Fertilisation normally occurs as the ovum is entering the oviduct. The accessory coverings (egg-white, shell membrane and shell) are added in the oviduct, and by the time of oviposition the developing embryo has reached the *blastoderm* stage comprising $\approx 60\,000$ cells (Fig. 12.2b), about 3 mm diameter. Only about 500 of these cells will contribute to the embryo proper, the remainder forming extraembryonic structures. The anterio-posterior polarity of the blastoderm is determined by gravity, the highest point of the blastoderm as it is situated in the uterus will become the future posterior side, and the lowest point the anterior one (Eyal-Giladi, 1993).

The blastoderm comprises two layers of cell, the upper layer or *epiblast* known as the *ectoderm*, and the lower layer or *hypoblast* known as the *endoderm* (Fig. 12.3a). *Gastrulation* begins with the formation of the middle layer, known as the *mesoderm*, and it is from blocks of mesoderm that the *somites* arise and the segmentation pattern is developed. Segmentation

involves the building of repeating homologous structures. These segments are more evident in the embryo than in the adult, where the discontinuities are not easily seen. Vertebrae, ribs, muscle and the dermal layer of the skin are derived from the paired blocks of mesodermal cells or somites. The mesoderm arises from ingression of cells of the epithelial epiblast beginning at the posterior end, extending anteriorly (Fig. 12.3b) and giving rise to the *primitive streak* and *Henson's node*. Epiblast cells continue migrating through Henson's node, building up as a strip of mesodermal cells that differentiates into the *notochord*. Henson's node is not static but moves in a posterior direction leaving the notochord cells on either side. The neural folds develop from the ectoderm in response to signals from the underlying mesoderm. Mesodermal cells, on either side of the notochord, condense in pairs to form the somites (Fig. 12.3c). The somites form sequentially at about 30 h (stages 7–14). Before the somites are clearly visible, the pattern of condensation of cells has been

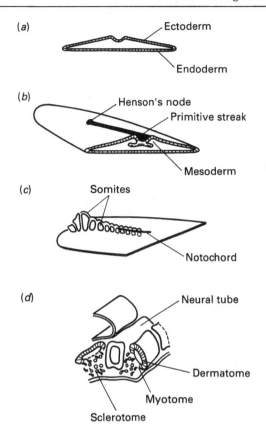

(a) Ectoderm
Endoderm

(b) Henson's node
Primitive streak
Mesoderm

(c) Somites
Notochord

(d) Neural tube
Dermatome
Myotome
Sclerotome

Fig. 12.3. Stages in the development of the chick embryo. (a) Blastoderm before invagination of the mesoderm. (b) Blastoderm after the formation of the primitive streak and Henson's node, mesoderm cells forming from the epiblast. (c) Neural fold stage. (d) Differentiation of the somites into scleroderm, myotome and dermatome.

determined in what is sometimes refered to as a 'pre-pattern' which, at the posterior ends, makes up the segmental plate. The somites differentiate into dermatome, myotome and sclerotome cells, which are the progenitors of the connective tissue, muscle and skeleton, respectively (Fig. 12.3d).

12.3 Segmentation and homeotic genes

The expression of segmentation genes is activated by concentration gradients set up in the developing embryo, and their role is to define increasingly restricted regions of the embryo. The segments of the embryo then have a pattern and a polarity. Once the segmental structure has been defined by the segmentation genes, the homeotic genes direct the development within the segmental structures in the

epidermis, mesoderm and nervous system. A number of different types of segmentation gene have been identified in *Drosophila*, and they can be classified according to the size of unit that they affect into either Gap genes, Pair rule genes or Segment polarity genes (Lewin, 1994). However, far less is known about their counterparts in avian species. A gene designated *pax-QNR*, characterised in quail, is homologous to the *paired* genes in *Drosophila* (Carriere *et al.*, 1993). It expresses five proteins, all of which can be phosphorylated *in vivo*. *pax-QNR* is expressed throughout the neuroretina, the neural tube and the pancreas during embryonic development. The functions of the five isoforms are not yet clear.

Homeotic mutants have been recognised since the early part of this century; the term homeosis was first used by Bateson (1894). The molecular basis of their expression has only been appreciated since nucleotide sequencing has been possible. A homeotic mutant is one causing a developmental abnormality, such that one part of an organism develops the characteristics of a different part. They have been most extensively studied in *Drosophila*, where there are eight homeotic genes located in two groups; antennapedia ANT-C and the bithorax BX-C. The order of the genes within the BX-C cluster along the chromosome corresponds with the position along the anterior–posterior axis of the animal in which they are expressed (McGinnis & Krumlauf, 1992; Scott, 1992). The ANT-C genes control the development of the more anterior parts of the fly, and most of the genes are also arranged along the chromosome in the anterior–posterior order. There are a larger number of homeotic genes in mammals and birds. Over 30 have been identified in the domestic fowl, arranged in four clusters (Sasaki, Yamamoto & Kuroiwa, 1992; Fainsod & Gruenbaum, 1994). Homeotic genes have been detected in a wide range of species. The ancestral *Hox* gene family probably contained five or six genes, ancestors to *Abd-B*, *Ant/Ubx*, *Dfd*, *pb* and *lab*. Groups 5–8 in mammals and birds, and *abd-A*, *Ubx*, *Antp* and *Scr* in *Drosophila* (Fig. 12.4) probably originated from independent duplications of a single ancestor (Mavilio, 1993). Initially, it was not feasible to have a systematic nomenclature for homeotic genes, but once a sufficient number had been characterised and homologies between them had been found, a more systematic nomenclature was adopted (Scott, 1992). This is illustrated in Fig. 12.4. Homology between the *Drosophila* genes and those of mammals and birds are shown by the vertical double-headed arrows. Genes are normally abbreviated to three letters written in italics, in this case *Hox*. The

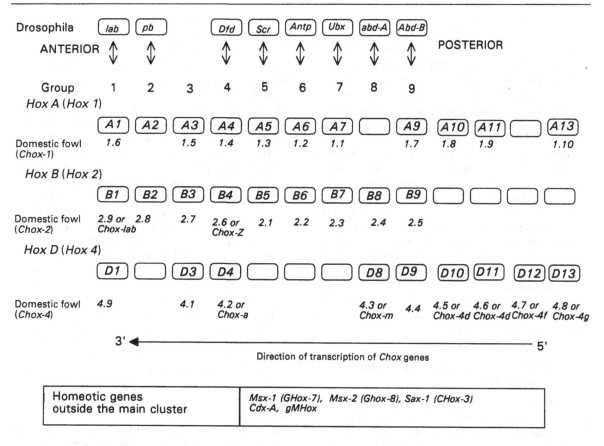

Fig. 12.4. Nomenclature of homeobox genes in domestic fowl. The recommended nomenclature (Scott, 1992) for domestic fowl homeobox genes is shown in the rectangular boxes, with earlier synonyms shown beneath. The nomenclature is based on homology with homeobox genes in *Drosophila* shown in the top row of boxes. Homeobox genes outside the cluster are shown at the bottom.

four mammalian clusters are given the letters A,B,C, and D, e.g. *HoxA*. The genes within a cluster are numbered, starting with the gene which is expressed nearest the anterior end of the developing embryo. The four clusters are believed to have arisen by gene duplication, and the corresponding genes in each cluster are considered to be paralogous (see Section 9.1). Genes from the domestic fowl have a *C* prefixed, and those from quail, *Q*. In addition to genes within the clusters, there are other genes that have been detected as having homeoboxes, but which lie outside the clusters. These are designated, where possible, using the first two letters of the homologous *Drosophila* gene, to indicate their lineage specification, followed by the letter *x*. For example, *Msx-1* is related to the *Drosophila* gene *Msh* (*muscle segment homeobox*) and *Cdx-A* is related to *caudal* (*cad*). This nomenclature will be adopted throughout this chapter, but some papers use the earlier nomen-

clature, which is also indicated in Fig. 12.4.

Homeotic genes contain a consensus sequence of 180 bp referred to as the *homeobox*. The homeobox encodes 60 amino acid residues known as the homeodomain, which usually occurs near the C-terminus. NMR and X-ray analyses of homeodomain-containing proteins show that they contain four helices, the second and third of which make a helix–turn–helix structure, and the ninth residue in helix-3 interacts with DNA bases. Homeodomain-containing proteins are sequence-specific DNA-binding proteins that act as transcriptional regulators (Gehring, 1992). They bind strongly to DNA ($K_d \approx 10^{-9}$–10^{-10} M), and residue nine of helix-3 determines the sequence specificity for DNA binding.

A number of homeotic genes have been charac-terised and sequenced in the domestic fowl and quail. These include *Chox-A11*, *Chox-D11*, *Qhox-A7*, *Msx-1*, *Msx-2*, *Chox-D12*, (Coelho *et al.*, 1991a,b; Mackem &

Mahon, 1991; Yokouchi *et al.*, 1991; Rogina *et al.*, 1992; Xue, Xue & le Douarin, 1993). A number of homeotic genes are expressed as multiple isoforms (Fainsod & Gruenbaum, 1989; Rangini *et al.*, 1989; Carriere *et al.*, 1993), some lacking the homeodomain. The mechanism for this is at present unclear. The expression of homeotic genes has been studied in a number of embryonic tissues, either using Northern blotting to detect transcription or by using antibodies to detect homeotic proteins. The *Hox* genes begin to be expressed during gastrulation in the most posterior regions of the ectoderm and mesoderm. As morphogenesis proceeds, expression of the *Hox* genes moves anteriorly, becoming restricted along the anterio-posterior axis. Differential expression of the multiple *Hox* genes occurs throughout the embryonic axis, so that each rhombomere has a unique pattern of expression. *Rhombomeres* are bulges that relate to the segmental pattern of the hindbrain. This differential expression is sometimes referred to as the Hox code and provides the molecular definition of position within the embryonic axis.

12.4 Homeobox-containing genes and limb development

The development of limb buds is one of the most studied areas in avian development. Two aspects will be considered in this and the next section, namely the expression of homeobox-containing genes and the role of the morphogen retinoic acid. Limb bud development is initiated by the release of specialised mesenchymal cells from the somatic layer of the lateral plate, followed by their migration laterally and accumulation under the epithelial tissue. These mesenchymal cells then induce the ectodermal cells to elongate and form an *apical ectodermal ridge* at the distal tip of the limb bud (Fig. 12.5). The apical ectodermal ridge directs the outgrowth of mesodermal cells of the limb. This allows the formation of various skeletal elements by limb mesoderm in their appropriate sequence along the proximodistal axis, by interacting with the mesenchymal cells beneath. The positional information, along the anterio-posterior limb axis, originates from the *zone of polarising activity* localised at the posterior margin of the limb bud (Fig. 12.5), which is the likely source of the morphogen retinoic acid. The zone of polarising activity acts in conjunction with the apical ectodermal ridge. Several homeobox genes have a role in pattern formation including *Msx-1* and *Msx-2* (formerly *GHox-7* and *Ghox-8*), and the *Chox-D* cluster.

Msx-1 and *Msx-2* encode similar homeodomain proteins, but they have different spatial and temporal expression. For example, in the leg bud at stage 19, *Msx-1* transcripts are localised in the mesodermal region underneath the apical ectodermal ridge and are fairly uniformly distributed in the dorsal and ventral mesoderm (Fig. 12.5). At this same stage, *Msx-2* transcripts are present in high amounts underneath the apical ectodermal ridge and in the anterodorsal mesoderm, but little is present in the posterioventral mesoderm (Fig. 12.5) (Coelho *et al.*, 1991a; Nohno *et al.*, 1992). By stage 26, the distribution of the two transcripts is very similar. Coelho *et al.* (1991b) looked for the presence of *Msx-1* and *Msx-2* transcripts in developing limb buds from normal and *limbless* chick embryos. The latter form limb buds but never develop an apical ectodermal ridge and so fail to undergo normal elongation. By stage 18, *Msx-1* expression is very much reduced in the mesoderm of the *limbless* compared to normal, but *Msx-2* expression was very similar in both. However, expression of *Msx-2* in the ectoderm of *limbless* is severely reduced. Coelho *et al.* (1991b) suggest that expression of *Msx-2* may be regulated by the product of the *limbless* gene. *Msx-2* expressed by the apical ectodermal ridge may amplify directly or indirectly the expression of *Msx-1*. Two polydactylous mutants (*talpid* and *diplopodia-5*), in which extra digits develop, also show changes in the zones in which both *Msx-1* and *Msx-2* are expressed (Coelho, Upholt & Kosher, 1993). Limb bud development not only involves growth and differentiation of cells but also, in shaping the developing wing bud, programmed cell death (apoptosis), in which cells that have fulfilled their roles are removed. Both *Msx-1* and *Msx-2* are expressed in the necrotic zone (Fig. 12.5b,c) of the limb bud and are thought to participate in programmed cell death (Coelho *et al.*, 1991a; 1993).

Other homeobox-containing genes are expressed in developing limb buds. The two paralogous genes *Chox-A11* and *Chox-D11* have very similar structures, their homeodomains differing by three amino acid residues, but they show different spatial expression during limb bud development (Rogina *et al.*, 1992). Whereas *Chox-D11* is expressed in high amounts in the posterior half of the limb mesoderm at stages 20–22 but is absent from the anterior half, *Chox-A11* is expressed in high amount throughout the limb bud. *Chox-D12* is also expressed in the posterior zone of the limb bud, suggesting it is involved in anterior–posterior formation (Mackem & Mahon, 1991). Much of the *Chox-D* cluster is expressed in overlapping domains during limb bud development from stages 20–29 (Fig. 12.6) (Nohno *et al.*, 1991). As

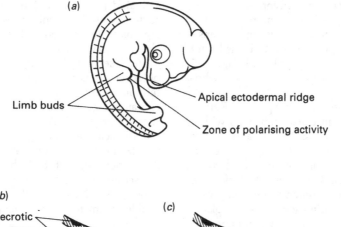

Fig. 12.5. Homeotic gene expression in the developing limb buds. (a) Chick embryo at stage 22, showing the limb buds. (b) Expression of *Msx-1* in the limb bud. (c) Expression of *Msx-2* in the limb bud.

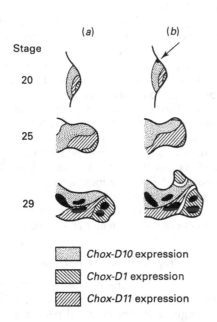

Fig. 12.6. Expression of *Chox-D1, Chox-D11* and *Chox-D10* during stages 20–36 of (a) normal wing bud development, and (b) digit duplication induced by retinoic acid (Nohno *et al.*, 1991).

the bud grows, the pattern of overlapping domains is extended along the proximodistal axis. Chuong (1993) has studied the expression of *Chox-C6* and *Chox-D4* during feather bud development. Using antibodies against *Chox-C6* and *Chox-D4* proteins, he classified the distribution of their microgradients along the anterio-posterior axis of feather buds into four basic patterns. The type of microgradient for the two homeoproteins differed in different regions of the body. If each of the proteins can have one of four types of microgradient, then two homeoproteins can give rise to $4 \times 4 = 16$ combinations. It is, therefore, possible to see how a complex spatial pattern can arise from a combination of a few genes.

12.5 Retinoic acid and limb bud development

Limb bud development is a good system in which to study the role of retinoids as morphogens. Retinoids are natural and synthetic compounds related to retinol and retinoic acid. Retinoic acid is formed by the oxidation of retinol (see Section 2.4.1), and both retinoic acid and the related metabolite 3,4-didehydroretinoic acid are present in the normal limb buds of chick embryos (Scholfield, Rowe & Brickell, 1992), the latter at substantially higher concentration (Mavilio, 1993). When locally applied to the anterior margin of chick limb buds, retinoic acid induces duplications of the digits. The zone of polarising activity is located at the posterior margin of the limb bud (Fig. 12.5), and this is thought to release a morphogen, whose concentration gradient across the anterio-posterior axis results in specifying the tissue pattern. Retinoic acid has been detected in chick limb buds with an overall concentration of $\approx$ 25 nM and a peak concentration of $\approx$ 60 nM at the posterior end. This and other evidence favours retinoic acid as the morphogen (see Ragsdale & Brockes, 1991). However, when the results of experiments are analysed in which (i) the zone of polarising activity is transplanted to the anterior of the limb, and (ii) when exogenous retinoic acid is added to limb buds, the data are not consistent with a simple retinoic acid diffusion model or a more complex autocatalytic model (Noji et al., 1991; Papageorgiou & Almirantis, 1992). Although retinoic acid is assumed to have a role in pattern formation, it may act synergistically with other morphogens, or there may be several steps to the process.

Retinoic acid is a hydrophobic molecule able to diffuse across the plasmalemma. Its mechanism of action is similar to that of the steroid hormones (see Section 10.1). It binds to nuclear receptors, which act as transcriptional activators. Two families of retinoic acid and retinoid receptors, RAR and RXR, have been identified in mammals and domestic fowl. The RAR family includes RAR-α, RAR-β, RAR-γ and RAR-δ, and the RXR includes RXR-α, RXR-β and RXR-γ (Smith & Eichele, 1991; Gudas, 1994). Each gene is able to form multiple transcripts, probably by alternative splicing; for example, RAR-γ forms seven transcripts, and RAR-β three (3.2, 3.4 and 4.6 kb). The proteins of both families have six domains: A–F. Each family has a DNA binding-domain (C) and a ligand-binding domain (E). The amino acid sequences are very similar in the B, C, D and E domains of RAR-α, RAR-β and RAR-γ, but diverge in A and F. RXR shows sequence similarity to RAR-α, RAR-β and RAR-γ only in the C domain (Rowe, Richman & Brickell, 1991). A further difference is that all-*trans*-retinoic acid specifically binds and activates the RAR subfamily and is probably the physiological ligand, whereas it only has a very low affinity for RXRs (Thaller, Hofmann & Eichele, 1993). 9-*cis*-Retinoic acid is able to bind both RARs and RXR; the latter with high affinity and, therefore, it is probably the physiological ligand. RARs and RXRs can exist as homodimers or RAR–RXR heterodimers. The ligand-activated homo- or heterodimers bind the retinoic acid-response elements present in the promoter regions of target genes. The retinoic acid-response elements have two copies of the motif 5'-$^A/_GG^G/_T$ TCA-3', separated by 5 bp (Gudas, 1994). The range of target genes having promoters that recognise the retinoic acid receptors is not yet clear, but homeobox-containing genes are the most likely candidates. In support of this, Yokouchi et al. (1991) found that retinoic acid decreased the concentration of Msx-1 mRNA in the anterior margin of developing limb bud. They suggest that this repression of Msx-1 transcription releases the cells in the anterior necrotic zone (Fig. 12.5) from their necrotic mode. This enables them to reprogramme as cartilage-forming cells with potential for digit duplication. They suggest that retinoic acid binds retinoic acid receptors and this leads to the repression of Msx-1 expression. Transcription of the retinoic acid-receptor gene itself is activated by retinoic acid. The exogenous application of retinoic acid to the zone of polarising activity causes digit duplication. This is preceded by an extension of the expression domains for Chox genes, as illustrated for the Chox-D cluster (Fig. 12.6b), which is assumed to initiate the digit duplication (Nohno et al., 1991). The action of retinoic acid is generally antineoplastic, i.e. it promotes differentiation and suppression of cell division, but not all cell types are similarly affected.

During early limb bud development (stages 20–27), transcripts of the α- and γ-retinoic acid receptors are fairly uniformly distributed throughout the bud, whereas the β-receptors are restricted to the proximal region of the bud, and highest levels are in the region that contributes to the shoulder. Grafting experiments show that this expression is position dependent (Schofield et al., 1992). RXR-γ is expressed in wing buds at stage 20. Both all-*trans*- and 9-*cis*-retinoic acid, when applied to wing buds, induce pattern duplications, but 9-*cis*-retinoic acid is about 25-fold more potent. All-*trans*-retinoic acid is converted to 9-*cis*-retinoic acid in the wing bud, and this may account for the effects of exogenously added all-*trans*-retinoic acid (Thaller et al., 1993).

12.6 Development of the nervous system

The development of the nervous system involves many complex interactions between cells, and a large number of genes must be involved in its development. A description of its development requires a knowledge of the cell migrations that lead to its organisation. A number of methods have been used to produce fate maps, including the use of fluorescent vital dye to label groups of cells and follow their subsequent migration (Lumsden, Sprawson & Graham, 1991), and the use of chick–quail chimaeras (see Section 12.1). Using neurochimaeras, it has been possible to study, for example, the ontogenesis of the cerebellum, the establishment of functional synapses between neurones and to construct a fate map of the early neural primordium (le Douarin, 1993).

The expression of the *engrailed* genes in the developing nervous system has been studied. The *engrailed* gene in *Drosophila* is one of the segmentation genes controlling segment polarity. It is essential for the establishment and maintenence of segment structure and is expressed at the posterior of each segment. Segmentation is also important in vertebrate development, although the segmentation is less obvious. Evidence for the common arrangement in both vertebrates and *Drosophila* is that homeobox-containing genes show similar expression along the anterio-posterior axis in both. The rhombomeres in the chick are the likely counterparts of parasegmentation in *Drosophila*, in each case they are demarkation boundaries of gene expression. In spite of these similarities and the sequence conservation between *engrailed* genes in invertebrates and vertebrates, there is little evidence to suggest a role for *engrailed* genes in segmentation in vertebrates, such as the domestic fowl. The *En* genes are a subfamily of homeobox-containing genes and are highly conserved. Two have been identified in the domestic fowl, *En-1* and *En-2*, using antibodies raised against En proteins. They are present in the developing brain, mandibular arch, spinal cord, dermatome and ventral limb bud ectoderm (Gardner & Barald, 1992). Using transplant surgery, Gardner & Barald (1991) have examined whether the cellular environment regulates the expression of *engrailed*-like protein. At stage 10–12, En protein is expressed in the mesencephalon and metencephalon but not in the prosencephalon. When reciprocal transplantations between these regions were carried out, the transplants took on the character of the host tissue with regard to En expression. Using an antibody specific for En-1 protein, and a second which recognises both En-1 and En-2, Gardner & Barald (1992) found that expression of *En-2* lagged behind *En-1* in the head region (prosencephalon to rhombomere 2). They concluded that *En* genes play a role in neurogenesis and the regionalisation of early cranial neuroepithelium.

The role in development of many of the homeodomain-containing genes is not known, and much of the work recently carried out is aimed at a temporal

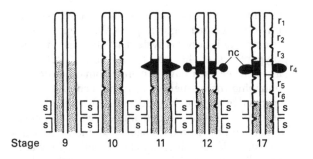

Fig. 12.7. Expression of *Ghox-lab* during development of the hindbrain from stages 9 to 17 (Sundin & Eichele, 1990). s, segment; r, rhombomere; nc, neural crest.

and spatial description of homeobox-containing gene expression, as the first step towards understanding their role. Most homeobox-containing genes are first expressed about the time of neurulation or later, some are expressed during gastrulation, e.g. *Ghox-lab*. In many cases, the anterior boundary of their expression is within the rhombomeres of the hindbrain or the cervical region of the spinal cord. The lack of expression in the fore- and midbrain may be because the hindbrain is a modified extension of the spinal cord, with the fore- and midbrain newly added structures. *Ghox-lab* expression has been studied in detail in the developing hindbrain from stages 7–17 (Sundin & Eichele, 1990). *Ghox-lab* is considered to be homologous to the *labial* gene in *Drosophila* and *Hox-1.6* or *Hox-2.9* in mouse. When more information is available as to its sequence, it may become renamed as *Chox-A1* or *Chox-B1* (see Fig. 12.4). Figure 12.7 shows how the spatial expression of *Ghox-lab* changes during the hindbrain development and rhombomere formation. From being generally expressed from the hindbrain areas anteriorly at stage 7, its expression becomes focused in rhombomere 4. More detailed examination of its expression in rhombomere 4 at stage 17 shows that it is expressed in all cells, except the floor plate and roof plate. Sundin & Eichele (1990) suggest that *Ghox-lab* is concerned with segment identity and is the vertebrate equivalent of the homeotic selector gene; it could, therefore, programme the neural crest of rhombomere 4 to form the glial anlage of the facial and acoustic ganglia.

Msx-1 and *Msx-2* genes, thought to be involved in short-range epithelial–mesenchymal interactions in developing limb buds (Section 12.4), are also differentially expressed during chick craniofacial development (Nishikawa *et al.*, 1994). *Msx-2* has a broader, more intensive expression in the lateral choroid plexus and cranial skull than *Msx-1*. The homeobox-containing gene *pax-QNR* is expressed in the embryonic nervous system. It is expressed as five polypeptides of 48, 46, 43, 33 and 32 kDa. The largest three each have a paired domain (a domain containing three α-helices with the second and third separated by eight amino acid residues), homeodomain, carboxyl domain and a domain between the homeodomain and the paired domain. The smaller two lack the paired domain. It is not clear whether they arise from alternative splicing. They are expressed along the entire anterio-posterior axis of the neural tube. In the neuroretina, the amounts of the 33 and 32 kDa proteins decrease with neuroretina differentiation (Carriere *et al.*, 1993). The promoter region of *pax-QNR* has been characterised as having a TATA-like box and CAAT box as well as several *cis*-regulatory elements. The 46 kDa protein product of *pax-QNR* expression encodes a transcriptional activator which is able to *trans*-activate its own promoter (Plaza, Dozier & Saule, 1993). This property of autoregulation has been found with several homeobox-containing genes in *Drosophila*.

Quox-1 is another homeobox-containing gene expressed in the nervous system. It has a high sequence similarity to mouse *Hox-1.1*, and thus should perhaps be designated *Qhox-A7* (Fig. 12.4). It is unusual, when compared with other homeobox genes, in that it is expressed in the central nervous system from the rostral part through all regions of the brain and spinal cord (Xue, Gehring & le Douarin, 1991), where most other homeobox-containing genes are not expressed anterior of the mesencephalon. As the embryonic nervous system develops, *Quox-1* mRNA becomes increasingly abundant in the brain and less abundant in the spinal cord as the rate of proliferation decreases. In the early stages of development, it is detectable in the neural primordium at the presomite stages, and later in all segmented somites. After the somites differentiate into sclerotome, dermatome and myotome (Fig. 12.3*d*), *Quox-1* expression is restricted to the myotome (Xue *et al.*, 1993).

Msx-1 is widely expressed in the stage 6 embryo, including the primitive streak. At stage 11, it is detected in the neural crest at the midbrain level (Suzuki *et al.*, 1991).

12.7 Cell adhesion molecules

The previous sections have been concerned with describing the temporal and spatial expression of homeobox-containing genes, but not with how expression is initiated and what are the downstream molecules of the homeodomain proteins. These molecules are important in determining the phenotype of the differentiated cells, and so far little is known about their identity. One group of molecules that are likely to be important are cell adhesion molecules and their receptors. The development of a multicellular organism from a single fertilised egg involves cell proliferation, cell differentiation, cell motility and cell adhesion. As proliferation and differentiation occurs, the cells have to move relative to one another and make connections with adjacent cells, giving rise to the organised tissues and organs of the fully developed animal. In normal differentiated tissues, cells having the same phenotype are most frequently adjacent to one another, and the boundaries with different cell types are usually sharp. In what is now regarded as a classical series of experiments, Townes & Holtfretter (1955) dissociated embryonic cells and mixed them together in random fashion. They found that cells of the same phenotype adhered to one another and only infrequently attached to cells of different phenotypes. If the cells mixed together came from more than one species, they adhered to cells of the same tissue type rather than the same species. These findings suggested that cells of the same phenotype recognised one another, and it seemed most likely that the recognition arises from cell surface molecules.

When embryonic development has reached the blastula stage, only a single cell type resembling epithelia exists, but later during gastrulation three germinal layers form: the upper layer or ectoderm, the middle layer or mesoderm and the lower layer or endoderm. The cells of both the endoderm and ectoderm are epithelial in character, forming a continuous layer in which cells are held to one another by specialised sites on the cell surfaces known as *adherens junctions*, and with a basement membrane at their base. The mesodermal cells are more loosely associated and, therefore, more mobile and are surrounded by an extracellular matrix. Overall two types of adhesive contact occur, cell–cell adhesion and cell–matrix adhesion, the former being more important in the endoderm and ectoderm and the latter in the mesoderm, although both types occur in each.

Several groups of molecules are now known to be involved in these processes, some of which are cell-surface proteins and others are proteins secreted to form the extracellular matrix. The principal groups of adhesion molecules are summarised in Table 12.1. The basement membrane is the first extracellular matrix to form during embryonic development. It is

Table 12.1. *Properties of adhesion molecules*

Protein	Structural characteristics	Location	Interactions	Functions
Cadherins	Transmembrane glycoprotein, 120–140 kDa	Plasmalemma	Ca^{2+}-dependent homophilic binding to extracellular domain, cytoskeletal binding to cytosol domain	Maintenance of cell-cell interactions
Integrins	Transmembrane $\alpha\beta$-heterodimer α_1, 165 kDa; β, 90–120 kDa	Plasmalemma	Binding to laminins, fibronectin, collagen; specificity depends on $\alpha\beta$ composition	Maintenance of cell to basement membrane and extracellular matrix binding
Laminin	Trimer 850 kDa A-chain 400 kDa B1-chain 215 kDa B2-chain 205 kDa	Extracellular matrix	Binds collagen, entactin, heparin and integrin	Principal component of extracellular matrix
Fibronectin	550 kDa homodimer (2 × 275 kDa) linked by two disulphide bridges	Extracellular matrix	Binds collagen, heparin and integrin	
Entactin	Sulphated glycoprotein, dumbell shaped, 150 kDa single chain	Extracellular matrix	Binds laminin	
Tenascin	190–230 kDa	Extracellular matrix		Regulation of cell–matrix attachments
Collagen IV	550 kDa, two $\alpha1$(IV) and one $\alpha2$(IV); chains disulphide bonded	Extracellular matrix	Binds laminin and fibronectin	Structural component of basement membranes
Heparin sulphate proteoglycan	Low-density form 600–700 kDa; single polypeptide with three heparin sulphate chains	Extracellular matrix	Binds laminin and fibronectin	Structural component of basement membrane
Heparin sulphate proteoglycan	High-density form 130 kDa; peptide core (5–30 kDa) + 3–4 heparin sulphate chains	Extracellular matrix	Binds laminin and fibronectin	Structural component of basement membrane

Source: Martin & Timpl (1987); Takeichi (1991); Hynes (1992); Rajaraman (1993); Tucker *et al.* (1994).

thought to create the barriers that allow embryonic cells to segregate and differentiate to form tissues (Martin & Timpl, 1987). There are three main classes of transmembrane molecule involved in cell–cell adhesion, and cell–matrix adhesion, namely proteins of the immunoglobulin superfamily, integrins and cadherins (Takeichi, 1991; Chuong, 1993). In addition, there are the proteins that constitute the extracellular matrix, namely laminins, fibronectins, entactin/nidogen, tenascin, collagen type IV and heparin sulphate proteoglycan. The best known of the immunoglobulin superfamily is *N-CAM*. It is widely distributed, and

undergoes Ca^{2+}-independent homophilic binding. The second group, *integrins*, recognise principally extracellular matrices but are also capable of intercellular adhesion. They are heterodimers, composed of an α-chain (120–180 kDa), and a β-chain (90–110 kDa). Both polypeptides are transmembrane glycoproteins, each having a single membrane domain, a short cytoplasmic domain and a large extracellular domain. They are capable of binding proteins such as collagen, laminin, fibronectin, vitronectin and osteopontin via their extracellular domain (Hynes, 1992). The integrin family includes a number of different integrins. At least 14 α-subunits are known and eight β-subunits, giving rise to a potentially large number of dimers. The specificity of the integrin depends on the combination of subunits, e.g. $\alpha_3\beta_1$ recognises fibronectin, laminin and collagens, whereas $\alpha_6\beta_1$ binds only laminin (Bronner-Fraser *et al.*, 1992).

The third group are the *cadherins*, which show Ca^{2+}-dependent homophilic binding. This last group are thought to be crucial for sorting different cell types in vertebrates. The importance of Ca^{2+} can be demonstrated, since suitable chelating agents cause disruption of the tissues. The cadherins have been divided into subclasses, based mainly on their tissue distribution. The three principal subclasses are E-cadherins, N-cadherins and P-cadherins. In avian species, E-cadherin has also been called L-CAM and uvomorulin. Each cadherin has a unique spatial and temporal expression. The tissue distribution of cadherins so far found in the domestic fowl is shown in Table 12.2. Cadherins are transmembrane proteins, most of which have 723–748 amino acid residues and $M_r \approx 120\,000$–$140\,000$. They have a single transmembrane domain, with the N-terminal part forming the extracellular domain and the C-terminal part the cytoplasmic domain (Fig. 12.8). The N-terminal part has an internal repeat sequence and contains the Ca^{2+}-binding sites and the homophilic binding sites. Their interactions are predominantly homophilic, but weaker interactions have been observed between cells expressing two different cadherins. The cytoplasmic domains of cadherins and integrins are believed to interact with bundles of actin microfilaments, via membrane-bound plaque structures. The exact molecular structures of these plaques have still to be elucidated, but they contain the proteins vinculin and talin. *Vinculin* is present in all adherens junctions, but *talin* is only present in cell–matrix junctions (Duband & Thiery, 1990). T-cadherin is unusual in lacking cytoplasmic sequences and having a lower M_r, 90 000 (Ranscht & Bronner-Fraser, 1991). The extracellular domain is N-glycosylated. From a number of experiments in which chimaeric cadherins were constructed and in which amino acids were replaced by site-directed mutagenesis, it has been deduced that the first 113 amino acid residues are important in determining specificity and binding. The cytoplasmic domain is also important for cadherin function, and its deletion causes them to lose their cell–cell binding function. Their binding to cytoskeletal proteins such as actin is essential for functioning.

During embryonic development, different cadherins are temporarily and spatially expressed. During the blastula stage, the chick embryo epiblast cells only expresses E-cadherin, but when gastrulation begins the mesodermal cells separating from the epiblast lose E-cadherin and begin to express N-cadherin. In the process of neural tube formation, the neural plate that originally expressed E-cadherin gradually loses it and acquires N-cadherin. Generally, when a population of cells separates from its parent cells it loses the parent type cadherin and acquires a different cadherin, but when cells of different lineages become connected, they acquire the same cadherins (Takeichi, 1991). N-cadherin gene is expressed as four mRNAs (8.0, 4.7, 3.8 and 3.3 kb), probably resulting from alternative splicing. The 8.0 kb mRNA is only expressed in brain, the 3.8 kb is only expressed in heart, but the 4.7 kb is the most abundant in both heart and brain (Dalseg, Andersson & Bock, 1990). Neural crest cells migrate extensively during embryonic development to defined sites where they differentiate. The neural crest cells from trunk somites differentiate into dermatome, sclerotome and myotome (Fig. 12.3d). T-cadherin is expressed preferentially in the caudal half of each sclerotome (Ranscht & Bronner-Fraser, 1991). At this stage, the neural crest cells are migrating into the rostral part of the sclerotome, but no migration occurs in the caudal part. Ranscht & Bronner-Fraser (1991) suggest that T-cadherin influences the pattern of migration of the neural crest cells in a negative way by preventing them from entering the caudal part of the sclerotome, and thus maintains somite polarity. Neural crest cell membranes are rich in cell adhesion molecules such as fibronectin, laminin, tenascin, collagen, and they also have several integrin receptors, which promote the attachment of cells to the extracellular matrix (Bronner-Fraser, 1993). The migration of melanocyte precursor cells from neural crest cells is described in Section 12.8.

B-cadherin has been cloned and sequenced from a chick brain cDNA library, and it shows extensively homologues with E-cadherin, P-cadherin and L-CAM, more so than with N-cadherin. It is expressed in a variety of chick embryo tissues at stage 39, particularly in the epithelial cells lining the choroid plexus. On the basis of its pattern of expression, it is thought to be important in neurogenesis (Napolitano *et al.*, 1991). In addition to the cadherins, other proteins involved in cell adhesion, such as tenascin, fibronectin, laminin and integrin, are differentially expressed during development (Chuong, 1993). The spatial and temporal distribution of vinculin and talin during early embryo development is similar to that of N-cadherin, L-CAM and integrin receptors (Duband & Thiery, 1990). *Laminin* is the most abundant protein in the basement membrane. It is a large

Table 12.2. *Cadherins*

Cadherin type	Alternative names	Tissue distribution
N-cadherin (nerve)	A-CAM, N-cal-CAM	Lens, nerves, muscle, ear, early embryo (gastrulation)
E-cadherin (epithelial)	L-CAM, uvomorulin	Early embryo (blastula epiblast cells)
R-cadherin (retina)		Retina
B-cadherin (brain)		Early embryo, liver, eye, heart, muscle, intestine
T-cadherin (truncated)	K-CAM (near identity with B-cadherin)	Early embryo
P-cadherin		No avian homologue yet identified; embryo, skin, liver, heart, kidney

Source: Napolitano *et al.* (1991); Sorkin *et al.* (1991); Geiger & Ayalon (1992).

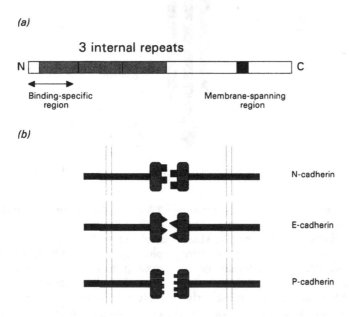

Fig. 12.8. (*a*) Cadherin structure. (*b*) Diagramatic representation of homophilic binding of N-, E- and P-cadherins.

protein comprising three polypeptide chains, A, B1 and B2, and having binding sites for collagen type IV, heparin sulphate proteoglycan and entactin (Fig. 12.9). These proteins bind together to form the integrated structure of the extracellular matrix. Fibronectin, laminin, entactin and tenascin have been studied in extracellular matrix during development. Laminin is the first extracellular matrix protein to be detected during embryonic development. It forms a very stable complex with entactin ($K_d < 10\,\text{nM}$). *Entactin* is found both in basement membranes and in the extracellular matrix of mesenchymal tissues. It has been detected in the epiblast and hypoblast of developing chick embryo and is abundant in the neural plate and in many developing tissues. It is found together with laminin and is thought to have a functional role in the directed cell migration of the early embryo (Zagris, Stavridis & Chung, 1993). *Fibronectin* is another cell–matrix binding protein present in the extracellular matrix. It has binding domains for heparin, collagen and integrins. The integrin cell recognition site on fibronectin is the sequence Arg–Gly–Asp present in the middle portion of the molecule. This binds the cell to the extracellular matrix, but it is reinforced by the lower-specificity heparin-binding sites (Ruoslahti, 1988).

Both laminin and fibronectin promote adhesion and neurite outgrowth, whereas tenascin generally has an opposing effect in acting as a barrier for growing axons (Perez & Halfter, 1993). *Tenascin* is a

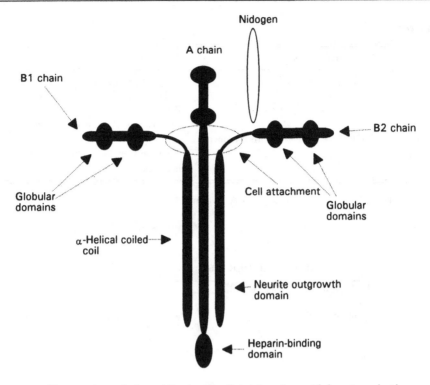

Fig. 12.9. Arrangement of laminin chains A, B1 and B2 showing their interactions with heparin and nidogen.

multimeric glycoprotein and exists in multiple forms as the result of alternative mRNA splicing. The principal forms in the domestic fowl are 230 kDa, 200 kDa and 190 kDa (Tucker *et al.*, 1994). Integrin distribution has been studied throughout development, using antibodies against particular α- and β-subunits. The β_1-subunit has wide distribution in most early embryo cells. However β_1 is capable of forming heterodimers with different α-subunits, and the binding capacity of all the different $\alpha\beta$ heterodimers has not been determined. Bronner-Fraser *et al.* (1992) found the α_6-subunit is developmentally regulated in the chick embryo and is prominent in the developing nervous system at stage 30, and also in myoblast precursors. Its distribution generally overlaps with that of β_1-subunit and laminin; $\alpha_6\beta_1$ binds laminin specifically. The α_5-subunit is also widely distributed in early embryos, associated with both endothelial and mesenchymal cells and persists in later embryos (beyond stage 36), but it is absent from most adult cells (Muschler & Horwitz, 1991). The heterodimer $\alpha_3\beta_1$ integrin, which binds fibronectin, laminin and collagen, is abundant in chick embryo fibroblasts (Nermut *et al.*, 1993).

12.8 Melanocyte differentiation

Melanin is the principal pigment contributing to plumage colours of birds. It occurs in feathers in two main forms, eumelanin and phaeomelanin. The former is black and the latter red. Both are produced by mature melanocytes. The plumage pattern depends on the spatial distribution of melanocytes throughout the skin. This section is principally concerned with the differentiation of the pluripotent neural crest cells into fully differentiated melanocytes. For details of pigment production and of the genetic basis of plumage pattern, see Veevers (1982) and Stevens (1991a), respectively. The differentiation of neural crest cells to form melanocytes is a good experimental system in which to study the mechanisms of cell lineage segregation. In order to do this, it is necessary to identify the cells from which the melanocytes are derived and determine the stages at which they migrate, eventually to the epidermis of the skin. During this process, their developmental potential becomes progressively more restricted as they become committed melanocytes.

The neural crest cells are the pluripotent parent cells of a number of different cell lineages, including the peripheral nervous system, endocrine cells such as those of the adrenal medulla, connective tissue, muscle, bone and cartilage. The neural crest cells in the trunk region of the developing embryo migrate mainly in two pathways, the dorsolateral pathway, giving rise to skin melanocytes, and the ventral pathway, giving rise to sensory and sympathetic neurons, such as glial and Schwann cells and adrenal chromaffin cells (Artinger & Bronner-Fraser, 1992). A number of techniques have been used to trace the course of melanocyte differentiation and the restriction of their developmental potential. These include the use of monoclonal antibodies to detect melanocyte-specific markers, liposoluble cyanine dye to follow cell migration and the chick–quail chimaera system (Kitamura et al., 1992; Bronner-Fraser, 1993; Nataf et al., 1993), and in vitro analysis of explants of epidermal ectoderm to examine their developmental potential (Richardson & Sieber-Blum, 1993). Monoclonal antibodies that have been used are: (i) MEBL-1, which detects a 135 kDa protein exclusively in melanocyte precursors before distinct melanosomes structures are formed and before DOPA can be detected (stage 16), (ii) MelEM, which first detects a 26 kDa protein in melanoblast as soon as they have reached the subectodermal mesenchyme, and (iii) Mel1 and Mel2, which detect glycoproteins (123 kDa and 85 kDa, respectively) present in melanosomes and promelanosomes. Melanosomes are the subcellular organelles that produce melanin. Neither of the glycoproteins (123 kDa or 85 kDa) correspond to tyrosinase (70 kDa), which plays a key role in melanin biosynthesis. Melanocyte differentiation progresses cephalocaudally, and MEBL-1 positive cells can be detected in the cephalic somites at stage 16, in the vagal somites at stage 17 and in the wing somites at stage 18 (Kitamura et al., 1992). The restriction of the developmental potential of neural crest cells has been demonstrated by removing explants of neural crest cells, dissociating them and then culturing them. Neural crest cells from stage 13–15 embryos developed in culture into both pigment-producing cells and adrenergic cells, but those from stage 20–21 embryos only produced pigment cells (Artinger & Bronner-Fraser, 1992). Recent work suggests that melanin synthesis begins in the dermis before the terminal migration to the epidermis has occurred (Hulley, Stander & Kidson, 1991). There is much information on the path of migration from the neural crest to the epidermis, and some information on proteins expressed at different stages, but the functions of the proteins are not yet clear, nor the control of their expression understood.

12.9 Protein expression during embryonic muscle development

During muscle development, unicellular myoblasts, derived from the precursor cells in the somites, fuse to form the multinuclear mature skeletal muscle cells or myofibres. The process involves specific myoblast–myoblast fusion. Two cell-surface glycoproteins have been implicated in this process, namely N-cadherin and N-CAM (Soler & Knudsen, 1991). The latter is a member of the immunoglobulin superfamily (see Section 12.8). Both N-cadherin and N-CAM are expressed in clusters periodically distributed in the plasma membrane of avian skeletal myoblasts. N-CAM is expressed as at least three principal isoforms (115 kDa, 145 kDa and 120 kDa) generated by alternative splicing. All three forms are expressed between E5 and E18 in embryonic muscle tissue but with different proportions of each at different stages. The largest (155 kDa) is expressed transiently between E11 and E14, which coincides with the period of extensive myotube formation suggesting that it has an important role in muscle differentiation (Yoshima et al., 1993). Myoblasts arise in the myotome, which is the source of skeletal muscle of both trunk and limbs. The myotome itself arises from the dorsolateral cells of the somites. The somite is initially composed of columnar epithelial-like cells arranged radially around a small lumen. Various markers become expressed as the myogenic programme is initiated, and these include a family of myogenic transcription factors of the myoD family, and the protein desmin, which is localised at the Z-line in striated muscle (Borman, Urlakis & Yorde, 1994). Desmin expression precedes that of other muscle-specific proteins. The myoD family includes myoD, myf5, myogenin and mrf4/herculin. The earliest myogenic transcription factor expressed in somites prior to myotome formation (stage 12) is CMD1 in domestic fowl, and qmf1 (the quail homologue of myoD) in quail. This is followed in quail by expression of qmf3 (equivalent to myf5), qmf2 (equivalent to myogenin) and genes giving rise to other muscle-specific proteins such as troponin I (Coutinho et al., 1993).

12.10 The role of growth factors in embryonic development

In the previous sections of this chapter, emphasis has been placed on the importance of spatial gene expression, the role of controlling genes such as homeobox genes, cell movement and the importance of cell adhesion molecules in establishing the positions of cells relative to one another. For all these processes to occur, intercellular signalling is necessary. One group of morphogens, the retinoids, has been discussed in connection with limb development (Section 12.5), but it is becoming clear that several intercellular signals may be necessary for the intricate processes involved. This section considers the role of polypeptide growth factors.

Early experiments, in which cells and tissues were transplanted from one position to another within developing amphibian and avian embryos showed that most frequently transplanted tissue developed or became transformed to become the same tissue as that in which it was transplanted. This suggested that intercellular signals from the recipient region exerted a controlling influence on differentiation. Until the early 1980s, it was generally felt that hormones were unlikely to play an important role in the early stages of development, at least prior to the appearence of at least rudimentary endocrine glands (de Pablo & Roth, 1990). The discovery of large numbers of paracrine and autocrine polypeptide growth factors, together with the detection of such growth factors in early embryos, changed the position. For example, insulin is present in unfertilised eggs, and both insulin and insulin mRNA are present at E2 (de Pablo & Roth, 1990). Many of the polypeptide growth factors were first discovered as the result of trying to define the growth medium required to support vertebrate cells in culture. The growth factors are generally classified, on the basis of their structures and gene sequence into five classes: epidermal (EGF), transforming (TGF), heparin-binding (HBGF), platelet-derived (PDGF) and insulin-like (IGF). The names usually refer to the function of the first member of the family to be discovered.

The blastoderm consists of $\approx 60\,000$ cells and comprises an upper layer or epiblast and a lower layer or hypoblast (Figs. 12.2 and 12.3). The hypoblast develops by ingression of small groups of cells and the embryonic axis starts to form. If at this stage, the epiblast is detached from the hypoblast and cultured separately, it will form non-axial mesoderm. The mesoderm will develop, but it will be disorganised in not having an axis (Mitrani et al., 1990). Therefore, it appears that the hypoblast may provide the factors necessary for axial development. Mitrani et al. (1990) found that if activin B, a member of the TGF family, was added to the cultured epiblasts it induced the formation of axial structures including notochord and segmented somites. Using a cDNA probe, they were able to demonstrate by Northern blotting that activin B is transcribed most actively in the hypoblast,

and they suggest that it is the endogenous inducer of the body axis. Sanders and Prasad (1991) detected TGF-β_1 in the mesoderm and primitive streak of the gastrulating embryo. They also found that if TGF-β_1 was added to cultures of cells having epithelial characteristics, and grown on fibronectin-coated coverslips, the cells adopted the appearance of mesenchymal cells. Further transcripts of the TGF family have been detected in the early chick embryos; these include TGF-β_2, TGF-β_3 and Vgc1 (Harris et al., 1993). From these collective observations, Sanders, Hu & Wride (1994) suggest that TGFs have a role in embryonic axis formation and blastoderm expansion. All of these together with bFGF (basic fibroblast growth factor) may be necessary in different combinations during early development. The mesodermal cells arising by ingression from the epiblast layer move away from the midline using the basement membrane of the overlying epiblast and the extracellular matrix. Brown and Sanders (1991) have examined the role of fibronectin and laminin in directing this movement. They microinjected fibronectin and laminin into the tissue space adjacent to the primitive streak. Their results suggest that fibronectin is definitely involved as a mediator of cell attachment, and laminin may also be. Fibronectin probably uses the integrin $\alpha_3\beta_1$ receptor, abundant in chick embryo fibroblasts.

A number of polypeptide growth factors have been detected in the chick embryo, either by using antibodies to detect the polypeptides themselves, or by detection of the gene transcripts using Northern blotting. Their temporal and spatial expression has then been studied, but in most cases the evidence for their function in development is circumstantial, i.e. it generally relates their expression to particular developmental events. It is not always clear whether their presence is causing a particular event or whether it is a consequence of a developmental change. IGF-I mRNA is expressed in most embryonic tissues except liver. IGF-I concentration in the serum increases from E6 to E15 and then declines until hatching. IGF-I receptors are also widespread in chick embryo from gastrulation onwards, but the number of receptors per cell decreases during late development (Yang et al., 1993). PDGF and FGF-2 are thought to play roles in chick limb bud development (Potts & Carrington, 1993; Savage et al., 1993). PDGFα receptor mRNA has been detected in both mesoderm and ectoderm throughout the development of the limb bud, and FGF-2 has been found in highest concentrations in the centre of the limb bud. Two putative growth factors, Quek1 and Quek2 (quail endothelial kinases), which are related to the PDGF family, have been detected in endothelial cells of developing quail embryos using monoclonal antibodies. Quek1 is also expressed in the mesoderm from the onset of gastrulation (Eichmann *et al.*, 1993).

Many of the growth factors, including FGF, IGF, PDGF and Quek, have tyrosine kinase domains. Phosphorylation of tyrosine residues in key proteins is probably an important control mechanism in development. Patstone & Maher (1993) have measured the general level of phosphorylation of tyrosine residues in developing embryos using antiphosphotyrosine antibodies. They found highest levels at E7 to E11 and in cells starting to differentiate rather than those most actively proliferating. Thyroid hormone receptors are differentially expressed during chick retinal development, suggesting they have stage-specific functions in neuroretinal development (Sjöberg, Vennström & Forrest, 1992). Polypeptide growth factors believed to be involved in cell–cell signalling are the *wnt* group of proto-oncogenes. These were first identified in mouse mammary tumours. The family, which includes over 12 members, encodes a cysteine-rich protein having a signal peptide. The protein is glycosylated and binds tightly to the extracellular matrix. Two members of the *wnt* family, *wnt-5a* and *wnt-7a* show different patterns of expression in developing limb buds. *wnt-5a* is expressed in the apical ectodermal ridge and in distinct domains along the proximodistal axis, whereas *wnt-7a* is expressed exclusively in the dorsal ectoderm of the limb bud. They may, therefore, have roles in establishing patterning in the limbs (Dealy *et al.*, 1993).

12.11 Conclusions

The study of avian development can be divided into three overlapping phases. The first is the full description of the morphological events that occur during development. The avian embryo is a particularly suitable organism for this study, since each of the stages is readily accessible. Fate maps, which form part of this study, show the extent to which the developmental programme is committed at each successive stage. The second stage is to describe the molecular changes that underly the morphological changes in development. Much has been learnt in recent years about the homeotic genes, several of the proteins involved in cell adhesion and the importance of paracrine and autocrine secretions; but there is still much to do. The third phase is to account for the sequences observed and to try to understand the causal events. This applies both to the morphological and molecular sequences. The importance of concentration gradients in initiating positional events is being actively studied. Work on other organisms such as *Drosophila, Xenopus*, nematodes and the mouse shows many common features in developmental biology.

13

The molecular basis of avian immunology

13.1 Introduction

The immune system constitutes the principal defence mechanism of an organism. Defence mechanisms exist in most organisms from protozoa to vertebrates. In protozoans, such as amoeba, a nucleus may be transplanted into an enucleated syngeneic strain and the recipient will survive, but if the nucleus is from disparate strains, then the chance the recipient has of survival falls with increasing disparity between the strains, i.e. a basic form of recognition of self and non-self exists. In invertebrates, a phagocytic system exists for engulfing foreign microorganisms; however, it is only in vertebrates that the refined and highly discriminating system based on the production of humoral antibodies exists (Manning & Turner, 1976). Almost all vertebrates are able to produce immunoglobulins having the two heavy and two light chains (H_2L_2) as their principal antibody, the only exception being in the Agnatha, where the lamprey antibody consists of heavy chains linked only by non-covalent bonds.

There are number of important characteristics of the avian immune system, discussed in this chapter, some of which are common to vertebrates in general and others that have not been found outside the class Aves. Most of this chapter is devoted to two topics, namely, the generation of avian antibodies and the nature and diversity of avian histocompatibility antigens. There are a huge number of different immunoglobulins generated to interact specifically with antigenic groups. The mechanism by which the pluripotent immature lymphocyte becomes committed to the production of a single species of antibody has been one of the most exciting problems arising from the central dogma (DNA→RNA→protein). There are two different antibody-producing systems: the humoral system in which lymphocytes secrete soluble immunoglobulins into the plasma, and the cellular antibodies that are present on the membranes of lymphocytes. The antibody-producing system has a memory. If an organism encounters a foreign antigen for the first time, this initiates what is referred to as the primary response. If it subsequently encounters the same antigen, the secondary response is both swifter and larger in magnitude. Organisms have the ability to distinguish self and non-self. This is apparent in tissue grafts but was first recognised as a result of blood transfusions. The major histocompatibility complex is the system of cell-surface antigens that gives the organism its almost unique identity, and it also plays a key role in the immune response. In mammals, it is known that the range of potential antigens that the foetus is exposed to through the placenta generates immunological tolerance to those antigens, if encountered post-partum. What happens in egg-laying vertebrates; at what stage in their development do they develop an immunological response? The cell surface antigens that have been most intensively studied in the domestic fowl are those on the erythrocyte membrane, and these are discussed in Section 13.6.

13.2 Avian immunoglobulin structure

Most work on avian immunoglobulins has used the domestic fowl, but some work has also been carried out on other species, including pheasant, quail, duck, turkey, pigeon, goose, hawk and cormorant (Table 13.1). Immunoglobulins (Ig) are the proteins that constitute the antibodies. All are made up of heavy (H) and light (L) chains linked by disulphide bridges (Fig. 13.1). In the early studies of the structure of immunoglobulins, it was found that when immunoglobulin G (IgG) was incubated with papain, it was cleaved into two fragments, known as F_{ab}, the fragment containing the antibody-combining site, and F_c (fragment crystalline), which could be crystallised

Table 13.1. *Properties of avian immunoglobulins as typified by the domestic fowl*

Class	Overall structure (size in kDa)	Heavy chain structure (size in kDa)	Light chain structure (size in kDa)	Carbohydrate (%)	Location	Other species in which studied
IgY (IgG)	H_2L_2 (173)	V-C-C-C-C (71)	V-C (21–25)	5–6	Serum mainly, egg-yolk, cropmilk	Turkey, pheasant, quail, pigeon, duck
IgM	$(H_2L_2)_5$, (860) $(H_2L_2)_4$? (680)	V-C-C-C (63–70) J-chain (19)	V-C (22–24)		Serum, cropmilk	Pigeon
IgA	H_2L_2 (170) and $(H_2L_2)_2$ linked by J-chain; $(H_2L_2)_3$? in bile	V-C-C-C-C, present in polymeric form (66) J-chain (19)	V-C (19)		Minor component of serum (monomer), gut-associated lymphoid tissue, gall bladder, Harderian gland, cropmilk (dimer)	
IgY(ΔF_c)	H_2L_2 (118)	V-C-C (35)	V-C (23)	0.6	Serum	Ducks, geese and waterfowl
IgX	$(H_2L_2)_5$? (890)				Bile and intestine	Duck

Source: Kobayashi & Hirai (1980); Ch'ng & Benedict (1981); Butler (1983); Higgins, Shortridge & Ng (1987); Parvari *et al.* (1988); Liu & Higgins (1990); Piquer *et al.* (1991); Engberg *et al.* (1992); Magor *et al.* (1992); Mansikka (1992); Olah *et al.* (1992).

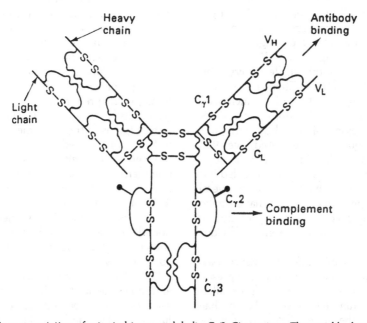

Fig. 13.1. Diagrammatic representation of a typical immunoglobulin G (IgG) structure. The variable domains are indicated as V_L and V_H for the light and heavy chains, respectively, and the constant domains as C_L and $C_\gamma 1$, $C_\gamma 2$ and $C_\gamma 3$ for the light and heavy chains, respectively. -S-S-, disulphide bridge.

(Porter, 1973). The F_c fragment contains the complement-binding site. Each light and heavy chain has a constant (C) region at the N-terminal end, and a variable (V) region at the C-terminal. The constant region is so called because its sequence is relatively constant when comparing different immunoglobulins within the same class. It contains the complement-fixation site. In contrast, there is much sequence variation in the V region of different immunoglobulins within the same class. The V region contains the antibody-binding domain, which with each different immunoglobulin recognises a specific epitope on a particular antigen. Since a typical vertebrate may have the potential to recognise several million epitopes, there are an enormous number of different antibodies (immunoglobulins). Within the V region there are three hypervariable regions, and these are located at the ends of regions of β-structure (Fig. 13.2), positioned to bind the epitope on a particular antigen.

With the large number of different immunoglobulins, a form of classification is necessary. There are three different levels of classification of immunoglobulins; isotypes or classes, allotypes, and idiotypes. *Isotypes* or classes are based on the differences in the C region of the heavy chains. In mammals there are at least five different classes: IgG, IgM, IgA, IgD and IgE. Of these, only the first three are so far known in avian species (Butler, 1983), but there is some evidence for IgD (Chen, Lehmeyer & Cooper, 1982). In birds, the heavy chains corresponding with IgM and IgA immunoglobulins are designated μ and α, and that for IgG is designated ν (Parvari *et al.*, 1988) to distinguish it from the smaller heavy chain in mammalian IgG designated γ. There are also two types of light chain, designated κ and λ, which occur in all three classes. However in the domestic fowl and turkey, the avian species in which antibody structures have been most thoroughly investigated, over 95% of the light chains are λ (Grant, Sanders & Hood, 1971). The major classes are sometimes divided into subclasses. The subclasses have the same overall composition of polypeptide chains but have smaller differences in the sequence of the heavy chains. All classes and subclasses occur in every normal individual within a species.

A further division is into *allotypes*, which have smaller differences from one another and represent alternative forms. These are the result of allelic differences in the constant regions of the heavy chain genes. For example, in the domestic fowl, 14 allotypes of IgG-1 and five of the IgM-1 subclasses have been defined (Ch'ng & Benedict, 1981), and some of these

are also present in turkey and golden pheasant. An individual can only have one (homozygous) or two (heterozygous) allotypes. Both the isotypes and allotypes result from differences in the C regions of the immunoglobulins. The immunoglobulin IgG-1.7 exists as an allotype in the domestic fowl, present only in certain genotypes, but it appears to be present in all pheasant and turkey sera, which fits the definition of isotype (Lamont & Dietert, 1990). *Idiotypes* (Greek *idio* individual), however, result from differences in the V region, the antigenic determinants of the antibody. They are limited to either a single immunoglobulin or a small number of very closely related immunoglobulins.

Of the three classes of immunoglobulin, IgG is the principal one present in domestic fowl serum. Originally this IgG was referred to as IgY, because it is distinctive in being about 20 kDa heavier than IgG present in mammals. This is also the case with IgG from ring-necked pheasant, Japanese quail, and duck (Ch'ng & Benedict, 1981; Zimmerman, Shalatin, & Grey, 1971) and probably applies to all birds. However, since it is the functional equivalent of IgG from mammals, some workers prefer to retain the IgG designation. The larger mass of this immunoglobulin (20 kDa) is accounted for by the heavy chains comprising $4C + 1V$, compared with mammalian heavy chains which are $3C + 1V$ (Parvari *et al.*, 1988). This seems a sufficient difference to warrant the IgY designation that is used throughout this chapter. Three different subclasses of IgY have been characterised in domestic fowl (Butler, 1983). IgM is the next most concentrated in serum, and it is detectable earlier on during embryonic development. IgA, although present in the lowest concentration in serum, is the predominant form in external mucosal secretions (Table 13.2). In plasma, approximately one fifth of IgA is in the monomeric state (H_2L_2) and the rest is polymerised, sedimenting with a similar sedimentation coefficient to that of IgM. In the polymeric form, the monomers are linked through a J chain of 19 kDa, as is the case with IgM (Kobayashi & Hirai, 1980). In external secretions it is mainly present as the dimer. The sequence of its H chain is homologous with the H chain of IgA from mammalian sources, but it has four complete C_H domains compared with only three in mammals (Mansikka, 1992). The IgA class has a major role as the antibodies against antigens present in external secretions. It has been detected in the intestine, bile, lungs, oviduct and the Harderian glands (ocular gland in the anterior portion of the conjunctiva that maintains moisture and an oily secretion in birds and

Table 13.2. *Secretion of immunoglobulin in avian species*

Species	Ig class	Serum concentration (mg/ml)	Other secretions (mg/ml)	Effect of age
Domestic fowl	IgY	Adult, 5–6 E7, 0.025 E21, 1 Maternal, 16	Egg-yolk, 8 Bile, 0.2 Saliva, 0.28 Tears, 0.87	First expressed in bursa at E16
Domestic fowl	IgM	1.7–2.6	Bile, 0.28 Egg-white, 0.15	First expressed in bursa at E13
Domestic fowl	IgA	0.3–5	Bile, 3.2 Saliva, 0.09 Tears, 0.15 Egg-white, 0.7	First expressed in bursa at E18
Turkey	IgA	0.00–0.002	Bile, 0.55–1.47 Jejunum, 0.05–0.5	Linear increase in jejunum from 1–29 days
Duck	IgY(ΔF$_c$)			Predominant Ig in duck serum
Pigeon	IgY	7	Egg-yolk, 5.37 Cropmilk, 0.34	
Pigeon	IgA	0.18	Egg-yolk, 0.02 Egg-white, 1.56 Cropmilk, 1.45	

Source: Lebacq-Verheyden, Vaerman & Heremans (1974); Rose *et al.* (1974); Kowalczyk *et al.* (1985); Cooper & Burrows (1989); Piquer *et al.* (1991); Engberg *et al.* (1992); Shimizu, Nagashima & Hashimoto (1993).

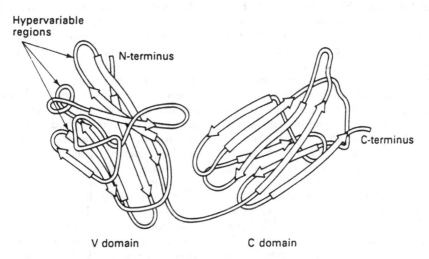

Fig. 13.2. The domain structure of light chain immunoglobulin. Note the two separate C and V domains and the hypervariable regions (or CDR). The arrows represent β-structure, which is part of the framework regions.

other classes of vertebrate where the lacrimal gland is either absent or poorly developed). The IgA concentration in bile is much higher than in plasma (Table 13.2), and it has been suggested that it is actively transported from blood to bile. When secreted into the gut it is protected from proteolysis by a secretory component (Piquer *et al.*, 1991). It is also found in the cropmilk of pigeons. The cropmilk gland is a thickening of the oesophageal wall and is lined with squamous epithelium. The latter are not usually associated with IgA secretion. By intravenous injection of labelled IgA, Goudswaard et al. (1979) have shown that IgA becomes detectable in cropmilk, and the most probable route seems to be via saliva. The IgA in cropmilk probably provides passive immunity for the developing squabs.

Immunoglobulins are present in egg-yolk and in very much smaller concentrations in egg-white. Immunoglobulin present in egg-yolk is almost entirely IgY, whereas the much smaller amounts present in egg-white (Table 13.2) are mainly IgA with a small amount of IgM (Rose, Orlans & Buttress, 1974). These are of maternal origin and provide passive immunity to the developing embryo until it is able to produce fully functional antibodies. Oocytes sequester immunoglobulins throughout their maturation, and in the last 3 days before ovulation the rate of uptake may be as high as 45 mg/day. The final concentration in egg-yolk is similar to that in maternal serum (Table 13.2). During embryonic development, IgY is transported from the yolk to the developing embryo. In the last 3 days of embryonic life the transport is most rapid ($\approx 600\ \mu g$/day), accounting for the rise in the E21 embryo (Table 13.2) (Kowalczyk, et al., 1985). There has been much interest in the presence of IgY in egg-yolk in recent years as a potential source of antibodies (Larsson et al., 1993). The amounts that can be extracted from egg-yolk compare favourably with those obtainable from an immunised rabbit. For example, a hen laying six eggs per week can produce 5–10 times as much immunoglobulin as can be obtained from a rabbit in a week after repeated bleeding. Whilst almost all of the IgY found in the embryo is taken up from the egg-yolk, the IgA and IgM in the embryo is derived from the egg-white. The IgA and IgM, present in the egg-white, is assumed to be taken up from oviduct secretions as the egg-white is laid down in the developing egg (Rose et al., 1974).

Ducks produce, in addition to the normal IgY, a smaller antibody with $M_r \approx 120\,000$, sometimes referred to as IgY(ΔF_c). In the normal adult Pekin duck and mallard, IgY and IgY(ΔF_c) are present in serum in roughly equal amounts, whereas in the Muscovy duck, IgY comprises about 70% and IgY(ΔF_c) about 30% (Grey, 1967). Following immunisation, however, IgY(ΔF_c) becomes the predominant immunoglobulin (Grey, 1967; Liu & Higgins, 1990). Duck antibodies have been shown to be deficient in biological effector mechanisms, e.g. complement fixation or skin sensitisation. They are capable of binding antigens efficiently, but the antibody–antigen complexes do not precipitate (Toth & Norcross, 1981). This may be because of the lack of a 'normal' F_c region on the predominant immunoglobulin, IgY(ΔF_c). Magor et al. (1992) have investigated the genetic and structural relationship between IgY and IgY(ΔF_c) in the duck. IgY(ΔF_c), which is present in ducks, geese and other waterfowl, has heavy chains with the structure 2C + 1V. IgY(ΔF_c) is not a proteolytic cleavage product of IgY. The most likely relationship is that of alternative splicing of the primary transcript from a single gene, although two separate genes encoding IgY and IgY(ΔF_c) cannot be ruled out.

13.3 Antibody biosynthesis

13.3.1 Generation of antibody diversity

It is estimated that a typical vertebrate is capable of generating at least 10^9 different antibodies. This far exceeds the total number of genes in the genome, which is estimated at between 10^5 and 2×10^5. Antibodies are highly specific in their recognition of antigens. The specificity is largely determined by the V region of each immunoglobulin. The light chain of domestic fowl IgG includes a 92 amino acid residue variable sequence, a 13 residue linking region and a 103 residue constant region (McCormack, Tjoelker & Thompson, 1991a). Within the variable region there are three *hypervariable* regions (also referred to as complementarity determining regions, *CDR*) flanked by four framework regions (Fig. 13.2). The three hypervariable regions are quite small, each comprises between 7 and 15 amino acid residues. This is sufficient to account for a very large number of potentially different antibodies. For example, if in a sequence of 10 amino acid residues, each residue could be replaced by any of the other 19 different amino acids, this would generate 20^{10} (10^{12}) different sequences. If one gene were required to encode each different antibody, then an enormous number of different genes would be required, far exceeding the DNA content of the genome.

Long before any immunoglobulins had been sequenced, Burnet (1959) proposed his *clone selection theory*. Soluble antibodies are produced by mature lymphocytes. Burnet proposed that undifferentiated stem cells, which mature into antibody-secreting cells, are initially pluripotent and potentially able to produce a wide variety of antibodies. During the cell divisions that lead to their maturation, they become unipotent, i.e. each cell produces only one type of

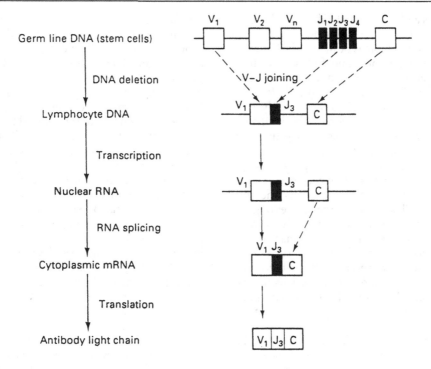

Fig. 13.3. The assembly of an antibody light chain. V_1, V_2.....V_n are variable domains, J_1, J_2, J_3 and J_4 are joining domains and C is the constant domain.

antibody. If a circulating antigen interacted with one of these cells, the cell could be stimulated to divide and secrete the particular antibody that would be able to bind the antigen. He postulated that a wide variety of cells is formed, but only those selected by their ability to combine with an antigen would proliferate and survive. The main tenets of Burnet's theory have proved correct.

To account for the large number of different unipotent antibody-producing cells generated from the multipotent stem cells requires a mechanism for reorganisation of the genes encoding these antibodies. A number of mechanisms have been suggested, and there is sufficient evidence from immunoglobulin gene sequences in mammalian species to confirm that two main mechanisms operate, namely somatic mutation and somatic recombination. The relative importance of these varies between species. *Somatic mutation* is the process in which the DNA of the stem cells is rapidly mutating, whilst cells are proliferating by mitotic division. *Somatic recombination* suggests that there are several separate regions (DNA sequences) on the chromosomes that code for parts of these genes; these regions can join together in a large number of different combinations, thereby creating a

large variety of genes. The details of such recombinations have been worked out in a number of different species. The genes for the light chains comprise joining regions (J) in addition to C and V. Heavy chains also contain a diversity region (D), which increases still further the number of recombinations possible. The way in which the large variety of light chains is generated is illustrated in Fig. 13.3, and a similar mechanism operates for the heavy chain. Based on the estimated number of different V, D and J genetic elements for the light and heavy chains in mouse and using a somatic recombination model such as that in Fig. 13.3 with junctional diversity, it is possible to account for 1.25×10^9 antibodies (Williamson & Turner, 1987). Somatic mutation is more difficult to quantify, but it would increase the number still further.

13.3.2 *The role of B cells, T cells and the bursa of Fabricius*

Both circulating antibodies and the cellular immune system originate from the lymphocytes. The two types of cell involved are T cells and B cells, named after the organs in which they were originally found, namely thymus and the bursa of Fabricius. The bursa is unique to the class Aves; no counterpart has been found in reptiles, amphibians or mammals. However, lymphocytes serving the same function as those found in the bursa are present in mammalian foetal liver and spleen. Since the bursa has a dedicated immunological function, it is easier to study some aspects of antibody generation in birds than in mammals, where foetal liver has diverse functions. The lymph tissues are divided into 'primary' and 'secondary' organs. The preliminary phase of establishing surface receptors on stem cells occurs in the primary organs, of which the bursa is the most important in birds. The stem cells themselves originate in the mesenchyme tissue, whence they seed the primary organs (Fig. 13.4). From the primary organs, they migrate to the secondary sites, such as the spleen, lymph nodes and caecal tonsils. That secondary sites are dependent on the primary organs for their source of lymph cells can be demonstrated in birds by surgical removal of the primary organ at the appropriate stage of development.

B and T lymphocytes cannot be distinguished on the basis of their size or morphology, but they can be distinguished by chemical or immunological methods. The B cells are responsible for the humoral response and, therefore, produce soluble immunoglobulins. The T cells are more heterogeneous in function and can be subdivided into three groups. The *cytotoxic* or killer cells (T_c) have surface receptors similar in structure to the soluble antibodies. These surface receptors can bind foreign cells and set in train the steps leading to their destruction. They are, therefore, responsible for cellular immunity. The second type of T cell, the T_h or *helper cells*, assist the B cells and certain other T cells by stimulating them to produce antibodies. The third type, T_s or *suppressor T cells*, suppress B cell activity and T_c activity. They are particularly important in suppressing responses to self-antigens. This ensures that antibodies are only produced against foreign antigens. Defective or deficient T_s cells lead to autoimmune diseases, in which the immune system destroys part of its own tissues. Examples of this in domestic fowl are spontaneous autoimmune thyroiditis (resembling Hashimoto's thyroiditis in humans) and delayed amelanosis (Wick *et al.*, 1989) The former occurs in the obese strain (OS) domestic fowl, in which thyroid autoantibodies occur spontaneously and the thyroid undergoes progressive destruction associated with a chronic inflammatory lesion.

The development of the bursa prior to hatching is important in understanding avian antibody generation.

The *bursa* is an oval sac-like structure in the dorsal region of the cloaca at its junction with the colon. It is relatively small, less than 0.5% of body weight. The bursa rudiment develops as a diverticulum of the cloaca and can be seen in the chick embryo at E5 as a proliferation bud of epithelial cells. Lymphocyte stem cells begin to colonise the epithelial buds between E8 and E11 to form bursal follicles (McCormack *et al.*, 1991a). These stem cells arise in blood islands of the yolk sac (Cooper & Burrows, 1989). Each follicle is initially colonised by two to three stem cells and continues to grow by rapid cell proliferation, until two to four weeks post-hatching. Within the bursa, B cells expand in number from $3-5 \times 10^4$ Ig-positive progenitor cells to $1-2 \times 10^9$ mature cells (Thompson, 1992). IgM molecules can be detected on the surface of developing lymphocytes at E12, and by E20 90% of the bursal cells can be detected having IgM. IgM production is followed by IgY production and then IgA (Cooper & Burrows, 1989 and Table 13.2). During the last few days of embryonic life, the lymphocytes begin to migrate from the bursa to the secondary lymphoid tissues (Fig. 13.4). This process continues after hatching, so that by about five weeks most of the stem cells have left the bursa for the periphery, where they continue to proliferate B cells for the rest of the fowl's life. The bursa reaches its maximum weight of about 3 g between 4 and 12 weeks, when it contains 10^4-10^5 discrete follicles, each containing $\approx 10^5$ lymphocytes. It then begins to involute and atrophy as the bird reaches sexual maturity.

13.3.3 Generation of the antibody repertoire in the domestic fowl

Although the domestic fowl is the only avian species in which antibody formation has been studied in detail, it seems probable that similar mechanisms occurs in all avian species. The rearrangement of the light chain genes has been examined in mallard duck, pigeon, quail, turkey, cormorant and hawk and appears to be essentially the same as in the domestic fowl, although there are differences in the Muscovy duck (McCormack *et al.*, 1989a). The unusual features of antibody generation in birds, compared with mammals are: (i) the role of the bursa, (ii) B cell precursors are produced during a short period of development in birds, whereas in mammals B cell precursors are produced continuously from bone marrow, and (iii) the antibody diversity in birds is generated from a much more limited repetoire of V, J and D genes (McCormack *et al.*, 1991a).

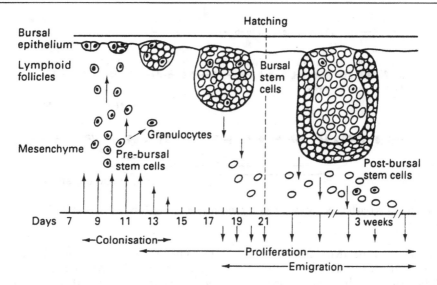

Fig. 13.4. Schematic representation of bursal B cell morphogenesis in the domestic fowl.

The importance of the bursa in antibody production is evident from experiments in which the whole or part of the bursa is ablated at different stages of development. The bursa comprises two cell types: epithelial cells formed at the site of the bursa and lymphoid cells that seed the embryonic bursa. The most critical stage of removal of the bursa is E18. This leaves the bird profoundly immunodeficient, because at this stage all the B cells and B cell precursors are present in the bursa. At stages later than E18, many B cells have left the bursa for secondary lymph glands. If the lymphoid cells, but not the epithelial cells, are specifically ablated at E15–E17, then both soluble and cell surface immunoglobulins fail to develop. However, ablation at E5 does not affect the number of circulating B cells but has a pronounced effect on the repertoire, which is much more limited. A similar effect is observed when bursectomy is performed at 60 h. The chickens are incapable of specific antibody responses, because only a few B cell precursors differentiate into mature immunoglobulin-producing cells.

The generation of the range of different light and heavy chains, which accounts for the diversity of soluble antibodies, occurs in two stages (Mastellar & Thompson, 1994). First, exon domains containing the VJ elements for light chains, and the VDJ elements for heavy chains, are generated. This process occurs before the stem cells migrate to the bursa. Second, expansion and rearrangement of the immunoglobulin repetoire occurs in the bursa. The latter process occurs by gene conversion. The pre-bursal cells that enter the bursa between E8 and E14 have already rearranged their H and L chains. The bursa provides the microenvironment for their expansion. On average, two stem cells enter each follicle, each of which has an immunoglobulin with a particular V–J and V–D–J arrangement of its L and H chains (Thompson, 1992). In the absence of a bursa, an alternative site or sites may allow limited B cell differentiation to take place. There is some evidence that differentiation can take place in the Harderian gland (Mansikka et al., 1990). The bursa thus influences the development of diversity.

Once the immunoglobulin gene structures of the light and heavy chains in the domestic fowl were established (Reynaud, Dahan & Weill, 1983; Parvari et al., 1988), it was apparent that their organisation differs significantly from those in mammals (Fig. 13.5). The light-chain gene consists of only single copies of $C_L J_L$ and V_L gene segments. The heavy-chain gene consists of only single copies of C_μ, J_H and V_H together with a cluster of ≈ 16 D gene segments (Reynaud, Anquez & Weill, 1991). Because 15 of the D elements are very similar, they contribute little to the diversity. This differs from the mammalian organisation, where there are multiple copies of the V and J gene segments (see Fig. 13.3). Other birds examined, including mallard duck, pigeon, quail, turkey, olivaceus cormorant and great black hawk, also have single V_L gene segments. Only the Muscovy duck has so far been found to have more

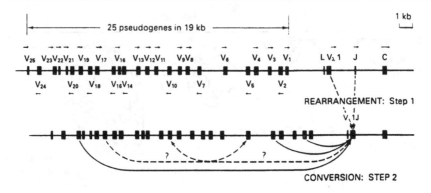

Fig. 13.5. Organisation and generation of diversity by reorganisation of the light chain (Reynaud *et al.*, 1987b).

than one V_L gene segment (McCormack *et al.*, 1989a). The other unusual feature of the domestic fowl immunoglobulin genes is the presence, upstream of the V_L gene segment, of a cluster of 25 V_L-homologous gene segments, and upstream of the V_μ, a cluster of 90 V_μ gene segments. These homologous gene segments are generally referred to as pseudogenes, but Reynaud, Dahan & Weill (1987b) point out that 'partial gene' might be a more appropriate term. These homologous gene segments or pseudogenes are arranged in both transcriptional orientations (Fig. 13.5). They are truncated at the 5' end and lack both the leader exon and the promoter region. At the 3' end, they lack a functional recombination signal required for V–J rearrangement.

In forming the diverse antibody genes, the V gene segment and the J segment first combine. This involves the excision of a piece of DNA, which circularises using the CACGTG sequence at each end (Fig. 13.6). This circular episome has been detected between E10 and E18, most readily in haemopoietic tissues such as spleen and liver, but with a little in the bursa (Mastellar & Thompson, 1994). It lacks a centromere and is, therefore, not propagated during B cell development (McCormack *et al.*, 1989b). Its formation is accompanied by the addition of a single random base to both the J and V joining regions, giving rise to some diversity of V–J junctions. In the heavy chain, a D gene segment has to be added. The formation of the V_H–DJ_H chimaeras is also accompanied by the excision of a circular episome. The latter can be detected about E14–E15, indicating coordination in the formation of the diverse L and H chains (Pickel *et al.*, 1993). Most of the diversity arises from the incorporation of ψV gene segments into the V_L and V_μ segments by the process of gene conversion. (Pseudogenes, such as ψV, are genes

resembling known genes, in this case V genes, but which are non-functional because of deletions or additions that prevent their transcription and/or translation.) The evidence for this came from comparing the cDNA sequences of IgL cDNA clones from E18 and from three-week old chicks. Reynaud *et al.* (1987a) noticed that the sequence differences between the two age groups could be accounted for by a mechanism in which a sequence replacement occurred between V_L and the ψV genes. The blocks replaced varied in length but were generally between 10 and 120 bp. The pattern of replacement that emerged was as follows: (i) the proximal ψV_L genes were more frequently substituted than the distal ψV genes, (ii) truncated ψV_L were rarely used, (iii) ψ_L genes in the antisense orientation were more frequently used than those in the sense orientation, (iv) the greater the length of homology between V_L and ψV_L, the more likely that ψV_L will act as a donor (ψVs with less than 150 bp sequence similarity with V_{L1} do not appear to be used), and (v) when successive substitutions occur, the earlier substitutions influence the pattern of later substitutions. The reasons for (iii) and (iv) are probably that the ψgene folds back to a position antiparallel with the V_L before segment replacement.

The mechanism known for this type of segment replacement is *gene conversion* (Fig. 13.7). It involves a unidirectional transfer of the pseudogene fragments, which occurs in the bursa. It is estimated that each B cell will undergo 4–10 intrachromosomal gene conversion events before migrating from the bursa (Thompson, 1992). This is equivalent to 0.05–0.1 per cell generation (Reynaud *et al.*, 1987a). This could account for a repertoire of at least 10^{11} distinct antibody molecules. Conversion does not occur with equal frequency throughout the V_L gene segment

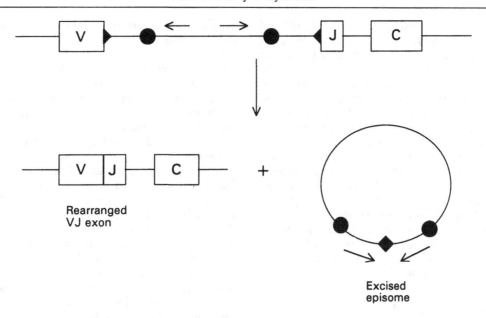

Fig 13.6. The recombination of V and J genetic elements to form a single exon results in the expulsion of an episome (Mastellar & Thompson, 1994). The square and triangle are recombination signal sequences.

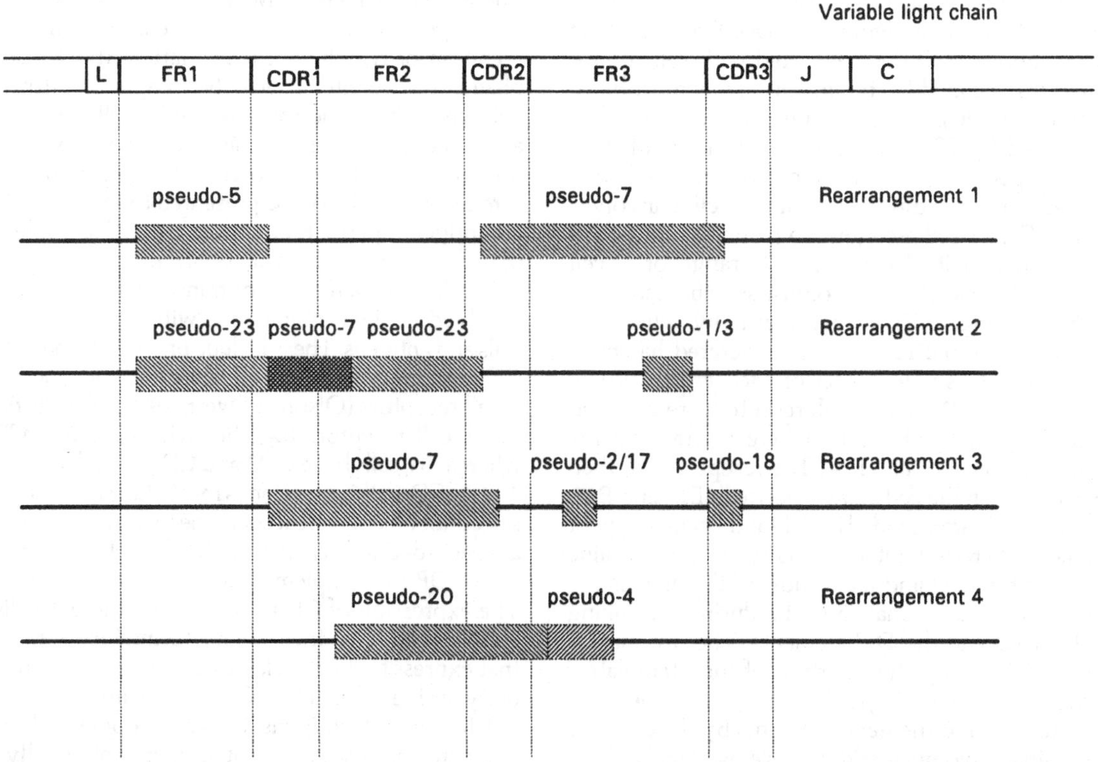

Variable light chain

| | L | FR1 | CDR1 | FR2 | CDR2 | FR3 | CDR3 | J | C | |

pseudo-5   pseudo-7   Rearrangement 1

pseudo-23 pseudo-7 pseudo-23   pseudo-1/3   Rearrangement 2

pseudo-7   pseudo-2/17 pseudo-18   Rearrangement 3

pseudo-20   pseudo-4   Rearrangement 4

Fig. 13.7. The rearrangement of a variable sequence by gene conversion. The top line shows a variable light chain comprising three CDRs and three framework regions (FR). Below are shown four rearrangments of this sequence in which pseudogenes have replaced different sections of the original sequence (Mastellar & Thompson, 1994).

but occurs most frequently in the CDR1, CDR3 and FR2/CDR2 boundary regions. Extrachromosomal circular DNA, similar to that mentioned above in connection with V–J joining, has also been found in E18 bursas, containing V_L-ψV chimaeras. Kondo *et al.* (1993) propose that it arises from excision–deletion involved in the diversification of IgL gene repertoire. The products of two recombination activating genes (RAG-1 and RAG-2) are thought to be involved in V(D)J recombination process in mammals. In the domestic fowl RAG-2, but not RAG-1 is expressed in the bursa at E15 to E18, and it has been suggested that it plays a role in the gene conversion process (Carlson *et al.*, 1991). However, in mutant lymphocytes grown in culture and deficient in RAG-2 immunoglobulin light chain, gene conversion still occurs, suggesting that RAG-2 does not have an essential role in this process (Takeda *et al.*, 1992).

13.4 The role of T cells

The bursa is concerned primarily with the formation of soluble immunoglobulins that will be able to combine with soluble foreign antigens present in the plasma. The T cells, however, produce T cell receptors that recognise peptides associated with cell surface molecules encoded by the major histocompatibility complex (MHC). Thus foreign antigens may be recognised by B cells or T cells by means of their antigen-specific cell surface receptor. In B cells this involves an immunoglobulin, but in T cells it involves the MHC. A T cell recognises a particular antigen in the context of its MHC. A wide range of T cell receptors is generated to recognise specific histocompatibility antigens. The mechanism by which the diversity of T cell receptors is generated has been more recently studied (McCormack *et al.*, 1991b; Pickel *et al.*, 1993). The T cell receptors are detected exclusively in the thymus, where rearrangements occur between E12 and E14. The receptors can then be detected on the cell surface between E14 and E17. The receptors are encoded in V, D and J gene segment families. There is limited diversity in the germline elements for V, D and J, but most of the diversity is generated by junctional diversity, during the linking of the V–D and the D–J elements. The diversity is created in part by the insertion of 'non-templated' nucleotides in the joining regions. These are nucleotides that are added to the gene segments by the catalysis of terminal deoxynucleotidyltransferase. Unlike DNA and RNA polymerase, this enzyme does not require a template but adds nucleotides randomly. The number of 'non-templated' nucleotides inserted increases

during development between E18 and four weeks post-hatching, and varies between 1 and 15 nucleotides (McCormack *et al.*, 1991b). The mechanism for generating multiple T cell receptors in birds is primarily by diversification during V(D)J joining and, therefore, is very similar to that used by mammals. It contrasts with that of avian B cells, which is primarily dependent on gene conversion.

In Sections 13.3.1 and 13.3.3, the mechanism for generating the wide range of antibodies to interact with foreign antigens was outlined. How does this lead to the destruction and removal of the antigenic entities? Soluble antigens, recognised by the antibodies secreted by B cells, form antigen–antibody complexes. The resulting change in conformation of the antibodies enables them to combine with circulating complement proteins, which are then able to trigger off a series of lytic reactions. However, the B cells are present in the secondary lymph organs and many antigens are not readily transported there. The cell-mediated response rather than the humoral response is the most important for dealing with foreign cells such as bacteria. Cell-mediated response is brought about by T cells. There are two main pathways. The first is that T cells react specifically with the antigen, activating the secretion of lymphokines. The latter attract macrophages to the site, which then engulf the foreign antigen. The second route is that T_c cells react with antigen bound to cells causing lysis. For this, T cells do not recognise the foreign antigen alone. The foreign antigen has to be presented to the host cell as an antigenic peptide bound to the MHC molecule (Arstila, Vainio & Lassila, 1994).

T cells have a number of transmembrane proteins involved in their interactions with other T cells, B cells and antigens. These include proteins of the MHC, CD proteins (CD3, CD4, CD8 are the main ones) and T cell receptors (Olson & Ewert, 1994). T_c cells have the T cell receptors together with CD3 and CD8, whereas T_h cells have CD3 and CD4 (Golub & Green, 1991). (CD, in this case stands for *cluster of differentiation*, since antibodies that react with the same marker antigen are regarded as a cluster. It should not be confused with CDR for complementarity determining region.) The expression of CD4 and CD8 on these T cells is closely related to their ability to interact with cells that express particular classes of major histocompatibility antigen (Fig. 13.8). CD4 augments the binding of MHC class II antigens. CD4 from domestic fowl is a transmembrane glycoprotein that is physically associated with a tyrosine protein kinase. CD4 enhances the affinity between MHC and the T cell receptor, and it is also involved in signal transduction through

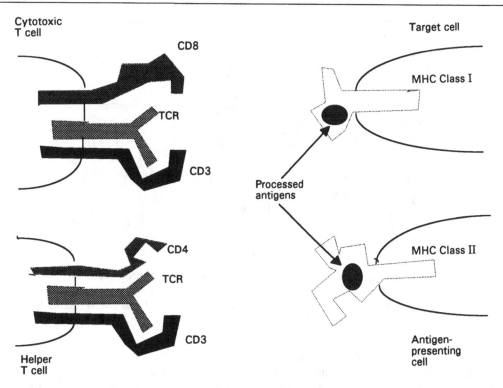

Fig. 13.8. A cytotoxic T cell and a helper T cell combining with a target cell and an antibody presenting cell, respectively. The processed antigens are presented combined with MHC classes I and II. TCR, T cell receptor.

the T cell membrane, in which the tyrosine kinase is assumed to participate (Luhtala *et al.*, 1993). CD8 is a disulphide-linked dimer, also associated with a tyrosine kinase. The two chains of the CD8 protein may be expressed either as a homodimer $\alpha\alpha$ or a heterodimer $\alpha\beta$. T cells can also be classified into groups according to the CD proteins they express. CD4$^+$ cells make up 45% of the T cells circulating in avian plasma, CD8$^+$ 15%, and a third class that do not express either, known as $\gamma\delta$ T cells, make up 50%. The function of the last group is not known (Arstila *et al.*, 1994).

13.5 The major histocompatibility complex

The MHC is a multiallelic gene complex, encoding principally proteins found on the plasmalemma, which are important in many aspects of the immune response. Much of the early work relating to MHC centred around the graft rejection of foreign tissue. In the 1930s, it was shown in mice that whether an implanted tumour grew or was rejected depended on the class of antigens they possessed. However much earlier, the discovery of the ABO blood group system by Landsteiner in 1900 could be regarded as

the first evidence of histocompatibility antigens, although restricted to one type of tissue, namely erythrocytes. Blood group antigens were first identified in domestic fowl by Landsteiner & Miller (1924). The term histocompatability antigens was first used in 1948 to describe the antigens responsible for tissue rejection. The domestic fowl MHC was the second to be discovered, after that of mice and before that of humans (Kroemer *et al.*, 1990). It has been most studied in these three species. In mice and humans there are three classes of MHC. In birds, notably the domestic fowl, there is an additional class of MHC, class IV. Class I are surface antigens found in all nucleated cells throughout the body, including erythrocytes. Class II are surface antigens of the primary immune system expressed in B cells, plasma cells, macrophages and activated T cells; their role is in antigen presentation to T lymphocytes. Class III genes encode complement proteins. These are well characterised in mammals, but the existence of an equivalent set of genes in the domestic fowl is doubtful. Class IV have a distinct structure from class II (Fig. 13.9) and were originally thought to be restricted to erythrocytes. They have only been

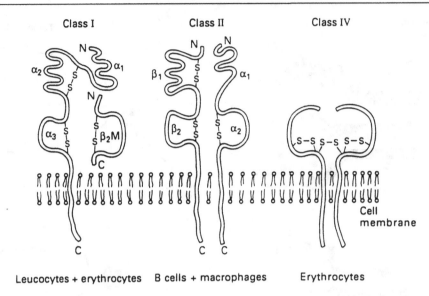

Fig. 13.9. The structure of class I (coded at the *B-F* locus), II (coded at the *B-L* locus) and IV (coded at the *B-G* locus) histocompatibility antigens in the domestic fowl.

detected in birds. Salomonsen *et al.* (1991) have shown, using monoclonal antibodies, that they are also present in thrombocytes, peripheral B and T cells, bursal B cells and thymocytes and epithelial cells of the bursa, thymus and caecal tonsils. Simonsen (1981) proposed a fifth class (class V) for genes that help to determine quantitative heritable traits concerned with growth and reproduction, e.g. embryonic mortality, hatchability, body weight and egg production, which are linked to the MHC genotype. However, this class has not generally been adopted.

Graft rejection is mainly a cell-mediated response, as opposed to a humoral response, and is brought about by T_c cells, although soluble antibodies are also produced. Class I and II MHC are most important in this respect. From the importance in graft–host rejection, it became apparent that MHC was the major system distinguishing self and non-self, playing a key role in the immune response. Tissue grafts between donor and acceptors are normally rejected unless the individuals are very closely related. For a system to be so discriminating as the MHC, it must be highly polymorphic. MHC is both multiallelic and multigenic and is the most highly polymorphic genetic system known.

13.5.1 Structure of avian histocompatibility antigens

The properties of the three classes of MHCs that have been studied in domestic fowl are summarised in Table 13.3. All three encode transmembrane proteins, the structures of which are illustrated in Fig. 13.9. The most important in determining tissue compatibility is class I, which comprises two chains α- and β_2-microglobulin. The α-chain (45 kDa) is polymorphic and consists of three extracellular domains, α_1, α_2 and α_3 of 90 residues each, a transmembrane domain of 43 residues and an intracellular domain. The α_1- and α_2-chains are the main sites of polymorphism. The smaller chain β_2-microglobin (14 kDa) has an invariant sequence. Partial sequences from domestic fowl and turkey β_2-microglobulin are highly conserved (see Lamont, 1993). X-ray crystallographic studies show that the two polymorphic domains, α_1 and α_2 at the external part of the molecule, form a groove that is important in relation to the immune response. Class II MHCs have two chains α and β. The β-chain is polymorphic, but the α-chain appears monomorphic (Lamont, 1993). Both α- and β-chains are N-glycosylated.

Class IV has two polypeptide chains linked by disulphide bonds, either as homodimers or heterodimers. Estimates of the size of the polypeptide chains are between 30 and 55 kDa, and their size varies between class IV antigens from different haplotypes (Kaufman *et al.*, 1990; Kline *et al.*, 1991). Those isolated from erythrocytes are not N-glycosylated, which is unusual for this type of transmembrane protein. The variation in size results from differences in the lengths of the cytoplasmic

Table 13.3. *Histocompatibility antigens in the domestic fowl*

Type	Region of genome	Number of genes identified	Spread on the genome (kb)	Cells in which expressed	Polypeptide structure (size in kDa)
Class I	B-F	6	≈ 80	All nucleated cells	α_1-α_2-α_3 domains (45), β_2 (12)
Class II	B-L	5	≈ 220	B cells, monocytes and dendritic cells	α_1-α_2 domains (32–34), β_1-β_2 domains (27–29)
Class IV	B-G	≈ 20	250–600	Erythrocytes, thrombocytes, lymphocytes, thymocytes and epithelial cells of intestine, caeca and liver	homo- & heterodimer (each chain 35–55)

Classes III and V have not been included since their existence and characteristics in avian species are in doubt.

region of the chain, which comprises an α-helical coil composed of heptad repeats of amino acid residues (Kaufman *et al.*, 1991). The size of the extracellular domain is constant, having both inter- and intrapolypeptide disulphide bonds. The extracellular domain has a structure similar to that of the variable region of immunoglobulins, with most polymorphism occurring in the complementarity determining regions (CDR). The variation in length of the cytoplasmic domain may result from alternative splicing or post-translational modification, such as proteolysis. Using monoclonal antibodies against the class IV heterodimer from the domestic fowl, Morgan *et al.* (1990) were unable to detect cross-reacting material of the surfaces of erythrocytes of Japanese quail, Bob-white quail, pigeon and duck. This could be because these species lack the class IV heterodimer, but more probably because they are sufficiently different antigenically.

13.5.2 Gene organisation of avian histocompatibility antigens

The three genomic regions that encode the three classes of histocompatibility antigen are *B-F*, *B-L* and *B-G*. Each of these is multiallelic. Each individual bird will have a particular *haplotype*, or particular combination of alleles at these loci. The different haplotypes have generally been defined by the antibodies raised in response to their gene products, a process known as *serotyping*. As sequence data become available, the genes can be defined by their nucleotide sequences. Briles & Briles (1982) have identified 27 different haplotypes in domestic fowl at the MHC locus, and it is possible that more exist. Each of these can be distinguished by a unique pattern of reactions with

15 different antisera, or by using monoclonal antibodies (Lamont, Warner & Nordskog, 1987). A number of haplotypes are associated with certain biological traits, e.g. susceptibility/resistance to lethal infections such as Marek's disease, lymphoid leukosis, caecal coccidiosis, fowl cholera, sarcoma induction by Rous sarcoma virus and spontaneous autoimmune thyroiditis (Kroemer *et al.*, 1990). So far a minimum of 14 haplotypes have been identified in a study of 30 families of the ring-necked pheasant (Jarvi & Briles, 1992), but there are probably more. Antigens to the domestic fowl MHC show cross-reactivity with that of the ring-necked pheasant. The molecular basis underlying the different haplotypes is being studied using restriction fragment length polymorphism (RFLP; see Avise, 1994) and nucleotide sequencing. Juul-Madsen *et al.* (1993) analysed seven domestic fowl haplotypes by RFLP using a single restriction endonuclease, *Bgl*II, and cDNA probes specific for *B-F* and *B-L*, and they were able to obtain a distinct pattern with each haplotype. Lamont *et al.* (1990) used a different combination of haplotypes and found that some RFLPs were shared between haplotypes. Miller, Abplanalp & Goto (1988) were able to distinguish 15 out of 17 genotypes using the restriction endonuclease *Pvu*II and a *B-G* cDNA clone for hybridisation. The power of discrimination of RFLP analysis, therefore, depends on the particular haplotypes involved and the choice of restriction endonucleases. RFLP analysis has revealed further potential complexity in the histocompatibility genes. Haplotypes that appear homogeneous on the basis of serotyping sometimes show variable RFLP patterns. This led Briles *et al.* (1993) to discover a second independently segregating polymorphic *Mhc*-like locus, designated *Rfp-Y*. This second locus shows a

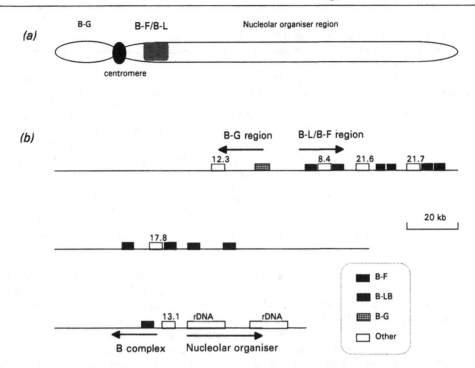

Fig. 13.10. Gene organisation of the major histocompatibility antigens in domestic fowl. (*a*) Diagram of microchromosome 16 ($\approx$ 8000 kb length) containing the MHC genes and the nucleolar organiser region. (*b*) Larger-scale topography of three non-overlapping regions within $\approx$ 400 kb of microchromosome 16, containing the *B-F*, *B-LB* and *B-G* genes with other interspersed genes. Arrows indicate the orientation of the genes. (Kroemer *et al.*, 1990; Kaufman *et al.*, 1991).

degree of cross-hybridisation with *Mhc*, suggesting that it has a related structure, but it is not clear at present what its function is, or whether it is expressed.

The genetic locus containing *B-F*, *B-L* and *B-G* is known as the *B-locus*. It was originally designated as blood group type B, together with the other domestic fowl blood groups A, C and D (see Briles, 1984). Schierman & Nordskog (1961) showed that it was distinct from the other groups. When they carried out skin homografts (grafts between different individuals of the same species) some were accepted, others rejected. They found that rejection occurred only when the particular blood group B present in the donor was not present in the host. It, thus, became apparent that the locus of the blood group B was strongly linked to, or identical with, the MHC locus. Pink *et al.* (1977) showed that the B locus was complex, comprising at least the three subregions *B-F*, *B-L* and *B-G*. Each of these encoded a different class of MHC. Initially, the class IV antigens were considered to be simple blood group antigens, especially as their tissue distribution was, until recently, believed to be restricted to erythrocytes.

Evidence has accumulated in the 1980s that places them as one of the classes of histocompatibility antigen (Kaufman, Skjødt & Salomonsen, 1991). The evidence supporting this is: (i) class IV are transmembrane plasmalemma proteins showing a high degree of polymorphism, (ii) they are now known to have wide tissue distribution (Table 13.3), and (iii) they exert what is described as the 'adjuvant effect'. Antigens that would otherwise be poorly immunogenic are much improved by the presence of allogenic B-G proteins; this is known as the adjuvant effect.

Since 1988, much of the B locus has been cloned and many of the subregions identified (Fig. 13.10*b*). Eventually, when each of the haplotypes can be distinguished by their gene sequence, the molecular genetic basis for their association with the biological traits may be clarified. The B locus has proved to be much more compact than the MHC loci in mice and humans. It is present on microchromosome 16 and is estimated to occupy at most 2000 kb; the bulk of the microchromosome encodes the nucleolar organiser region (Fig. 13.10*a*) (Kroemer *et al.*, 1990). The *B-L* and *B-F* genes encoding class I and II antigens together occupy about 250 kb, compared with 1500 kb for class I and 900 kb for class II in mammalian systems. There are five *B-LB* gene loci and six *B-F* gene loci very closely linked, such that no crossovers between them have yet been detected. The five *B-LB* loci encode the β-chain of class I histocompatibility antigens and are divided into two families (*B-LBI* and *B-LBII* in one, and *B-LBIII*, *B-LBIV* and *B-LBV* in the other) on the basis of their relatedness. No *B-LA* gene has so far been described (Zoorob *et al.*, 1990). A single gene might be expected since the α-chain is monomorphic. It seems unlikely that class III genes equivalent to those found in mammals exist in domestic fowl. The close proximity of the *B-L* and *B-F* genes in domestic fowl precludes this. There is, however, preliminary evidence for a class III gene linked to the MHC (Spike & Lamout, 1995).

There are at least 20 *B-G* gene loci, starting 12 kb from the *BL/BF* locus. Very few recombinants between the *BF/BL* region and the *BG* region have been found, and a recombinantion frequency 0.05% has been estimated. This is consistent with the strong linkage disequilibrium observed between *B-F* and *B-G* alleles in outbred flocks of domestic fowl (Kaufman *et al.*, 1991). Estimates of the span of the *B-G* region are between 250 kb and 600 kb (Kroemer *et al.*, 1990; Kaufman *et al.*, 1991).

Besides the six *B-F* and five *B-LB* genes, there are other genes within the *Mhc* region (Fig. 13.10*b*) that have been detected using cDNA probes and which have been generated from mRNAs expressed in various tissues. Their functions are unknown and they are simply named according to their size. The 12.3 gene encodes a protein resembling the β-subunit of GTP-binding protein. GTP-binding proteins are involved in signal transduction and lymphocyte activation. The 8.4 gene is an immunoglobulin-like membrane-bound protein, and the 17.8 gene a transmembrane protein of the animal lectin superfamily (Bernot, Zoorob & Auffray, 1994).

There has been few studies of MHC in other avian species, but there is evidence for linkage between MHC and blood groups in turkeys and ring-necked pheasants (Lamont & Dietert, 1990). Four turkey class II genotypes have been identified, using RFLP and a domestic fowl DNA probe. Turkey erythrocytes appear either not to express class I proteins, or to express only very low levels (Emara *et al.*, 1992). At least three class II *B* haplotypes and, therefore, six genotypes exist in ring-necked pheasant (Wittzell *et al.*, 1994).

Table 13.4. Blood groups in the domestic fowl

Group	Number of alleles	Chromosome	Comments
Ea-A	8	1	53–55 kDa, closely linked to *Ea-E*, increasing expression from E3 to maturity
Ea-B	30	16(micro)	Encodes MHC, also present on lymphocytes
Ea-C	8	1	Also present on lymphocytes, closely linked to a minor histocompatibility complex
Ea-D	5	1	On the long arm
Ea-E	11	1	
Ea-H	3	1	
Ea-I	8	1	On the long arm
Ea-J	2	1	
Ea-K	4	?	
Ea-L	2	?	
Ea-N	2	?	
Ea-P	10	1	
Ea-R	2	?	

13.6 Blood groups in the domestic fowl

Blood groups arise from the different pattern of cell surface antigens on erythrocytes. The first blood groups in the domestic fowl were described by Briles, McGibbon & Irvin (1950) and the history of their discovery is reviewed by Briles (1984). Table 13.4 lists the different groups that have been identified. Group B has received the most attention, since it is responsible for host graft rejection etc. and is present not only on erythrocytes but also on many other types of cell. It turned out to be the avian counterpart of the MHC (described above). Most of the other groups have not yet been studied by molecular biologists. A number have been mapped in relation to other gene loci using the classical method of determining recombinant frequencies. All show codominance. Of the 13 groups, eight (*Ea-A*, *Ea-B*, *Ea-D*, *Ea-E*, *Ea-H*, *Ea-I*, *Ea-J* and *Ea-P*) have been mapped with varying degrees of precision (Bitgood *et al.*, 1980; 1991). Monoclonal antibodies have been raised that are able to differentiate some of the EaA antigens (Fulton, Briles & Lamont, 1990). Using these, Fulton *et al.* (1990) precipitated EaA2 and EaA4 antigens, and their M_r were determined as 53 000 and

54 500, respectively. Both appear to be glycosylated with intrachain disulphide bonds. Contrary to earlier reports, Fulton *et al.* (1990) could not detect EaA on lymphocytes. Groups B and C are also present on white blood cells (Lamont & Dietert, 1990). Species other than domestic fowl have not been extensively studied, but seven alloantigens have been found in turkey, three in Japanese quail, six to seven in Muscovy, Pekin and mallard ducks, and three in ring-necked pheasants (Lamont & Dietert, 1990). Biochemical and molecular biological studies have yet to be made of most of the blood group determinants.

Appendix: English common names of birds cited in the text

The names used are those given in *A Complete Checklist of the Birds of the World*, 2nd edn (Howard & Moore, 1991).

Adélie penguin	*Pycoscelis adeliae*
American coot	*Fulica americana*
American crow	*Corvus brachyrhynchos*
American goldfinch	*Carduelis tristis*
American robin	*Turdus migratorus*
American tree sparrow	*Spizella arborea*
Andean cock of the rock	*Rupicola peruviana*
Andean condor	*Vultur gryphus*
Andean goose	*Chloephage melanoptera*
Arctic tern	*Sterna paradisaea*
Bald eagle	*Haliaeetus leucocephalus*
Bar-headed goose	*Anser indicus*
Bay-breasted warbler	*Dendroica castanea*
Belted kingfisher	*Megacerle alcyon*
Blackbird	*Turdus merula*
Black-footed albatross	*Diomedea nigripes*
Black-headed gull	*Larus ridibundus*
Black jacobin	*Melanotrochilius fuscus*
Black kite	*Milvus migrans*
Black-necked grebe	*Podiceps nigrocollis*
Black stork	*Ciconia nigra*
Black vulture	*Coragyps atratus*
Boat-tailed grackle	*Quiscalus major*
Budgerigar	*Melopsittacus undulatus*
Catbird	*Dumetella carolinensis*
Cinereous vulture	*Aegypius monachus*
Common eider	*Somateria mollissima*
Common peafowl	*Pavo cristatus*
Common starling	*Sturnus vulgaris*
Cow bird	*Molothrus bonariensis*
Dark-eyed junco	*Junco hyemalis*
Dove	*Streptopelia senegalensis*
Eastern white pelican	*Pelecanus oncrotalus*
Emperor penguin	*Aptenodytes patagonicus*
Emu	*Dromaius novaehollandiae*
European robin	*Erithacus rubecula*
Eurasian buzzard	*Buteo buteo*
Eurasian tree sparrow	*Passer montanus*

Garden warbler	*Sylvia borin*
Gentoo penguin	*Pygoscelis papua*
Glaucous-winged gull	*Larus glaucescens*
Goldcrest	*Regulus regulus*
Golden eagle	*Aquila chrysaetos*
Golden oriole	*Oriolus oriolus*
Golden pheasant	*Chrysolophus pictus*
Great black hawk	*Buteogallus urubitinga*
Great blue heron	*Ardea herodias*
Great bustard	*Otis tarda*
Great crested grebe	*Podiceps cristatus*
Great cormorant	*Phalacrocorax carbo*
Great horned owl	*Bubo virginianus*
Great tit	*Parus major*
Green jungle fowl	*Gallus varius*
Grey-backed camaroptera	*Camaroptera brevicauda*
Grey heron	*Ardea cinerea*
Grey jungle fowl	*Gallus sonneratii*
Greylag goose	*Anser anser*
Griffon vulture	*Gyps fulvus*
Guinea-fowl	*Numida meleagris*
Herring gull	*Larus argentatus*
Hoatzin	*Opisthocomus hoazin*
Hoopoe	*Upupa epops*
Horned lark	*Eremophila alpestris*
House finch	*Carpodacus mexicanus*
House sparrow	*Passer domesticus*
Imperial eagle	*Aquila heliaca*
Javanese cormorant	*Phalacrocorax niger*
Kelp gull	*Larus dominicanus*
King penguin	*Aptenodytes patagonicus*
King vulture	*Sarcorhamphus papa*
Lady Amherst pheasant	*Chrysolophus amherstiae*
Lapwing	*Vanellus vanellus*
Laughing gull	*Larus atricilla*
Lesser black-backed gull	*Larus fuscusta*
Little egret	*Egretta garzetta*
Little green heron	*Butorides striatus virescens*
Macaw	*Ara* spp.
Marsh harrier	*Circus aeruginosus*
Mew gull	*Larus canus*
Mocking bird	*Mimus theuca*

Mute swan	*Cygnus olor*	Ruffed grouse	*Bonasa umbellus*
Night heron	*Nycticorax nycticorax*	Rufous hummingbird	*Selasphorus rufus*
Northern bullfinch	*Pyrrhula pyrrhula*	Rüppell's griffon	*Gyps rueppelli*
Northern goshawk	*Accipter gentilis*	Sanderling	*Calidris alba*
Olivaceus cormorant	*Phalacrocorax olivaceus*	Sarus crane	*Grus antigone*
Osprey	*Pandion haliaetus*	Sedge warbler	*Acrocephalus schoenbaenus*
Ostrich	*Struthio camelis*	Spoonbill	*Platalea leucorodia*
Parakeet	*Melopsittacus undulatus*	Spruce grouse	*Dendragapus canadensis*
Pied flycatcher	*Ficedula hypoleuca*	Stone-curlew	*Burhinus oedicnemus*
Red-billed quelea	*Quelea quelea*	Tufted duck	*Aythya fuligula*
Red grouse	*Lagopus lagopus*	Turkey vulture	*Cathartes aura*
Red jungle fowl	*Gallus gallus*	Ural owl	*Strix uralensis*
Red-necked phalarope	*Phalaropus lobatus*	Western marsh harrier	*Circus aeruginosus*
Red-tailed hawk	*Buteo jamaicensis*	White-crowned sparrow	*Zonotrichia leucophrys gambelii*
Reed bunting	*Emberiza schoeniclus*	White-headed vulture	*Aegypius occipitalis*
Reed warbler	*Acrocephalus scirpaceus*	White stork	*Ciconia ciconia*
Rhea	*Rhea americana*	Willow grouse	*Lagopus lagopus*
Ring-necked pheasant	*Phasianus colchicus torquatus*	Wood pigeon	*Columba palumbus*
Rockhopper penguin	*Eudyptes chrysocome*	Yellow finch	*Sicalis luteola*
Rock ptarmigan	*Lagopus mutus*	Yellowhammer	*Emberiza citrinella*
Royal tern	*Thalasseus maximus*	Yellow-vented bulbul	*Pycnotus goiavier*

References

Abbadie, C., Kabrun, N., Bouali, F. *et al.* (1993). High-levels of c-*rel* expression are associated with programmed cell-death in the developing avian embryo and in bone-marrow cells *in vitro. Cell*, **75**, 899–912.

Abramova, E.B., Mil'man, L.S. & Kuznetsov, A.A. (1992). Ontogenic studies of chicken liver pyruvate kinase. *J. Evol. Biochem. Physiol.*, **27**, 323–30.

Adamo, M.L. & Hazelwood, R.L. (1990). Characterization of liver and cerebellar binding sites for avian pancreatic polypeptide. *Endocrinology*, **126**, 434–40.

Adamo, M.L. & Hazelwood, R.L. (1992). The effect of C-terminus and N-terminus iodination on avian pancreatic polypeptide (APP) binding to its chicken brain receptor. *Neuropeptides*, **21**, 225–30.

Adamo, M.L., Simon, J., Rosebrough, R.W., McMurthy, J.P., Steele, N.C. & LeRoith, D. (1987). Characterisation of the chicken muscle insulin receptor. *Gen. Comp. Endocrinol.*, **68**, 456–65.

Adams, R.L.P., Knowler, J.T. & Leader, D.P. (1992) *The Biochemistry of the Nucleic Acids*, 11th edn, London: Chapman & Hall.

Agell, N. & Mezquita, C. (1988). Cellular content of ubiquitin and formation of ubiquitin conjugates during chicken spermatogenesis. *Biochem. J.*, **250**, 883–9.

Aguilera, E., Moreno, J. & Ferrer, M. (1993). Blood chemistry values in three *Pygoscelis* penguins. *Comp. Biochem. Physiol.*, **105A**, 471–3.

Ahn, J.Y., Hong, S.O., Kwak, K.B. *et al.* (1991). Developmental regulation of proteolytic activities and subunit pattern of 20 S proteasome in chick embryonic muscle. *J. Biol. Chem.*, **266**, 15746–9.

Ajuyah, A.O., Lee, K.H., Hardin, R.T. & Sim, J.S. (1991). Changes in the yield and in the fatty acid composition of whole carcass and selected meat portions of broiler chickens fed full-fat oil seeds. *Poult. Sci.*, **70**, 2304–14.

Akhayat, O., Infante, A.A., Infante, D., Martins de Sa, S., Grossi de Sa, M.-F. & Scherrer, K. (1987). A new type of prosome-like particle, composed of small cytoplasmic RNA and multimers of a 21 kDa protein, inhibits protein synthesis *in vivo. Eur. J. Biochem.*, **170**, 23–33.

Alerstam, T. & Lindström, Å. (1990). Optimal bird migration: the relative importance of time, energy, and safety. In *Bird Migration: Physiology and Ecophysiology*, ed. E. Gwinner, pp.

331–51. Berlin: Springer Verlag.

Alessi, D., MacDougall, L.K., Sola, M.M. & Cohen, P. (1992). The control of protein phosphatase-1 by targetting subunits. The major myosin protein phosphatase in avian smooth muscle is a novel form of protein phosphatase-1. *Eur. J. Biochem.*, **210**, 1023–35.

Alexander, C. & Day, C.E. (1973). Distribution of serum lipoproteins of selected vertebrates. *Comp. Biochem. Physiol.*, **46B**, 295–312.

Allegretto, E.A., Pike, J.W. & Haussler, M.R. (1987). Immunochemical detection of unique proteolytic fragments of the chick 1,25-dihydroxyvitamin D_3. *J. Biol. Chem.*, **262**, 1312–9.

Almquist, H.J. (1952). Amino acid requirements of chicken and turkeys – a review. *Poult. Sci.*, **31**, 966–81.

Alonso, J., Alonso, J.C., Munoz-Pulido, R. *et al.* (1990). Hematology and blood chemistry of free-living young great bustards (*Otis tarda*). *Comp. Biochem. Physiol.*, **97A**, 611–13.

Alonso, J.C., Huecas, V., Alonso, J.A., Abelenda, M., Munoz-Pulido, R. & Puerta, M.L. (1991). Hematology and blood chemistry of adult white storks (*Ciconia ciconia*). *Comp. Biochem. Physiol.*, **98A**, 395–7.

Altman, R.B. & Kirmayer, A.H. (1976). Diabetes mellitus in the avian species. *J. Am. Hosp. Assoc.*, **12**, 531–7.

Aminlari, M. & Shahbazi, M. (1994). Rhodanese (thiosulfate:cyanide sulfurtransferase) distribution in the digestive tract of chicken. *Poult. Sci.*, **73**, 1465–9.

Anderson, S., Bankier, A.T., Barrell, B.G. *et al.* (1981). Sequence and organization of the human mitochondrial genome. *Nature*, **290**, 457–65.

Andreeva, I.E., Livanova, N.B., Eronina, T.B., Silonova, G.V. & Poglazov, B.F. (1986). Phosphorylase kinase from chicken skeletal muscle: quaternary structure, regulatory properties and partial proteolysis. *Eur. J. Biochem.*, **158**, 99–106.

Anjanayaki, E.A., Tsolas, O. & Horecker, B.L. (1977). Crystalline fructose 1,6-bisphosphatase from chicken breast muscle. *Arch. Biochem. Biophys.*, **183**, 48–56.

Annison, E.F. (1983). Lipid metabolisn. In *Physiology and Biochemistry of the Domestic Fowl*, vol. 4, ed. B.M. Freeman, pp. 165–74. London: Academic Press.

Ansari, H.A., Takagi, N. & Sasaki, M. (1986). Interordinal conservatism of chromosomal banding patterns in *Gallus domesticus* (Galliformes) and *Melopsittacus undulatus* (Psittaciformes). *Cytogenet. Cell Genet.*, **43**, 6–9.

Ansari, H.A., Takagi, N. & Sasaki, M. (1988). Morphological differentiation of sex chromosomes in three species of ratite birds. *Cytogenet. Cell Genet.*, **47**, 185–8.

Aoyama, H., Asamoto, K., Nojyo, Y. & Kinutani, M. (1992). Monoclonal antibodies specific to quail embryo tissues: their epitopes in the developing quail embryo and their application to identification of quail cells in quail-chick chimeras. *J. Histochem. Cytochem.*, **40**, 1769–77.

Apostol, I., Giletto, A., Komiyama, T., Zhang, W.-L. & Laskowski, M. (1993). Amino acid sequences of ovomucoid third domains from 27 additional species of birds. *J. Prot. Chem.*, **12**, 419–33.

Arnold, J.E. & Gevers, W. (1990). Auto-ubiquitination of ubiquitin-activating enzymes from chicken breast muscle. *Biochem. J.*, **267**, 751–7.

Arstila, T.P., Vainio, O. & Lassila, O. (1994). Central role of CD4$^+$ T cells in avian immune response. *Poult. Sci.*, **73**, 1019–26.

Arthur, R.R. & Straus, N.A. (1983). DNA sequence organization and transcription of the chicken genome. *Biochim. Biophys. Acta*, **741**, 171–9.

Artinger, K.B. & Bronner-Fraser, M. (1992). Partial restriction in the developing potential of late emigrating avian neural crest cells. *Dev. Biol.*, **149**, 149–57.

Asotra, K. (1986). Glucose-6-phosphatase activity in normal and denervated developing chick gastrocnemii: reappraisal of glycogenolytic and glycolytic metabolism in skeletal muscle. *Exp. Path.*, **29**, 103–12.

Attwood, P.V. & Graneri, B.D.L.A. (1992). Bicarbonate-dependent ATP cleavage by pyruvate carboxylase in the absence of pyruvate. *Biochem. J.*, **287**, 1011–17.

Auer, H., Mayr, B., Lambrou, M. & Schleger, W. (1987). An extended chicken karyotype, including the NOR chromosome. *Cytogenet. Cell Genet.*, **45**, 218–21.

Avise, J.C. (1994). *Molecular Markers, Natural History and Evolution*. New York: Chapman & Hall.

Avise, J.C., Ankney, C.D., & Nelson, W.S. (1990). Mitochondrial gene trees and the evolutionary relationship of Mallard and Black ducks. *Evolution*, **44**, 1109–19.

Avise, J.C., Arnold, J.A., Ball, R.M. *et al.* (1987). Intraspecific phylogeography: the mitochondrial DNA bridge between population genetics and systematics. *Annu. Rev. Ecol. System.*, **18**, 489–522.

Avise, J.C. & Nelson, W.S. (1989). Molecular genetic relationships of the extinct dusky seaside sparrow. *Science*, **243**, 646–8.

Bachmann, K., Harrington, B.A. & Craig, J.P. (1972). Genome size in birds. *Chromosoma*, **37**, 405–16.

Back, D.W., Goldman, M.J., Fish, J.E., Ochs, R.S. & Goodridge, A.G. (1986). The fatty-acid synthase gene in avian liver – 2 messenger RNAs are expressed in parallel by feeding, primarily at the level of transcription. *J. Biol. Chem.*, **261**, 4190–7.

Bailey, E. & Horne, J.A. (1972). Formation and utilization of acetoacetate and 3-hydroxybutyrate by various tissues of adult pigeon (*Columba livea*). *Comp. Biochem. Physiol.*, **42B**, 659–67.

Baker, D.H. (1991). Partitioning of nutrients for growth and other metabolic functions: efficiency and priority considerations. *Poult. Sci.*, **70**, 1797–805.

Balash, J., Musquera, S., Palacios, L., Jimenez, M. & Palameque, J. (1976). Comparative hematology of some Falconiformes. *Condor*, **78**, 258–73.

Baldwin, E. (1967). *Dynamic Aspects of Biochemistry*, 5th edn, Ch. 13. Cambridge: Cambridge University Press.

Balinsky, B.I. (1981). *An Introduction to Embryology*. New York: Saunders.

Ball, G.F. & Balthazart, J. (1990). Steroid modulation of muscarinic cholinergic and α_2-adrenergic receptor density in the nucleus intercollicularis of the Japanese quail. *Eur. J. Neurosci.*, **2**, 828–35.

Ballantyne, J.S., John, T.M. & George, J.C. (1988). The effects of glucagon on hepatic mitochondrial metabolism in the pigeon, *Columbia livia*. *Gen. Comp. Endocrinol.*, **72**, 130–5.

Ballard, F.J., Johnson, R.J., Owens, P.C. *et al.* (1990). Chicken insulin-like growth factor-1: amino acid sequence, radioimmunoassay, and plasma levels between strains and during growth. *Gen. Comp. Endocrinol.*, **79**, 459–68.

Bamberger, M.J. & Lane, M.D. (1990). Possible role of the Golgi apparatus in the assembly of very low density lipoprotein. *Proc. Natl. Acad. Sci. USA*, **87**, 2390–4.

Banaszak, L., Sharrock, W. & Timmins, P. (1991). Structure and function of a lipoprotein: lipovitellin. *Annu. Rev. Biophys. Biophysical Chem.*, **20**, 221–46.

Bandman, E. (1992). Contractile protein isoforms in muscle development. *Dev. Biol.*, **154**, 273–83.

Baniahmad, A., Muller, M., Steiner, C. & Renkawitz, R. (1987). Activity of two different silencer elements of the chicken lysozyme gene can be expressed by enhancer elements. *EMBO J.*, **6**, 2297–303.

Baniahmad, A., Steiner, C., Köhne, A.C. & Renkawitz, R. (1990). Modular structure of a chicken lysozyme silencer: involvement of an unusual thyroid hormone receptor binding site. *Cell*, **61**, 505–14.

Bannister, D.W. (1984). Activation of phosphofructokinase and pyruvate kinase by 6-phosphogluconate in chicken liver (*Gallus domesticus*): No evidence for a regulatory role. *Int. J. Biochem.*, **16**, 895–9.

Barber, D.L., Sanders, E.J., Aebersold, R. & Schneider, W.J. (1991). The receptor for yolk lipoprotein deposition in the chicken oocyte. *J. Biol. Chem.*, **266**, 18761–70.

Barré, H., Nedergaard, J.& Cannon, B. (1986). Increased respiration in skeletal muscle mitochondria from cold-acclimated ducklings: uncoupling effects of free fatty acids. *Comp. Biochem. Physiol.*, **85B**, 343–8.

Bartlett, G.R. (1982). Developmental changes of phosphates in red cells of the emu and the rhea. *Comp. Biochem. Physiol.*, **73A**, 129–34.

Bartlett, G.R. & Borgese, T.A. (1976). Phosphate compounds in red cells of the chicken and duck embryo and hatchling. *Comp. Biochem. Physiol.*, **55A**, 207–10.

Bate, A.J. & Dickson, A.J. (1986). The importance of ketone bodies as tissue specific fuels in the development of the chick embryo neonate. *Biochem. Soc. Trans.*, **14**, 712–13.

Bateson, W. (1894). *Materials for the Study of Variation*. London:

MacMillan.

Baumann, R., Goldback, E., Haller, E.A. & Wright, P. (1984). Organic phosphates increase the solubility of avian haemoglobin D and embryonic chick haemoglobin. *Biochem. J.*, **217**, 767–71.

Baverstock, P.R., Adams, M., Polkinghorne, R.W. & Gelder, M. (1982). A sex-linked enzyme in birds – Z chromosome conservation but no dosage compensation. *Nature*, **296**, 763–6.

Becker, K.J., Geyer, H., Eigenbrodt, E. & Schoner, W. (1986). Purification of pyruvate kinase isoenzymes type M_1 and M_2 from dog (*Canis familiaris*) and comparison of their properties with those from chicken and rat. *Comp. Biochem. Physiol.*, **83B**, 823–9.

Beekman, J.M., Wijnholds, J., Schippers, I.J., Pot, W., Gruber, M. & AB, G. (1991). Regulatory elements and DNA-binding proteins mediating transcription from the chicken very-low-density apolipoprotein II gene. *Nucl. Acids Res.*, **19**, 5371–7.

Beers, K.W., Nejad, H. & Bottje, W.G. (1992). Aflatoxin and glutathione in domestic fowl (*Gallus domesticus*) I. Glutathione elevation and attenuation by high dietary methionine. *Comp. Biochem. Physiol.*, **101C**, 239–44.

Behe, M.J. (1990). Histone deletion mutants challenge the molecular clock hypothesis. *Trends Biochem. Sci.*, **15**, 374–6.

Ben-Dror, I., Havazelet, N. & Vardimon, L. (1993). Developmental control of glucocorticoid transcriptional activity in embryonic retina. *Proc. Natl. Acad. Sci. USA*, **90**, 1117–21.

Bensadoun, A. & Rothfield, A. (1972). The form of absorption of lipids in the chicken, *Gallus domesticus*. *Proc. Soc. Exp. Biol. Med.*, **141**, 814–17.

Bentley, P.J. (1982). *Comparative Vertebrate Endocrinology*, 2nd edn. Cambridge: Cambridge University Press.

Benveniste-Schrode, K., Doering, J.L., Hauck, W.W., Schrode, J., Kendra, K.L. & Drexler, B.K. (1985). Evolution of chick type I procollagen genes. *J. Mol. Evol.*, **22**, 209–19.

Berdnikov, V.A., Gorel, F.L. & Svinarchuk, F.P. (1978). Study of evolutionary changes in subfraction composition of histone H1 in birds. *Biokhimiya*, **43**, 830–7.

Bergsma, D.J., Grichnik, J.M., Gossett, L.M.A. & Schwartz, R.J. (1986). Delimitation and characterization of *cis*-acting DNA sequences required for the regulated expression and transcriptional control of the chicken skeletal α-actin gene. *Mol. Cell. Biol.*, **6**, 2462–75.

Berko-Flint, Y., Levkowitz, G. & Vardimon, L. (1994). Involvement of c-*jun* in the control of glucocorticoid receptor transcriptional activity during development of chicken retinal tissue *Eur. J. Mol. Biol.*, **13**, 646–54.

Berkowitz, E.A., Chu, W.W. & Evans, M.I. (1993). Insulin inhibits the estrogen-dependent expression of the chicken very low density apolipoprotein II gene in Leghorn male hepatoma cells. *Mol. Endocrinol.*, **7**, 507–14.

Bernardi, G. (1993). The isochore organization of the human genome and its evolutionary history – a review. *Gene*, **135**, 57–66.

Bernardi, G., Olofsson, B., Filipski, J. *et al.* (1985). The mosaic genome of warm-blooded vertebrates. *Science*, **228**, 953–7.

Bernot, A., Zoorob, R. & Auffray, C. (1994). Linkage of a new member of the lectin supergene family to chicken *Mhc*

genes. *Immunogenetics*, **39**, 221–9.

Bertland, L.H. & Kaplan, N.O. (1970). Studies on the conformations of multiple forms of chicken heart aspartate aminotransferase. *Biochemistry*, **9**, 2653–65.

Bhattacharyya, N. & Banerjee, D. (1993). Transcriptional regulation of the gene encoding apolipoprotein AI in chicken LMH cells. *Gene*, **137**, 315–20.

Bhattacharyya, N., Chattapadhyay, R., Hirsch, A. & Banerjee, D. (1991). Isolation, characterization and sequencing of the chicken apolipoprotein-AI-encoding gene. *Gene*, **104**, 163–8.

Bhattacharyya, N., Chattapadhyay, R., Oddoux, C. & Banerjee, D. (1993). Characterization of the chicken apolipoprotein A-I gene 5'-flanking region. *DNA Cell Biol.*, **12**, 597–604.

Bickmore, W.A. & Sumner, A.T. (1989). Mammalian chromosome banding – an expression of genome organization. *Trends Genet.*, **5**, 144–8.

Birkhold, S.G. & Samms, A.R. (1994). Concurrent identification of calpains I and II from chicken breast muscle. *Comp. Biochem. Physiol.*, **107B**, 519–23.

Bishop, J.M. (1983). Cellular oncogenes and retroviruses. *Annu. Rev. Biochem.*, **52**, 301–54.

Bitgood, J.J., Kendall, R.L., Briles, R.W. & Briles, W.E. (1991). Erythrocyte alloantigen loci *Ea-D* and *Ea-I* map to chromosome 1 in the chicken. *Anim. Genet.*, **22**, 449–54.

Bitgood, J.J., Shoffner, R.N., Otis, J.S. & Briles, W.E. (1980). Mapping of the genes for pea comb, blue egg, barring, silver, and blood groups A, E, H, and P in the domestic fowl. *Poult. Sci.*, **59**, 1686–93.

Bitgood, J.J. & Somes, R.G. (1990). Linkage relationships and gene mapping. In *Poultry Breeding and Genetics*, ed. R.D. Crawford, pp. 469–495. Amsterdam: Elsevier.

Blackburn, E.H. (1991). Structure and function of telomeres. *Nature*, **350**, 569–73.

Blem, C.R. (1990). Avian energy storage. In *Current Ornithology*, vol. 7, ed. D.M. Power, pp. 59–113. New York: Plenum.

Blomstrand, E., Challis, R.A.J., Cooney, G.J. & Newsholme, E.A. (1983). Maximum activities of hexokinase, 6-phosphofructokinase, oxoglutarate dehydrogenase, and carnitine palmitoyltransferase in rat and avian muscles. *Biosci. Rep.*, **3**, 1149–53.

Bloom, K.S. & Anderson, J.N. (1982). Hormonal regulation of the conformation of the ovalbumin gene in chick oviduct chromatin. *J. Biol. Chem.*, **257**, 13018–27.

Bloom, S.E. (1981). Detection of normal and aberrant chromosomes in chicken embryos and in tumor cells. *Poult. Sci.*, **60**, 1355–61.

Bloom, S.E. & Bacon, L.D. (1985). Linkage of the major histocompatibility (B) complex and the nucleolar organizer in the chicken. *J. Hered.*, **76**, 146–54.

Bode, W. & Huber, R. (1992). Natural proteinase-inhibitors and their interaction with proteinases. *Eur. J. Biochem.*, **204**, 433–51.

Bodemer, C.W. (1968). *Modern Embryology*. New York: Holt, Rinehart & Winston.

Boguski, M.S., Birkmeier, E.H., Elshourbagy, N.A., Taylor, J.M. & Gordon, J.I. (1986). Evolution of apoproteins. *J. Biol. Chem.*, **261**, 6398–407.

Bohnet, S., Rogers, L., Sasaki, G. & Kolattukudy, P.E. (1991). Estradiol induces proliferation of peroxisome-like microbodies and the production of 3-hydroxy fatty acid diesters, the female pheromones, in the uropygial glands of male and female Mallards. *J. Biol. Chem.*, **266**, 9795–804.

Boland, R., Minghetti, P.P., Lowe, K.E. & Norman, A.W. (1991). Sequences near the CCAAT region and putative 1,25 dihydroxyvitamin D_3 response element and further upstream novel regulatory sequences of calbindin D_{28K} promoter show DNAse I footprinting protection. *Mol. Cell. Endocrinol.*, **75**, 57–63.

Bond, U. & Schlesinger, M.J. (1986). The chicken ubiquitin gene contains a heat shock promoter and expresses an unstable mRNA in heat shock cells. *Mol. Cell. Biol.*, **6**, 4602–10.

Bondok, A.A., Botros, K.G. & El-Mohandes, E.I. (1988). Fluorescence histochemical study of the localisation and distribution of β-adrenergic receptor sites in the spinal cord and cerebellum of the chicken. *J. Anat.*, **160**, 167–74.

Bonifer, C., Vidal, M., Grosveld, F. & Sippel, A.E. (1990). Tissue specific and position independent expression of the complete domain for chicken lysozyme in transgenic mice. *EMBO J.*, **9**, 2843–8.

Bork, P. (1992). The modular architecture of vertebrate collagens. *FEBS Lett.*, **307**, 49–54.

Borman, W.H., Urlakis, K.J. & Yorde, D.E. (1994). Analysis of the *in vivo* myogenic status of chick somites by desmin expression *in vitro*. *Dev. Dynamics*, **199**, 268–79.

Bosca, L., Rousseau, G.G. & Hue, L. (1985). Phorbol 12-myristate 13-acetate and insulin increase the concentration of fructose-2,6-bisphosphate and stimulate glycolysis in chicken embryo fibroblasts. *Proc. Natl. Acad. Sci. USA*, **82**, 6440–4.

Boyd, C.D., Song, J., Kniep, A.C. *et al.* (1991). A restriction fragment length polymorphism in the pigeon pro $\alpha2(I)$ collagen gene: lack of an allelic association with an atherogenic phenotype in pigeons genetically susceptible to the development of spontaneous atherosclerosis. *Connective Tissue Res.*, **26**, 187–97.

Boyer, P.L., Colmenares, C., Stavnezer, E. & Hughes, S.H. (1993). Sequence and biological activity of chicken *snoN* cDNA clones. *Oncogene*, **8**, 457–66.

Bradfield, P.M. & Baggott, G.K. (1993). The effect of water loss upon the urate, urea and ammonia content of the egg of the Japanese quail *Coturnix coturnix japonica*. *Comp. Biochem. Physiol.*, **106A**, 187–93.

Brady, L.J., Romsos, D.R. & Leveille, G.A. (1979). Gluconeogenesis in isolated chicken (*Gallus domesticus*) liver cells. *Comp. Biochem. Physiol.*, **63B**, 193–8.

Branden, C. & Tooze, J. (1991). *Introduction to Protein Structure*. New York: Garland.

Bray, M.M. (1993). Effect of ACTH and glucocorticoids on lipid metabolism in the Japanese quail, *Coturnix coturnix japonica*. *Comp. Biochem. Physiol.*, **105A**, 689–96.

Breitbart, R.E., Andreadis, A., Nadal-Ginard, B. (1987). Alternative splicing: a ubiquitous mechanism for the generation of multiple protein isoforms from single genes. *Annu. Rev. Biochem.* **56**, 467–96.

Brewitt, B., Talian, J.C. & Zelenka, P.S. (1992). Cell cycle synchrony in the developing chicken lens epithelium. *Dev. Biol.*, **152**, 315–22.

Brewton, R.G., Ouspenskala, M.V., Vanderrest, M. & Mayne, R.

(1992). Cloning of the chicken $\alpha3(IX)$ collagen chain completes the primary structure of type IX collagen. *Eur. J. Biochem.*, **205**, 443–9.

Brice, A.T. (1992). The essentiality of nectar and arthropods in the diet of Anna's hummingbird (*Calypte anna*). *Comp. Biochem. Physiol.*, **101A**, 151–5.

Brigham, R.M. & Trayhurn, P. (1994). Brown fat in birds? A test for the mammalian BAT-specific mitochondrial protein in common poorwills. *Condor*, **96**, 208–11.

Briles, W.E. (1984). Early chicken blood group investigations. *Immunogenetics*, **20**, 217–26.

Briles, W.E. & Briles, R.W. (1982). Identification of haplotypes of the chicken major histocompatibility complex. *Immunogenetics*, **15**, 449–59.

Briles, W.E., Goto, R.M., Auffray, C. & Miller, M.M. (1993). A polymorphic system related to but genetically independent of the chicken major histocompatibility complex. *Immunogenetics*, **37**, 408–14.

Briles, W.E., McGibbon, W.H. & Irvin, M.R. (1950). On multiple alleles effecting cellular antigens in the chicken. *Genetics*, **33**, 633–52.

Brittain, T. (1991). Cooperativity and allosteric regulation in non-mammalian vertebrate haemoglobins. *Comp. Biochem. Physiol.*, **99B**, 731–40.

Brock, M.K. & White, B.N. (1991). Multifragment alleles in DNA fingerprints of the parrot, *Amazona ventralis*. *J. Hered.*, **82**, 209–12.

Broders, F., Zahraoui, A. & Scherrer, K. (1990). The chicken α-globin gene domain is transcribed into a 17-kilobase polycistronic RNA. *Proc. Natl. Acad. Sci. USA*, **87**, 503–7.

Bronner-Fraser, M. (1993). Neural crest cell migration in the developing embryo. *Trends Cell Biol.*, **3**, 392–7.

Bronner-Fraser, M., Artinger, M., Muschler, J. & Horwitz, A.F. (1992). Developmentally regulated expression of α_6 integrin in avian embryos. *Development*, **115**, 197–211.

Brotherton, T.W., Reneker, J. & Ginder, G.D. (1990). Binding of HMG 17 to mononucleosomes of the avian β-globin gene cluster in erythroid and non-erythroid cells. *Nucl. Acids Res.*, **18**, 2011–15.

Brotherton, T., Zenk, D., Kahanic, S. & Reneker, J. (1991). Avian nuclear matrix proteins bind very tightly to cellular DNA of β-globin gene enhancer in a tissue-specific fashion. *Biochemistry*, **30**, 5845–50.

Brown, A.J. & Sanders, E.J. (1991). Interactions between mesoderm cells and the extracellular matrix following gastrulation in the chick embryo. *J. Cell Biol.*, **99**, 431–41.

Brown, G.W. (1966). Studies in comparative evolution 1. Avian liver arginase. *Arch. Biochem. Biophys.*, **114**, 184–94.

Brown, G.W. (1970). Nitrogen metabolism in birds. In *Comparative Biochemistry of Nitrogen Metabolism.*, vol. 2, ed. J.W. Campbell, pp. 711–93. London: Academic Press.

Brown, T.A. & DeLuca, H.F. (1990). Phosphorylation of 1,25-dihydroxyvitamin D_3 receptor. *J. Biol. Chem.*, **265**, 10025–9.

Brown, W.M., George, M. & Wilson, A.C. (1979). Rapid evolution of animal mitochondrial DNA. *Proc. Natl. Acad. Sci. USA*, **76**, 1967–71.

Browne, D.L. & Dodgson, J.B. (1993). The gene encoding

chicken chromosomal protein HMG-14a is transcribed into multiple mRNAs. *Gene*, **124**, 199–206.

Brush, A.H. (1983). Self-assembly of avian Φ-keratins. *J. Prot. Chem.*, **2**, 63–75.

Brygier, J., de Bruin, S.H., van Hoof, P.H.K.B. & Rollema, H.S. (1975). The interaction of organic phosphates with human and chicken haemoglobin. *Eur. J. Biochem.*, **60**, 379–83.

Buckle, R., Balmer, M., Yenidunya, A.A. & Allan, J. (1991). The promoter and enhancer of the inactive chicken β-globin gene contains precisely positioned nucleosomes. *Nucl. Acids Res.*, **19**, 1219–26.

Buckner, J.S. & Kolattukudy, P.E. (1975). Lipid biosynthesis in sebaceous glands: regulation of the synthesis of *n*- and branched fatty acids by malonyl coenzyme A decarboxylase. *Biochemistry*, **14**, 1768–73.

Buckner, J.S. & Kolattukudy, P.E. (1976a). One step purification and properties of a two-peptide fatty acid synthase from the uropygial gland of the goose. *Biochemistry*, **15**, 1948–57.

Buckner, J.S. & Kolattukudy, P.E: (1976b). Biochemistry of bird waxes. In *Chemistry and Biochemistry of Natural Waxes*. ed. P.E. Kolattukudy, pp. 147–99. Amsterdam: Elsevier.

Bullfield, G., Isaacson, J.H. & Middleton, R.J. (1988). Biochemical correlates of selection for weight-for-age in chicken: twenty-fold higher muscle ornithine decarboxylase levels in modern broilers. *Theoret. Appl. Genet.*, **75**, 432–7.

Bumstead, N. & Palyga, J. (1992). A preliminary linkage map of the chicken genome. *Genomics*, **13**, 690–7.

Bungert, J., Waldschmidt, R., Kober, I. & Seifart, K.H. (1992). Transcription factor-IIA is inactivated during terminal differentiation of avian erythroid-cells. *Proc. Natl. Acad. Sci. USA*, **89**, 11678–82.

Burke, T. & Bruford, M.W. (1987). DNA fingerprinting in birds. *Nature*, **327**, 149–52.

Burke, T., Hanotte, O., Bruford, M.W. & Cairns, E. (1991). Multilocus and single locus minisatellite analysis in population biological studies. In *DNA Fingerprinting: Approaches and Applications*, ed. T. Burke, G. Dolf., A.J. Jeffreys & R. Wolff, pp. 217–229. Basel: Birkhauser Verlag.

Burley, R.W., Evans, A.J. & Pearson, J.A. (1993). Molecular aspects of the synthesis and deposition of hens' egg yolk with special reference to low density lipoprotein. *Poult. Sci.*, **72**, 850–5.

Burley, R.W. & Vadehra, D.V. (1989). *Avian Egg: Chemistry and Biology.*, Chichester: Wiley-Interscience.

Burnet, F.M. (1959). *The Clonal Selection Theory of Acquired Immunity*. Cambridge: Cambridge University Press.

Burr, G.O. & Burr, M.M. (1929). A new deficiency disease produced by the rigid exclusion of fat from the diet. *J. Biol. Chem.*, **82**, 345–55.

Butler, E.J. (1983). Plasma proteins. In *Physiology and Biochemistry of the Domestic Fowl*, vol. 4, ed. B.M Freeman, pp. 321–38. London: Academic Press.

Buyse, J., Decuypere, E., Leenstra, F.R. & Scanes, C.G. (1992). Abdominal adipose tissue from broiler chickens selected for body weight or for food efficiency differ in *in vitro* lipolytic sensitivity to glucagon and to chicken growth hormone, but not to dibutyryl cAMP. *Br. Poult. Sci.*, **33**, 1069–75.

Byrne, B.M., Gruber, M. & Ab, G. (1989). The evolution of egg yolk proteins. *Prog. Biophys. Mol. Biol.*, **53**, 33–69.

Byrnes, L., Luo, C.-C., Li, W.-H., Yang, C. & Chan, L. (1987). Chicken apolipoprotein A-I: cDNA sequence, tissue expression and evolution. *Biochem. Biophys. Res. Commun.*, **148**, 485–92.

Cadenas, E. (1989). Biochemistry of oxygen toxicity. *Annu. Rev. Biochem.*, **58**, 79–110.

Cahaner, A., Nitsan, Z. & Nir, I. (1986). Weight and fat content of adipose and nonadipose tissues in broilers selected for or against abdominal adipose tissue. *Poult. Sci.*, **65**, 215–22.

Calkhoven, C.F., AB, G. & Wijnholds, J. (1992). cC/EBP, a chicken transcription factor of the leucine zipper C/EBP family. *Nucl. Acids Res.*, **20**, 4093.

Calogeraki, I., Barnier, J.V., Eychène, A., Felder, M.P., Calothy, G., Marx, M. (1993). Genomic organization and nucleotide-sequence of the coding region of the chicken c-Rmil (B-*raf*) protooncogene. *Biochem. Biophys. Res. Commun.*, **193**, 1324–31.

Calomenopoulou, M., Kaloyianni, M. & Beis, I.D. (1989). Purification and regulatory properties of pigeon erythrocyte pyruvate kinase. *Comp. Biochem. Physiol.*, **93B**, 697–706.

Campbell, R. & Langslow, D.R. (1978). The effect of insulin administration *in ovo* on glucose-6-phosphatase in chick embryos. *Biochem. Soc. Trans.*, **6**, 149–52.

Campbell, R.M. & Scanes, C.G. (1987). Growth hormone inhibition of glucagon- and cAMP-induced lipolysis by chicken adipose tissue *in vitro*. *Proc. Soc. Exp. Biol. Med.*, **184**, 456–60.

Canfield, W.M. & Kornfeld, S. (1989). The chicken liver cation independent mannose-6-phosphate receptor lacks the high affinity binding site for IGF-II. *J. Biol. Chem.*, **264**, 7100–3.

Cantos, F.J., Alonso-Gómez, A.L. & Delgado, M.J. (1994). Seasonal changes in fat and protein reserves of the black-headed gull, *Larus ridibundus*, in relation to migration. *Comp. Biochem. Physiol.*, **108A**, 117–22.

Carey, C., Marsh, R.L., Bekoff, A., Johnston, R.M. & Olin, A.M. (1988). Enzyme activities in muscles of seasonally acclimatized house finches. In *Physiology of Cold Adaptation in Birds*, ed. C. Bech & R.E. Reinertsen, pp. 95–103. New York: Plenum.

Carlson, L.M., Oettinger, M.A., Schatz, D.G. *et al.* (1991). Selective expression of RAG-2 in chicken B-cells undergoing immunoglobulin gene conversion. *Cell*, **64**, 201–8.

Carpenter, K.J. & Sutherland, B. (1994). Eijkman's contribution to the discovery of vitamins. *J. Nutr.*, **125**, 155–63.

Carriere, C., Plaza, S., Martin, P., Quatannens, B., Bailly, M., Stehelin, D. & Saule, S. (1993). Characterization of quail Pax-6 (Pax-QNR) proteins expressed in the neuroretina. *Mol. Cell. Biol.*, **13**, 7257–66.

Caspersson, T., Farber, S., Foley, G.E. *et al.* (1968). Chemical differentiation among metaphase chromosomes. *Exp. Cell Res.*, **49**, 219–22.

Catlett, R.H., Walters, T.W. & Dutro, P.A. (1978). The effect of flying and not flying on the myoglobin content of heart muscle of pigeon *Columbia livia domestica*. *Comp. Biochem. Physiol.*, **59A**, 401–2.

Cazals-Hatem, D.L., Louie, D.C., Tanaka, S. & Reed, J.C. (1992). Molecular cloning and DNA sequence analysis of cDNA encoding chicken homologue of the Bcl-2 oncoprotein.

Biochim. Biophys. Acta, **1132**, 109–13.

Cerra, M.C., Canonaco, M. & Tota, B. (1993). ANF binding sites in the heart of the quail (*Coturnix coturnix japonica*). *Peptides*, **14**, 913–18.

Cervantes-Oliver, P., Durieutrautman, O., Delavierklutchko, C. & Strosberg, A.D. (1985). The oligosaccharide moiety of the β_1-adrenergic receptor from turkey erythrocytes has a biantennary N-acetyllactosamine-containing structure. *Biochemistry*, **24**, 3765–70.

Chandra, H.S. (1994). Proposed role of W chromosome inactivation and the absence of dosage compensation in avian sex determination. *Proc. R. Soc. Lond.*, **258B**, 79–82.

Chang, K.S., Zimmer, W.E., Bergsma, D.J., Dodgson, J.B. & Schartz, R.J. (1984). Isolation and characterization of six different chicken actin genes. *Mol. Cell. Biol.*, **4**, 2498–508.

Chang, L.-H., Chuang, L.-F., Tsai, C.-P., Tu, C.-P.D. & Tam, M.F. (1990). Characterization of glutathione S-transferase from day old chick livers. *Biochemistry*, **29**, 744–50.

Chang, L.-H., Wang, L.-Y. & Tam, M.F. (1991). The single residue on an alpha family chick liver glutathione S-transferase CL 3-3 is not functionally important. *Biochem. Biophys. Res. Commun.*, **180**, 323–8.

Chapman, B.S., Hood, L.E. & Tobin, A.J. (1982a). Amino acid sequences of the ε and α^E globins of HbE, a minor early embryonic hemoglobin of the chicken. *J. Biol. Chem.*, **257**, 643–50.

Chapman, B.S., Hood, L.E. & Tobin, A.J. (1982b). Minor early embryonic chick hemoglobin H: amino acid sequences of ε and α^D chains. *J. Biol. Chem.*, **257**, 643–50.

Chapman, B.S., Tobin, A.J. & Hood, L.E. (1980). Complete amino acid sequence of the major early embryonic α-like globins of the chicken. *J. Biol. Chem.*, **255**, 9051–9.

Chapman, B.S., Tobin, A.J. & Hood, L.E. (1981). Complete amino-acid sequence of major early embryonic β-like globin in chickens. *J. Biol. Chem.*, **256**, 5524–31.

Chaudhuri, C.R. & Chatterjee, I.B. (1969). L-Ascorbic acid synthesis in birds: phylogenetic trend. *Science*, **164**, 435–6.

Chen, C.-L.H., Lehmeyer, J.E. & Cooper, M.D. (1982). Evidence for an IgD homologue on chicken lymphocytes. *J. Immunol.*, **129**, 2580–5.

Chen, J., Zong, C.S. & Wang, L.-H. (1994). Tissue and epithelial cell-specific expression of chicken proto-oncogene c-*ros* in several organs suggests that it may play roles in their development and mature functions. *Oncogene*, **9**, 773–80.

Chen, S. & Zhou, D. (1992). Functional domains of aromatase cytochrome P450 inferred from comparative analyses of amino acid sequences and substantiated by site directed mutagenesis experiments. *J. Biol. Chem.*, **267**, 22587–94.

Chen, Z.-Q., Lin, C.C. & Hodgetts, R.B. (1989). Cloning and characterization of a tandemly repeated DNA sequence in the crane family (Gruidae). *Genome*, **32**, 646–54.

Cherel, Y., Charrassin, J.-B. & Handrich, Y. (1993). Comparison of body reserve buildup in prefasting chicks and adults of king penguins (*Aptenodytes patagonicus*). *Physiol. Zool.*, **66**, 750–70.

Cherel, Y. & Freby, F. (1994). Daily body-mass loss and nitrogen excretion during molting fast of macaroni penguins. *Auk*, **111**, 492–5.

Cherel, Y., Robin, J.-P. & Le Maho, Y. (1988). Physiology and biochemistry of long-term fasting in birds. *Canad. J. Zool.*, **66**, 159–66.

Chieri, R.A., Basabe, J.C., Farina, J.M.S. & Foglia, V.G. (1972). Studies on the carbohydrate metabolism in penguins (*Pygocellis papua*). *Gen. Comp. Endocrinol.*, **18**, 1–4.

Chim, T.Y. & Quebbeman, A.J. (1978). Quantitation of renal uric acid synthesis in the chicken. *Am. J. Physiol.*, **234**, F446–51.

Chiou, S.H., Hung, C.C. & Lin, C.W. (1992). Biochemical characterization of crystallin from pigeon lens. *Biochim. Biophys. Acta*, **1160**, 317–24.

Chiou, S.-H., Lee, H.-J. & Chang, G.-G. (1990). Kinetic analysis of duck ε-crystallin, a lens structural protein with lactate dehydrogenase activity. *Biochem. J.*, **267**, 51–8.

Chiou, S.-H., Lo, C.-H., Chang, C.-Y., Itoh, T., & Samejima, T. (1991). Ostrich crystallins: structural characterization of δ-crystallin with enzyme activity. *Biochem. J.*, **273**, 295–300.

Ch'ng, L.K. & Benedict, A.A. (1981). The phylogenetic relationships of immunoglobulin allotypes and 7S immunoglobulin isotypes of chickens and other Phasianoids (turkey, pheasant, quail). *Immunogenetics*, **12**, 541–54.

Chou, W.-Y., Huang, S.-M., Liu, Y.H. & Chang, G.-G. (1994). Cloning and expression of pigeon liver cytosolic NADP$^+$-dependent malic enzyme cDNA and some of its abortive mutants. *Arch. Biochem. Biophys.*, **310**, 158–66.

Christidis, L. (1990). Aves. In *Animal Cytogenetics*. vol. 4, ed, B. John, pp. 101–14. Berlin: Gebruder Borntraeger.

Christidis, L., Shaw, D.D. & Schodde, R. (1991). Chromosomal evolution in parrots, lorikeets and cockatoos (Aves: Psittaciformes). *Heriditas*, **114**, 47–56.

Christie, W.W. & Moore, J.H. (1972). The lipid components of the plasma, liver and ovarian follicles of the domestic chicken (*Gallus gallus*). *Comp. Biochem. Physiol.*, **41B**, 287–95.

Chuong, C.-M. (1993). The making of a feather: homeoproteins, retinoids and adhesion molecules. *BioEssays*, **15**, 513–21.

Cleer, W.F. & Coughlan, M.P. (1975). Avian xanthine dehydrogenases. I. Isolation and characterization from turkey kidney. *Comp. Biochem. Physiol.*, **50B**, 311–22.

Coates, M.E. (1984). Metabolic role of the vitamins. In *Physiology and Biochemistry of the Domestic Fowl*, vol. 5, ed. B.M. Freeman, pp. 27–37. London: Academic Press.

Cobb, J.A., Manning, D., Kolatkar, P.R., Cox, D.J. & Riggs, A.F. (1992). Deoxygenation-linked association of a tetrameric component of chicken hemoglobin. *J. Biol. Chem.*, **267**, 1183–9.

Cochet, M., Gannon, F., Hen, R., Marteaux, L., Perrin, F & Chambon, P. (1979). Organization and sequence studies of the 17-piece chicken conalbumin gene. *Nature*, **282**, 567–74.

Coelho, C.N.D., Krabbenhoft, K.M., Upholt, W.B., Fallon, J.F. & Kosher, R.A. (1991b). Altered expression of the chicken homeobox-containing genes GHox-7 and GHox-8 in the limb buds of *limbless* mutant chick embryos. *Development*, **113**, 1487–93.

Coelho, C.N.D., Sumoy, L., Rodgers, B.J. *et al.* (1991a). Expression of the chicken homeobox-containing gene GHox-8 during embryonic limb development. *Mechanisms. Dev.*, **34**, 143–54.

Coelho, C.N.D., Upholt, W.B. & Kosher, R.A. (1993). The expression pattern of the chicken homeobox-containing gene GHox-7 in developing polydactylous limb buds suggests its involvement in apical ectodermal ridge-directed outgrowth of limb mesoderm and in programmed cell death. *Differentiation*, **52**, 129–37.

Cohen, P. (1992). Signal integration at the level of protein kinases, protein phosphatase and their substrates. *Trends Biochem. Sci.*, **17**, 408–13.

Colella, R., Johnson, A. & Bird, J.W.C. (1991). Steady-state cystatin mRNA levels in chicken tissues in response to oestrogen. *Biomed. Biochim. Acta*, **50**, 607–11.

Colella, R., Sakaguchi, Y., Nagase, H. & Bird, J.W.C. (1989). Chicken egg white cystatin. *J. Biol. Chem.*, **264**, 17164–9.

Connolly, E., Nedergaard, J.& Cannon, B. (1988). Shivering and nonshivering thermogenesis in birds: a mammalian view. In *Physiology of Cold Adaptation in Birds*, ed. C. Bech & R.E. Reinertsen, pp. 37–48. New York: Plenum.

Constans, T., Chevalier, B., Derouet, M. & Simon, J. (1991). Insulin sensitivity and liver insulin receptor structure in ducks from two genera. *Am. J. Physiol.*, **261**, R882–90.

Conway, C.J., Eddleman, W.R. & Simpson, K.L. (1994). Seasonal changes in fatty acid composition of the wood thrush. *Condor*, **96**, 791–4.

Cooney, A.J., Leng, X.H., Tsai, S.Y., O'Malley, B.W. & Tsai, M.J. (1993). Multiple mechanisms of chicken ovalbumin upstream promoter transcription factor-dependent repression of transactivation by the vitamin-D, thyroid-hormone, and retinoic acid receptors. *J. Biol. Chem.*, **268**, 4152–60.

Cooper, D.A., Lu, S.-C., Viswanath, R., Freiman, R.N. & Bensadoun, A. (1992). The structure and complete nucleotide sequence of the avian lipoprotein lipase gene. *Biochim. Biophys. Acta*, **1129**, 166–71.

Cooper, D.A., Stein, J.C., Strielman, P.J. & Bensadoun, A. (1989). Avian adipose lipoprotein lipase: cDNA sequence and reciprocal regulation of mRNA levels in adipose and heart. *Biochim. Biophys. Acta*, **1008**, 92–101.

Cooper, J.A., Reiss, N.A., Schwartz, R.J. & Hunter, T. (1983). Three glycolytic enzymes are phosphorylated at tyrosine in cells transformed by Rous sarcoma virus. *Nature*, **302**, 218–23.

Cooper, M.D. & Burrows, P.D. (1989). B-cell differentiation. In *Immunoglobulin Genes*, ed. T. Honjo, F.W. Alt & T.H. Alt, pp. 1–22. London: Academic Press.

Cooper, T.A. & Ordahl, C.P. (1985). A single cardiac troponin T gene generates embryonic and adult isoforms via developmentally regulated alternate splicing. *J. Biol. Chem.*, **260**, 11140–8.

Corradino, R.A. (1993). Calbindin D_{28K} regulation in precociously matured chick egg shell gland *in vitro*. *Gen. Comp. Endocrinol.*, **91**, 158–66.

Coutinho, L.L., Morris, J., Marks, H.L., Buhr, R.J. & Ivarie, R. (1993). Delayed somite formation in a quail line exhibiting myofiber hyperplasia is accompanied by delayed expression of myogenic regulatory factors and myosin heavy chain. *Development*, **117**, 563–9.

Coux, O., Camoin, L., Nothwang, H.-G. *et al.* (1992a). The protein of M_r 21 000 constituting the prosome-like particle of duck erythroblasts is homologous to apoferritin. *Eur. J. Biochem.*, **207**, 823–32.

Coux, O., Nothwang, H.G., Scherrer, K. *et al.* (1992b). Structure and RNA content of prosomes. *FEBS Lett.*, **300**, 49–55.

Cracraft, J. (1974). Phylogeny and evolution of the ratite birds. *Ibis*, **116**, 494–521.

Cramb, G. & Langslow, D.R. (1984). The endocrine pancreas: control of secretions and actions of the hormones. In *Physiology and Biochemistry of the Domestic Fowl*, vol. 5, ed. B.M. Freeman, pp. 94–124. London: Academic Press.

Cramb, G., Langslow, D.R. & Phillips, J.H. (1982). The binding of pancreatic hormones to isolated chicken hepatocytes. *Gen. Comp. Endocrinol.*, **46**, 297–309.

Craven, R.C., Bennett, R.P. & Wills, J.W. (1991). Role of the avian retroviral protease in the activation of reverse transcriptase during virion assembly. *J. Virol.*, **65**, 6205–17.

Crawford, C., Willis, A.C. & Gagnon, J. (1987). The effects of autolysis on the structure of chicken calpain II. *Biochem. J.*, **248**, 579–88.

Creighton, T.E. & Darby, N.J. (1989). Functional evolutionary divergence of proteolytic enzymes and their inhibitors. *Trends Biochem. Sci.*, **14**, 319–24.

Crewther, W.G., Fraser, R.D.G., Lennox, F.G. & Lindley, H. (1965). The chemistry of keratins. *Adv. Prot. Chem.*, **20**, 191–346.

Crittenden, L.B., Salter, D.W. & Federspiel, M.J. (1989). Segregation, viral phenotype, and proviral structure of 23 avian leukosis virus inserts in the germ line of chickens. *Theoret. Appl. Genet.*, **77**, 505–15.

Crosby, M.J. (1994). Avian classification. *Trends Ecol. Evol.*, **9**, 307–8.

Cross, M., Mangeldorf, J., Wedel, A. & Renkawitz, R. (1988). Mouse lysozyme: isolation, characterization and expression studies. *Proc. Natl. Acad. Sci. USA*, **85**, 6232–6.

Cvekl, A. & Paces, V. (1992). Interactions between protein bound to the duck β^A-globin gene promoter and enhancer detected by the DNAseI footprinting. *Gene*, **110**, 225–8.

D'Andrea, R.J., Coles, L.S., Lesnikowski, C., Tabe, L. & Wells, J.R.E. (1985). Chromosomal organisation of chicken histone genes: preferred associations and inverted duplications. *Mol. Cell. Biol.*, **5**, 3108–15.

D'Mello, J.P.F. (1988). Dietary interactions influencing amino acid utilisation by poultry. *World's Poult. Sci. J.*, **44**, 92–102.

Dalseg, A.-M., Andersson, A.-M. & Bock, E. (1990). Characterization of N-cadherin mRNA in chicken brain and heart by means of oligonucleotide probes. *FEBS Lett.*, **262**, 234–6.

Dalton, S., Robins, A.J., Harvey, R.P. & Wells, J.R.E. (1989). Transcription from the intron-containing chicken histone H2A$_f$ gene is not S-phase regulated. *Nucl. Acids Res.*, **17**, 1745–56.

Dalton, S. & Wells, J.R.E. (1988). A gene-specific promoter element is required for optimal expression of the histone H1 gene in S-phase. *EMBO J.*, **7**, 49–56.

Dam, H. (1935). The antihaemorrhagic vitamin of the chick. Occurrence and chemical nature. *Nature*, **135**, 652–3.

Das, M., Dixit, R., Seth, P.K. & Mukhtar, H. (1981). Glutathione S-transferase activity in the brain: species, sex, regional and age differences. *J. Neurochem.*, **36**, 1439–42.

Davison, T.F. & Langslow, D.R. (1975). Changes in plasma glucose and liver glycogen following the administration of gluconeogenic precursors to the starving fowl. *Comp.*

Biochem. Physiol., **52A**, 645–9.

Dawson, T.J. & Herd, R.M. (1983). Digestion in the emu: low energy and nitrogen requirements of this large ratite bird. *Comp. Biochem. Physiol.*, **75A**, 41–5.

Dawson, W.R. & Marsh, R.L. (1988). Metabolic acclimatization to cold and season in birds. In *Physiology of Cold Adaptation in Birds*, ed. C. Bech & R.E. Reinertsen, pp. 83–94. New York: Plenum.

Dawson, W.R., Marsh, R.L. & Yacoe, M.E. (1983). Metabolic adjustments of small passerine birds for migration and cold. *Am. J. Physiol.*, **245**, R755–67.

Dayhoff, M.O. (1978). *Atlas of Protein Sequence and Structure*, vol. 5, suppl. 3. Silver Spring, MD: National Biomedical Research Foundation.

de Gennaro, L.D. (1982). The glycogen body. In *Avian Biology*, vol. 6, ed. J.R. King & K.C. Parkes, pp. 341–71. New York: Academic Press.

de la Sena, C.A., Fechheimer, N.S. & Nestor, K.E. (1991). Variability of C-banding patterns in Japanese quail chromosomes. *Genome*, **34**, 993–7.

de Leon, F.A.P. & Burt, D. (1993). Physical map of the chicken: an update. *Manipulation of the Avian Genome.*, September 1993, ed. A. Gibbins & R. Etches. Guelph, Ontario: Department of Poultry Science.

de Pablo, F. & Roth, J. (1990). Endocrinization of the early embryo: an emerging role for hormones and hormone-like factors. *Trends Biochem. Sci.*, **15**, 339–42.

de Pomerai, D. (1990). *From Gene to Animal*, 2nd edn. Cambridge: Cambridge University Press.

Deák, F. Argraves, W.S., Kiss, I., Sparks, K.J. & Goetinck, P.F. (1985). Primary structure of the telopeptide and a portion of the helical domain of chicken type II procollagen as determined by DNA sequence analysis. *Biochem. J.*, **229**, 189–96.

Dealy, C.N., Roth, A., Ferrari, D., Brown, A.M.C. & Kosher, R.A. (1993). *Wnt-5a* and *Wnt-7a* are expressed in the developing chick limb bud in a manner suggesting roles in pattern formation along the proximodistal and dorsoventral axes. *Mech. Dev.*, **43**, 175–86.

Decuypere, E., Leenstra, F.R., Buyse, J., Huybrechts, L.M., Buomomo, F.C. & Berghman, L.R. (1993). Plasma levels of growth hormone and insulin-like growth factor-I and -II from 2 to 6 weeks of age in meat-type chickens selected for 6-week body weight or for feed conversion and reared under high or normal environmental temperature conditions. *Reprod. Nutr. Dev.*, **33**, 361–72.

Deeley, R.G., Burtch-Wright, R.A., Grant, C.E., Hoodless, P.A., Ryan, A.K. & Schrader, T.J. (1992). Synthesis and deposition of egg proteins. In *Manipulation of the Avian Genome*, ed. R.J. Etches & A.M. Verrinder Gibbins, pp. 205–22, Boca Raton, FL: CRC Press.

Degnan, S.M. & Moritz, C. (1992). Phylogeography of mitochondrial DNA in two species of White-eyes in Australia. *Auk*, **109**, 800–11.

Desjardins, P. & Morais, R. (1990). Sequence and gene organisation of the chicken mitochondrial genome. *J. Mol. Biol.*, **212**, 599–634.

Desjardins, P. & Morais, R. (1991). Nucleotide sequence and evolution of coding and noncoding regions of a quail mitochondrial genome *J. Mol. Evol.* **32**, 153–61.

Dickerson, R.E. & Geis, I. (1983). *Hemoglobin: Structure, Function, Evolution, and Pathology*. Menlo Park: Benjamin/Cummings.

Dickson, A.J. & Langslow, D.R. (1978). Hepatic gluconeogenesis in chickens. *Mol. Cell Biochem.*, **22**, 167–81.

Didier, R., Rémésy, C. & Demigne, C. (1981). Effet du jeûne sur les principaux métabolites sanguins et hépatiques de la néoglucogenèse et de la cetogenèse chez la caille domestique adulte (*Coturnix coturnix* japonica). *Reprod. Nutr. Dév.*, **21**, 421–8.

Dierich, A., Gaub, M.-P., LePennec, J.-P., Astinotti, D. & Chambon, P. (1987). Specificity of the chicken ovalbumin and conalbumin promoters. *EMBO J.*, **6**, 2305–12.

Dieterlen-Lièvre, F. & le Douarin, N. (1993). The use of avian chimeras in developmental biology. In *Manipulation of the Avian Genome*, ed. R.J. Etches & A.M.V. Gibbins, pp. 103–19, Boca Raton, FL: CRC Press.

Dixon, J.L., Chattapadhyay, R., Huima, T., Redman, C.M. & Banerjee, D. (1992). Biosynthesis of lipoprotein: location of nascent ApoAI and ApoB in the rough endoplasmic reticulum of chicken hepatocytes. *J. Cell Biol.*, **117**, 1161–9.

Dolan, M., Sugarman, B.J., Dodgson, J.B. & Engel, J.D. (1981). Chromosomal arrangement of the chicken β-type globin genes. *Cell*, **4**, 669–77.

Dölle, A. & Strätling, W.H. (1990). Genomic footprinting of proteins interacting with the chicken lysozyme promoter. *Gene* **95**, 187–93.

Dolnik, V.R. (1988). Bird migration across arid and mountainous regions of Middle Asia and Kasakhstan. In *Bird Migration: Physiology and Ecophysiology*, ed. E. Gwimmer, pp. 368–86. Berlin: Springer Verlag.

Domenget, C., Leprince, D., Pain, B. *et al*. (1992). The various domains of v-*myb* and v-*ets* oncogenes of E26 retrovirus contribute differently, but cooperatively, in transformation of hematopoietic lineages. *Oncogene*, **7**, 2231–41.

Dominguez-Steglich, M., Jelsch, J.-M., Garnier, J.-M. & Schmid, M. (1992). *In situ* mapping of the chicken progesterone receptor gene and the ovalbumin gene. *Genomics*, **13**, 1343–4.

Dominguez-Steglich, M., Meng, G., Bettecken, T., Muller, C.R. & Schmid, M. (1990). The dystrophin gene is automatically located on a microchromosome in chicken. *Genomics*, **8**, 536–40.

Dominguez-Steglich, M., Robbins, J. & Schmid, M. (1993). Mapping of the chicken N-CAM gene and a myosin heavy chain gene: avian michromosomes are not genetically inert reserves of DNA. *J. Exp. Zool.*, **265**, 295–300.

Donaldson, W.E. & Christensen, V.L. (1991). Dietary carbohydrate and glucose metabolism in turkey poults. *Comp. Biochem. Physiol.*, **98A**, 347–50.

Donaldson, W.E. & Christensen, V.L. (1992). Effects of injections of L-alanine, L-glucose and L-ascorbic acid in newly hatched turkey poults on glucose metabolism. *Comp. Biochem. Physiol.*, **101A**, 849–52.

Donaldson, W.E., Christensen, V.L. & Kreuger, K.K. (1991). Effects of stressors on blood glucose and hepatic glycogen in turkey poults. *Comp. Biochem. Physiol.*, **100A**, 945–7.

Downing, D.T. (1986). Preen gland and scent gland lipids. In

Biology of the Integument 2, ed. J. Bereiter-Hahn, A.G. Matoltsy & K.S. Richards, pp. 833–40. Berlin: Springer-Verlag.

Dozin, B., Quarto, R., Rossi, F. & Cancedda, R. (1990). Stabilization of the messenger-RNA follows transcriptional activation. *J. Biol. Chem.* **265**, 7216–20.

Duband, J.-L. & Thiery, J.P. (1990). Spatio-temporal distribution of the adherens junction-associated molecules vinculin and talin in the early avian embryo. *Cell Diff. Dev.*, **30**, 55–76.

Dublet, B., Oh, S., Sugrue, S.P. *et al.* (1989). The structure of avian typeXII collagen- α-1(XII) chains contain 190 kDa non-triple helical amino-terminal domains and form homotrimeric molecules. *J. Biol. Chem.*, **264**, 13150–6.

Duchamp, C., Barré, H., Rouanet, J.-L. *et al.* (1991). Nonshivering thermogenesis in king penguin chicks. I. Role of skeletal muscle. *Am. J. Physiol.*, **261**, R1438–45.

Duclos, M.J., Chevalier, B., Le Marchand-Brustel, Y., Tanti, J.F., Goddard, C. & Simon, J. (1993). Insulin-like growth factor-I-stimulated glucose transport in myotubes derived from chicken muscle satellite cells. *J. Endocrinol.*, **137**, 465–72.

Dudek, H., Tantravahi, R.V., Rao, V.N., Reddy, E.S.P. & Reddy, E.P. (1992). *Myb* and *ets* proteins cooperate in transcriptional activation of the *mim*-1 promoter. *Proc. Natl. Acad. Sci. USA*, **89**, 1291–5.

Dufour, E., Obled, A., Valin, C., Bechet, D., Ribadeaudumas, B. & Huet, J.C. (1987). Purification and amino-acid sequence of chicken liver cathepsin-L. *Biochemistry*, **26**, 5689–95.

Dunn, M.A., Johnson, N.E., Liew, M.Y.B. & Ross, E. (1993). Dietary aluminium chloride reduces the amount of intestinal calbindin D-28K in chicks fed low calcium or low phosphorus diets. *J. Nutr.*, **123**, 1786–93.

Dunnington, A.E., Gal, O., Siegel, P.B. *et al.* (1991). Deoxyribonucleic acid fingerprint comparisons between selected populations of chickens. *Poult. Sci.*, **70**, 463–7.

Durairaj, G. & Martin, E.W. (1975). The effect of temperature and diet on the fatty acid composition of the Japanese quail. *Comp. Biochem. Physiol.*, **50**, 237–48.

Edson, N.L., Krebs, H.A. & Model, A. (1936). The synthesis of uric acid in the avian organism: hypoxanthine as an intermediary metabolite. *Biochem. J.*, **30**, 1380–5.

Egana, M., Trueba, M. & Sancho, M.J. (1986). Some special characteristics of glycogen synthase from chicken liver. *Comp. Biochem. Physiol.*, **83B**, 771–4.

Eguchi, Y., Nakashima, Y., Oshiro, M. & Takei, H. (1990). Complete nucleotide sequence of a pigeon α-globin cDNA. *Nucl. Acids Res.*, **18**, 7135.

Eguchi, Y., Takei, H., Oshiro, M., Toda, T. & Nakashima, Y. (1991). Protein structure and cDNA nucleotide sequence of the Japanese quail α^A globin. *Biol. Chem. Hoppe-Seyler*, **372**, 113–18.

Ehrich, M. & Larsen, C. (1983). Drug metabolism in adult white Leghorn hens – response to enzyme inducers. *Comp. Biochem. Physiol.*, **74C**, 383–6.

Eichmann, A., Marcelle, C., Bréant, C. & le Douarin, N.M. (1993). Two molecules related to the VEGF receptor are expressed in early endothelial cells during avian embryonic development. *Mech. Dev.*, **42**, 33–48.

Eigenbrodt, E. & Schoner, W. (1977). Purification and properties of the pyruvate kinase isoenzymes type L and M₂ from chicken liver. *Physiol. Chem. Hoppe-Seyler*, **358**, 1033–46.

Elbrecht, A. & Smith, R.G. (1992). Aromatase enzyme activity and sex determination in chickens. *Science*, **255**, 467–70.

Elkin, R.G. (1987). A review of duck nutrition research. *World's Poult. Sci. J.*, **43**, 84–106.

Ellegren, H. (1991). Fingerprinting bird's DNA with a synthetic polynucleotide probe (TG)$_n$. *Auk*, **108**, 956–8.

Ellis, D.L. & Danzo, B.J. (1989). Identification of an androgen receptor in the adult chicken oviduct. *J. Steroid Biochem.*, **33**, 1081–6.

Emara, M.G., Nestor, K.E., Foster, D.N. & Lamont, S.J. (1992). The turkey major histocompatibility complex: identification of Class II genotypes by restriction fragment length polymorphism analysis of deoxyribonucleic acid. *Poult. Sci.*, **71**, 2083–9.

Emmanuel, B. & Gilanpour, H. (1978). Studies on enzymes of nitrogen metabolism, and nitrogen end products in the developing chick (*Gallus domesticus*) embryo. *Comp. Biochem. Physiol.*, **61B**, 287–9.

Empie, M.W. & Laskowski, M. (1982). Thermodynamics and kinetics of single residue replacements in avian ovomucoid third domains: effect on inhibitor interactions with serine proteinases. *Biochemistry*, **21**, 2274–84.

Engberg, R.M., Kaspers, B., Schranner, I., Kösters, J. & Lösch, U. (1992). Quantification of the immunoglobulin classes IgG and IgA in the young and adult pigeon (*Columbia livia*). *Avian Path.*, **21**, 409–20.

Engstrom, L., Ekman, P., Humble, E. & Zetterqvist, O. (1987). Pyruvate kinase. In *The Enzymes*, 3rd edn, vol. 18, ed. P.D. Boyer & E.G. Krebs, pp. 47–76. New York: Academic Press.

Epple, A. & Brinn, J.E. (1987). *The Comparative Physiology of the Pancreatic Islets*. New York: Springer-Verlag.

Epplen, J.T., Leipoldt, M., Engel, W. & Schmidtke, J. (1978). DNA sequence organization in avian genomes. *Chromosoma*, **69**, 307–321.

Erisman, M.D. & Astrin, S.M. (1988). The *myc* gene. In *The Oncogene Handbook*, ed. E.P. Reddy, A.M. Skalka & T. Curran. pp. 341–77. Amsterdam: Elsevier.

Eshdat, Y., Chapot, M.P. & Strosberg, A.D. (1989). Turkey β-adrenergic-receptor. *FEBS Lett.*, **246**, 166–70.

Espinet, C., Bartrons, R. & Carreras, J. (1988). Effects of fructose 2,6-bisphosphate and glucose 1,6-bisphosphate on phosphofructokinase from chicken erythrocytes. *Comp. Biochem. Physiol.*, **90B**, 453–7.

Espinet, C., Bartrons, R. & Carreras, J. (1989). The effect of fructose 2,6-bisphosphate and glucose metabolization by chicken erythrocytes. *FEBS Lett.*, **258**, 143–6.

Evans, A.J. (1977). The growth of fat. In *Growth and Poultry Meat Production*, ed. K.N. Boorman & B.J. Wilson, pp. 29–64. Edinburgh: British Poultry Science.

Evans, C.-O., Healey, J.F., Greene, Y. & Bonkovsky, H.L. (1991). Cloning, sequencing and expression of cDNA for chick liver haem oxygenase. *Biochem. J.*, **273**, 659–66.

Evans, M.I., O'Malley, P.J., Krust, A. & Burch, J.B.E. (1987). Developmental regulation of the estrogen receptor and the estrogen responsiveness of five yolk protein genes in the avian liver. *Proc. Natl. Acad. Sci. USA*, **84**, 8493–7.

Eyal-Giladi, H. (1993). Early determination and morphogenetic processes in birds. In *Manipulation of the Avian Genome*, ed. R.J. Etches & A.M.V. Gibbins, pp. 29–37. Boca Raton, FL: CRC Press.

Eychène, A., Barnier, J.V., Dézélée, P., Marx, M., Laugier, D., Calogeraki, I. & Calothy, G. (1992). Quail neuroretina c-Rmil(B-*raf*) proto-oncogene cDNAs encode two proteins of 93.5 and 95 kDa resulting from alternative splicing. *Oncogene*, **7**, 1315–23.

Fagan, J.M., Wajnberg, E.F., Culbert, L. & Waxman, L. (1992). ATP depletion stimulates calcium-dependent protein breakdown in chick skeletal muscle. *Am. J. Physiol.*, **262**, E637–43.

Fagan, J.M. & Waxman, L. (1989). A novel ATP-requiring protease from skeletal muscle that hydrolyses non-ubiquitinated proteins. *J. Biol. Chem.*, **264**, 17864–8.

Fainsod, A. & Gruenbaum, Y. (1989). A chicken homeo box gene with developmentally regulated expression. *FEBS Lett.*, **250**, 381–5.

Fainsod, A. & Gruenbaum, Y. (1994). Homeobox genes in avian development. *Poult. Sci. Rev.*, **6**, in press.

Faraci, F.M. (1991). Adaptations to hypoxia in birds: how to fly high. *Annu. Rev. Physiol.*, **53**, 59–70.

Farquharson, C., Whitehead, C., Rennie, S., Thorp, B. & Loverage, N. (1992). Cell proliferation and enzyme activities associated with the development of avian tibial dyschondroplasia: an *in situ* biochemical study. *Bone*, **13**, 59–67.

Fawcett, D.H. & Bulfield, G. (1990). Molecular cloning, sequence analysis and expression of putative chicken insulin-like growth-I cDNAs. *J. Mol. Endocrinol.*, **4**, 201–11.

Featherston, W.R. & Horn, G.W. (1973). Dietary influences on the activities of enzymes involved in branched-chain amino acid catabolism in the chick. *J. Nutr.*, **103**, 757–65.

Feeney, R.E. & Osuga, D.T. (1991). Biologically active proteins in eggs. In *Food Enzymology*, vol. 2, ed. P.F. Fox, pp. 265–85. London: Elsevier.

Ferreira, A., Litthauer, D., Saayman, H., Oelofsen, W., Crabb, J & Lazure, C. (1991). Purification and primary structure of glucagon from ostrich pancreas splenic lobes. *Int. J. Peptide Res.*, **38**, 90–5.

Fielding Hejtmancik, J., Thompson, M., Wiastow, G. & Piatigorsky, J. (1986). cDNA deduced protein sequence for βB-1 crystallin polypeptide of the chicken lens. *J. Biol. Chem.*, **261**, 982–7.

Fischer, P.W.F. & Goodridge, A.G. (1978). Coordinate regulation of acetylCoA carboxylase and fatty acid synthase in liver cells of the developing chick *in vivo* and in culture. *Arch. Biophys. Biochem.*, **190**, 332–44.

Fisher, C. (1980). Protein deposition in poultry. In *Protein Deposition in Animals*, ed. P.J. Buttery & D.B. Lindsay, pp. 251–70. London: Butterworths.

Fisher, H. (1972). The nutrition of birds. In *Avian Biology*, vol. 2, ed. D.S. Farner & J.R. King, pp. 431–69. New York: Academic Press.

Fisher, R.B. (1954). *Protein Metabolism*. London: Methuen.

Fister, P., Eigenbrodt, E., Presek, P., Reinacher, M. & Schoner, W. (1983). Pyruvate kinase type M_2 is phosphorylated in the intact chicken liver cell. *Biochem. Biophys. Res. Commun.*, **115**, 409–14.

Flannery, A.V., Easterby, J.S. & Beynon, R.J. (1992). Turnover of glycogen phosphorylase in the pectoralis muscle of broiler and layer chickens. *Biochem. J.*, **286**, 915–22.

Fleenor, D.E., Hickman, K.H., Lindquester, G.J. & Devlin, R.B. (1992). Avian cardiac tropomyosin gene produces tissue-specific isoforms through alternative RNA splicing. *J. Muscle Res. Cell Motility*, **13**, 55–63.

Foglia, T.A., Cartwright, A.L., Gyurik, R.J. & Philips, J.G. (1994). Fatty acid turnover rates in the adipose tissues of the growing chicken (*Gallus domesticus*). *Lipids*, **29**, 497–502.

Foltzer, C., Harvey, S. & Mialhe, P. (1987). Pancreatic somatostatin, glucagon and insulin during post-hatch growth in the duck (*Anas platyrhynchos*). *J. Endocrinol.*, **113**, 65–70.

Foos, G., Grim, S. & Klempnauer, K.H. (1992). Functional antagonism between members of the *myb* family – B-*myb* inhibits v-*myb*-induced gene activation. *EMBO J.*, **11**, 4619–29.

Forryschaudies, S. & Hughes, S.H. (1991). The chicken tropomyosin-1 gene generates 9 messenger-RNAs by alternative splicing. *J. Biol. Chem.*, **266**, 13821–7.

Foster, R.J., Bonsall, R.F., Poulose, A.J. & Kolattukudy, P.E. (1985). Interaction of S-acyl fatty acid synthase thioester hydrolase with fatty acid synthase. *J. Biol. Chem.*, **260**, 1386–9.

Freeman, B.M. (1983). Adrenal glands. In *Physiology and Biochemistry of the Domestic Fowl*. vol. 4, ed. B. M. Freeman, pp. 407–24. London: Academic Press.

Fritschi, S. & Stranzinger, G. (1985). Fluorescent chromosome banding in inbred chicken: quinacrine bands, sequential chromomycin and Dapi bands. *Theoret. App. Genet.*, **71**, 408–412.

Fruton, J.S. (1972). *Molecules and Life*. pp. 379–85. New York: Wiley Interscience.

Fujisawasehara, A., Nabeshima, Y., Komiya, T., Uetsuki, T., Asakura, A. & Nabeshima, Y. (1992). Differential transactivation of muscle-specific regulatory elements including the myosin light chain box by chicken Myod, myogenin, and Mrf4. *J. Biol. Chem.*, **267**, 10031–8.

Fukui, Y., Saltiel, A.R. & Hanafusa, H. (1991). Phosphatidylinositol-3 kinase is activated in v-*src*, v-*yes*, and v-*fps* transformed chicken embryo fibroblasts. *Oncogene*, **6**, 407–11.

Fulton, J.E., Briles, R.W. & Lamont, S.J. (1990). Monoclonal antibody differentiates chicken A system alloantigens. *Anim. Genet.*, **21**, 39–45.

Furita, S. & Hashimoto, T. (1982). Pyruvate dehydrogenase complex from pigeon breast muscle. *Meth. Enzymol.*, **89**, 414–20.

Furuse, M., Murai, A., Kita, K., Asakura, K. & Okumura, J.-I. (1991). Lipogenesis depending on sexual maturity in female Japanese quail (*Coturnix coturnix*). *Comp. Biochem. Physiol.*, **100B**, 343–5.

Gadow, H.F. (1892) On the classification of birds. *Proc. Zool. Soc. Lond.*, **1892**, 229–56.

Gallo, C.V.D., Vassetzky, Y.S., Targa, F.R., Georgiev, G.P., Scherrer, K. & Razin, S.V. (1991). The presence of sequence-specific protein binding sites correlate with replication activity and matrix binding in a 1.7 kb-long DNA fragment of the chicken α-globin gene domain. *Biochem. Biophys. Res. Commun.*, **179**, 512–19.

Gapp, D.A. (1987). Endocrine and related factors in the control of metabolism in nonmammalian vertebrates. In *Fundamentals of Comparative Vertebrate Endocrinology*, ed. I. Chester-Jones, P.M. Ingleton & J.G. Phillips, pp. 623–60, New York: Plenum Press.

Garcia, F.J., Pons, A., Alemany, M. & Palou, A. (1986). Tissue glycogen and lactate handling by the developing domestic fowl. *Comp. Biochem. Physiol.*, **85A**, 155–9.

Gardner, C.A. & Barald, K.F. (1991). The cellular environment controls the expression of *engrailed*-like protein in the cranial neuroepithelium of quail-chick embryos. *Development*, **113**, 1037–48.

Gardner, C.A. & Barald, K.F. (1992). Expression patterns of *engrailed*-like proteins in the chick embryo. *Dev. Dynamics*, **193**, 370–88.

Gasaway, W.C. (1976a). Cellulose digestion and metabolism by captive rock ptarmigan. *Comp. Biochem. Physiol.*, **54A**, 179–82.

Gasaway, W.C. (1976b). Seasonal variation in diet, volatile fatty acid production and size of the cecum of rock ptarmigan. *Comp. Biochem. Physiol.*, **53A**, 109–14.

Gasaway, W.C. (1976c). Volatile fatty acids and metabolizable energy derived from cecal fermentation in the willow ptarmigan. *Comp. Biochem. Physiol.*, **53A**, 115–21.

Gehring, W.J. (1992). The homeobox in perspective. *Trends Biochem. Sci.*, **17**, 277–80.

Geiger, B. & Ayalon, O. (1992). Cadherins. *Annu. Rev. Cell Biol.*, **8**, 307–32.

George, F.W. & Wilson, J.D. (1982). Developmental pattern of increased aromatase activity in the Sebright bantam chicken. *Endocrinology*, **110**, 1203–7.

George, R., Barber, D.L. & Schneider, W.J. (1987). Characterization of the chicken oocyte receptor for low and very low density lipoproteins. *J. Biol. Chem.*, **262**, 16838–47.

Gerondakis, S. & Bishop, J.M. (1986). Structure of the protein encoded by the chicken proto-oncogene c-*myb*. *Mol. Cell. Biol.*, **6**, 3677–84.

Giardina, B., Corda, M., Pellegrini, M.G. Condo, S.G. & Brunori, M. (1985). Functional properties of the hemoglobin system of two diving birds (*Podiceps nigricollis* and *Phalacrocorax carbosinensis*). *Mol. Physiol.*, **7**, 281–92.

Gill, K.G. (1989). *Ornithology*. pp. 243–58. New York: Freeman.

Ginsberg, B.H., Kahn, C.R. & Roth, J. (1976). The insulin receptor of the turkey erythrocyte: characterization of the membrane-bound receptor. *Biochim. Biophys. Acta*, **443**, 227–42.

Givol, I., Tsarfaty, I., Resau, J. *et al.* (1994). Bcl-2 expressed using a retroviral vector is localized primarily in the nuclear membrane and the endoplasmic reticulum of chicken embryo fibroblasts. *Cell Growth Diff.*, **5**, 419–29.

Glick, B. (1988). The bursa of Fabricius: development, growth, modulation and endocrine function. *CRC Crit. Rev. Poult. Biol.*, **1**, 107–32.

Glover, I.D., Barlow, D.J., Pitts, J.E. (1985) Conformational studies on the pancreatic polypeptide hormone family. *Eur. J. Biochem.*, **142**, 379–85.

Godovac-Zimmermann, J., Kösters, J., Braunitzer, G. & Göltenboth, R. (1988). Structural adaptation of bird hemoglobins to high-altitude respiration and the primary sequences of black-headed gull (*Larus ridibundus*, Charadriiformes) α^- and β/β'-chains. *Biol. Chem. Hoppe-Seyler*, **369**, 341–8.

Golden, A. & Brugge, J.S. (1988). The *src* gene. In *The Oncogene Handbook*, ed. E.P. Reddy, A.M. Skalka & T. Curran, pp. 149–73. Amsterdam: Elsevier.

Golden, S., Riesenfeld, G. & Katz, J. (1982). Carbohydrate metabolism of hepatocytes of starved Japanese quail. *Arch. Biochem. Biophys.*, **213**, 118–26.

Golub, E.S. & Green, D.R. (1991). *Immunology: A Synthesis*. Sunderland, USA: Sinauer.

Goodridge, A.G. (1968a). Citrate-cleavage enzyme, malic enzyme and certain dehydrogenases in embryonic and growing chicks. *Biochem. J.*, **108**, 663–6.

Goodridge, A.G. (1968b). Lipolysis in vitro in adipose tissue from embryonic and growing chicks. *Am. J. Physiol.*, **214**, 902–18.

Goodridge, A.G., Jenik, R.A., McDevitt, M.A., Morris, S.M. & Winberry, L.K. (1984). Malic enzyme and fatty acid synthase in the uropygial gland and liver of embryonic and neonatal ducklings – tissue specific regulation of gene expression. *Archs. Biochem. Biophys.*, **230**, 82–92.

Gorney, E. & Yom-Tov, Y. (1994). Fat, hydration condition, and moult of Steppe Buzzards *Buteo buteo vulpinus* on spring migration. *Ibis*, **136**, 185–92.

Goudswaard, J., van der Donk, J.A., van der Gaag, I. & Noordzij, A. (1979). Peculiar IgA transfer in the pigeon from mother to squab. *Dev. Comp. Immunol.*, **3**, 307–19.

Graber, G. & Baker, D.H. (1973). The essential nature of glycine and proline for growing chicks. *Poult. Sci.*, **52**, 892–6.

Grajal, A., Strahl, S.D., Parra, R., Dominguez, M.G. & Neher, A. (1989). Foregut fermentation in the hoatzin, a neotropical leaf-eating bird. *Science*, **245**, 1236–8.

Grajer, K.-H., Hörlein, A. & Igo-Kemenes, T. (1993). Hepatic nuclear factor 1α (HNF-1α) is expressed in the oviduct of hens and interacts with regulatory elements of the lysozyme gene. *Biol. Chem. Hoppe-Seyler*, **374**, 319–26.

Grant, J.A., Sanders, B. & Hood, L. (1971). Partial amino acid sequences of chicken and turkey immunoglobulin light chains. Homology with mammalian λ chains. *Biochemistry*, **10**, 3123–32.

Gray, M.W. (1989). Origin and evolution of mitochondrial DNA. *Annu. Rev. Cell Biol.*, **5**, 25–50.

Green, S. & Chambon, P. (1991). The oestrogen receptor: from perception to mechanism. In *Nuclear Hormone Receptors*, ed. M.G. Parker, pp. 15–38, London: Academic Press.

Greer, J.L. & Hargis, P.S. (1992). Cytosolic modulators of acyl-CoA: cholesterol acyltransferase (ACAT) activity in hepatic tissue of the chicken (*Gallus domesticus*). *Comp. Biochem. Physiol.*, **103B**, 9–13.

Gregg, K. & Rogers, G.E. (1986). Feather keratin: composition, structure and biogenesis. In *Biology of the Integument*, vol. 2, ed. J. Bereiter-Hahn, A.G. Matoltsy & K.S. Richards, pp. 666–94. Berlin: Springer Verlag.

Grewal, T., Thiesen, M., Borgmeyer, U. *et al.* (1992). The −6.1-kilobase chicken lysozyme enhancer is a multifactorial complex containing several cell-type-specific elements. *Mol. Cell. Biol.*, **12**, 2339–50.

Grey, H.M. (1967). Duck immunoglobulins. I. Structural studies on a 5.7S and 7.8S γ-globulin. *J. Immunol.*, **98**, 811–19.

Griffin, H.D. (1992). Manipulation of eggyolk cholesterol: a physiologist's view. *World's Poult. Sci. J.*, **48**, 101–12.

Griffin, H.D., Guo, K., Windsor, D. & Butterwith, S.C. (1992). Adipose tissue lipogenesis and fat deposition in leaner broiler chickens. *J. Nutr.*, **122**, 363–8.

Griffin, H.D. & Hermier, D. (1988). Plasma lipoprotein metabolism and fattening in poultry. In *Leanness in Domestic Birds*, ed.B. Leclercq & C.C. Whitehead, pp. 175–201. London: Butterworths.

Griffin, H.D., Perry, M.M. & Gilbert, A.B. (1984). Yolk Formation. In *Physiology and Biochemistry of the Domestic Fowl*, vol. 5, ed. B.M. Freeman, pp. 345–80. London: Academic Press.

Griffin, H.D., Windsor, D. & Zammit, V.A. (1990). Regulation of carnitine palmitoyltransferase I in chick liver. *Biochem. Soc. Trans.*, **18**, 981–2.

Grimes, H.L., Szente, B.E. & Goodenow, M.M. (1992). c-ski cDNAs are encoded by 8 exons, 6 of which are closely linked within the chicken genome. *Nucl. Acids Res.*, **20**, 1511–16.

Griminger, P. (1986). Lipid metabolism. In *Avian Physiology*, 4th edn, ed. P.D. Sturckie, pp. 345–58. New York: Springer Verlag.

Gröschel-Stewart, U. & Zuber, C. (1990). Is glycogen a major energy source in avian gizzard smooth muscle contraction. *Experientia*, **46**, 686–8.

Groscolas, R. (1978). Study of molt fasting followed by an experimental forced fasting in the emperor penguin *Aptenodytes forsteri*: relationship between feather growth, body weight loss, body temperature and plasma fuel levels. *Comp. Biochem. Physiol.*, **61A**, 287–95.

Groscolas, R. & Cherel, Y. (1992). How to molt while fasting in the cold: the metabolic and hormonal adaptations of Emperor and King penguins. *Ornis Scand.*, **23**, 328–34.

Groscolas, R. & Rodriguez, A. (1981). Glucose metabolism in fed and fasted emperor penguins (*Aptenodytes forsteri*). *Comp. Biochem. Physiol.*, **70**, 191–8.

Gross, D.S. & Garrard, W.T. (1988). Nuclease hypersensitive sites in chromatin. *Annu. Rev. Biochem.*, **57**, 159–97.

Gu, Z.-W., Xie, Y.-H., Yang, M. *et al.* (1993). Primary structure of Beijing duck apolipoprotein A-1. *J. Prot. Chem.*, **12**, 585–91.

Gudas, L.J. (1994). Retinoids and vertebrate development. *J. Biol. Chem.*, **269**, 15399–402.

Guha, B. & Ghosh, A. (1978). A cytomorphological study of the endocrine pancreas of some Indian birds. *Gen. Comp. Endocrinol.*, **34**, 38–44.

Guha, B. & Ghosh, A. (1992). Diabetes and avian resistance. *Curr. Sci.*, **62**, 564–8.

Gullberg, A., Tegelstrom, H. & Gelter, H.P. (1992). DNA fingerprinting reveals multiple paternity in families of Great and Blue Tits (*Parus major* and *P. caeruleus*). *Hereditas*, **117**, 103–8.

Gunawardana, V.K. (1977). Stages of spermatids in the domestic fowl: a light microscope study using araldite sections. *J. Anat.*, **123**, 351–60.

Gyllensten, U.B., Jakobson, S., Temrin, H. & Wilson, A.C. (1989). Nucleotide sequence and genomic organization of bird minisatellites. *Nucl. Acids Res.*, **17**, 2203–14.

Haberfield, A., Dunnington, E.A. & Siegel, P.B. (1992). Genetic distances estimated from DNA fingerprints in crosses of White Plymouth Rock chickens. *Animal Genet.*, **23**, 167–73.

Haché, R.J.G. & Deeley, R.G. (1988). Organization, sequence and nuclease sensitivity of repetitive elements flanking the chicken apoVLDLII gene: extended similarity to elements flanking the chicken vitellogenin gene. *Nucl. Acids Res.*, **16**, 97–113.

Hadjiisky, P., Hermier, D., Truffert, J., De Gennes, J.-L. & Grosgogeat, Y. (1993). Effect of pravastatin, a HMG CoA reductase inhibitor, on blood lipids and aortic lipidosis in cholesterol-fed White Carneau pigeons. *Biochim. Biophys. Acta*, **1181**, 279–86.

Hagerman, P.J. (1990). Sequence-directed curvature of DNA. *Annu. Rev. Biochem.*, **59**, 755–81.

Haggblom, L., Terwilliger, R.C. & Terwilliger, N.B. (1988). Changes in myoglobin and lactate dehydrogenase in muscles of a diving bird, the pigeon guillemot during maturation. *Comp. Biochem. Physiol.*, **91B**, 273–7.

Haldane, J.B.S. (1932). *The Causes of Evolution*. London: Longmans & Green. (Reprinted by Princeton University Press, Princeton, 1990.)

Hamada, Y., Yanagisawa, M., Katsuragawa, Y. *et al.* (1990). Distinct vascular and intestinal smooth muscle myosin heavy chain mRNA are encoded by a single-copy gene in the chicken. *Biochem. Biophys. Res. Commun.*, **170**, 53–8.

Hamburger, V. & Hamilton, H.L. (1951). A series of normal stages in the development of the chick embryo. *J. Morphol.*, **88**, 49–92. (Reprinted in 1993 in *Dev. Dynamics*, **195**, 231–72.)

Hamer, M.J. & Dickson, A.J. (1987). Developmental changes in hepatic fructose 2,6-bisphosphate content and phosphofructokinase-1 activity in the transition of chicks from embryonic to neonatal nutritional environment. *Biochem. J.*, **245**, 35–9.

Hamer, M.J. & Dickson, A.J. (1989). Influence of developmental stage on glycogenolysis and glycolysis in hepatocytes isolated from chick embryos and neonates. *Biochem. Soc. Trans.*, **17**, 1107–8.

Hamer, M.J. & Dickson, A.J. (1990). Control of glycolysis in cultured chick embryo hepatocytes. Fructose 2,6-bisphosphate content and phosphofructokinase-1 activity are stimulated by insulin and epidermal growth factor. *Biochem. J.*, **269**, 685–90.

Han, P.F. & Johnson, J. (1982). Fructose- 1,6-bisphosphatase from turkey liver. *Meth. Enzymol.*, **90**, 334–40.

Hanafusa, H. (1988). The fps/fes oncogene. In *The Oncogene Handbook*, ed. E.P. Reddy, A.M. Skalka & T. Curran, pp. 39–57. Amsterdam: Elsevier.

Hanotte, O., Burke, T., Armour, J.A.L. & Jeffreys, A.J. (1991). Hypervariable minisatellite DNA sequences in the Indian Peafowl *Pavo cristatus*. *Genomics*, **9**, 587–97.

Hanson, H.C. (1962). The dynamics of condition factors in Canada Geese and their relation to seasonal stresses. *Arctic Inst N. Am. Technical Paper*, **12**, 1–68.

Harata, M., Ouchi, K., Ohata, S., Kikuchi, A. & Mizuno, S. (1988). Purification and characterization of W-protein. *J. Biol. Chem.*, **263**, 13952–61.

Hardie, D.G. (1991). *Biochemical Messengers*. London: Chapman

& Hall.

Hardison, R.C. (1991). Evolution of globin gene families. In *Evolution at the Molecular Level*, ed. R.K. Selander, A.G. Clark & T.S. Whittam, pp. 272–90. Sunderland, MA: Sinauer.

Harel, R., Sharma, Y.D., Aguilera, A. *et al.* (1992). Cloning and developmental expression of the α3-chain of chicken type-IX collagen. *J. Biol. Chem.*, **267**, 10070–6.

Harold, F.M. (1986). *The Vital Force*. New York: Freeman.

Harris, I., Mizrahi, L., Ziv, T., Thomsen, G. & Mitrani, E. (1993). Identification of TGF-β-related genes in the early chick embryo. *Roux's Arch. Dev. Biol.*, **203**, 159–63.

Harris, J.R. (1983). Ultrastructure and biochemistry of the erythrocyte. In *Physiology and Biochemistry of the Domestic Fowl*, vol. 4, ed. B.M. Freeman, pp. 236–312. London: Academic Press.

Harris, L.L., Talian, J.C. & Zelenka, P.S. (1992). Contracting patterns of c-*myc* and N-*myc* expression in proliferating, quiescent, and differentiating cells of the embryonic chicken lens. *Development*, **115**, 813–20.

Harvey, A.L. & Marshall, I.G. (1986). Muscle. In *Avian Physiology*, 4th edn, ed. P.D. Sturkie, pp. 74–86. New York: Springer-Verlag.

Harvey, S., Scanes, C.G. & Brown, K.I. (1986). Adrenals. In *Avian Physiology*, 4th edn, ed. P.D. Sturkie, pp. 479–93. New York: Springer-Verlag.

Hasegawa, S., Terazono, K., Nata, K., Takada, T., Yamamoto, H. & Okamoto, H. (1990). Nucleotide sequence determination of chicken glucagon precursor cDNA: chicken preproglucagon does not contain glucagon-like peptide II. *FEBS Lett.*, **264**, 117–20.

Hastings, K.E.M., Bucher, E.A. & Emerson, C.P. (1985). Generation of troponin T isoforms by alternative RNA splicing in avian skeletal muscle. *J. Biol. Chem.*, **260**, 13699–703.

Hastings, K.E.M., Koppe, R.I., Marmor, E., Bader, D., Shimada, Y. & Toyota, N. (1991). Structure and developmental expression of troponin-I isoforms – cDNA clone analysis of avian cardiac troponin-I messenger RNA. *J. Biol. Chem.*, **266**, 19659–65.

Hathaway, D.R., Werth, D.K. & Haeberle, J.R. (1982). Limited autolysis reduces the Ca^{2+} requirement of smooth muscle Ca^{2+} activated protease. *J. Biol. Chem.*, **257**, 9072–7.

Hauser, H., Graf, T., Beug, H. *et al.* (1981). In *Haematology and Blood Transfusion*, vol. 26, ed. R. Neth, R.C. Gallo, T. Graf, K. Mannweiller, & K. Winkler, pp. 175–178. Berlin: Springer Verlag.

Haussler, M.R., Myrtle, J.F. & Norman, A.W. (1968). The association of a metabolite of vitamin D_3 with intestinal mucosa chromatin in vivo. *J. Biol. Chem.*, **243**, 4055–64.

Hayashi, K., Nimpf, J. & Schneider, W.J. (1989). Chicken oocytes and fibroblasts express different apolipoprotein-B-specific receptors. *J. Biol. Chem.*, **264**, 3131–9.

Hayashi, K., Tomita, Y., Maeda, Y., Shimagawa, Y., Inoue, K. & Hashizume, T. (1985). The rate of degradation of myofibrillar proteins of skeletal muscle in broiler and layer chickens estimated by N-methylhistidine in excreta. *Br. J. Nutr.*, **54**, 157–63.

Hayashi, S., Goto, K., Okada, T.S. & Kondoh, H. (1987). Lens-specific enhancer in the third intron regulates expression of the chicken δ1-crystallin gene. *Genes Dev.*, **1**, 818–28.

Hayman, A.R., Köppel, J. & Trueb, B. (1991). Complete structure of the chicken α2(VI) collagen gene. *Eur. J. Biochem.*, **197**, 177–84.

Hayman, M.J. (1987). erb-B: growth factor receptor turned oncogene. In *Oncogenes and Growth Factors*, ed. R.A. Bradshaw & S. Prentis, pp. 84–9. Amsterdam: Elsevier.

Hazelwood, R.L. (1984). Pancreatic hormones, insulin/glucagon ratios, and somatostatin as determinants of avian carbohydrate metabolism. *J. Exp. Zool.*, **232**, 647–52.

Hazelwood, R.L. (1986a). Pancreas. In *Avian Physiology*, 4th edn, ed. P.D. Sturkie, pp. 494–500. New York: Springer-Verlag.

Hazelwood, R.L. (1986b). Carbohydrate metabolism. In *Avian Physiology*, 4th edn, ed. P.D. Sturkie, pp. 303–25. New York: Springer-Verlag.

Hazelwood, R.L. (1993a). From avian pancreatic polypeptide to mammalian neuropeptides: carbohydrate metabolism implications. In *Avian Endocrinology*, ed. P.J. Sharp, pp. 189–200. Bristol: Society of Endocrinology.

Hazelwood, R.L. (1993b). The pancreaticpolypeptide (PP-fold) family: gastrointestinal, vascular, and feeding behavioral implications. *Proc. Soc. Exp. Biol. Med.*, **202**, 44–63.

Head, M.W., Peter, A. & Clayton, R.M. (1991). Evidence for the extralenticular expression of members of the β-crystallin gene family in the chick and a comparison with δ-crystallin during differentiation and transdifferentiation. *Differentiation*, **48**, 147–56.

Hebda, C.A. & Nowak, T. (1982). The purification, characterization and activation of phosphoenolpyruvate carboxykinase. *J. Biol. Chem.*, **257**, 5503–14.

Hecht, A., Berkenstamm, A., Strömstedt, P.-E., Gustafsson, J.-A. & Sippel, A.E. (1988). A progesterone responsive element maps to the far upstream steroid dependent DNAse hypersensitive site of the chicken lysozyme chromatin. *EMBO J.*, **7**, 2063–73.

Heidenreich, K.A. (1991). Insulin in the brain. What is its role? *Trends Endocrinol. Met.*, **2**, 9–12.

Heitlinger, E., Peter, M., Häner, M., Lustig, A., Aebi, U. & Nigg, E.A. (1991). Expression of chicken lamin B_2 in *Escherichia coli*: characterization of its structure, assembly, and molecular interactions. *J. Cell Biol.*, **113**, 485–95.

Helm-Bychowski, K.M. & Wilson, A.C. (1986). Rates of nuclear DNA evolution in pheasant-like birds: evidence from restriction maps. *Proc. Natl. Acad. Sci. USA*, **83**, 688–92.

Hendriks, W., Mulders, J.W.M., Bibby, M.A., Slingsby, C., Bloemendal, H. & de Jong, W.W. (1988). Duck lens-crystallin and lactate dehydrogenase B4 are identical: selective pressure from two distinct functions on a single copy gene. *Proc. Natl. Acad. Sci. USA*, **85**, 7114–18.

Hennekes, H., Peter, M., Weber, K. & Nigg, E.A. (1993). Phosphorylation on protein kinase C sites inhibits nuclear import of lamin B_2. *J. Cell Biol.*, **120**, 1293–304.

Herault, Y., Chatelain, G., Brun, G. & Michel, D. (1992). v-*src*-induced-transcription of the avian clusterin gene. *Nucl. Acids Res.*, **20**, 6377–83.

Hermier, D., Chapman, M.J. & Leclercq, B. (1984). Plasma lipoprotein profile in fasted and refed chickens selected for high or low adiposity. *J. Nutr.*, **114**, 1112–21.

Hermier, D., Forgez, P. & Chapman, M.J. (1985). A density gradient study of the lipoprotein and apolipoprotein distribution in the chicken, *Gallus domesticus*. *Biochim. Biophys. Acta*, **836**, 105–18.

Hermier, D., Salichon, M.R. & Whitehead, C.C. (1991). Relationships between plasma lipoproteins and glucose in fasted chickens selected for leanness or fatness by three criteria. *Rep. Nutr. Dev.*, **31**, 419–29.

Herschman, H.R. (1991). Primary response genes induced by growth factors and tumor promoters. *Annu. Rev. Biochem.*, **60**, 281–319.

Hershko, A. & Ciechanover, A. (1982). Mechanisms of intracellular protein breakdown. *Annu. Rev. Biochem.*, **51**, 335–64.

Hershko, A. & Ciechanover, A. (1992). The ubiquitin system for protein degradation. *Annu. Rev. Biochem.*, **61**, 761–807.

Hiebl, I., Braunitzer, G. & Schneeganss, D. (1987). The primary structures of the major and minor hemoglobin-components of adult Andean goose (*Chloephaga melanoptera*, Anatidae): the mutation leu→ser in position 55 of the β-chains. *Biol. Chem. Hoppe-Seyler*, **368**, 1559–69.

Hiebl, I., Schneeganss, D. & Braunitzer, G. (1986). The primary structures of the α^D-chains of the bar-headed goose (*Anser indicus*), the greylag goose (*Anser anser*) and the Canada goose (*Branta canadensis*). *Biol. Chem. Hoppe-Seyler*, **367**, 591–9.

Hiebl, I., Schneeganss, D., Grim, F., Kösters, J. & Braunitzer, G. (1987). The primary structure of the major and minor hemoglobin component of adult European black vulture (*Aegypius monachus*, Aegypiinae). *Biol. Chem. Hoppe-Seyler*, **368**, 11–8.

Hiebl, I., Weber, R.E., Schneeganss, D. & Braunitzer, G. (1989). The primary structures and functional properties of the major and minor hemoglobin components of the adult white-headed vulture (*Trigonoceps occiptalis*, Aegypiinae). *Biol. Chem. Hoppe-Seyler*, **370**, 699–706.

Hiebl, I., Weber, R.E., Schneeganss, D., Kösters, J. & Braunitzer, G. (1988). Structural adaptations in the major and minor hemoglobin components of adult Rüppell's griffon (*Gyps rueppellii*, Aegypiinae): a new molecular pattern for hypoxic tolerance. *Biol. Chem. Hoppe-Seyler*, **369**, 217–32.

Higgins, D.A., Shortridge, K.F. & Ng, P.L.K. (1987). Bile immunoglobulin of the duck (*Anas platyrhynchos*) – II. Antibody response in influenza A virus infections. *Immunol.*, **62**, 499–504.

Hill, R.W. & Wyse, G.A. (1989). *Animal Physiology*, 2nd edn, pp. 58–75. Cambridge, MA: Harper & Row.

Hillgartner, F.B., Williams, A.S., Flanders, J.A. Morin, D. & Hanson, R.J. (1981). Myofibrillar protein degradation in the chicken. *Biochem. J.*, **196**, 591–601.

Hinman, L.M., Ksiezak-Reding, H., Baker, A.C. & Blass, J.P. (1986). Pigeon liver phosphoprotein phosphatase: an effective activator of pyruvate dehydrogenase in tissue homogenates. *Arch. Biochem. Biophys.*, **246**, 381–90.

Hirai, H. & Varmus, H.E. (1990). Mutations in *src* homology regions 2 and 3 of activated chicken c-*src* that result in preferential transformation of mouse and chicken cells. *Proc. Natl. Acad. Sci. USA*, **87**, 8592–6.

Hiremath, L.S., Kessler, P.M., Sasaki, G.C. & Kolattukudy, P.E.

(1992). Estrogen induction of alcohol dehydrogenase in the uropygial gland of mallard ducks. *Eur. J. Biochem.*, **203**, 449–57.

Hodin, J. & Wistow, G. (1993). 5'Race PCR of messenger RNA for 3 taxon-specific crystallins – for each gene one promoter controls both lens and nonlens expression. *Biochem. Biophys. Res. Commun.*, **190**, 391–6.

Holmes, W.N., Al-Ghawas, S.C., Cronshaw, J. & Rohde, K.E. (1991). The structural organization and the steroidogenic responsiveness in vitro of adrenal gland tissue from the neonatal mallard duck (*Anas platyrhynchos*). *Cell Tissue Res.*, **263**, 557–66.

Holmes, W.N., Butler, D.G. & Phillips, J.G. (1961). Observations on the effects of maintaining glaucous-winged gulls (*Larus glaucescens*) on fresh water and sea water for long periods. *J. Endocrinol.*, **23**, 53–61.

Hood, L., Campbell, J.H., & Elgin, S.C.R. (1975). The organization, expression, and evolution of antibody genes and other multigene families. *Annu. Rev. Genet.*, **9**, 305–53.

Hoodless, P.A., Ryan, A.K., Schrader, T.J. & Deeley, R.G. (1992). Characterization of liver-enriched proteins binding to a developmentally demethylated site flanking the avian apoVLDLII gene. *DNA Cell Biol.*, **11**, 755–65.

Horwitz, J. & Piatigorsky, J. (1980). Evolutionary and developmental differences in δ-crystallins from birds and reptile lenses. *Biochim. Biophys. Acta*, **624**, 21–9.

Howard, A. & Pelc, S.R. (1953). Synthesis of desoxyribonucleic acid in normal and irradiated cells and its relation to chromosome breakage. *Heredity*, **6** (suppl.), 261–73.

Howard, R. & Moore, A. (1991). *A Complete Checklist of Birds of the World*, 2nd edn. London: Academic Press.

Hruz, P.W. & Miziorko, H.M. (1992). Avian 3-hydroxy-3-methylglutaryl-CoA lyase: sensitivity of enzyme activity to thiol/disulfide exchange and identification of provimal reactive cysteines. *Prot. Sci.*, **1**, 1144–53.

Huckaby, C.S., Conneely, O.M., Beattie, W.G., Dobson, A.D.W., Tsai, M.-J. & O'Malley, B.W. (1987). Structure of the chromosomal chicken progesterone receptor gene. *Proc. Natl. Acad. Sci. USA*, **84**, 8380–4.

Hughes, M., Sehgal, A., Hadman, M. & Bos, T. (1992). Heterodimerization with c-Fos is not required for cell transformation of chicken embryo fibroblasts by Jun. *Cell Growth Diff.*, **3**, 889–97.

Hughes, S.H., Stubblefield, E., Payvar, F. *et al.* (1979). Gene localization by chromosome fractionation: globin genes are on at least two chromosomes and three estrogen inducible genes on three chromosomes. *Proc. Natl. Acad. Sci. USA*, **76**, 1348–52.

Hulley, P.A., Stander, C.S. & Kidson, S.H. (1991). Terminal migration and early differentiation of melanocytes in embryonic chick skin. *Dev. Biol.*, **145**, 182–94.

Humphries, P.N. (1973). Behavioural and morphological changes in the Mallard and their relationship to plasma oestrogen concentrations. *J. Endocrinol.*, **58**, 353–4.

Husain, M.M., Kumar, A. & Mukhtar, H. (1984). Hepatic mixed-function oxidase activities of wild pigeons. *Xenobiotica*, **14**, 761–6.

Hutchison, R.E., Wozniak, A.W. & Hutchison, J.B. (1992).

Regulation of female brain aromatase activity during the reproductive cycle of the dove. *J. Endocrinol.*, **134**, 385–96.

Hynes, R.O. (1992). Integrins – versatility, modulation and signalling in cell adhesion. *Cell*, **69**, 1–25.

Inagami, T. (1989). Atrial natriuretic factor. *J. Biol. Chem.* **264**, 3043–6.

Iñarrea, P., Villareal, E., Moya, I., Aguirre, P. & Palacios, J. (1992). Development of hepatic activity of 3-hydroxy-3-methyl-glutaryl-coenzyme A reductase and cholesterol 7α-hydroxylase in young chick. *Comp. Biochem. Physiol.*, **103A**, 417–20.

Ingermann, R.L., Stock, M.K., Metcalfe, J. & Bissonnette, J.M. (1985). Monosaccharide uptake by erythrocytes of embryonic and adult chicken. *Comp. Biochem. Physiol.*, **80A**, 369–72.

Ingram, V.M. (1961). Gene evolution and the haemoglobins. *Nature*, **189**, 704–8.

Inoue, A., Yanagisawa, M. & Masaki, T. (1992). Differential tissue expression of multiple genes for chicken smooth muscle/nonmuscle myosin regulatory light chains. *Biochim. Biophys. Acta*, **1130**, 197–202.

Inui, A., Okita, M., Miura, M. *et al.* (1990). Characterization of the receptors for peptide-YY and avian pancreatic polypeptide in chicken and pig brains. *Endocrinology*, **127**, 934–41.

Iodice, A.A., Leong, V. & Weinstock, I.M. (1966). Separation of cathepsins A and D of skeletal muscle. *Arch. Biochem. Biophys.*, **117**, 477–86.

Iruela-Arispe, M. & Sage, E.H. (1991). Expresssion of Type VIII collagen during morphogenesis of the chicken and mouse heart. *Dev. Biol.*, **144**, 107–18.

Isaacks, R.E., Harkness, D.R., Sampsell, R.N., Adler, J.L., Kim, C.Y. & Goldman, P.H. (1976). Studies on avian erythrocyte metabolism-IV. Relationship between the major phosphorylated metabolic intermediates and oxygen affinity of whole blood in adults and embryos in several Galliformes. *Comp. Biochem. Physiol.*, **55A**, 29–33.

Isaacks, R.E, Harkness, D., Sampsell, R. *et al.* (1977). Studies on avian erythrocyte metabolism. *Eur. J. Biochem.*, **77**, 567–74.

Isaacks, R.E., Kim, C., Liu, H.L., Goldman, P., Johnson, A. & Harkness, D. (1983). Studies on avian erythrocyte metabolism. XIII. Changing organic phosphate composition in age-dependent density populations of chicken erythrocytes. *Poult. Sci.*, **62**, 1639–46.

Ishiura, S., Murofushi, H., Suzuki, K. & Imahori, K. (1978). Studies of a calcium-activated protease from chicken skeletal muscle. *J. Biochem.*, **84**, 225–30.

Ishiura, S., Tsuji, S., Murofushi, H. & Suzuki, K. (1982). Purification of an endogenous 68000-dalton inhibitor of Ca^{2+}–activated neutral protease from chicken skeletal muscle. *Biochim. Biophys. Acta*, **701**, 216–23.

Jacob, J. (1976). Bird waxes. In *Chemistry and Biochemistry of Natural Waxes*, ed. P.E. Kolattukudy, pp. 93–146. Amsterdam: Elsevier.

Jacob, J., Balthazart, J. & Schoffeniels, E. (1979). Sex differences in the chemical composition of uropygial gland waxes in domestic ducks. *Biochem. System. Ecol.*, **7**, 149–53.

Jacob, J. & Ziswiler, V. (1982). The uropygial gland. In *Avian Biology*, vol. 6, ed. D.S. Farner, J.R. King & K.C. Parkes, pp. 199–324. New York: Academic Press.

James, N.T. (1972). A study in the concentration and function of mammalian and avian myoglobin in type I skeletal muscle fibres. *Comp. Biochem. Physiol.*, **41B**, 457–60.

Jarvi, S.I. & Briles, W.E. (1992). Identification of the major histocompatibility complex in the ring-necked pheasant, *Phasianus colchicus. Animal Genet.*, **23**, 2110–20.

Jaussi, R., Behra, R., Giannattasio, S., Flura, T. & Christen, P. (1987). Expression of cDNAs encoding the precursor and the mature form of chicken mitochondrial aspartate aminotransferase in *Escherichia coli. J. Biol. Chem.*, **262**, 12434–7.

Jeffrey, I.W. & Pain, V.M. (1993). Stimulation by insulin of protein synthesis in cultured chick embryo fibroblasts. *Biochimie*, **75**, 791–6.

Jeffreys, A.J., Wilson, V. & Thein, S.L. (1985). Hypervariable 'minisatellite' regions in human DNA. *Nature*, **314**, 67–73.

Jeltsch, J.M., Krozowski, Z., Quirin-Stricker, C. *et al.* (1986). Cloning the chicken progesterone receptor. *Proc. Natl. Acad. Sci. USA*, **83**, 5424–8.

Jenni-Eiermann, S. & Jenni, L. (1992). High plasma triglyceride levels in small birds during migratory flight: a new pathway for fuel supply during endurance locomotion at very high mass-specific metabolic rates. *Physiol. Zool.*, **65**, 112–23.

Jensen, L.S. (1989). Relationship between protein and amino acid requirements of poultry. In *Proc. Georgia Nutr. Conf., Atlanta Georgia*. pp. 8–15.

Jérôme, V., Vourc'h, C., Baulieu, E.-E. & Catelli, M.-G. (1993). Cell cycle regulation of the chicken hsp90α expression. *Exp. Cell Res.*, **205**, 44–51.

Jessen, T.-H., Weber, R.E., Fermi, G., Tame, J. & Braunitzer, G. (1991). Adaptation of bird haemoglobins to high altitudes: demonstration of molecular mechanism by protein engineering. *Proc. Natl. Acad. Sci. USA*, **88**, 6519–22.

Johansen, K., Berger, M., Bicudo, J.E.P.W., Ruschi, A. & de Almeida, P.J. (1987). Respiratory properties of blood and myoglobin in hummingbirds. *Physiol. Zool.*, **60**, 269–78.

John, B. (1988). The biology of heterochromatin. In *Heterochromatin: Molecular and Structural Aspects*, ed. R.S. Verma, pp. 1–147, Cambridge: Cambridge University Press.

John, T.M., Pilo, B., Hazelwood, R.L. & George, J.C. (1989). Circulating levels of insulin and avian pancreatic polypeptide in pigeons under different experimental conditions. *Gen. Comp. Endocrinol.*, **75**, 71–7.

Johnson, P., Parkes, C. & Barrett, A.J. (1984). Characterization of calpains and inhibitors from chicken gizzard smooth–muscle. *Biochem. Soc. Trans.*, **12**, 1106–7.

Jones, M. (1991). Muscle protein loss in laying House Sparrows *Passer domesticus. Ibis*, **133**, 383–92.

Joshi, V.C. & Aranda, L.P. (1979). Hormonal regulation of the terminal enzyme of microsomal stearyl CoA desaturase in cultured avian liver explants. *J. Biol. Chem.*, **254**, 11779–82.

Jost, J.P., Saluz, H.P., McEwan, I. *et al.* (1990). Tissue specific expression of avian vitellogenin gene is correlated with DNA hypomethylation and *in vivo* specific-DNA interactions. *Philos. Trans. R. Soc. Lond*, **326B**, 231–40.

Juul-Madsen, H.R., Hedemand, J.E., Salamonsen, J. & Simonsen, M. (1993). Restriction fragment length polymorphism analysis of the chicken *B-F* and *B-L* genes and their association with serologically defined B haplotypes. *Animal Genet.*, **24**, 243–7.

Kajimoto, Y. & Rotwein, P. (1989). Structure and expression of a chicken insulin-like growth factor I precursor. *Mol. Endocrinol.*, **3**, 1907–13.

Kameda, K. & Goodridge, A.G. (1991). Isolation and partial characterisation of the gene for goose fatty acid synthase. *J. Biol. Chem.*, **266**, 419–26.

Kanayama, K., Wada, K., Negami, A., Yamamura, H. & Tanabe, T. (1985). Purification of protein phosphatase from hen oviduct. *FEBS Lett.*, **184**, 78–81.

Kaneko, J.J. (1980). *Clinical Biochemistry of Domestic Animals*, p. 27. New York: Academic Press.

Kang, C.W., Sunde, M.L. & Swick, R.W. (1985). Growth and protein turnover in the skeletal muscles of broiler chicks. *Poult. Sci.*, **64**, 370–9.

Karasawa, Y. & Maeda, M. (1994). Role of caeca in the nitrogen nutrition of the chicken fed on a moderate protein diet or a low protein diet plus urea. *Brit. Poult. Sci.*, **35**, 383–91.

Karmann, H., Rideau, N., Zorn, T., Malan, A. & Le Maho, Y. (1992). Early insulin response after food intake in geese. *Am. J. Physiol.*, **263**, R782–4.

Kashuba, V.I., Kavsan, V.M., Ryndich, A.V. *et al.* (1993). Complete nucleotide sequence of Rous sarcoma virus varient adapted to duck cells. *Mol. Biol.*, **27**, 269–78.

Katanuma, N. (1989). Possible regulatory mechanisms of intracellular protein catabolism through lysosyme. In *Intracellular Proteolysis: Mechanisms and Regulations*, ed. N. Katanuma & E. Kominami, pp. 3–26. Tokyo: Japanese Scientific Societies Press.

Kato, I., Kohr, W.J. & Laskowski, M. (1978). Evolution of ovomucoids. *Proc. 11th FEBS Meeting*, vol. 47, pp. 197–206. Oxford: Pergamon Press.

Kato, S., Itoh, S., Miura, Y., Naito, H. & Noguchi, T. (1990). Characterization of insulin receptors in primary cultures of quail (*Coturnix coturnix Japonica*) oviduct cells. The level of insulin receptor is regulated by steroid and peptide hormones. *Comp. Biochem. Physiol.*, **97B**, 783–91.

Kaufman, J., Salomonsen, J., Skjødt, K. & Thorpe, D. (1990). Size polymorphism of the chicken major histocompatibility complex-encoded B-G molecules is due to length variation in the cytoplasmic heptad repeat region. *Proc. Natl. Acad. Sci. USA*, **87**, 8277–81.

Kaufman, J., Skjødt, K. & Salomonsen, J. (1991). The B-G multgene family of the chicken major histocompatibility complex. *Crit. Rev. Immunol.*, **11**, 113–43.

Kawamoto, S. & Adelstein, R.S. (1991). Chicken nonmuscle myosin heavy-chains–differential expression of messenger-RNAs and evidence for 2 different polypeptides. *J. Cell Biol.*, **112**, 915–24.

Kawashima, S., Nomoto, M., Hayashi, M., Inomata, M., Nakamura, M. & Imahori, K. (1984). Comparison of calcium-activated neutral proteases from skeletal muscle of rabbit and chicken. *J. Biochem.*, **95**, 95–101.

Khochbin, S. & Wolffe, A.P. (1994). Developmentally regulated expression of linker-histone variants in vertebrates. *Eur. J. Biochem.*, **225**, 501–10.

Kim, J.W., Fletcher, D.L., Campion, D.R., Gaskins, H.R. & Dean, R.G. (1992). Effect of genetic background on the developmental expression of c-*fos* and c-*myc* in chicken. *Mol. Biol. Rep.*, **16**, 85–90.

Kim, R.Y. & Wistow, G.J. (1993). Expression of the duck α-enolase/TAU crystallin gene in transgenic mice. *FASEB J.*, **7**, 464–9.

Kimmel, J.R., Pollock, H.G. & Hazelwood, R.L. (1968). Isolation and characterisation of chicken insulin. *Endocrinology*, **83**, 1323–30.

Kimura, M. (1983). *The Neutral Theory of Molecular Evolution.* Cambridge: Cambridge University Press.

King, A.S. & McLelland, J. (1984). *Birds: Their Structure and Function.* London: Ballière Tindall.

Kiss, R.S., Ryan, R.O., Hicks, L.D., Oikawa, K. & Kay, C.M. (1993). Physical properties of apoprotein A-1 from chicken (*Gallus domesticus*). *Biochemistry*, **32**, 7872–8.

Kitamura, K., Takiguchi-Hayashi, K., Sezaki, M., Yamamoto, H. & Takeuchi, T. (1992). Avian crest cells express a melanogenic trait during early migration from the neural tube: observations with new monoclonal antibody, MEBL-1. *Development*, **114**, 367–78.

Klandorf, H., Clarke, B.L., Scheck, A.C. & Brown, J. (1986). Regulation of glucokinase in the domestic fowl. *Biochem. Biophys. Res. Commun.*, **139**, 1086–93.

Klasing, K.C. & Jarrell, V.L. (1985). Regulation of protein degradation in chick muscle by several hormones and metabolites. *Poult. Sci.*, **64**, 694–9.

Kline, K., Bacon, L.D., Dietert, R.R., Lillehoj, H.S., Morgan, T.J. & Sanders, B.G. (1991). MHC coded B-G homodimer and heterodimer heterogeneity among different chicken lines. *J. Hered.*, **82**, 31–6.

Knapp, L.W., Shames, R.B., Barnes, G.L. & Sawyer, R.H. (1993). Region-specific patterns of β-keratin expression during avian development. *Dev. Dynamics*, **196**, 283–90.

Knezetic, J.A. & Felsenfeld, G. (1989). Identification and characterization of a chicken α-globin enhancer. *Mol. Cell. Biol.*, **9**, 893–901.

Knezetic, J. A. & Felsenfeld, G. (1993). Mechanism of developmental regulation of απ, the embryonic α globin gene. *Mol, Cell Biol.*, **13**, 4632–9.

Kobayashi, K. & Hirai, H. (1980). Studies on the subunit components of chicken polymeric immunoglobulins. *J. Immunol.*, **124**, 1695–704.

Kodama, H., Saitoh, H., Tone, M., Kuhara, S., Sakaki, Y. & Mizuno, S. (1987). Nucleotide sequences and unusual electrophoretic behaviour of the W chromosome-specific repeating DNA units of the domestic fowl, *Gallus gallus domesticus*. *Chromosoma*, **96**, 18–25.

Koelkebeck, K.W., Baker, D.H., Han, Y. & Parsons, C.M. (1991). Effect of excess lysine, methionine, threonine or tryptophan on production performance of laying hens. *Poult. Sci.*, **70**, 1651–3.

Kohen, R., Yamamoto, Y., Cundy, K.C. & Ames, B.N. (1988). Antioxidant activity of carnosine, homocarnosine and anserine present in muscle and brain. *Proc. Natl. Acad. Sci. USA*, **85**, 3175–9.

Kohler, C.D., Schaadt, C.P. & Vekemans, M.J.J. (1989). Chromosomal banding studies of the osprey *Pandion haliaetus*

(Aves, Falconiformes). *Genome*, **32**, 1037–40.

Koizumi, I., Suzuki, Y. & Kaneko, J.J. (1991). Studies on the fatty acid composition of intramuscular lipids of cattle, pigs and birds. *J. Nutr. Sci. Vit.*, **37**, 545–54.

Kok, K., Snippe, L., AB, G. & Gruber, M. (1985). Nuclease-hypersensitivity in chromatin of the estrogen-inducible apoVLDLII gene of the chicken. *Nucl. Acids Res.*, **13**, 5189–202.

Kolattukudy, P.E. & Rogers, L. (1987). Biosynthesis of 3-hydroxy fatty acids, the pheromone components of female mallard ducks, by cell-free preparations from the uropygial gland. *Arch. Biochem. Biophys.*, **252**, 121–9.

Kolattukudy, P.E., Rogers, L. & Flurkey, W. (1985). Suppression of a thioesterase gene expression and the disappearance of short chain fatty acids in the preen gland of the Mallard duck during eclipse, the period following postnuptial molt. *J. Biol. Chem.*, **260**, 10789–93.

Komiyama, T., Bigler, T.L., Yoshida, N., Noda, K. & Laskowski, M. (1991). Replacement of P1 Leu18 by Glu18 in the reactive site of turkey ovomucoid third domain converts it into a strong inhibitor of glu-specific *Streptomyces-griseus* proteinase (GLUSGP). *J. Biol. Chem.*, **266**, 10727–30.

Kondo, T., Arakawa, H., Kitao, H., Hirota, Y. & Yamagishi, H. (1993). Signal joint immunoglobulin Vλ1-Jλ and novel joints of chimeric V pseudogenes on extrachromosomal circular DNA from chicken bursa. *Eur. J. Immunol.*, **23**, 245–9.

Kono, N. & Uyeda, K. (1973). Chicken liver phosphofructokinase. *J. Biol. Chem.*, **248**, 8592–602.

Kornberg, A. (1989). *For the Love of Enzymes: The Odyssey of a Biochemist*. Cambridge, MA: Harvard University Press.

Kornberg, R.D. & Lorch, Y. (1992). Chromatin structure and transcription. *Annu. Rev. Cell Biol.*, **8**, 563–87.

Kouba, M., Bernard-Griffiths, M.-A. & Lemarchal, P. (1993). Liver stearyl-CoA desaturase activity and fatness in birds. *In vitro* studies in the growing turkey and chicken. *Comp. Biochem. Physiol.*, **105A**, 359–62.

Kouba, M., Catheline, D. & Leclercq, B. (1992). Lipogenesis in turkeys and chickens: a study of body composition and liver lipogenic enzyme activities. *Br. Poult. Sci.*, **33**, 1003–14.

Kovács, K.J., Westphal, H.M. & Péczely, P. (1989). Distribution of glucocorticoid receptor-like immunoreactivity in the brain, and its relation to CRF and ACTH immunoreactivity in the hypothalamus of the japanese quail, *Coturnix coturnix japonica*. *Brain Res.*, **505**, 239–45.

Kowalczyk, K., Daiss, J., Halpern, J & Roth, T.F. (1985). Quantitation of maternal-fetal IgG transport in the chicken. *Immunology*, **54**, 755–62.

Kraft, H.J., Hendriks, W., de Jong, W.W., Lubsen, N.H. & Schoenmakers, J.G.G. (1993). Duck lactate dehydrogenase B/ε-crystallin gene: lens recruitment of a GC-promoter. *J. Mol. Biol.*, **229**, 849–59.

Krebs, H.A. & Henseleit, K. (1932). Untersuchungen über die Harnstoffbildung in Tierkörper. *Hoppe-Seyler Z. Physiol. Chem.*, **210**, 33–43.

Krebs, H.A. & Lowenstein, J.M. (1960). The tricarboxylic acid cycle. In *Metabolic Pathways*, vol. 1, ed. D.M. Greenberg, pp. 129–203. New York: Academic Press.

Krek, W. & Nigg, E.A. (1989). Structure and developmental expression of the chicken *CDC2* kinase. *EMBO J.*, **8**, 3071–8.

Krementz, D.G. & Ankney, C.D. (1988). Changes in lipid and protein reserves and in diet of breeding house sparrows. *Canad. J. Zool.*, **66**, 950–6.

Kretsovali, A., Müller, M.M., Weber, F. *et al.* (1987). A transcriptional enhancer located between adult β-globin and embryonic ε-globin genes in chicken and duck. *Gene*, **58**, 167–75.

Kroemer, G., Bernot, A., Behar, G. *et al.* (1990). Molecular genetics of the chicken MHC: current status and evolutionary aspects. *Immunol. Rev.*, **113**, 119–45.

Kruski, A.W. & Scanu, A.M. (1975). Properties of rooster serum high density lipoproteins. *Biochim. Biophys. Acta*, **409**, 26–38.

Krust, A., Green, S., Argos, P. *et al.* (1986). The chicken oestrogen receptor sequence: homology with v-*erb*A and the human oestrogen and glucocorticoid receptors. *EMBO J.*, **5**, 891–7.

Kudzma, D.J., Swaney, J.B. & Ellis, E.N. (1979). Effects of estrogen administration on the lipoproteins and apoproteins of the chicken. *Biochim. Biophys. Acta*, **572**, 257–68.

Kuhnlein, U. & Zadworny, D. (1990). Molecular aspects of poultry breeding. *Proc. 4th World Congr. Genet. Appl. Livestock Prod.*, **16**, 21–30.

Kuhnlein, U., Zadworny, D., Dawe, Y., Fairfull, R.W. & Gavora, J.S. (1990). Assessment of inbreeding by DNA fingerprinting: development of a calibration curve using defined strains of chickens. *Genetics*, **125**, 161–5.

Kunnas, T.A., Wallén, M.J. & Kulomaa, M.S. (1993). Induction of chicken avidin and related mRNAs after bacterial infection. *Biochim. Biophys. Acta*, **1216**, 441–5.

Kutlu, H.R. & Forbes, J.M. (1993). Changes in growth and blood parameters in heat-stressed broiler chicks in response to dietary ascorbic acid. *Livestock Prod. Sci.*, **36**, 335–50.

Lackner, R., Challis, R.A.J., West, D. & Newsholme, E.A. (1984). A problem in the radiochemical assay of glucose 6-phosphatase in muscle. *Biochem. J.*, **218**, 649–51.

Lamont, S.J. (1993). The major histocompatibility complex in chickens. In *Manipulation of the Avian Genome*, ed. R.J. Etches & A.M.V. Gibbins, pp. 185–203, Boca Raton, FL: CRC Press.

Lamont, S.J. & Dietert, R.R. (1990). Immunogenetics, In *Poultry Breeding and Genetics*, ed. R.D. Crawford, pp. 497–541, Amsterdam: Elsevier.

Lamont, S.J., Gerndt, B.M., Warner, C.M. & Bacon, L.D. (1990). Analysis of restriction length polymorphisms of the major histocompatibility complex of 15I$_5$B-congenic chicken lines. *Poult. Sci.*, **69**, 1195–203.

Lamont, S.J., Warner, C.M. & Nordskog, A.W. (1987). Molecular analysis of the chicken major histocompatibility complex gene and gene products. *Poult. Sci.*, **66**, 819–24.

Landon, E.J. & Carter, C.E. (1960). The preparation, properties, and inhibition of hypoxanthine dehydrogenase of avian kidney. *J. Biol. Chem.*, **235**, 819–24.

Landschulz, W.H., Johnson, P.F. & McKnight, S.L. (1988). The leucine zipper: a hypothetical structure common to a new class of DNA binding proteins. *Science*, **240**, 1759–64.

Landsteiner, K. & Miller, C.P. (1924). On individual differences in chicken blood. *Proc. Soc. Exp. Biol. Med.*, **22**, 100–2.

Langelier, M., Connelly, P. & Subbiah, M.T.R. (1975). Plasma lipoprotein profile and composition in White Carneau and Show Racer breeds of pigeons. *Canad. J. Biochem.*, **54**, 27–31.

Langslow, D.R. (1978). Gluconeogenesis in birds. *Biochem. Soc. Trans.*, **6**, 1148–52.

Langslow, D.R., Kimmel, J.R. & Pollock, H.G. (1973). Studies of the distribution of a new avian pancreatic polypeptide and insulin among birds, reptiles, amphibians and mammals. *Endocrinology*, **93**, 558–65.

Lapennas, G.N. & Reeves, R.B. (1983). Oxygen affinity of blood of adult domestic fowl and red jungle fowl. *Resp. Physiol.*, **52**, 27–39.

Larrabee, M.G. (1989). The pentose cycle (hexose monophosphate shunt). *J. Biol. Chem.*, **264**, 15875–9.

Larsson, A., Balow, R.-M., Lindahl, T.L. & Forsberg, P.-O. (1993). Chicken antibodies: taking advantage of evolution – a review. *Poult. Sci.*, **72**, 1807–12.

Laskowski, M., Apostol, I., Ardelt, W. *et al.* (1990). Amino acid sequences of ovomucoid third domain from 25 additional species of birds. *J. Prot. Chem.*, **9**, 715–26.

Laskowski, M., Kato, I., Ardelt, W. *et al.* (1987). Ovomucoid third domain from 100 avian species: isolation, sequences, and hypervariability of enzyme inhibitor contact residues. *Biochemistry*, **26**, 202–21.

Latshaw, J.D. (1993). Dietary lysine concentrations from deficient to excessive and the effects on broiler chicks. *Br. Poult. Sci.*, **34**, 951–8.

Lauderdale, J.D. & Stein, A. (1992). Introns of the chicken ovalbumin gene promote nucleosome alignment *in vitro*. *Nucl. Acids Res.*, **20**, 6589–96.

Laurin, D.E. & Cartwright, A.L. (1993). Factors affecting insulin responsiveness of triglyceride synthesis in incubated chicken hepatocytes. *Proc. Soc. Exp. Biol. Med.*, **202**, 476–81.

Lavin, S., Cuenca, R., Marco, I., Velarde, R. & Vinas, L. (1992). Hematology and blood chemistry of the marsh harrier (*Circus aeruginosus*). *Comp. Biochem. Physiol.*, **103A**, 493–5.

Law, D.J. & Titball, J.G. (1992). Identification of a putative collagen-binding protein from chicken skeletal muscle as glycogen phosphorylase. *Biochim. Biophys. Acta*, **1122**, 225–33.

le Douarin, N.M. (1993). Embryonic chimaeras in the study of brain development. *Trends Neurosci.*, **16**, 64–72.

le Maho, Y., Vu van Kha, H., Koubi, H. *et al.* (1981). Body composition, energy expenditure, and plasma metabolites in long-term fasting geese. *Am. J. Physiol.*, **241**, E342–54.

Lebacq-Verheyden, A.-M., Vaerman, J.-P. & Heremans, J.F. (1974). Quantification and distribution of chicken immunoglobulins IgA, IgM and IgG in serum and secretions. *Immunol.*, **27**, 683–92.

Leclercq, B. (1984). Adipose tissue metabolism and its control in birds. *Poult. Sci.*, **63**, 2044–54.

Lee, D.C., Kim, R.Y. & Wistow, G.J. (1993). An avian αB-crystallin: non-lens expression and sequence similarities with both small (hsp27) and large (hsp70) heat shock proteins. *J. Mol. Biol.*, **232**, 1221–6.

Lefkowitz, R.J. & Caron, M.G. (1988). Adrenergic receptors. *J. Biol. Chem.*, **263**, 4993–6.

Legrand, P. & Lemarchal, P. (1992). Stearyl-CoA desaturase activity and triglyceride secretion in isolated and cultured hepatocytes from genetically lean and fat lines. *Comp. Biochem. Physiol.*, **102B**, 371–5.

Lehner, C.F., Stick, R., Eppenberger, H.M. & Nigg, E.A. (1987). Differential expression of nuclear lamin proteins during chicken development. *J. Cell Biol.*, **105**, 577–87.

LeMeur, M., Glanville, N., Mandel, J.L., Gerlinger, P., Palmiter, R. & Chambon, P. (1981). The ovalbumin gene family: hormonal control of the X and Y gene transcription and mRNA accumulation. *Cell*, **23**, 561–71.

Leshin, M., Baron, J., George, F.W. & Wilson, J.D. (1981). Increased estrogen formation and aromatase activity in fibroblasts cultured from the skin of chickens with henny feathering trait. *J. Biol. Chem.*, **256**, 4341–4.

Lewin, B. (1994). *Genes V*. Oxford: Oxford University Press.

Li, L., Pan, L. & Xu, G.-J. (1991). Properties of chicken liver phosphofructokinase-2. *Science in China*, **34**, 916–22.

Li, W.-H. & Graur, D. (1991). *Fundamentals of Molecular Evolution*. Sunderland, MA: Sinauer.

Li, X., Zelenka, P.S. & Piatigorsky, J. (1993). Differential expression of the two δ-crystallin genes in lens and non-lens tissues: shift favouring δ_2 expression from embryonic to adult chickens. *Dev. Dynamics*, **196**, 114–23.

Libri, D., Mouly, V., Lemonnier, M. & Fiszman, M.Y. (1990). A nonmuscle tropomyosin is encoded by the smooth/skeletal β-tropomyosin gene and its RNA is transcribed from an internal promoter. *J. Biol. Chem.*, **265**, 3471–3.

Libri, D., Piseri, A. & Fiszman, M.Y. (1991). Tissue-specific splicing *in vivo* of the β-tropomyosin gene-dependence on an RNA secondary structure. *Science*, **252**, 1842–5.

Lila, D., Susana, Z. & Ricardo, S. (1994). Induction of a calbindin-D_{9K}-like protein in avian muscle cells by 1,25-dihydroxy-vitamin D_3. *Biochem. Mol. Biol. Int.*, **32**, 859–67.

Lillie, F.R. (1919). *The Development of the Chick*. New York: Holt.

Lin, C.-W. & Chiou, S.-H. (1992). Sequence analysis of pigeon δ-crystallin gene and its deduced primary structure: comparison of δ crystallin with and without endogenous argininosuccinate lyase activity. *FEBS Lett.*, **311**, 276–80.

Linares, A., Caamaño, G.J., Diaz, R., Gonzalez, F.J. & Garcia-Pelegrin, E. (1993). Changes in ketone body utilization by chick liver, duodenal mucosa and kidney during embryonic and postnatal development. *Comp. Biochem. Physiol.*, **105B**, 277–82.

Lindquist, S. (1986). The heat-shock response. *Annu. Rev. Biochem.*, **55**, 1151–91.

Linsenmayer, T.F., Gibney, E., Igoe, F. *et al.* (1993). Type V collagen: molecular structure and fibrillar organization of the chicken α1(V) NH_2-terminal domain, a putative regulator of corneal fibrillogenesis. *J. Cell Biol.*, **121**, 1181–9.

Lipstein, B., Boer, P. & Sperling, O. (1978). Regulation of *de novo* purine synthesis in chick liver: role of phosphoribosyl-pyrophosphate availability and of salvage purine nucleotide biosynthesis. *Biochim. Biophys. Acta*, **521**, 45–54.

Liu, L.-F., Wu, S.-H. & Tam, M.F. (1993). Nucleotide sequence of class-α glutathione *S*-transferases from chicken liver. *Biochim. Biophys. Acta*, **1216**, 332–4.

Liu, S.S. & Higgins, D.A. (1990). Yolk-sac transmission and post-hatching ontogeny of serum immunoglobulins in the duck (*Anas platyrhynchos*). *Comp. Biochem. Physiol.*, **97B**, 637–44.

Liu, W.-H., Means, G.E. & Feeney, R.E. (1971). The inhibitory properties of avian ovoinhibitors against proteolytic enzymes. *Biochim. Biophys. Acta*, **229**, 176–85.

Lois, R. & Martinson, H.G. (1989). Chicken globin gene-transcription is cell lineage specific during the time of the switch. *Biochemistry*, **28**, 2281–7.

Lonberg, N. & Gilbert, W. (1983). Primary structure of chicken muscle pyruvate kinase mRNA. *Proc. Natl. Acad. Sci. USA*, **80**, 3661–5.

Lonberg, N. & Gilbert, W. (1985). Intron/exon structure of the chicken pyruvate kinase gene. *Cell*, **40**, 81–90.

Longmire, J.L., Ambrose, R.E., Brown, N.C. *et al.* (1991). Use of sex-linked minisatellite fragments to investigate genetic differentiation and migration of North American populations of peregrine falcon (*Falco peregrinus*). In *DNA Fingerprinting: Approaches and Applications*, ed. T. Burke, G. Dolf, A.J. Jeffreys & R. Wolff, pp. 217–229. Basel: Birkhauser Verlag.

Lucas, P.C. & Granner, D.K. (1992). Hormone response domains in gene transcription. *Annu. Rev. Biochem.*, **61**, 1131–73.

Luhtala, M., Salamonsen, J., Hirota, Y., Onodera, T., Toivanen, P. & Vainio, O. (1993). Analysis of chicken CD4 by monoclonal antibodies indicates evolutionary conservation between avian and mammalian species. *Hybridoma*, **12**, 633–45.

Lumsden, A., Sprawson, N. & Graham, A. (1991). Segmental origin and migration of neural crest cells in the hindbrain region of the chick embryo. *Development*, **113**, 1281–91.

Lundgren, B.O. (1988). Catabolic enzyme activities in the pectoralis muscle of migratory and non-migratory Goldcrests, Great Tits, and Yellowhammers. *Ornis Scand.*, **19**, 190–4.

Lundgren, B.O. & Kiessling, K.H. (1985). Seasonal variation in catabolic enzyme activities in breast muscle of some migratory birds. *Oecologia* (Berlin), **66**, 468–71.

Lundgren, B.O. & Kiessling, K.H. (1986). Catabolic enzyme activities in pectoralis muscle of premigratory and migratory Reed Warblers *Acrocephalus scirpaceus* (Herm.). *Oecologia* (Berlin), **68**, 529–32.

Lutz, P.L., Longmuir, I.S. & Schmidt-Nielsen, K. (1974). Oxygen affinity of bird blood. *Resp. Physiol.*, **20**, 325–30.

Luxembourg, A., Hekman, M. & Ross, E.M. (1991). Turkey erythrocyte β-adrenergic-receptor-selective proteolysis of both domains. *FEBS Lett.*, **283**, 155–8.

MacGregor, J.S., Annamalai, A.E., van Tol, A., Black, W.J. & Horecker, B.L. (1982). Fructose-1,6-bisphosphatase from chicken and rabbit muscle. *Meth. Enzymol.*, **90**, 340–5.

Mack, A.L., Gill, F.B., Colburn, R. & Spolsky, C. (1986). Mitochondrial DNA: a source of genetic markers for the studies of similar passerine species. *Auk*, **103**, 676–81.

Mackem, S. & Mahon, K.A. (1991). Ghox 4.7: a chick homeobox gene expressed primarily in limb buds with limb-type differences in expression. *Development*, **112**, 791–806.

Macleod, K., LePrince, D. & Stehelin, D. (1992). The *ets* gene family. *Trends Biochem. Sci.*, **17**, 251–6.

Maeda, Y., Hayashi, K., Hashiguchi, T. & Okamoto, S. (1986). Genetic studies on the muscle protein turnover rate of Coturnix quail. *Biochem. Genet.*, **24**, 207–16.

Maeda, Y., Matsuoka, S., Furuichi, N., Hayashi, K. & Hashiguchi, T. (1987). The effect of the *dw* gene on the muscle protein turnover rate in chickens. *Biochem. Genet.*, **25**, 253–8.

Maeda, Y., Okamoto, S. & Hashiguchi, T. (1992). Comparison of the rates of muscle protein metabolism between the domestic and the wild Coturnix quail. *Biochem. Genet.*, **30**, 503–6.

Maggini, S., Stoecklintschan, F.B., Morikoferzwez, S. & Walter, P. (1992). New kinetic-parameters for rat-liver arginase measured at near-physiological steady-state concentrations of arginine and Mn^{2+}. *Biochem. J.*, **283**, 653–60.

Maginniss, L.A. (1985). Red cell organic phosphates and Bohr effects in house sparrow blood. *Resp. Physiol.*, **59**, 93–103.

Magor, K.E., Warr, G.W., Middleton, D., Wilson, M.R. & Higgins, D.A. (1992). Structural relationship between the two IgY of the duck, *Anas platyrhynchos*: molecular genetic evidence. *J. Immunol.*, **149**, 2627–33.

Maki, Y., Bos, T.J., Davis, C., Starbuck, M. & Vogt, P.K. (1987). Avian sarcoma virus 17 carries the *jun* oncogene. *Proc. Natl. Acad. Sci. USA*, **84**, 2848–52.

Malcarney, H.L., Martinez del Rio, C. & Apanius, V. (1994). Sucrose intolerance in birds: simple nonlethal diagnostic methods and consequences for assimilation of complex carbohydrates. *Auk*, **111**, 170–7.

Mannervik, B. (1985). The isoenzymes of glutathione transferase. *Adv. Enzymol. Mol. Biol.*, **57**, 357–417.

Manning, M.J. & Turner, R.J. (1976). *Comparative Immunobiology*. Glasgow: Blackie.

Mansikka, A. (1992). Chicken IgA H chains: implications concerning the evolution of H chain genes. *J. Immunol.*, **149**, 855–61.

Mansikka, A., Jalkanen, S., Sandberg, M., Granfors, K., Lassila, O. & Toivanen, P. (1990). Bursectomy of chicken embryos at 60 hours of incubation leads to an oligoclonal B cell compartment and restricted Ig diversity. *J. Immunol.*, **145**, 3601–9.

Mapes, J.P. & Krebs, H.A. (1978). Rate-limiting factors in urate synthesis and gluconeogenesis in avian liver. *Biochem. J.*, **172**, 193–203.

Marchand, M.J., Maisin, L., Hue, L. & Rousseau, G.G. (1992). Activation of 6-phosphofructo-2-kinase by pp60$^{v\text{-}src}$ is an indirect effect. *Biochem. J.*, **285**, 413–17.

Marcu, K.B., Bossone, S.A. & Patel, A.J. (1992). *myc* function and regulation. *Annu. Rev. Biochem.*, **61**, 809–60.

Maresh, C.G., Kwan, T.H. & Kalman, S.M. (1969). Carbamyl phosphate synthetase in the chick. *Canad. J. Biochem.*, **47**, 61–3.

Maridor, G. & Nigg, E.A. (1990). cDNA sequences of chicken nucleolin/C23 and NO38/B23, two major nucleolar proteins. *Nucl. Acids Res.*, **18**, 1286.

Marsh, R.L. (1981). Catabolic enzyme activities in relation to premigratory fattening and muscle hypertrophy in the gray catbird (*Dumetella carolinensis*). *J. Comp. Physiol.*, **141**, 417–23.

Marsh, R.L. & Dawson, W.R. (1982). Substrate metabolism in seasonally acclimatized American goldfinches. *Am. J. Physiol.*, **242**, R563–9.

Marsh, R.L., Dawson, W.R., Camilliere, J.J. & Olson, J.M. (1990). Regulation of glycolysis in the pectoralis muscles of seasonally acclimatized American goldfinches exposed to cold. *Am. J. Physiol.*, **258**, R711–7.

Marston, S.B. & Redwood, C.S. (1991). The molecular anatomy of caldesmon. *Biochem. J.*, **279**, 1–16.

Martin, G.R. & Timpl, R. (1987). Laminin and other basement membrane components. *Annu. Rev. Cell Biol.*, **3**, 57–85.

Martin, G.S. (1970). Rous sarcoma virus: a function required for the maintainence of the transformed state. *Nature*, **227**, 1021–3.

Martin, S.J., Green, D.R. & Cotter, T.G. (1994). Dicing with death: dissecting the components of the apoptosis machinery. *Trends Biochem. Sci.*, **19**, 26–30.

Martinerie, C., Viegas-Pequignot, E., Guenard, I. *et al.* (1992). Physical mapping of human loci homologous to the chicken *nov* proto-oncogene. *Oncogene*, **7**, 2529–34.

Martinez, A., Lopez, J. & Sesma, P. (1993). Development of the diffuse endocrine system in the chicken proventriculus. *Cell Tissue Res.*, **271**, 107–13.

Martinez del Rio, C. (1990). Dietary, phylogenetic, and ecological correlates of intestinal sucrase and maltase activity in birds. *Physiol. Zool.*, **63**, 987–1011.

Maruyama, K., Sunde, M.L. & Swick, R.W. (1978). Growth and muscle turnover in the chick. *Biochem. J.*, **176**, 573–82.

Mason, S.L. & Ward, L.C. (1981). Branched chain amino acid metabolism in two avian species: *Coturnix coturnix Japonica* and *Gallus domestica*. *Comp. Biochem. Physiol.*, **69B**, 265–72.

Mastellar, E.L. & Thompson, C.B. (1994). B cell development in the chicken. *Poult. Sci.*, **73**, 998–1011.

Mathew, A., Grdisa, M. & Johnstone, R.M. (1993). Nucleosides and glutamine are primary energy substrates for embryonic and adult chicken red cells. *Biochem. Cell Biol.*, **71**, 288–95.

Mathews, C.K. & van Holde, K.E. (1990). *Biochemistry*. New York: Benjamin/Cummings.

Mathieu-Costello, O. (1993). Comparative aspects of muscle capillary supply. *Annu. Rev. Physiol.*, **55**, 503–25.

Matsumine, H., Herbst, M.A., Ou, S., Wilson, J.D. & McPhaul, M.J. (1991). Aromatase messenger RNA in the extragonadal tissues of chickens with the henny-feathering trait is derived from a distinctive promoter structure that contains a segment of a retroviral long terminal repeat – functional-organization of the Sebright, Leghorn, and Campine aromatase genes. *J. Biol. Chem.*, **266**, 19900–7.

Matsuo, I. & Yasuda, K. (1992). The cooperative interaction between 2 motifs of an enhancer element of the chicken αA-crystallin gene, α-CE1 and α-CE2, confers lens-specific expression. *Nucl. Acids Res.*, **20**, 3701–12.

Matthews, D.M. (1975). Intestinal absorption of peptides. *Physiol. Rev.*, **55**, 537–608.

Matzke, M.A., Varga, F., Berger, H. *et al.* (1990). A 41–42 bp tandemly repeated sequence isolated from nuclear envelopes of chicken erythrocytes is located predominantly on microchromosomes. *Chromosoma*, **99**, 131–7.

Maurice, D.V., Lightsey, S.F., Kuo-Tung, H. & Rhoades, J.F. (1991). Comparison of glutathione *S*-transferase activity in the rat and birds: tissue distribution and rhythmicity in chicken (*Gallus domesticus*) liver. *Comp. Biochem. Physiol.*, **100B**, 471–4.

Mavilio, F. (1993). Regulation of vertebrate homeobox-containing genes by morphogens. *Eur. J. Biochem.*, **212**, 273–88.

Mayne, R., Sanderson, R.D., Wiedemann, H., Fitch, J.M. & Linsenmayer, T.F. (1983). The use of monoclonal-antibodies to fragments of chicken type-IV collagen in structural and localization studies. *J. Biol. Chem.*, **258**, 5794–7.

Mayr, B., Lambrou, M., Kalat, M., Schleger, W. & Bigelbach, A. (1990). Characterization of heterochromatin by sequential counterstain-enhanced fluorescence in three domestic species: *Meleagris gallopavo*, *Columba livia domestica*, and *Anser anser* L. *J. Hered.*, **81**, 468–75.

Mayr, B., Lambrou, M. & Schleger, W. (1989). Further resolution of the quail karyotype and characterization of micro-chromosomes by counterstain-enhanced fluorescence. *J. Hered.*, **80**, 147–50.

McCaffrey, J. & Hamilton, J.W. (1994). Developmental regulation of basal and hormone-inducible phosphoenolpyruvate carboxykinase gene expression in chick embryo liver *in vivo*. *Arch. Biochem. Biophys.*, **309**, 10–17.

McCarrey, J.B., Abplanalp, H. & Abbott, U.K. (1981). Studies in the H-W (H-Y) antigen in chickens. *J. Hered.*, **72**, 169–71.

McCormack, W.T., Carlson, L.M., Tjoelker, L.W. & Thompson, C.B. (1989a). Evolutionary comparison of the avian Ig_L locus: combinatorial diversity plays a role in the generation of the antibody repetoire in some avian species. *Int. Immunol.*, **1**, 332–41.

McCormack, W.T., Tjoelker, L.W., Carlson, L.M., Petryniak, B., Barth, C.F., Humpries, E.H. & Thompson, C.B. (1989b). Chicken Ig_L gene rearrangement involves deletion of a circular episome and addition of single non-random nucleotides to both coding segments. *Cell*, **56**, 785–91.

McCormack, W.T., Tjoelker, L.W., Stella, G., Postema, C.E. & Thompson, C.B. (1991b). Chicken T-cell receptor β-chain diversity: an evolutionarily conserved $D_β$-encoded glycine turn within the hypervariable CDR3 domain. *Proc. Natl. Acad. Sci. USA*, **88**, 7699–703.

McCormack, W.T., Tjoelker, L.W. & Thompson, C.B. (1991a). Avian B-cell development: generation of an immunoglobulin repertoire by gene conversion. *Annu. Rev. Immunol.*, **9**, 219–41.

McDermott, J.B., Peterson, C.A. & Piatigorsky, J. (1992). Structure and lens expression of the gene encoding chicken βA3/A1-crystallin. *Gene*, **117**, 193–200.

McDonald, M.L. & Swick, R.W. (1981). The effect of protein depletion and repletion on muscle-protein turnover in the chick. *Biochem. J.*, **194**, 811–19.

McDonnell, D.P., Mangelsdorf, D.J., Pike, J.W., Haussler, M.R. & O'Malley, B.W. (1987). Molecular cloning of complementary DNA encoding the avian receptor for vitamin D. *Science*, **235**, 1214–17.

McEwan, I.J., Saluz, H.-P. & Jost, J.-P. (1991). *In vivo* and *in vitro* protein–DNA interactions at the distal oestrogen response element of the chicken vitellogenin gene: evidence for the same protein binding to this sequence in hen and rooster liver. *J. Steroid Biochem. Mol. Biol.*, **38**, 275–83.

McFarland, D.C., Ferrin, N.H., Gilkerson, K.K. & Pesall, J.E. (1992). Tissue distribution of insulin-like growth factor in the turkey. *Comp. Biochem. Physiol.*, **103B**, 601–7.

McGinnis, W. & Krumlauf, R. (1992) Homeobox genes and axial patterning. *Cell*, **68**, 283–302.

McMurthy, J.P., Rosebrough, R.W. & Steele, N.C. (1987). Insulin metabolism and its effect on blood electrolytes and glucose in the turkey hen. *Comp. Biochem. Physiol.*, **86A**, 309–13.

McNabb, F.M.A., McNabb, R.A., Prather, I.D., Conner, R.N. & Adkisson, C.S. (1980). Nitrogen excretion by turkey vultures. *Condor*, **82**, 219–23.

McPhaul, M.J., Noble, J.F., Simpsom, E.R., Mendelson, C.R. & Wilson, J.D. (1988). The expression of a functional cDNA encoding the chicken cytochrome P-450$_{arom}$ (aromatase) that catalyses the formation of estrogen from androgen. *J. Biol. Chem.*, **263**, 16358–63.

Meng, A., Carter, R.E. & Parkin, D.T. (1990). The variability of DNA fingerprints in three species of swan. *Heredity*, **64**, 73–80.

Mezquita, J., Lopez-Ibor, B., Pau, M. & Mezquita, C. (1993). Intron and intronless transcription of the chicken polyubiquitin gene UbII. *FEBS Lett.*, **319**, 244–8.

Mezquita, J. & Mezquita, C. (1991). Characterization of a chicken polyubiquitin gene preferentially expressed during spermatogenesis. *FEBS Lett.*, **279**, 69–72.

Mezquita, J. & Mezquita, C. (1992). A cluster of tRNA genes is present in the 5'-flanking region of chicken ubiquitin gene UbII. *Nucl. Acids Res.*, **20**, 5477.

Miksik, I. & Hodny, Z. (1992). Glycated hemoglobin in mute swan (*Cygnus olor*) and rook (*Corvus frugilegus*). *Comp. Biochem. Physiol.*, **103B**, 553–5.

Millar, C.D., Lambert, D.M., Bellamy, A.R., Stapleton, P.M. & Young, E.C. (1992). Sex-specific restriction fragments and sex ratios revealed by DNA fingerprinting in the Brown Skua. *J. Hered.*, **83**, 350–5.

Miller, M.M., Abplanalp, H. & Goto, R. (1988). Genotyping chickens for the *B-G* subregion of the major histocompatibility complex using restriction fragment length polymorphisms. *Immunogenetics*, **28**, 374–9.

Mills, G.L. & Taylaur, C.E. (1971). The distribution and composition of serum lipoproteins in eighteen animals. *Comp. Biochem. Physiol.*, **40B**, 489–501.

Mimura, H., Cao, X., Ross, F.P., Chiba, M. & Teitelbaum, S.L. (1994). 1,25-Dihydroxyvitamin D-3 transcriptionally activates the β_3-integrin subunit gene in avian osteoclast precursors. *Endocrinology*, **134**, 1061–6.

Minick, M.C. & Duke, G.E. (1991). Simultaneous determination of plasma insulin, glucose and glucagon levels in fasting, previously stressed but convalescing bald eagles, *Haliaeetus leucocephalus* Linnaeus. *Comp. Biochem. Physiol.*, **99A**, 307–11.

Minie, M., Clark, D., Trainor, C. *et al.* (1992a). Developmental regulation of globin gene expression. *J. Cell Sci.* **16s**, 15–20.

Minie, M.E., Kimura, T. & Felsenfeld, G. (1992b). The developmental switch in embryonic ρ-globin expression is correlated with erythroid lineage-specific differences in transcription factor levels. *Development*, **115**, 1149–64.

Mirsky, A.E. & Ris, H. (1951). The desoxyribonucleic acid content of animal cells and its evolutionary significance. *J. Gen. Physiol.*, **34**, 451–63.

Mitrani, E., Ziv, T., Thomsen, G., Shimoni, Y., Melton, D.A. & Bril, A. (1990). Activin can induce the formation of axial structures and is expressed in the hypoblast of the chick. *Cell*, **63**, 495–501.

Miyata, A., Minamino, N., Kangawa, K. & Matsuo, H. (1988). Identification of a 29-amino acid natriuretic peptide in chicken heart. *Biochem. Biophys. Res. Commun.*, **155**, 1330–7.

Mölders, H., Jenuwein, T., Adamkiewicz, J. & Möller, R. (1987). Isolation and structural analysis of a biologically active chicken c-*fos* cDNA: identification of evolutionarily conserved domains in *fos* protein. *Oncogene*, **1**, 377–85.

Monod, J., Wyman, J. & Changeux, J.-P. (1965). On the nature of allosteric transitions: a plausible model. *J. Mol. Biol.*, **12**, 88–118.

Montarras, D., Fiszman, M. & Gros, F. (1982). Changes in tropomyosin during development of chick embryonic skeletal muscles *in vivo* and during differentiation of chick muscle cells in vitro. *J. Biol. Chem.*, **257**, 545–8.

Mooers, A. & Cotgreave, P. (1994). Sibley and Alquist's tapestry dusted off. *Trends Evol. Ecol.*, **9**, 458–9.

Moore, B.E. & Bose, H.R. (1989). Expression of the c-*rel* and c-*myc* proto-oncogenes in avian tissues. *Oncogene*, **4**, 845–52.

Moore, L.A., Tidyman, W.E., Arrizubieta, M.J. & Bandman, E. (1993). The evolutionary relationship of avian and mammalian myosin heavy-chain genes. *J. Mol. Evol.*, **36**, 21–30.

Moran, C. (1993). Microsatellite repeats in pig (*Sus domestica*) and chicken (*Gallus domesticus*). *J. Hered.*, **84**, 274–80.

Moran, E.T. (1987). Protein requirement, egg formation and hen's ovulatory cycle. *J. Nutr.*, **117**, 612–18.

Morgan, T.J., Lillehoj, H.S., Sanders, B.G. & Kline, K. (1990). Characterization and developmental expression of the chicken B–G heterodimer. *Dev. Comp. Immunol.*, **14**, 425–37.

Morgenstern, R., Lundqvist, G., Anderson, G., Balk, L. & DePierre, J.W. (1984). The distribution of microsomal glutathione transferase among different organelles, different organs, and different organisms. *Biochem. Pharmacol.*, **33**, 3609–14.

Mori, G. & George, J.C. (1978). Seasonal changes in serum levels of certain metabolites, uric acid, and Ca^{2+} in migratory Canada geese (*Branta canadensis interior*). *Comp. Biochem. Physiol.*, **59B**, 263–9.

Morris, S.M., Nilson, J., Jenik, R.A., Winberry, L.K., McDevitt, M.A. & Goodridge, A.G. (1982). Molecular cloning of gene sequences for fatty acid synthase and evidence for nutritional regulation of fatty acid synthase mRNA concentration. *J. Biol. Chem.*, **257**, 3225–9.

Morris, S.M., Winberry, L.K., Fisch, J.E., Back, D.W. & Goodridge, A.G. (1984). Developmental and nutritional regulation of the messenger RNAs for fatty acid synthase, malic enzyme and albumin in the livers of embryonic and newly-hatched chicks. *Mol. Cell. Biochem.*, **64**, 63–8.

Mortensen, A. & Tindall, A.R. (1981). Caecal decomposition of uric acid in captive and free ranging willow ptarmigan (*Lagopus lagopus lagopus*). *Acta Physiol. Scand.*, **111**, 129–33.

Moun, T., Johansen, S., Erikstad, K.E. & Piatt, J.F. (1994). Phylogeny and evolution of the auks (subfamily Alcinae) based on mitochondrial DNA sequences. *Proc. Nat. Acad. Sci. USA*, **91**, 7912–6.

Müller, A.R. & Gerstberger, R. (1992). The α_2-adrenergic receptor system in the hypothalamus of the Pekin duck. *Cell Tissue Res.*, **268**, 99–107.

Murphy, M.E. (1993a). The essential amino acid requirements for maintenance in the White-crowned Sparrow, *Zonotrichia leucophyrys gambelii*. *Canad. J. Zool.*, **71**, 2121–30.

Murphy, M.E. (1993b). The protein requirement for maintenance in the White-crowned Sparrow, *Zonotrichia leucophrys gambelii*. *Canad. J. Zool.*, **71**, 2111–20.

Murphy, M.E. (1994). Amino acid compositions of avian eggs and tissues: nutritional implications. *J. Avian Biol.*, **25**, 27–38.

Murphy, M.E. & King, J.R. (1982). Amino acid composition of the plumage of the White-crowned sparrow. *Condor*, **84**, 435–8.

Murphy, M.E. & King, J.R. (1985). Diurnal variation in liver and muscle glutathione pools of molting and nonmolting white-crowned sparrows. *Physiol. Zool.*, **58**, 646–54.

Murphy, M.E. & King, J.R. (1992). Energy and nutrient use during moult by White-crowned sparrows *Zonotrichia leucophrys gambelii*. *Ornis Scand.*, **23**, 304–13.

Murray, A. & Hunt, T. (1993). *Cell Cycle: an Introduction*. Oxford: Oxford University Press.

Muscarella, D.E., Vogt, V.M. & Bloom, S.E. (1985). The ribosomal RNA gene cluster in aneuploid chickens: evidence for increased gene dosage and regulation of gene expression. *J. Cell Biol.*, **101**, 1749–56.

Muschler, J.L. & Horwitz, A.F. (1991). Down-regulation of the chicken $\alpha_5\beta_1$ integrin fibronectin receptor during development. *Development*, **113**, 327–37.

Nabeshima, Y., Fujii-Kuriyama, Y., Muramatsu, M. & Ogata, K. (1984). Alternative transcription and two modes of splicing result in two myosin light chains from one gene. *Nature*, **308**, 333–8.

Nagahara, N., Nishino, T., Kanisawa, M. & Tsuhima, K. (1987). Effect of protein on purine nucleoside phosphorylase and xanthine dehydrogenase activities of liver and kidney in chicken and pigeon. *Comp. Biochem. Physiol.*, **88B**, 589–93.

Nah, H.-D., Barembaum, M. & Upholt, W.B. (1992). The collagen $\alpha 1$(XI) collagen gene is widely expressed in embryonic tissues. *J. Biol. Chem.*, **267**, 22581–6.

Nakamura, D., Tiersch, T.R., Douglas, M. & Chandler, R.W. (1990). Rapid identificaton of sex in birds by flow cytometry. *Cytogenet. Cell Genet.*, **53**, 201–5.

Nakano, T. & Graf, T. (1991). Goose-type lysozyme gene of the chicken: sequence, genomic organization and expression reveals major differences to chicken-type lysozyme gene. *Biochim. Biophys. Acta*, **1090**, 273–6.

Nakao, M., Mizutani, T., Bhakta, A., Ribarac-Stepic, N. & Moudgil, V.K. (1992). Phosphorylation of chicken oviduct progesterone receptor by cAMP-dependent protein kinase. *Arch. Biochem. Biophys.*, **298**, 340–8.

Nakayama, T., Takechi, S. & Takami, Y. (1993). The chicken histone gene family. *Comp. Biochem. Physiol.*, **104B**, 635–9.

Nanda, I. & Schmid, M. (1994). Localization of telomeric (TTAGGG)n sequence in chicken (*Gallus domesticus*) chromosomes. *Cytogenet. Cell Genet.*, **65**, 190–3.

Napolitano, E.W., Venstrom, K., Wheeler, E.F. & Reichardt, L.F. (1991). Molecular cloning and characterization of B-cadherin, a novel chick cadherin. *J. Cell Biol.*, **113**, 893–905.

Nataf, V., Mercier, P., Ziller, C. & le Douarin, N. (1993). Novel markers of melanocyte differentiation in the avian embryo. *Exp. Cell Res.*, **207**, 171–82.

Neelin, J.M., Callaham, P.X., Lamb, D.C. & Murray, K. (1964). The histones of chicken erythrocyte nuclei. *Canad. J. Biochem.*, **42**, 1743–52.

Nehlig, A., Crone, M.C. & Lehr, P.R. (1980). Variations of 3-hydroxybutyrate dehydrogenase activity in brain and liver mitochondria of the developing chick. *Biochim. Biophys. Acta*, **633**, 22–32.

Nemere, I., Dormanen, M.C., Hammond, M.W., Okamura, W.H. & Norman, A.W. (1994). Identification of a specific binding protein for $1\alpha 25$ dihydroxyvitamin D_3 in basal-lateral membranes of chick intestinal epithelium and the relationship to transcalthachia. *J. Biol. Chem.*, **269**, 23750–6.

Nemere, I., Leathers, V.L., Thompson, B.S., Luben, R.A. & Norman, A.W. (1991). Redistribution of calbindin-D_{28K} in chick intestine in response to calcium transport. *Endocrinology*, **129**, 2972–84.

Nermut, M.V., Burt, J.S., Hirst, E.M.A. & Larjava, H. (1993). Distribution of avian integrin during the lifetime of chicken embryo fibroblasts *in vitro*: study by immunofluorescence and immuno electron microscopy. *Micron*, **24**, 363–75.

Nestor, P.V., Forde, R.C., Webb, P. & Gannon, F. (1994). The genomic organisation, sequence and functional analysis of the 5 flanking region of the chicken estrogen receptor gene. *J. Steroid Biochem. Molec. Biol.*, **50**, 121–30.

Newsholme, E.A. & Leech, A.R. (1983). *Biochemistry for Medical Sciences*. Chichester: Wiley.

Newton, P.A. & Hamer, M.J. (1989). Tissue-specific developmental changes in phosphofructokinase activity in embryonic and neonatal chick. *Biochem. Soc. Trans.*, **17**, 1106–7.

Nigg, E.A. (1993). Cellular substrates of p34^{cdc2} and its companion cyclin-dependent kinases. *Trends Cell Biol.*, **3**, 296–301.

Nikolaropoulis, S. & Sotiroudis, T.G. (1985). Phosphorylase kinase from chicken gizzard: partial purification and characterisation. *Eur. J. Biochem.*, **151**, 467–73.

Nikovits, W., Kuncio, G. & Ordahl, C.P. (1986). The chicken fast skeletal troponin I gene: exon organization and sequence. *Nucl. Acids Res.*, **14**, 3377–90.

Nimpf, J., Radosavljevic, M. & Schneider, W.J. (1989a). Specific postendocytic proteolysis of apoprotein B in oocytes does not abolish receptor recognition. *Proc. Natl. Acad. Sci. USA*, **86**, 906–10.

Nimpf, J., Radosavljevic, M.J. & Schneider, W.J. (1989b). Oocytes from the mutant restricted ovulator hen receptor for very low density lipoprotein. *J. Biol. Chem.*, **264**, 1393–8.

Nimpf, J. & Schneider, W.J. (1991). Receptor-mediated lipoprotein transport in laying hens. *J. Nutr.*, **121**, 1471–4.

Ninomiya, Y., Gordon, H., van der Rest, M., Schmid, T., Linsenmayer, T. & Olsen, B.R. (1986). The developmentally regulated Type X collagen gene contains a long open reading frame without introns. *J. Biol. Chem.*, **261**, 5041–50.

Nir, I. & Lin, H. (1982). The skeleton, an important site of lipogenesis in the chick. *Ann. Nutr. Alim.*, **26**, 100–5.

Nishikawa, K., Nakanishi, T., Aoki, C., Hattori, T., Takahashi, K. & Taniguchi, S. (1994). Differential expression of homeobox-containing genes *Msx-1* and *Msx-2* and homeoprotein *Msx-2* expression during craniofacial development. *Biochem. Mol. Biol. Int.*, **32**, 763–71.

Nishimura, I., Muragaki, Y. & Olsen, B.R. (1989). Tissue-specific forms of type IX collagen-proteoglycan arise from the use of two widely separate promoters. *J. Biol. Chem.*, **264**, 20033–41.

Nishimura,T., Kato, Y., Okitani, A. & Kato, H. (1991). Purification and properties of aminopeptidase C from chicken skeletal muscle. *Agri. Biol. Chem.*, **5**, 1771–8.

Nishimura, T. & Vogt, P.K. (1988). The avian cellular homologue of the oncogene *jun*. *Oncogene*, **3**, 659–63.

Nishizawa, M., Goto, N. & Kawai, S. (1987). An avian transforming retrovirus isolated from a nephroblastoma that carries the *fos* gene as the oncogene. *J. Virol.*, **61**, 3733–40.

Noble, R.C. & Connor, K. (1984). Lipid metabolism in the chick embryo. *World's Poult. Sci. J.*, **40**, 114–20.

Nohno, T., Noji, S., Koyama, E. *et al.* (1992). Differential expression of two *Msh*-related homeobox genes *Chox-7* and *Chox-8* during chick limb development. *Biochem. Biophys. Res. Commun.*, **182**, 121–8.

Nohno, T., Noji, S., Koyama, E. *et al.* (1991). Anteroposterior axial polarity during limb development. *Cell*, **64**, 1197–205.

Noji, S., Nohno, T., Koyama, E. *et al.* (1991). Retinoic acid induces polarizing activity but is unlikely to be a morphogen in the chick limb bud. *Nature*, **350**, 83–6.

Norbury, C. & Nurse, P. (1992). Animal cell cycles and their control. *Annu. Rev. Biochem.*, **61**, 441–70.

Nordlie, R.C. (1974). Metabolic regulation by multifunctional glucose-6-phosphatase. *Curr. Topics Cell Reg.*, **8**, 33–117.

Norman, A.W. (1987). Studies on the vitamin D endocrine system in the avian. *J. Nutr.*, **117**, 797–807.

Norman, A.W. & Hurwitz, S. (1993). The role of the vitamin D endocrine system in avian bone biology. *J. Nutr.*, **123**, 310–16.

Nothwang, H.G., Coux, O., Keith, G., Silva-Pereira, I. & Scherer, K. (1992). The major RNA in prosomes of Hela cells and duck erythroblasts is tRNA. *Nucl. Acids Res.*, **20**, 1959–65.

Nunez, G. & Clarke, M.F. (1994). The Bcl-2 family of proteins: regulators of cell death and survival. *Trends Cell Biol.*, **4**, 399–403.

Nys, Y. (1993). Regulation of plasma 1,25(OH)$_2$D$_3$, of osteocalcin and of intestinal and uterine calbindin in hens. In *Avian Endocrinology*, ed. P.J. Sharp, pp. 345–58. Bristol: Society of Endocrinology.

Ochoa, S., Mehler, A.H. & Kornberg, A. (1947). Reversible oxidative decarboxylation of malic acid. *J. Biol. Chem.*, **167**, 871–2.

Ochs, R.S. & Harris, R.A. (1978). Studies on the relationship between glycolysis, lipogenesis, gluconeogenesis and pyruvate kinase activity of rat and chicken hepatocytes. *Arch. Biochem. Biophys.*, **190**, 193–201.

Ochs, R.S. & Harris, R.A. (1980). Aminooxyacetate inhibits gluconeogenesis by isolated chicken hepatocytes. *Biochim. Biophys. Acta*, **632**, 260–9.

Ohyama, K., Tokimitsu, I. & Tajima, S. (1991). Mechanisms of stimulation of collagen synthesis during chick embryonic skin development. *J. Biochem.*, **110**, 257–60.

Oku, H., Ishikawa, M., Nagata, J., Toda, T. & Chinen, I. (1993). Lipoprotein and apoprotein profile of Japanese quail. *Biochim. Biophys. Acta*, **1167**, 22–8.

Olah, I., Scott, T.R., Gallego, M., Kendall, C. & Glick, B. (1992). Immunoglobulin-G develops an intimate relationship with central canal epithelium in the harderian gland of the chicken. *Poult. Sci.*, **71**, 664–76.

Oliva, R. & Dixon, G.H. (1989). Chicken protamine genes are intronless. The complete genomic sequence and organization of the two loci. *J. Biol. Chem.*, **264**, 12472–81.

Olmo, E., Capriglione, T. & Odierna, G. (1989). Genome size evolution in vertebrates: trends and constraints. *Comp. Biochem. Physiol.*, **92B**, 447–53.

Olofsson, B. & Bernardi, G. (1983). The distribution of CR1, and *Alu*-like family of interspersed repeats, in the chicken genome. *Biochim. Biophys. Acta*, **740**, 339–43.

Olson, W.C. & Ewert, D.L. (1994). A novel multilineage cell-surface antigen expressed on terminally differentiated chicken B cells in mucosal tissues. *Cell. Immunol.*, **154**, 328–41.

Orkin, S.H. (1990). Globin gene regulation and switching – circa 1990. *Cell*, **63**, 665–72.

Orlowski, M. (1990). The multicatalytic complex: a major extralysosomal proteolytic system. *Biochem.*, **29**, 10289–97.

Orti, E., Bodwell, J.E. & Munck, A. (1992). Phosphorylation of steroid hormone receptors. *Endocrine Res.*, **13**, 105–28.

Oscar, T.P. (1991). Glucagon-stimulated lipolysis of primary cultured broiler adipocytes. *Poult. Sci.*, **70**, 326–32.

Osley, M.A. (1991). The regulation of histone synthesis in the cell cycle. *Annu. Rev. Biochem.*, **60**, 827–61.

Page, S., Miller, J.B., DiMario, J.X., Hager, E.J., Moser, A. & Stockdale, F.E. (1992). Developmentally regulated expression of three slow isoforms of myosin heavy chain: diversity among the first fibers to form in avian muscle. *Dev. Biol.*, **154**, 118–28.

Pageaux, J.-F., Dufrene, L., Laugier, C., Perche, O. & Sandoz, D. (1989). Heterogeneity of progesterone receptor expression in epithelial cells of immature and differentiating quail oviduct. *Biol. Cell*, **67**, 135–40.

Pageaux, J.F., Joulain, C., Fayard, J.M., Lagarde, M. & Laugier, C. (1992). Changes in fatty-acid composition of plasma and oviduct lipids during sexual-maturation of Japanese quail. *Lipids*, **27**, 518–25.

Pages, T. & Planas, J. (1983). Muscle myoglobin and flying habits in birds. *Comp. Biochem. Physiol.*, **74A**, 289–94.

Palmiter, R.D. (1972). Regulation of protein synthesis in chick oviduct. *J. Biol. Chem.*, **247**, 6450–61.

Pałyga, J. (1990). Polymorphism of histone H1 in goose erythrocytes. *Biochem. Genet.*, **28**, 359–65.

Pałyga, J. (1991). A comparison of the histone H1 complements of avian erythrocytes. *Int. J. Biochem.*, **23**, 845–9.

Papageorgiou, S. & Almirantis, Y. (1992). Diffusion or autocatalysis of retinoic acid cannot explain pattern formation in the chick wing bud. *Dev. Dynamics*, **194**, 282–8.

Pardue, S.L. & Thaxton, J.P. (1986). Ascorbic acid in poultry. *World's Poult. Sci. J.*, **42**, 107–23.

Park, I.N., Shin, S. & Marquardt, R.R. (1991). Effects of niacin deficiency on the relative turnover rates of proteins in various tissues of Japanese quail. *Int. J. Biochem.*, **23**, 1005–12.

Parker, E.M. & Ross, E.M. (1991). Truncation of the extended carboxyl-terminal domain increases the expression and regulator activity of the avian β-adrenergic receptor. *J. Biol. Chem.*, **266**, 9987–96.

Parr, T., Waites, G.T., Patel, B., Millake, D.B., & Critchley, D.R. (1992). Chick skeletal-muscle α-actinin genes give rise to two alternatively spliced isoforms which differ in the EF-hand Ca^{2+} - binding domain. *Eur. J. Biochem.*, **210**, 801–9.

Parrish, J.W. & Martin, E.W. (1977). The effect of dietary lysine level on the energy and nitrogen balance of the dark-eyed junco. *Condor*, **79**, 24–30.

Parvari, R., Avivi, A., Lentner, F. *et al.* (1988). Chicken immunoglobulin γ-heavy chains: limited VH gene repetoire, combinatorial diversification by D gene segments and evolution of the heavy chain locus. *EMBO J.*, **7**, 739–44.

Patejunas, G. & Young, A.P. (1990). Constitutive and glucocorticoid-mediated activation of glutamine synthetase gene expression in developing chicken retina. *J. Biol. Chem.*, **265**, 15280–5.

Pato, M.D. & Adelstein, R.S. (1983a). Purification and characterization of a multisubunit phosphatase from turkey gizzard smooth muscle. *J. Biol. Chem.*, **258**, 7047–54.

Pato, M.D. & Adelstein, R.S. (1983b). Characterization of a Mg^{2+}-dependent phosphatase from turkey gizzard smooth muscle. *J. Biol. Chem.*, **258**, 7055–8.

Pato, M.D., Adelstein, R.S., Crouch, D., Safer, B., Ingebritsen, T.S. & Cohen, P. (1983). The protein phosphatases involved in cellular regulation. *Eur. J. Biochem.*, **132**, 283–7.

Patstone, G. & Maher, P.A. (1993). Phosphotyrosine-containing proteins are concentrated in differentiating cells during chicken embryonic development. *Growth Factors*, **9**, 243–52.

Patten, B.M. (1951). *Early Embryology of the Chick*, 4th edn. New York: McGraw-Hill.

Pazirandeh, M., Chirala, S.S. & Wakil, S.J. (1991). Site-directed mutagenesis studies on the recombinant thioesterase domain of chicken fatty acid synthase. *J. Biol. Chem.*, **266**, 20946–52.

Pearce, J. (1980). Comparative aspects of lipid metabolism in avian species. *Biochem. Soc. Trans.*, **8**, 295–6.

Pearce, J. (1983). Carbohydrate metabolism. In *Physiology and Biochemistry of the Domestic Fowl*, vol. 4, ed. B.M. Freeman, pp. 147–64. London: Academic Press.

Pearce, J. & Brown, W.O. (1971). Carbohydrate metabolism. In *Physiology and Biochemistry of the Domestic Fowl*, vol. 1, ed. D.J. Bell & B.M. Freeman, pp. 295–320. London: Academic Press.

Penner, C.G. & Davie, J.R. (1992). Multisubunit erythroid complexes binding to the enhancer element of the chicken histone H5 gene. *Biochem. J.*, **283**, 905–11.

Perez, R.G. & Halfter, W. (1993). Tenascin in the developing chick visual system: distribution and potential role as a modulator of the retinal axon growth. *Dev. Biol.*, **156**, 278–92.

Perler, F., Efstratiadis, A., Lomedico, P., Gilbert, W., Kolodner, R. & Dodgson, J. (1980). The evolution of genes: the chicken preproinsulin gene. *Cell*, **20**, 555–66.

Perry, M.M. & Sang, H.M. (1993). Transgenesis in chickens. *Transgenic Res.*, **2**, 125–33.

Peter, M., Nakagawa, J., Dorée, M., Labbé, J.C. & Nigg, E.A. (1990). In vitro disassembly of nuclear lamina and M phase-specific phosphorylation of lamins by cdc2 kinase. *Cell*, **61**, 591–602.

Petschow, D., Würdinger, I., Baumann, R., Duhm, J., Braunitzer, G. & Bauer, C. (1977). Causes of high blood O_2 affinity of animals at high altitude. *J. App. Physiol.*, **42**, 139–43.

Phillips, D.C. (1967). The hen egg-white lysozyme molecule.

Proc. Natl. Acad. Sci. USA, **57**, 484–95.

Piatigorsky, J., O'Brien, W.E., Norman, B.L. *et al.* (1988). Gene sharing by δ-crystallin and argininosuccinate lyase. *Proc. Natl. Acad. Sci. USA*, **85**, 3479–83.

Piatigorsky, J. & Wistow, G. (1991). The recruitment of crystallins: new functions precede gene duplication. *Science*, **252**, 1078–9.

Pickel, J.M., McCormack, W.T., Chen, C.-I.H., Cooper, M.D. & Thompson, C.B. (1993). Differential regulation of V(D)J recombination during development of avian B and T cells. *Int. Immunol.*, **5**, 919–27.

Piersma, T. (1988). Breast muscle atrophy and constraints on foraging during the flightless period of wing moulting Great Crested Grebes. *Ardea*, **76**, 96–106.

Pikaart, M., Irving, J. & Villeponteau, B. (1991). Decline in histone H5 phosphorylation during erythroid senescence in chick embryos. *Mech. Ageing Dev.*, **59**, 189–95.

Pike, J.W., Sleator, N.M. & Haussler, M.R. (1987). Chicken intestinal receptor for 1,25-dihydroxyvitamin D_3. *J. Biol. Chem.*, **262**, 1305–11.

Pinchasov, Y., Elmaliah, S. & Bezdin, S. (1994). Plasma apolipoprotein VLDL-II and egg production in laying hens: establishment of an ELISA method. *Reprod. Nutr. Dev.*, **34**, 361–9.

Pink, J.L.R., Droege, W., Hála, K., Miggiano, V.C. & Ziegler, A. (1977). A three locus model for the major histocompatibility complex. *Immunogenetics*, **5**, 203–16.

Piquer, F.J., Sell, J.L., Albatshan, H.A., Mallarino, E.G., Sotosalanova, M.F. & Angel, C.R. (1991). Posthatching changes in the immunoglobulin-A concentration in the jejunum and bile of turkeys. *Poult. Sci.*, **70**, 2476–83.

Plaza, S., Dozier, C. & Saule, S. (1993). Quail *PAX-6* (*PAX-QNR*) encodes a transcriptional factor able to bind and *trans*-activate its own promoter. *Cell Growth Diff.*, **4**, 1041–50.

Pochet, R., van Rampelbergh, J., Bastianelli, E. & van Eldik, L.J. (1994). Calmodulin, calbindin-D28K and calretinin in rat and chicken pineal glands: immunocytochemical and immunoblotting analysis. *Biochim. Biophys. Acta*, **1223**, 318–24.

Poletti, A., Conneely, O.M., McDonnell, D.P., Schrader, W.T., O'Malley, B.W. & Weigel, N.L. (1993). Chicken progesterone-receptor expressed in *Saccharomyces cerevisiae* is correctly phosphorylated at all 4 ser–pro phosphorylation sites. *Biochemistry*, **32**, 9563–9.

Polo, F.J., Celdrán, J., Viscor, G. & Palomeque, J. (1994). Blood chemistry of captive herons, egrets, spoonbill, ibis and gallinule. *Comp. Biochem. Physiol.*, **107A**, 343–7.

Porter, R.R. (1973). Structural studies of immunoglobulins. *Science*, **180**, 713–16.

Postnikov, Y.V., Shick, V.V., Belyavsky, A.V. *et al.* (1991). Distribution of high mobility group proteins 1/2,E and 14/17 and linker histone-H1 and histone H-5 on transcribed and non-transcribed regions of chicken erythrocyte chromatin. *Nucl. Acids Res.*, **19**, 717–25.

Potts, J.D. & Carrington, J.L. (1993). Selective expression of the chicken platelet-derived growth factor α (PDGFα) receptor during limb bud development. *Dev. Dynamics*, **198**, 14–21.

Presek, P., Remacher, M. & Eigenbrodt, E. (1988). Pyruvate kinase type M_2 is phosphorylated at tyrosine residues in cells transformed by Rous sarcoma virus. *FEBS Lett.*, **242**, 194–8.

Presland, R.B., Gregg, K., Molloy, P.L., Morris, C.P., Crocker, L.A. & Rogers, G.E. (1989a). Avian keratin genes. I. A molecular analysis of the structure and expression of a group of feather keratin genes. *J. Mol. Biol.*, **209**, 549–59.

Presland, R.B., Whitbread, L.A. & Rogers, G.E. (1989b). Avian keratin genes. II. Chromosomal arrangements and close linkage of three gene families. *J. Mol. Biol.*, **209**, 561–76.

Price, N.C. & Stevens, L. (1989). *Fundamentals of Enzymology*, 2nd edn. Oxford: Oxford University Press.

Proctor, N.S. & Lynch, P.J. (1993). *Manual of Ornithology*. New Haven: Yale University Press.

Puerta, M.L., Garcia del Campo, A.L., Huecas, V. & Abelenda, M. (1991). Hematology and blood chemistry of the white pelican (*Pelecanus onocrotalus*). *Comp. Biochem. Physiol.*, **98A**, 393–4.

Puerta, M.L., Munoz-Pulido, R., Huecas, V. & Abelenda, M. (1989). Hematology and blood chemistry of chicks of white and black storks (*Ciconia ciconia* and *Ciconia nigra*). *Comp. Biochem. Physiol.*, **94A**, 201–4.

Quasba, P.K. & Safaya, S.K. (1984). Similarity of the nucleotide sequences of rat α-lactalbumin and chicken lysozyme genes. *Nature*, **308**, 377–80.

Quéva, C., Leprince, D., Stéhelin, D. & Vandenbunder, B. (1993). p54$^{c\text{-}ets\text{-}1}$ and p68$^{c\text{-}ets\text{-}1}$, the two transcription factors encoded by the c-*ets*-1 locus, are differentially expressed during the development of the chick embryo. *Oncogene*, **8**, 2511–20.

Quéva, C., Ness, S.A., Grässer, F.A., Graf, T., Vandenbunder, B. & Stéhelin, D. (1992). Expression patterns of c-*myb* and of v-*myb* induced myeloid-1 (*min*-1) gene during the development of the chick embryo. *Development*, **114**, 125–33.

Rafestin-Oblin, M.E., Couette, B., Radanyi, C., Lombes, M. & Baulieu, E.-E. (1989). Mineralocorticoid receptor in chick intestine. *J. Biol. Chem.*, **264**, 9304–9.

Ragsdale, C.W. & Brockes, J.P. (1991). Retinoic acid receptors and vertebrate limb morphogenesis. In *Nuclear Hormone Receptors*, ed. M.G. Parker, pp. 269-95. London: Academic Press.

Rajaraman, R. (1993). Avian integrin α_v subunit associates with multiple β subunits and shows tissue-specific variation in its assocation. *Exp. Cell Res.*, **205**, 25–31.

Rajavashisth, T.B., Dawson, P.A., Williams, D.L., Shackelford, J.E., Lebherz, H. & Lusis, A.J. (1987). Structure, evolution and regulation of chicken apolipoprotein A-I. *J. Biol. Chem.*, **262**, 7058–65.

Ramenofsky, M. (1990). Fat storage and fat metabolism in relation to migration. In *Bird Migration: Physiology and Ecophysiology*, ed. E. Gwimmer, pp. 214–31. Berlin: Springer Verlag.

Ramirez, V., Savoie, P. & Morais, R. (1993). Molecular characterization of duck mitochondrial genome. *J. Mol. Evol.*, **37**, 296–310.

Rangini, Z., Frumkin, A., Shani, G. *et al.* (1989). The chicken homeo box genes *CHox1* and *CHox3*: cloning, sequencing and expression during embryogenesis. *Gene*, **76**, 61–74.

Ranscht, B. & Bronner-Fraser, M. (1991). T-cadherin expression alternates with migrating neural crest cells in the trunk of the avian embryo. *Development*, **111**, 15–22.

Rapp, U.R., Cleveland, J.L., Bonner, T.I., Storm, S.E. (1988) The *raf* oncogenes. In *The Oncogene Handbook*, ed. E.P. Reddy, A.M. Skalka & T. Curran, pp. 213–53. Amsterdam: Elsevier.

Rauen, K.A., Le Ciel, C.D.S., Abbott, U.K. & Hutchison, N.J. (1994). Localization of the chicken PgK gene to chromosome 4p by fluorescence in situ hybridization. *J. Hered.*, **85**, 147–50.

Rayment, I. & Holden, H.M. (1994). The three-dimensional structure of a molecular motor. *Trends Biochem. Sci.*, **19**, 129–34.

Raymoure, W.J., McNaught, R.W., Greene, G.L. & Smith, R.G. (1986). Receptor interconversion model of hormone action. *J. Biol. Chem.*, **261**, 17018–25.

Rayner, J.M.V. (1982). Avian flight energetics. *Annu. Rev. Physiol.*, **44**, 109–19.

Reddy, E.P. (1988). The *myb* oncogene. In *The Oncogene Handbook*, ed. E.P. Reddy, A.M. Skalka & T. Curran, pp. 327–39. Amsterdam: Elsevier.

Reddy, E.P., Skalka, A.M. & Curran, T. (1988). *The Oncogene Handbook*. Amsterdam: Elsevier.

Reddy, M.C., Grothvasselli, B., Sharon, A. & Farnsworth, P.N. (1993). Effect of catalytically active recruited crystallins on lens metabolism. *FEBS Lett.*, **335**, 285–8.

Reddy, S.R.R. & Campbell, J.W. (1970). Molecular weight of arginase from different species. *Comp. Biochem. Physiol.*, **32**, 499–509.

Reitman, M., Grasso, J.A., Blumenthal, R. & Lewit, P. (1993). Primary sequence, evolution, and repetitive elements of the *Gallus gallus* (chicken) β-globin cluster. *Genomics*, **18**, 616–26.

Retzek, H., Steyrer, E., Sanders, E.J., Nimpf, J. & Schneider, W.J. (1992). Molecular cloning and functional characterisation of chicken cathepsin D, a key enzyme for yolk formation. *DNA Cell Biol.*, **11**, 661–72.

Reynaud, C.-A., Anquez, V., Grimal, H. & Weill, J.-C. (1987a). A hyperconversion mechanism generates the chicken light chain preimmune repertoire. *Cell*, **48**, 379–88.

Reynaud, C.-A., Anquez, V. & Weill, J.-C. (1991). The chicken D locus and its contribution to the immunoglobulin heavy chain repertoire. *Eur. J. Immunol.*, **21**, 2661–70.

Reynaud, C.-A., Dahan, A. & Weill, J.-C. (1983). Complete sequence of a chicken λ light chain immunoglobulin derived from the nucleotide sequence of its mRNA. *Proc. Natl. Acad. Sci. USA*, **80**, 4099–103.

Reynaud, C.-A., Dahan, A. & Weill, J.-C. (1987b). A gene conversion program during the ontogenesis of chicken B cells. *Trends Genet.*, **3**, 248–51.

Rhyu, M.R., Nishimura, T., Okitani, A. & Kato, H. (1992). Purification and properties of aminopeptidase H from chicken skeletal muscle. *Eur. J. Biochem.*, **208**, 53–9.

Ricard, F.H., Leclercq, B., & Touraille, C. (1983). Selecting broilers for low or high abdominal fat: distribution of carcass fat and quality of meat. *Br. Poult. Sci.*, **24**, 511–16.

Rice, N.R. & Gilden, R.V. (1988). The *rel* oncogene. In *The Oncogene Handbook*, ed. E.P. Reddy, A.M. Skalka & T. Curran, pp. 495–512. Amsterdam: Elsevier.

Richardson, M.K. & Sieber-Blum, M. (1993). Pluripotent neural crest cells in the developing skin of the quail embryo. *Dev. Biol.*, **157**, 348–58.

Riddle, O. & Tange, M. (1928). Studies on the physiology of reproduction in birds. XXIV On the extirpation of the bursa of Fabricius in young doves. *Am. J. Physiol.*, **86**, 266–73.

Rideau, N. & Simon, J. (1992). Specificity of the permissive effect of D-glucose on insulin release in chicken pancreas. *Comp. Biochem. Physiol.*, **103A**, 739–45.

Rideau, N. & Simon, J. (1993). cAMP and/or acetyl choline permit an insulin response to fuel nutrients in chicken. *Comp. Biochem. Physiol.*,**106A**, 837–43.

Ridley, M. (1993). *Evolution*. Oxford: Blackwell.

Rifkind, A.B., Kanetoshi, A., Orlinick, J., Capdevila, J.H. & Lee, C. (1994). Purification and biochemical characterization of two major cytochrome P-450 isoforms induced by 2,3,7,8-tetrachlorodibenzo-*p*-dioxin in chick embryo liver. *J. Biol. Chem.*, **269**, 3387–96.

Rinaudo, J.A.S. & Zelenka, P.S. (1992). Expression of c-*fos* and c-*jun* mRNA in the developing chicken lens: relationship to cell proliferation, quiescence, and differentiation. *Exp. Cell Res.*, **199**, 147–53.

Rivett, A.J. (1993). Proteasomes: multicatalytic proteinase complexes. *Biochem. J.*, **291**, 1–10.

Riviere, J.L. (1980). T-Ethoxycoumarin O-dealkylase and cytochrome P-450 from grey partridge hepatic and duodenal microsomes. *Biochem. Biophys. Res. Commun.*, **97**, 546–9.

Robbins, K.R. & Baker, D.H. (1981). Kidney arginase activity in chicks fed diets containing deficient or excessive concentrations of lysine, arginine, histidine or total nitrogen. *Poult. Sci.*, **60**, 829–34.

Robbins, K.R., Baker, D.H. & Norton, H.W. (1977). Histidine status in the chick as measured by growth rate, plasma free histidine and breast muscle carnosine. *J. Nutr.*, **107**, 2055–61.

Rogers, C.M., Ramenofsky, M., Ketterson, E.D., Nolan, V. & Wingfield, J.C. (1993). Plasma corticosterone, adrenal mass, winter weather, and season in nonbreeding populations of Dark-eyed Juncos (*Junco hyemalis hyemalis*). *Auk*, **110**, 279–85.

Rogers, L., Kolattukudy, P.E. & de Renobales, M. (1982). Purification and characterization of S-acyl fatty acid synthase thioester hydrolase which modifies the product specificity of fatty acid synthase in the uropygial gland of Mallard. *J. Biol. Chem.*, **257**, 880–6.

Rogina, B., Coelho, C.N.D., Kosher, R.A. & Upholt, W.B. (1992). The pattern of expression of the chicken homolog of HOX1I in the developing limb suggests a possible role in the ectodermal inhibition of chondrogenesis. *Dev. Dynamics*, **193**, 92–101.

Rollema, H.S. & Bauer, C. (1979). The interaction of inositol pentaphosphate with the hemoglobins of highland and lowland geese. *J. Biol. Chem.*, **254**, 12038–43.

Rooney, T.A., Hager, R. & Thomas, A.P. (1991). β-Adrenergic receptor-mediated phospholipase-C activation independent of cAMP formation in turkey erythrocyte-membranes. *J. Biol. Chem.*, **266**, 15068–74.

Rose, M.E., Orlans, E. & Buttress, N. (1974). Immunoglobulin classes in the hen's egg: their segregation in yolk and white. *Eur. J. Immunol.*, **4**, 521–3.

Rosebrough, R.W. & McMurtry, J.P. (1992). Insulin, glucagon and catecholamine interactions in avian liver explants. *Comp. Biochem. Physiol.*, **103B**, 281–7.

Rosebrough, R.W., Mitchell, A.D. & Steele, N.C. (1990). Dietary protein and carbohydrate effects on the distribution of glycogen in chicken liver. *Nutr. Res.*, **10**, 1131–9.

Rosenburg, J., Pines, M. & Hurwitz, S. (1989). Inhibition of aldosterone secretion by atrial natriuretic peptide in chicken adrenocortical cells. *Biochim. Biophys. Acta*, **1014**, 189–94.

Rosser, B.W.C. & George, J.C. (1986). The avian pectoralis: histochemical characterization and distribution of muscle fiber types. *Canad. J. Zool.*, **64**, 1174–85.

Rowe, A., Richman, J.M. & Brickell, P.M. (1991). Retinoic acid treatment alters the distribution of retinoic acid receptor-β transcripts in the embryonic chick face. *Development*, **111**, 1007–16.

Royal, A., Garapin, A., Cami, B. *et al.* (1979). The ovalbumin gene region: common features in the organization of three genes expressed in chicken oviduct under hormonal control. *Nature*, **279**, 125–32.

Ruiz, N., Miles, R.D. & Harris, R.H. (1983). Choline, methionine and sulphate interrelationships in poultry nutrition. *World's Poult. Sci. J.*, **39**, 185–98.

Ruiz-Larrea, F. & Berrie, C.P. (1993). Characterization of a membrane-associated, receptor and G-protein responsive phosphoinositide-specific phospholipase C from avian erythrocytes. *FEBS Lett.*, **328**, 174–82.

Ruoslahti, E. (1988). Fibronectin and its receptors. *Annu. Rev. Biochem.*, **57**, 375–413.

Rupp, R.A.W., Nicolas, R.H., Borgmeyer, U. *et al.* (1988). TGGCA protein is present in erythroid nuclei and binds within the nuclease-hypersensitive sites 5' of the chicken β^H-globin and β^A-globin genes. *Eur. J. Biochem.*, **177**, 505–11.

Rushlow, C. & Warrior, R. (1992). The rel family of proteins. *Bioessays*, **14**, 89–95.

Saadoun, A. & Leclercq, B. (1983). Comparison of *in vivo* fatty acid synthesis of the genetically lean and fat chickens. *Comp. Biochem. Physiol.*, **75B**, 641–4.

Saadoun, A. & Leclercq, B. (1987). In vivo lipogenesis of genetically lean and fat chickens: effects of nutritional state and dietary fat. *J. Nutr.*, **117**, 428–35.

Saarela, S., Keith, J.S., Hohtola, E. & Trayhurn, P. (1991). Is the 'mammalian' brown fat-specific mitochondrial uncoupling protein present in adipose tissue of birds? *Comp. Biochem. Physiol.*, **100B**, 45–9.

Saitoh, H., Harata, M. & Mizuno, S. (1989). Presence of female-specific bent-repetitive DNA sequences in the genomes of turkey and pheasant and their interactions with W-protein of chicken. *Chromosoma*, **98**, 250–8.

Saitoh, H. & Mizuno, S. (1992). Distribution of XhoI and EcoRI family repetitive DNA sequences into separate domains in the chicken W chromosome. *Chromosoma*, **101**, 474–7.

Saitoh, Y., Saitoh, H., Ohtomo, K. & Mizuno, S. (1991). Occupancy of the majority of DNA in the chicken W chromosome by bent-repetitive sequences. *Chromosoma*, **101**, 32–40.

Sakhri, M., Jeacock, M.K. & Shepherd, D.A.L. (1992). Regulation of intracellular protein degradation in the isolated perfused liver of the chicken (*Gallus domesticus*). *Comp. Biochem. Physiol.*, **101B**, 17–21.

Salomonsen, J., Dunon, D., Skjødt, K., Thorpe, D., Vainio, O. &

Kaufman, J. (1991). Chicken major histocompatibility complex-encoded B-G antigens are found on many cell types that are important for the immune system. *Proc. Natl. Acad. Sci. USA*, **88**, 1359–63.

Sanders, E.J., Hu, N. & Wride, M.A. (1994). Expression of TGFβ1/β3 during early chick embryo development. *Anat. Rec.*, **238**, 397–406.

Sanders, E.J. & Prasad, S. (1991). Possible roles for TGFβ1 in the gastrulating chick embryo. *J. Cell Sci.*, **99**, 617–26.

Sang, H., Gribbin, C., Mather, C. Morrice, D. & Perry, M. (1993). Transfection of chick embryos maintained under *in vitro* conditions. In *Manipulation of the Avian Genome*, ed. R.J. Etches & A.M.V. Gibbins, pp. 121–33. Boca Raton, FL: CRC Press.

Sarkar, N.K. (1977). Implication of alanine and aspartate aminotransferases, glutamate dehydrogenase and phosphoenolpyruvate carboxykinase in glucose production in rats, chicken and pigs. *Int. J. Biochem.*, **8**, 427–32.

Sasaki, H., Yamamoto, M. & Kuroiwa, A. (1992). Cell type dependent transcription regulation by chick homeodomain proteins. *Mech. Development*, **37**, 25–36.

Sasse, C.E. & Baker, D.H. (1972). The phenylalanine and tyrosine requirements and their interrelationship for the young chick. *Poult. Sci.*, **51**, 1531–6.

Sato, S. & Shiratsuchi, A. (1990).'Chymotrypsin-like' activity of chicken liver multicatalytic proteinase resides in the smallest subunit. *Biochim. Biophys. Acta*, **1041**, 269–72.

Saunders, D.K. & Fedde, M.R. (1991). Physical conditioning: effect on the myoglobin concentration in skeletal and cardiac muscle of bar-headed geese. *Comp. Biochem. Physiol.*, **100A**, 349–52.

Saunders, D.K. & Klemm, R.D. (1994). Seasonal changes in the metabolic properties of muscle in blue-winged teal, *Anas discors*. *Comp. Biochem. Physiol.*, **107A**, 63–8.

Saunderson, C.L. & Leslie, S. (1983). $N^τ$-methyl histidine excretion by poultry: not all species excrete $N^τ$–methyl histidine quantitatively. *Br. J. Nutr.*, **50**, 691–700.

Saunderson, C.L. & Leslie, S. (1988). Muscle growth and protein degradation during early development in chicks of fast and slow growing strains. *Comp. Biochem. Physiol.*, **89A**, 333–7.

Saunderson, C.L. & Whitehead, C.C. (1987). $N^τ$ Methyl histidine excretion and [U-¹⁴C] amino acid oxidation in fully fed chickens from two lines for high and low body fat contents. *Comp. Biochem. Physiol.*, **86B**, 419–22.

Savage, M.P., Hart, C.E., Riley, B.B., Sasse, J., Olwin, B.B. & Fallon, J.F. (1993). Distribution of FGF–2 suggests it has a role in chick limb bud growth. *Dev. Dynamics*, **198**, 159–70.

Sawai, S., Kato, K., Wakamatsu, Y. & Kondoh, H. (1990). Organization and expression of the chicken N-*myc* gene. *Mol. Cell. Biol.*, **10**, 2017–26.

Scanes, C.G. (1987). The physiology of growth, growth hormone, and other growth factors in poultry. *Crit. Rev. Poult. Sci.*, **1**, 51–105.

Scanes, C.G., Campbell, R. & Griminger, P. (1987). Control of energy balance during egg production in the laying hen. *J. Nutr.*, **117**, 605–11.

Schierman, L.W. & Nordskog, A.W. (1961). Relationship of blood type to histocompatibility in chickens. *Science*, **134**, 1008–9.

Schip, A.D., Meijlink, F.C.P.W., Strijker, R., Gruber, M., Vliet, A.J., Klundert, J.A.M. & AB, G. (1983). The nucleotide sequence of the chicken apo very low density lipoprotein II gene. *Nucl. Acids Res.*, **11**, 2529–40.

Schip, F.D., Samallo, J., Broos, J., Ophuis, J., Mojet, M., Gruber, M. & AB, G. (1987). Nucleotide sequence of a chicken vitellogenin gene and derived amino acid sequence of the encoded yolk precursor protein. *J. Mol. Biol.*, **196**, 245–60.

Schjeide, O.A. & Schjeide, S.M. (1981). High-density and very high-density lipoproteins in nonmammalian vertebrates. In *High Density Lipoproteins*, ed. C.E. Day, pp. 400–18. New York: Dekker.

Schlinger, B.A. & Arnold, A.P. (1992). Plasma sex steroids and tissue aromatization in hatchling zebra finches: implications for the sexual differentiation of singing behaviour *Endocrinology*, **130**, 289–99.

Schneeganss, D., Braunitzer, G., Oberthr, W., Kösters, J. & Grimm, F. (1985). Die Hämoglobine des Feldspaerlings (*Passer montanus*, Passeriformes). *Biol. Chem. Hoppe-Seyler*, **366**, 893–9.

Schneider, W.J., Carroll, R., Severson, D.L. & Nimpf, J. (1990). Apolipoprotein VLDL-II inhibits lipolysis of triglyceride-rich lipoproteins in the laying hen. *J. Lipid Res.*, **31**, 507–13.

Schofield, J.N., Rowe, A. & Brickell, P.M. (1992). Position-dependence of retinoic acid receptor-β gene expression in the chick limb bud. *Dev. Biol.*, **152**, 344–53.

Schuchard, M., Subramanian, M., Ruesink, T. & Spelsberg, T.C. (1991). Nuclear matrix localization and specific matrix DNA binding by receptor binding factor 1 of the avian oviduct progesterone receptor. *Biochemistry*, **30**, 9516–22.

Schultz, G.A. & Church, R.B. (1972). DNA base sequence heterogeneity in the order Galliformes. *J. Exp. Zool.*, **179**, 119–128.

Schütz, H. & Gerstberger, R. (1990). Atrial natriuretic factor controls salt gland secretion in the pekin duck (*Anas platyrhynchos*) through interaction with high affinity receptors. *Endrocrinology*, **127**, 1718–26.

Schuur, E.R. & Baluda, M.A. (1991). Expression of MYB proteins in avian hemopoietic tissues. *Oncogene*, **6**, 1923–9.

Schwabe, J.W.R. & Rhodes, D. (1991). Beyond zinc fingers: steroid hormone receptors have a novel structural motif for DNA recognition. *Trends Biochem. Sci.*, **16**, 291–6.

Schwartz, D.E., Tizard, R. & Gilbert, W. (1983). Nucleotide sequence of Rous sarcoma virus. *Cell*, **32**, 853–69.

Scott, M.J., Huckaby, C.S., Kato, I. *et al.* (1987a). Ovoinhibitor introns specify functional domains as in the related and linked ovomucoid gene. *J. Biol. Chem.*, **262**, 5899–907.

Scott, M.J., Tsai, M.-J. & O'Malley, B.W. (1987b). Deoxyribonuclease I sensitivity of the ovomucoid-ovoinhibitor gene complex in oviduct nuclei and relative location of CR1 repetitive sequences. *Biochemistry*, **26**, 6831–40.

Scott, M.L., Holm, E.R. & Reynolds, R.E. (1963). Studies on the protein and methionine requirements of young Bobwhite quail and young Ringnecked pheasants. *Poult. Sci.*, **42**, 676–80.

Scott, M.L., Nesheim, M.C. & Young, R.J. (1982). *Nutrition of the Chicken*, 3rd edn. New York: M.L. Scott and Associates.

Scott, M.P. (1992). Vertebrate homeobox gene nomenclature.

Cell, **71**, 551–3.

Scotting, P., McDermott, H. & Meyer, R.J. (1991). Ubiquitin–protein conjugates and αB crystallin are selectively present in cells undergoing major cytomorphological reorganisation in early chicken embryos. *FEBS Lett.*, **285**, 75–9.

Senesi, S., Freer, G., Batoni, G. *et al.* (1992). Purine metabolism and B-lymphocyte development in the chicken bursa of Fabricius. *Dev. Comp. Immunol.*, **16**, 197–207.

Serrano, J., Bevins, C.L., Young, S.W. & De Pablo, F. (1989). Insulin gene expression in chicken ontogeny: pancreatic, extrapancreatic, and prepancreatic. *Dev. Biol.*, **132**, 410–18.

Shand, J.H., West, D.W., Noble, R.C. & Speake, B.K. (1994). The esterification of cholesterol in the liver of the chick embryo. *Biochim. Biophys. Acta*, **1213**, 224–30.

Shapira, E., Yarus, S. & Fainsod, A. (1991). Genomic organization and expression during embryogenesis of the chicken CR1 repeat. *Genomics*, **10**, 931–9.

Sharp, P.M. & Li, W.-H. (1987). Ubiquitin genes as a paradigm of concerted evolution of tandem repeats. *J. Mol. Evol.*, **25**, 58–64.

Shaw, E.M., Guise, K.S. & Shoffner, R.N. (1989). Chromosomal localization of chicken sequences homologous to the β-actin gene by in situ hybridization. *J. Hered.*, **80**, 475–8.

Sheldon, F.H. & Bledsoe, A.H. (1993). Avian molecular sytematics: 1970s to 1990s. *Annu. Rev. Ecol. System.*, **24**, 243–78.

Shen, C.S. & Mistry, S.P. (1979). Development of gluconeogenic, glycolytic, and pentose-shunt enzymes in the chicken kidney. *Poult. Sci.*, **58**, 663–7.

Sherwood, T.A., Alphin, R.L., Saylor, W.W. & White, H.B. (1993). Folate metabolism and deposition in eggs by laying hens. *Arch. Biochem. Biophys.*, **307**, 66–72.

Shields, C.E., Herman, Y.F. & Herman, R.H. (1964). 1-^{14}C glucose utilization of intact nucleated red blood cells of selected species. *Nature*, **203**, 935–6.

Shields, G.F. & Wilson, A.C. (1987). Calibration of mitochondrial DNA evolution in geese. *J. Mol. Evol.*, **24**, 212–17.

Shih, J.C.H., Pullman, E.P. & Kao, K.J. (1983). Genetic selection, general characterisation and histology of atherosclerosis-susceptible and resistant Japanese quail. *Atherosclerosis*, **49**, 41–53.

Shim, K.F. & Vohra, P. (1984). A review of the nutrition of Japanese Quail. *World's Poult. Sci. J.*, **40**, 261–74.

Shimizu, M., Nagashima, H. & Hashimoto, K. (1993). Comparative studies on molecular stability of immunoglobulin G from different species. *Comp. Biochem. Physiol.*, **106B**, 255–61.

Shinomura, T., Nishida, Y., Ito, K. & Kimata, K. (1993). cDNA cloning of PG-M, a large chondroitin sulfate proteoglycan expressed during chondrogenesis in chick limb buds. *J. Biol. Chem.*, **268**, 14461–9.

Shore, S.K. (1988). The *yes* oncogene. In *The Oncogene Handbook*, ed. E.P. Reddy, A.M. Skalka & T. Curran, pp. 175–81. Amsterdam: Elsevier.

Sibley, C.G. & Alquist, J.E. (1990). *Phylogeny and Classification of Birds: a Study in Molecular Evolution.* New Haven: Yale University Press.

Sibley, C.G. & Monroe, B.L. (1990). *Distribution and Taxonomy of Birds of the World.* New Haven: Yale University Press.

Sif, S., Capobianco, A.J. & Gilmore, T.D. (1993). The v-Rel oncoprotein increases expression from Sp1 site-containing promoters in chicken embryo fibroblasts. *Oncogene*, **8**, 2501–9.

Silva, R., Fischer, A.H. & Burch, J.B.E. (1989). The major and minor chicken vitellogenin genes are each adjacent to partially deleted pseudogene copies of the other. *Mol. Cell. Biol.*, **9**, 3557–62.

Silverman, M. (1991). Structure and function of hexose transporters. *Annu. Rev. Biochem.*, **60**, 757–94.

Simek, S. & Rice, N.R. (1988). Detection and characterization of the protein encoded by the chicken c-*rel* protooncogene. *Oncogene Res.*, **2**, 103–19.

Simon, J. (1989). Chicken as a useful species for the comprehension of insulin action. *Crit. Rev. Poult. Biol.*, **2**, 121–48.

Simon, J., Chevalier, B., Derouet, M. & Leclercq, B. (1991). Normal number and kinase activity of insulin receptors in liver of genetically fat chickens. *J. Nutr.*, **121**, 379–85.

Simon, J., Freychet, P. & Rosselin, G. (1977). A study of insulin binding sites in the chicken tissues. *Diabetologia*, **13**, 219–28.

Simon, J. & LeClercq, B. (1982). Longitudinal study of adiposity in chickens selected for high or low abdominal fat content: further evidence of a glucose-insulin imbalance in the fat line. *J. Nutr.*, **112**, 1961–73.

Simon, J. & Leroith, D. (1986). Insulin receptors of chicken liver and brain: characterization of α and β subunit properties. *Eur. J. Biochem.*, **158**, 125–32.

Simon, J. & Taouis, M. (1993). The insulin receptor in chicken tissues. In *Avian Endocrinology*, ed. P.J. Sharp, pp. 177–88, Bristol: Society of Endocrinology.

Simonsen, M. (1981). The major histocompatibility complex in a bird's-eye view. In *Immunobiology of the Major Histocompatibility Complex*, ed. M.B. Zaleski, C.J. Abeyounis & K. Kano, pp. 192–201, Basel: Karger.

Sitbon, G., Laurent, F., Mialhe, A. *et al.* (1980). Diabetes in birds. *Horm. Met. Res.*, **12**, 1–9.

Sjöberg, M., Vennström, B. & Forrest, D. (1992). Thyroid hormone receptors in chick retinal development: differential expression of mRNAs for α and N-terminal variant β receptors. *Development*, **114**, 39–47.

Skadhauge, E. (1983). Formation and composition of urine. In *Physiology and Biochemistry of the Domestic Fowl*, vol. 4, ed. B.M. Freeman, pp. 108–35. London: Academic Press.

Slater, E.P., Redeuilh, G. & Beato, M. (1991). Hormonal regulation of vitellogenin genes: an estrogen-responsive element in the *Xenopus* A2 gene and a multihormonal regulatory region in the chicken II gene. *Mol. Endocrinol.*, **5**, 386–96.

Smillie, L.B., Gologinska, K. & Reinach, F.C. (1988). Sequences of complete cDNAs encoding four variants of chicken skeletal muscle troponin T. *J. Biol. Chem.*, **263**, 18816–20.

Smith, C.W.J., Patton, J.G. & Nadal-Ginard, B. (1989). Alternative splicing in the control of gene expression. *Annu. Rev. Genet.*, **23**, 527–77.

Smith, D.F. & Toft, D.O. (1992). Composition, assembly and activation of the avian progesterone receptor. *J. Steroid Biochem. Mol. Biol.*, **41**, 201–7.

Smith, D.R., Vogt, P.K. & Hayman, M.J. (1989). The v-*sea* oncogene of avian erythroblastosis retrovirus S13: another member of the protein-tyrosine kinase gene family. *Proc. Natl. Acad. Sci. USA*, **86**, 5291–5.

Smith, J.N. (1968). The comparative metabolism of xenobiotics. *Adv. Comp. Physiol. Biochem.*, **3**, 173–232.

Smith, S.M. & Eichele, G. (1991). Temporal and regional differences in the expression pattern of distinct retinoic acid receptor-β transcripts in the chick embryo. *Development*, **111**, 245–52.

Sokolove, P.M. (1985). Altered membrane association of glycogen phosphorylase in the dystrophic chicken. *Biochim. Biophys. Acta*, **841**, 232–6.

Soler, A.P. & Knudsen, K.A. (1991). Colocalization of N-CAM and N-cadherin in avian skeletal myoblasts. *Dev. Biol.*, **148**, 389–92.

Söling, H.D. (1982). Species dependent compartmentation of guanine nucleotide metabolism in the vertebrate liver cell. In *Metabolic Compartmentation*, ed. H. Sies, pp. 123–49. London: Academic Press.

Söling, H.D., Kleine, J., Willms, B., Janson, G. & Kuhn, A. (1973). Relationship between the intracellular distribution of phosphoenolpyruvate carboxykinase, regulation of gluconeogenesis and energy cost of glucose formation. *Eur. J. Biochem.*, **37**, 233–43.

Söling, H.D. & Kleinicke, J. (1976). Species dependent regulation of hepatic gluconeogenesis in higher animals. In *Gluconeogenesis: Its Regulation in Mammalian Species*, ed. R.W. Hanson & M.A. Mehlman, pp. 369–462. New York: Wiley-Interscience.

Sollenberger, K.G., Kao, T.-L. & Taparowsky, E.J. (1994). Structural analysis of the chicken *max* gene. *Oncogene*, **9**, 661–4.

Somes, R.G. (1988). International Registry of Poultry Genetic Stocks. *Storrs Agri. Exp. Station Bull.*, 476.

Sommerville, J. (1986). Nucleolar structure and ribosome biogenesis. *Trends Biochem. Sci.*, **11**, 438–42.

Sorkin, B.C., Gallin, W.J., Edelman, G.M. & Cunningham, B.A. (1991). Genes for two calcium-dependent cell adhesion molecules have similar structures and are arranged in tandem in the chicken genome. *Proc. Natl. Acad. Sci. USA*, **88**, 11545–9.

Speake, B.K., Noble, R.C. & McCartney, R. (1993). Tissue-specific changes in lipid composition and lipoprotein lipase activity during the development of the chick embryo. *Biochim. Biophys. Acta*, **1165**, 263–70.

Speiss, J., Rivier, J.E., Rodkey, J.A., Bennett, C.D. & Vale, W. (1979). Isolation and characterization of somatostatin from pigeon pancreas. *Proc. Natl. Acad. Sci. USA*, **76**, 2974–8.

Spike, C.A. & Lamout, S.J. (1995). Genetic analysis of three loci homologous to human G9a: evidence for linkage of a class III gene with chicken MHC. *Animal Genet.*, **26**, 185–7.

Squires, M.W. & Naber, E.C. (1993). Vitamin profiles of eggs as indicators of nutritional status in the laying hen: riboflavin study. *Poult. Sci.*, **72**, 483–94.

Stanley, J.C., Dohm, G.L., McManus, B.S. & Newsholme, E.A. (1984). Activities of glucokinase and hexokinase in mammalian and avian livers. *Biochem. J.*, **224**, 667–71.

Stapel, S.O. & de Jong, W.W. (1983). Lamprey 48-kDa lens protein represents a novel class of crystallins. *FEBS Lett.*, **162**, 305–9.

Stapel, S.O., Zweers, A., Dodemont, H.J., Kan, J.H. & de Jong, W.W. (1985). ε-Crystallin, a novel avian and reptilian eye lens protein. *Eur. J. Biochem.*, **147**, 129–36.

Stavnezer, E. (1988). The *ski* gene. In *The Oncogene Handbook*, ed. E.P. Reddy, A.M. Skalka & T. Curran, pp. 393–401. Amsterdam: Elsevier.

Stavnezer, E., Brodeur, D. & Brennan, L.A. (1989). The v-*ski* oncogene encodes a truncated set of c–*ski* coding exons with limited sequence and structural relatedness to v-*myc*. *Mol. Cell. Biol.*, **9**, 4038–45.

Stehelin, D., Varmus, H.E., Bishop, J.M. & Vogt, P.K. (1976). DNA related to the transforming gene(s) of avian sarcoma viruses is present in normal avian DNA. *Nature*, **260**, 170–3.

Stein, J.P., Catterall, J.F., Kristo, P., Means, A.R. & O'Malley, B.W. (1980). Molecular cloning of the ovomucoid gene sequences from partially purified messenger RNA. *Cell*, **21**, 681–7.

Sternberg, D.W., Scholz, G., Fukui, Y. & Hanafusa, H. (1993). Activation of a histone H1 kinase by tyrosine phosphorylation in v-*src*-transformed fibroblasts. *EMBO J.*, **12**, 323–30.

Stevens, L. (1986). Gene structure and organisation in the domestic fowl (*Gallus domesticus*). *Worlds Poult. Sci. J.*, **42**, 232–42.

Stevens, L. (1991a). *Genetics and Evolution of the Domestic Fowl*. Cambridge: Cambridge University Press.

Stevens, L. (1991b). Egg white proteins. *Comp. Biochem. Physiol.*, **100B**, 1–9.

Steyrer, E., Barber, D.L. & Schneider, W.J. (1990). Evolution of lipoprotein receptors. *J. Biol. Chem.*, **265**, 19575–81.

Stifani, S., Barber, D.L., Aebersold, R. *et al.* (1991). The laying hen expresses two different low density lipoprotein receptor-related proteins. *J. Biol. Chem.*, **266**, 19079–87.

Stifani, S., George, R. & Schneider, W.J. (1988). Solubilization and characterization of the chicken oocyte vitellogenin receptor. *Biochem. J.*, **250**, 467–75.

Stock, A.D. & Bunch, T.D. (1982). The evolutionary implications of chromosome banding pattern homologies in the bird order Galliformes. *Cytogenet. Cell Genet.*, **34**, 136–48.

Strack, P.R., Wajnberg, E.F., Waxman, L. & Fagan, J.M. (1991). Purification of the multicatalytic proteinase from the nucleus and cytoplasm of chicken red blood cells. *Biomed. Biochim. Acta*, **50**, 4–6.

Strack, P.R., Wajnberg, E.F., Waxman, L. & Fagan, J.M. (1992). Comparison of the multicatalytic proteinases isolated from the nucleus and cytoplasm of chicken red blood cells. *Int. J. Biochem.*, **24**, 887–95.

Strandholm, J.J., Cardenas, J.M. & Dyson, R.D. (1975). Pyruvate kinase isozymes in adult and fetal tissues of chicken. *Biochemistry*, **14**, 2242–6.

Strijker, R., Assendelft, G.B., Dikkeschei, B.D., Gruber, M. & AB, G. (1986). Estradiol-dependent transcription initiation upstream from the chicken apoVLDLII gene coding for the very-low-density apolipoprotein II. *Gene*, **45**, 27–35.

Strittmatter, C.F. (1965). Studies on avian xanthine dehydrogenase. *J. Biol. Chem.*, **240**, 2557–64.

Stumph, W.E., Kristo, P., Tsai, M.-J. & O'Malley, B.W. (1981). A chicken middle-repetitive DNA sequence which shares homology with mammalian ubiquitous repeats. *Nucl. Acids Res.*, **9**, 5383–97.

Sturkie, P.D. (1986). Kidneys, extrarenal salt excretion, and urine.

In *Avian Physiology*, 4th edn, ed. P.D. Sturkie, pp. 359–82. New York: Springer-Verlag.

Subrahmanyan, S.C., Kasturi, R., Pazirandeh, M., Stolow, D.T., Huang, W.-Y. & Wakil, S.J. (1989). A novel cDNA extension procedure: isolation of chicken fatty acid synthase cDNA clones. *J. Biol. Chem.*, **264**, 3750–7.

Subramanian, M., Harris, S.A., Rasmussen, K. & Spelsberg, T.C. (1993). Rapid down-regulation of *c-jun* protooncogene transcription by progesterone in the avian oviduct. *Endocrinology*, **133**, 2049–54.

Sugano, T., Shiota, M., Khono, H. & Shimada, M. (1982a). Intracellular redox state and control of gluconeogenesis in perfused chicken liver. *J. Biochem.*, **91**, 1917–29.

Sugano, T., Shiota, M., Khono, H. & Shimada, M. (1982b). Stimulation of gluconeogenesis by glucagon and norepinephrine in the perfused chicken liver. *J. Biochem.*, **92**, 111–20.

Sugarman, B.J., Dodgson, J.B. & Engel, J.D. (1983). Genomic organization, DNA sequence, and expression of chicken embryonic histone genes. *J. Biol. Chem.*, **258**, 9005–16.

Sugden, P.H. & Fuller, S.J. (1991). Regulation of protein turnover in skeletal and cardiac muscle. *Biochem. J.*, **273**, 21–37.

Sumner, A.T. (1990). *Chromosome Banding*. London: Unwin Hyman.

Sun, J.-M., Hendzel, M.J. & Davie, J.R. (1992). Nuclear matrix proteins bind very tightly to specific regions of the chicken histone H5 gene. *Biochem. Cell Biol.*, **70**, 822–9.

Sun, J.-M., Penner, C.G. & Davie, J.R. (1992). Analysis of erythroid nuclear proteins binding to the promoter and enhancer elements of the chicken histone H-5 gene. *Nucl. Acids Res.*, **20**, 6385–92.

Sun, J.-M., Penner, C.G. & Davie, J.R. (1993). Repression of histone H5-gene expression in mature erythrocytes is correlated with reduced DNA-binding activities of transcription factors sp1 and GATA-1. *FEBS Lett.*, **331**, 141–4.

Sunde, M.L., Swick, R.W. & Kang, C.W. (1984). Protein degradation: an important consideration. *Poult. Sci.*, **63**, 2055–61.

Sundin, O.H. & Eichele, G. (1990). A homeo domain protein reveals the metameric nature of the developing chick hindbrain. *Genes Dev.*, **4**, 1267–76.

Sutrave, P. & Hughes, S.H. (1989). Isolation and characterization of three distinct cDNAs for the chicken c-ski gene. *Mol. Cell. Biol.*, **9**, 4046–51.

Suzuki, H.R., Padanilam, B.J., Vitale, E., Ramirez, F. & Solursh, M. (1991). Repeating developmental expression of G-Hox 7, a novel homeobox-containing gene in the chicken. *Dev. Biol.*, **148**, 375–88.

Suzuki, K. & Tsuji, S. (1982). Synergistic activation of calcium-activated neutral proteinase by Mn^{2+} and Ca^{2+}. *FEBS Lett.*, **140**, 16–18.

Suzuki, T., Maruyama, T. & Morohashi, M. (1994). Distribution of sulphydryl groups and disulphide linkages in tissue proteins of quail uropygial glands. *Br. Poult. Sci.*, **35**, 323–33.

Svacha, A., Weber, C.W. & Reid, B.L. (1970). Lysine, methionine and glycine requirements of Japanese quail to five weeks of age. *Poult. Sci.*, **49**, 54–9.

Swain, S.D. (1992a). Energy profiles during fasting in horned larks (*Eremophila alpestris*). *Physiol. Zool.*, **63**, 568–82.

Swain, S.D. (1992b). Flight muscle catabolism during overnight fasting in a passerine bird, *Eremophila alpestris*. *J. Comp.*

Physiol., **162B**, 383–92.

Swasdison, S., Mayne, P.M., Wright, D.W. *et al.* (1992). Monoclonal antibodies that distinguish avian type I and type III collagens: isolation, characterization and immunolocalization in various tissues. *Matrix*, **11**, 56–65.

Swick, R.W. (1982). Growth and protein turnover in animals. *Crit. Rev. Food Sci. Nutr.*, **16**, 117–26.

Swiderski, R.E. & Solursh, M. (1992a). Localization of type II collagen, long form α1(IX) collagen, and short form α1(IX) collagen transcripts in the developing chick notochord and axial skeleton. *Dev. Dynamics*, **194**, 118–27.

Swiderski, R.E. & Solursh, M. (1992b). Differential coexpression of long and short form type IX collagen transcripts during avian limb chondrogenesis in ovo. *Development*, **115**, 169–79.

Takahashi, K., Asao, T., Suzuki, N., Tashiro, M. & Kanamori, M. (1992). Purification and characterization of ovoinhibitor from Japanese quail egg white. *Nippon Nogeikagaku Kaishi*, **66**, 1757–64.

Takai, T., Yokoyama, C., Wada, K. & Tanabe, T. (1988). Primary structure of chicken liver acetylCoA carboxylase deduced from cDNA sequence. *J. Biol. Chem.*, **263**, 2651–7.

Takase, S. & Goda, T. (1990). Developmental changes in vitamin A level and lack of retinyl palmitate in chick lungs. *Comp. Biochem. Physiol.*, **96B**, 415–19.

Takeda, S., Masteller, E.L., Thompson, C.B. & Buerstedde, J.M. (1992). RAG-2 expression is not essential for chicken immunoglobulin gene conversion. *Proc. Natl. Acad. Sci. USA*, **89**, 4023–7.

Takeichi, M. (1991). Cadherin cell adhesion receptors as a morphogenetic regulator. *Science*, **251**, 1451–5.

Tamir, H. & Ratner, S. (1963). Enzymes of arginine metabolism in chicks. *Arch. Biochem. Biophys.*, **102**, 249–58.

Tan, K.O., Sater, G.R., Myers, A.M., Robson, R.M. & Huiatt, T.W. (1993). Molecular characterization of avian muscle titin. *J. Biol. Chem.*, **268**, 22900–7.

Tanaka, A., Salem, M., Eckrode, R.J. & Fujita, D.J. (1990). Characterization of avian retroviruses carrying activated transformed human c-src genes and of the steps involved in expression of activated src-PKinases *in vitro*. *Oncogene Res.*, **5**, 305–22.

Taouis, M., Derouet, M., Chevalier, B. & Simon, J. (1993). Corticosterone effect on insulin receptor number and kinase activity in chicken muscle and liver. *Gen. Comp. Endocrinol.*, **89**, 167–75.

Targa, F.R., Gallo, C.V.D., Huesca, M., Scherrer, K. & Marcaud, L. (1993b). Silencer and enhancer elements located at the 3'-side of the chicken and duck α-globin-encoding gene domains. *Gene*, **129**, 229–37.

Targa, F.R., Gallo, C.V.D. & Scherrer, K. (1993a). Analysis of the distribution of protein-binding DNA motives in the vicinity of the 3'-side chicken α-globin enhancer. *Biochem. Biophys. Res. Commun.*, **190**, 1163–72.

Targa, F.R., Huesca, M. & Scherrer, K. (1992). Preliminary characterization of a nuclear factor interacting with the silencer element at the 3'-side of the chicken α-globin gene domain. *Biochem. Biophys. Res. Commun.*, **188**, 416–23.

Tedford, B.L. & Meier, A.H. (1993). Daily rhythm of plasma insulin in Japanese quail (*Coturnix c. Japonica*) fed *ad libitum*.

Comp. Biochem. Physiol., **104A**, 143–5.

Tereba, A., McPhaul, M.J. & Wilson, J.D. (1991). The gene for aromatase (P450$_{arom}$) in the chicken is located on the long arm of chromosome 1. *J. Hered.*, **82**, 80–1.

Thaller, C. & Eichele, G. (1987). Identification and spatial distribution of retinoids in the developing chick limb bud. *Nature*, **237**, 625–8.

Thaller, C., Hofmann, C. & Eichele, G. (1993). 9-*cis*-retinoic acid, a potent inducer of digit pattern duplications in the chick wing bud. *Development*, **118**, 957–65.

Thampy, K.G. & Koshy, A.G. (1991). Purification, characterisation, and ontogeny of acetyl-CoA carboxylase isozyme of chick embryo brain. *J. Lipid Res.*, **32**, 1667–73.

Thompson, C.B. (1992). Creation of immunoglobulin diversity by intrachromosomal gene conversion. *Trends Genet.*, **8**, 416-22.

Thurston, R.J., Bryant, C.C. & Korn, N. (1993). The effects of corticosterone and catecholamine infusion on plasma glucose levels in chicken (*Gallus domesticus*) and turkey (*Meleagris gallapavo*). *Comp. Biochem. Physiol.*, **106C**, 59–62.

Tickle, C., Alberts, B., Wolpert, L. & Lee, J. (1982). Local application of retinoic acid to the limb bond mimics the action of the polarizing region. *Nature*, **296**, 564–6.

Tiersch, T.R. & Wachtel, S.S. (1991). On the evolution of genome size of birds. *J. Hered.*, **82**, 363–8.

Tikhonenko, A.T., Hartman, A.-R. & Linial, M.L. (1993). Overproduction of v-Myc in the nucleus and its excess over Max are not required for avian fibroblast transformation. *Mol. Cell. Biol.*, **13**, 3623–31.

Tinker, D.A., Brosnan, J.T. & Herzberg, G.R. (1986). Interorgan metabolism of amino acids, glucose, lactate, glycerol and uric acid in the domestic fowl (*Gallus domesticus*). *Biochem. J.*, **240**, 829–36.

Tixier-Bouchard, M. (1993). Current state of the art in poultry genome mapping. *Manipulation of the Avian Genome*, September 1993, Guelph, Ontario: Department of Poultry Science.

Tone, M., Nakano, N., Takao, E., Narisawa, S. & Mizuno, S. (1982). Demonstration of W chromosome-specific repetitive DNA sequences in the domestic fowl *Gallus gallus domesticus*. *Chromosoma*, **86**, 551–69.

Tone, M., Sakaki, Y., Hashiguchi, T., Mizuno, S. (1984). Genus specificity and extensive methylation of the W chromosome-specific repetitive DNA sequences from the domestic fowl, *Gallus gallus domesticus*. *Chromosoma*, **89**, 228–37.

Toth, T.E. & Norcross, N.L. (1981). Precipitating and agglutinating activity in duck anti-soluble protein immune sera. *Avian Dis.*, **25**, 338–52.

Townes, P. & Holtfretter, J. (1955). Directed movements and selective adhesion of embryonic amphibian cells. *J. Exp. Zool.*, **128**, 53–120.

Trueb, J. & Trueb, B. (1992). Molecular cloning of a novel Ras-like protein from chicken. *FEBS Lett.*, **306**, 181–4.

Tsai, M.-J. & O'Malley, B.W. (1991). Mechanisms of regulation of gene transcription by steroid receptors. In *The Hormonal Control Regulation of Gene Transcription*, ed. P. Cohen & J.G. Foulkes, pp. 101–16. Amsterdam: Elsevier.

Tsai, S.Y., Tsai, M.-J. & O'Malley, B.W. (1991). The steroid receptor superfamily: transactivators of gene expression. In *Nuclear Hormone Receptors*, ed. M.G. Parker, pp. 103–24. London: Academic Press.

Tsuji, S. (1983). Chicken ornithine transcarbamylase: purification and some properties. *J. Biochem.*, **94**, 1307–15.

Tucker, R.P., Spring, J., Baumgartner, S. *et al.* (1994). Novel tenascin variants with a distinctive pattern of expression in the avian embryo. *Development*, **120**, 637–47.

Tung, A.K. (1973). Biosynthesis of glucagon: evidence for a possible high molecular weight biosynthetic intermediate. *Horm. Met. Res.*, **5**, 416–24.

Uetsuki, T., Nabeshima, Y., Fujisawa-Sehara, A. & Nabeshima, Y.-I. (1990). Regulation of the chicken embryonic myosin light-chain (L23) gene: existence of a common regulatory element shared by myosin alkali light-chain genes. *Mol. Cell. Biol.*, **10**, 2562–9.

Ukhanova, M.V. & Leibush, B.N. (1988). Binding of glucagon by isolated chicken hepatocytes. *J. Evol. Biochem. Physiol.*, **24**, 392–7.

Unsicker, K. (1973). Fine structure and innervation of the avian adrenal gland. I. Fine structure of adrenal chromaffin cells and ganglion cells. *Z. Zellforsch. Mikrosk. Anat.*, **145**, 389–416.

Upholt, W.B. & Sandell, L.J. (1986). Exon/intron organization of the chicken type II procollagen gene: intron size distribution suggests a minimal intron size. *Proc. Natl. Acad. Sci. USA*, **83**, 2325–9.

Ureta, T., Reichberg, S.B., Radojkovic, J. & Slebe, J.C. (1973). Comparative studies on glucose phosphorylating isoenzymes of vertebrates. IV. *Comp. Biochem. Physiol.*, **45B**, 445–61.

van Hest, B.J., Molloy, P.L., Frankham, R. & Sheldon, B.L. (1994). Heat shock protein *HSP108* and a replication gene cluster are linked in the chicken. *Anim. Genet.*, **23**, 109–11.

van Schaftingen, E. & Hers, H.-G. (1986). Purification and properties of phosphofructokinase-2/fructose-2,6-phosphatase from chicken liver and pigeon muscle. *Eur. J. Biochem.*, **159**, 359–65.

van Wyk, E., van der Bank, F.H., Verdoorn, G.H. & Bouwman, H. (1993). Chlorinated hydrocarbon insecticide residues in the cape griffon vulture (*Gyps coprotheres*). *Comp. Biochem. Physiol.*, **104C**, 209–20.

Varmus, H. & Weinberg, R.A. (1993). *Genes and the Biology of Cancer*. New York: Scientific American Library.

Veevers, G. (1982). *The Colours of Animals*. London: Arnold.

Veiga, J.A.S., Roselino, E.S. & Migliorini, R.H. (1978). Fasting, adrenalectomy, and gluconeogenesis in the chicken and a carnivorous bird. *Am. J. Physiol.*, **234**, R115–21.

Velarde, F.L., Espinoza, D., Monge, C., de Muizon, C. & de Muizon, C. (1991). A genetic response to high altitude hypoxia: high hemoglobin-oxygen affinity in chicken (*Gallus gallus*) from the Peruvian Andes. *C. R. Acad. Sci. Paris*, **313**, 401–6.

Vennström, B. & Damm, K. (1988). The *erbA* and *erbB* oncogenes. In *The Oncogene Handbook*, ed. E.P. Reddy, A.M. Skalka & T. Curran, pp. 25–37. Amsterdam: Elsevier.

Venturini, G., Capanna, E. & Fontana, B. (1987). Size and structure of the bird genome II. Repetitive DNA and sequence organization. *Comp. Biochem. Physiol.*, **87B**, 975–9.

Venturini, G., D'Ambrogi, R. & Capanna, E. (1986). Size and

structure of the bird genome- 1. DNA content of 48 species of neognathae. *Comp. Biochem. Physiol.*, **85B**, 61–5.

Vermeer, C. (1990). γ-Carboxy glutamate-containing proteins and the vitamin K-dependent carboxylase. *Biochem. J.*, **266**, 625–36.

Vieira, A.V. & Schneider, W.J. (1993). Transport and uptake of retinol during chicken oocyte growth. *Biochim. Biophys. Acta*, **1169**, 250–6.

Vives, V., Sancho, J. & Gomezcap, J.A. (1981) Studies in vivo and in vitro of insulin effect on the metabolism of glucose in different chicken tissues. *Comp. Biochem. Physiol.*, **69B**, 479–485.

Vuorio, E. & de Crombrugghe, B. (1990). The family of collagen genes. *Annu. Rev. Biochem.*, **59**, 837–72.

Wada, K., Takai, T. & Tanabe, T. (1987). Amino acid sequence of chicken liver cathepsin L. *Eur. J. Biochem.*, **167**, 13–18.

Wada, K. & Tanabe, T. (1988). Purification and characterization of chicken liver cathepsin-B. *J. Biochem.*, **104**, 472–6.

Wakil, S.J., Stoops, J.K. & Joshi, V.C. (1983). Fatty acid synthesis and its regulation. *Annu. Rev. Biochem.*, **52**, 537–79.

Walchli, C., Koller, E., Trueb, J. & Trueb, B. (1992). Structural comparison of chicken genes for α-1(VI) and α-2(VI) collagen. *Eur. J. Biochem.*, **205**, 583–9.

Walchli, C., Trueb, J., Kessler, B., Winterhalter, K.H. & Trueb, B. (1993). Complete primary structure of chicken collagen-XIV. *Eur. J. Biochem.*, **212**, 483–90.

Wallner-Pendleton, E.A., Rogers, D. & Epple, A. (1993). Diabetes in a red-tailed hawk (*Buteo jamaicensis*). *Avian Path.*, **22**, 631–5.

Wals, P.A. & Katz, J. (1981). Glucokinase in bird liver: a membrane bound enzyme. *Biochem. Biophys. Res. Commun.*, **100**, 1543–8.

Wang, L.-H. (1988). The *ros* oncogene. In *The Oncogene Handbook*, ed. E.P. Reddy, A.M. Skalka & T. Curran, pp. 135–47. Amsterdam: Elsevier.

Watford, M. (1985). Gluconeogenesis in the chicken: regulation of phosphoenolpyruvate carboxylase gene expression. *Fed. Proc.*, **44**, 2469–74.

Watford, M. (1991). The urea cycle: a two-compartment system. *Essays Biochem.*, **26**, 49–58.

Watford, M., Hod, Y., Chiao, Y.-B., Utter, M. & Hanson, R.W. (1981). The unique role of the kidney in gluconeogenesis in the chicken. *J. Biol. Chem.*, **256**, 10023–7.

Watkins, B.A. (1991). Importance of essential fatty acids and their derivatives in poultry. *J. Nutr.*, **121**, 1475–85.

Watson, D.K., McWilliams, M.J. & Papas, T.S. (1988). Molecular organization of the chicken *ets* locus. *Virology*, **164**, 99–105.

Watson, J.D. & Crick, F.H.C. (1953). Molecular structure of nucleic acids: a structure for deoxyribose nucleic acid. *Nature*, **171**, 737–8.

Watson, J.D., Hopkins, N.H., Roberts, J.W., Steitz, J.A. & Weiner, A.M. (1987). *The Molecular Biology of the Gene*. California: Benjamin/Cummings.

Wei, D. & Andrews, G.K. (1988). Molecular cloning of chicken metallothionein. Deduction of the complete amino acid sequence and analysis of expression using cloned cDNA. *Nucl. Acids Res.*, **16**, 537–53.

Weiss, R., Teich, N., Varmus, H. & Coffin, J. (1982). *Molecular Biology of Tumor Viruses*. 2nd edn. Cold Spring Harbor,

New York: Cold Spring Harbor Laboratory Press.

Weldon, S.L., Rando, A., Kalonick, P. *et al.* (1990). Mitochondrial phosphoenolpyruvate carboxykinase from the chicken. *J. Biol. Chem.*, **265**, 7308–17.

Wenink, P.W., Baker, A.J. & Tilanus, M.G.J. (1993). Hypervariable-control-region sequences reveal global population structuring in a long-distance migrant shorebird, the Dunlin (*Calidris alpina*). *Proc. Natl. Acad. Sci. USA.*, **90**, 94–8.

Weston, K. (1992). Extension of the DNA-binding consensus of the chicken c-*myb* and v-*myb* proteins. *Nucl. Acids Res.*, **20**, 3043–9.

Wetmore, A. (1960). A classification for birds of the world. *Smithsonian Miscel. Collection*, **139**, 1–37.

Wetton, J.H., Carter, R.E., Parkin, D.T. & Walters, D. (1987). Demographic study of a wild house sparrow population by DNA fingerprinting. *Nature*, **327**, 147–9.

Whitbread, L.A., Gregg, K. & Rogers, G.E. (1991). The structure and expression of a gene encoding chick claw keratin. *Gene*, **101**, 223–9.

White, H.B. (1987). Vitamin-binding proteins in the nutrition of the avian embryo. *J. Exp. Zool.*, Suppl. **1**, 53–63.

White, H.B. & Merrill, A.H. (1988). Riboflavin-binding proteins. *Annu. Rev. Nutr.*, **8**, 279–99.

Whittow, G.C. (1986). Regulation of body temperature. In *Avian Physiology*, 4th edn, ed. P.D. Sturkie. p. 223. New York: Springer Verlag.

Wick, G., Brezinschek, H.P., Hála, K., Dietrich, H., Wolf, H. & Kroemer, G. (1989). The obese strain of chickens: an animal model with spontaneous autoimmune thyroiditis. *Adv. Immunol.*, **47**, 433–500.

Wiese, T.J., Lambeth, D.O. & Ray, P.D. (1991). The intracellular distribution and activities of phosphoenolpyruvate car-boxykinase isozymes in various tissues of several mammals and birds. *Comp. Biochem. Physiol.*, **100B**, 297–302.

Wiggins, D., Lund, P. & Krebs, H.A. (1982). Adaptation of urate synthesis in chicken liver. *Comp. Biochem. Physiol.*, **72B**, 565–8.

Will, B.H., Usui, Y. & Suttie, J.W. (1992). Comparative metabolism and requirement of vitamin K in chicks and rats. *J. Nutr.*, **122**, 2354–60.

Williams, D.L. (1979). Apoproteins of avian very low density lipoprotein: demonstration of a single high molecular weight apoprotein. *Biochemistry*, **18**, 1056–63.

Williams, J. (1968). A comparison of glycopeptides from the ovotransferrin and serum transferrin of the hen. *Biochem. J.*, **108**, 57–67.

Williamson, A.R. & Turner, M.W. (1987). *Essential Immunogenetics*. Oxford: Blackwell Scientific.

Wilson, A.C., Carlson, S.S. & White, T.J. (1977). Biochemical evolution. *Annu. Rev. Biochem.*, **46**, 573–639.

Wilson, E. & Stevens, L. (1995). Proteinase inhibitory properties of ovoinhibitor from Golden pheasant (*Chrysolophus pictus*). *Brit. Poult. Sci.*, **35**, 833–4.

Wilson, S.B., Back, D.W., Morris, S.M., Swierczynski, J. & Goodridge, A.G. (1986). Hormonal regulation of lipogenic enzymes in chick embryo hepatocytes in culture. *J. Biol. Chem.*, **261**, 15179–82.

Wilton, S.D., Crocker, L.A. & Rogers, G.E. (1985). Isolation and characterisation of keratin mRNA from the scale epidermis of the embryonic chick. *Biochim. Biophys. Acta*, **824**, 201–8.

Wingaarden, J.B. & Ashton, D.H. (1959). The regulation of activity of phosphoribosylpyrophosphate amidotransferase by purine ribonucleotides: a potential feedback control of purine biosynthesis. *J. Biol. Chem.*, **234**, 1492–6.

Wingfield, J.C., Matt, K.S. & Farner, D.S. (1984). Physiological properties of steroid hormone binding proteins in avian blood. *Gen. Comp. Endocrinol.*, **53**, 281–92.

Wistow, G. (1993). Lens crystallins: gene recruitment and evolutionary dynamism. *Trends Biochem. Sci.*, **18**, 301–6.

Wistow, G., Anderson, A. & Piatigorsky, J. (1990). Evidence for neutral and selective processes in the recruitment of enzyme-crystallins in avian lenses. *Proc. Natl. Acad. Sci. USA*, **87**, 6277–80.

Wistow, G.J., Mulders, J.W.M. & de Jong, W.W. (1987). The enzyme lactate dehydrogenase as a structural protein in avian and crocodilian lenses. *Nature*, **326**, 622–4.

Wistow, G.J. & Piatigorsky, J. (1988). Lens crystallins: the evolution and expression of proteins for a highly specialized tissue. *Annu. Rev. Biochem.*, **57**, 479–504.

Wistow, G.J. & Piatigorsky, J. (1990). Gene conversion and splice-site slippage in the argininosuccinate lyases/δ-crystallins of the duck lens: members of an enzyme superfamily. *Gene*, **96**, 263–70.

Wittenberg, B.A. & Wittenberg, J.B. (1989). Transport of oxygen in muscle. *Annu. Rev. Physiol.*, **51**, 857–78.

Wittzell, H., von Schantz, T., Zoorob, R. & Auffray, C. (1994). Molecular characterization of three *Mhc* class II *B* haplotypes in the ring-necked pheasant. *Immunogenet.*, **39**, 395–403.

Wolfe, F.H., Sathe, S.K., Goll, D.E., Kleese, W.C., Edmunds, T. & Duperret, S.M. (1989). Chicken skeletal muscle has three Ca^{2+}-dependent proteinases. *Biochim. Biophys. Acta*, **998**, 236–50.

Wolfes, R., Mathe, J. & Seitz, A. (1991). Forensics of birds of prey by DNA fingerprinting with ^{32}P-labelled oligonucleotide probes. *Electrophoresis*, **12**, 175–80.

Wu, C.-Y., Chen, S.-T., Choiu, S.-H. & Wang, K.T. (1992). Kinetic analysis of duck ε-crystallin with lactate dehydrogenase activity: determination of kinetic constants and comparison of substrate specificity. *Biochem. Biophys. Res. Commun.*, **186**, 874–80.

Wu, R.S. & Bonner, W.H. (1981). Separation of basal histone synthesis from S-phase histone synthesis in dividing cells. *Cell*, **27**, 321–30.

Wyles, J.S., Kunkel, J.G. & Wilson, A.C. (1983). Birds, behaviour and anatomical evolution. *Proc. Natl. Acad. Sci. USA*, **80**, 4394–7.

Xue, Z., Gehring, W.J. & le Douarin, N.M. (1991). Quox-1, a quail homeobox gene expressed in the embryonic central nervous system, including the forebrain. *Proc. Natl. Acad. Sci. USA*, **88**, 2427–31.

Xue, Z., Xue, X.J. & le Douarin, N.M. (1993). Quox-1, an *Antp*-like homeobox gene of the avian embryo: a developmental study using a Quox-1-specific antiserum. *Mech. Dev.*, **43**, 149–58.

Yacoe, M.E. & Dawson, W.R. (1983). Seasonal acclimatization in American goldfinches: the role of the pectoralis muscle.

Am. J. Physiol., **245**, R265–71.

Yamada, K., Tsuchiya, M., Mishima, K. & Shimoyama, M. (1992). p33, an endogenous target protein for arginine-specific ADP-ribosyltransferase in chicken polymorphonuclear leukocytes, is highly homologous to *min-1* protein (*myb*-induced myeloid protein-1). *FEBS Lett.*, **311**, 203–5.

Yamada, Y., Liau, G., Mudryj, M., Obici, S. & de Crombrugghe, B. (1984). Conservation of the sizes for one but not another class of exons in two chick collagen genes. *Nature*, **310**, 333–7.

Yamano, T., Yorita, K., Fujii, H. *et al.* (1988). Gluconeogenesis in perfused chicken kidney: effects of feeding and starvation. *Comp. Biochem. Physiol.*, **91B**, 701–6.

Yang, C.-Y., Huang, W.-Y., Chirala, S. & Wakil, S.J. (1988). Complete amino acid sequence of the thioesterase domain of chicken liver fatty acid synthase. *Biochemistry*, **27**, 7773–7.

Yang, Y.W.-H., Brown, D.R., Robcis, H.L., Rechler, M.M. & de Pablo, F. (1993). Developmental regulation of insulin-like growth factor binding protein–2 in chick embryo serum and vitreous humor. *Reg. Peptides*, **48**, 145-55.

Yarden, Y., Rodriguez, H., Wong, S.K.F. *et al.* (1985). The avian β-adrenergic-receptor – primary structure and membrane topology. *Proc. Natl. Acad. Sci. USA*, **83**, 6795–9.

Yasuda, K. & Okada, T.S. (1986). Structure and expression of the chicken crystallin genes. *Oxford Survey on Eukaryote Genes* **6**, 183–209.

Yau, J.-C., Denton, J.H., Bailey, C.A. & Sama, A.R. (1991). Customizing the fatty acid content of broiler tissues. *Poult. Sci.*, **70**, 167–72.

Yeung, T.C. & Gidari, A.S. (1980). Purification and properties of a chicken liver glutathione S-transferase. *Arch. Biochem. Biophys.*, **205**, 404–11.

Yokouchi, Y., Ohsugi, K., Sasaki, H. & Kuroiwa, A. (1991). Chicken homeobox gene *Msx-1*: structure, expression in limb buds and effect of retinoic acid. *Development*, **113**, 431–44.

Yorita, K., Yamano, T., Ikeda, K., Kobayashi, T., Shiota, M. & Sugano, T. (1987). Distribution of glycolysis and gluconeogenesis in perfused kidney. *Am. J. Physiol.*, **253**, R679–86.

Yoshima, T., Mimura, N., Aimoto, S. & Asano, A. (1993). Transitional expression of neural cell adhesion molecule isoforms during chicken embryonic myogenesis. *Cell Struct. Funct.*, **18**, 1–11.

Yoshimura, Y., Chang, C., Okamoto, T. & Tamura, T. (1993). Immunolocalization of androgen receptors in the small, preovulatory, and postovulatory follicles of laying hens. *Gen. Comp. Endocrinol.*, **91**, 81–9.

Yu, C.W. & Chiou, S.H. (1993). Facile cloning and sequence-analysis of goose δ-crystallin gene based on polymerase chain-reaction. *Biochem. Biophys. Res. Commun.*, **192**, 948–53.

Zagris, N., Stavridis, V. & Chung, A.E. (1993). Appearance and distribution of entactin in the early chick embryo. *Differentiation*, **54**, 67–71.

Zehner, Z.E., Li, Y., Roe, B.A., Paterson, B.M. & Sax, C.M. (1987). The chicken vimentin gene – nucleotide-sequence, regulatory elements, and comparison to the hamster gene. *J. Biol. Chem.*, **262**, 8112–20.

Zehner, Z.E. & Paterson, B.M. (1985). The chicken vimentin gene – aspects of organization and transcription. *Ann. New York Acad. Sci.*, **455**, 79–94.

Zhang, H.-Y. & Young, A.P. (1991). A single upstream glucocorticoid response element juxtaposed to an AP1/ATF/CRE-like site renders the chicken glutamine-synthetase gene hormonally inducible in transfected retina. *J. Biol. Chem.*, **266**, 24332–8.

Zhou, R.-P. & Duesberg, P.H. (1989). Avian proto-*myc* genes promoted by defective or nondefective retroviruses are single-hit transforming genes in primary cells. *Proc. Natl. Acad. Sci. USA*, **86**, 7721–5.

Zhu, X., Testori, A., Oh, S., Skinner, J.D. & Burgoyne, L.A. (1992). Avian species-specific tandem repeats contain nuclear protein binding sites. *Biochem. Biophys. Res. Commun.*, **182**, 447–51.

Zhuang, Y.-H., Landers, J.P., Schuchard, M.D. *et al.* (1993). Immunohistochemical localization of the avian progesterone receptor and its candidate receptor binding factor (RBF-1). *J. Cell. Biochem.*, **53**, 383–93.

Zimmerman, B., Shalatin, N. & Grey, H.M. (1971). Structural studies on the duck 5.7s and 7.8s immunoglobulins. *Biochemistry*, **10**, 482–8.

Zink, R.M. (1994). The geography of mitochondrial DNA variation, population structure, hybridization, and species limits in the Fox sparrow (*Passerella iliaca*). *Evolution*, **48**, 96–111.

Zoorob, R., Béhar, G., Kroemer, G. & Auffray, C. (1990). Organization of a functional chicken class II *B* gene. *Immunogenetics*, **31**, 179–87.

Zubay, G. (1993). *Biochemistry*, 3rd edn. Dubuque: W.C. Brown.

Zwaan, J. & Ikeda, A. (1968). Macromolecular events during differentiation of chicken lens. *Exp. Eye Res.*, **7**, 301–11.

Index

Printed in the United States
By Bookmasters